Water Science and Technology Library

Volume 126

The aim of the *Water Science and Technology Library* is to provide a forum for dissemination of the state-of-the-art of topics of current interest in the area of water science and technology. This is accomplished through publication of reference books and monographs, authored or edited. Occasionally also proceedings volumes are accepted for publication in the series. *Water Science and Technology Library* encompasses a wide range of topics dealing with science as well as socio-economic aspects of water, environment, and ecology. Both the water quantity and quality issues are relevant and are embraced by *Water Science and Technology Library*. The emphasis may be on either the scientific content, or techniques of solution, or both. There is increasing emphasis these days on processes and *Water Science and Technology Library* is committed to promoting this emphasis by publishing books emphasizing scientific discussions of physical, chemical, and/or biological aspects of water resources. Likewise, current or emerging solution techniques receive high priority. Interdisciplinary coverage is encouraged. Case studies contributing to our knowledge of water science and technology are also embraced by the series. Innovative ideas and novel techniques are of particular interest.

Comments or suggestions for future volumes are welcomed.

Vijay P. Singh, Department of Biological and Agricultural Engineering & Zachry Department of Civil and Environment Engineering, Texas A&M University, USA
Email: vsingh@tamu.edu

All contributions to an edited volume should undergo standard peer review to ensure high scientific quality, while monographs should also be reviewed by at least two experts in the field.

Manuscripts that have undergone successful review should then be prepared according to the Publisher's guidelines manuscripts: https://www.springer.com/gp/authors-editors/book-authors-editors/book-manuscript-guidelines

Umesh Chandra Chaube · Ashish Pandey ·
Vijay P. Singh

Canal Irrigation Systems in India

Operation, Maintenance, and Management

Springer

Umesh Chandra Chaube
Department of Water Resources
Development and Management
Indian Institute of Technology Roorkee
Roorkee, Uttarakhand, India

Ashish Pandey
Department of Water Resources
Development and Management
Indian Institute of Technology Roorkee
Roorkee, Uttarakhand, India

Vijay P. Singh
Department of Biological and Agricultural
Engineering
Texas A&M University
College Station, TX, USA

ISSN 0921-092X ISSN 1872-4663 (electronic)
Water Science and Technology Library
ISBN 978-3-031-42811-1 ISBN 978-3-031-42812-8 (eBook)
https://doi.org/10.1007/978-3-031-42812-8

This Springer imprint is published by the registered company Springer Nature Switzerland AG
The registered company address is: Gewerbestrasse 11, 6330 Cham, Switzerland

Paper in this product is recyclable.

Contents

About the Authors

Umesh Chandra Chaube is Former Professor of IIT Roorkee and has more than 48 years of experience in the water resource development and management sector covering field engineering, teaching, and research. He holds B.Tech. and M.Tech. degrees (Civil Engineering) from IIT Kanpur and Ph.D. degree (Water Resources) from IIT Delhi. He has supervised one post-doctoral research, 15 Ph.D.-level research, 95 Master degree-level dissertations, and 152 special problem studies. He has published 124 research papers in journals and conference proceedings.

Over a period of 33 years at the Department of Water Resource Development and Management of IIT Roorkee, he has been responsible for the capacity building of a large number of irrigation professionals from Afro-Asian countries. He has also served on several National Committees of the Government of India.

He is Life Member of the International Commission on Irrigation and Drainage and Life Fellow of the Institution of Engineers India, the Indian Water Resources Society, and the Indian Association of Hydrology. He has been Executive Vice President of the Indian Water Resources Society and Chief Editor of its Journal. He has served as Deputy Director/Assistant Director in Central Water Commission of Government of India and as Faculty at IIT Roorkee, IIT Indore, and Sri Vaishnav Vidyapeeth University, Indore. He was employed by United Nation-FAO as International Hydrologist and Water Management Specialist in Kabul (Afghanistan). He is the recipient of the Colombo Plan Scholarship, UNESCO Fellowship, and UNEP Fellowship, Emeritus Fellowship of IIT Roorkee and has been Professor Emeritus at Sri Vaishnav Vidyapeeth University at Indore.

Prof. Ashish Pandey is Bharat Singh Chair Professor of the Ministry of Water Resources, Government of India, and Head of the Department of Water Resources Development and Management, IIT Roorkee. He holds B.Tech. and M.Tech. (Gold Medal) from JNKVV, Jabalpur followed by a Ph.D. degree from IIT Kharagpur. His research interests include irrigation water management, soil, and water conservation engineering, remote sensing and GIS applications in water resources, etc. He supervised 13 Ph.D. and 88 M.Tech. dissertations. Dr. Pandey published 224 research papers in High Impact Journals/Conference and 37 book chapters. He has co-authored two textbooks on *Irrigation Engineering* and *Introductory Soil and Water Conservation Engineering*. He has also served as a lead editor of seven volumes of the Water Science and Technology Library book series for Springer Nature, Switzerland AG.

Currently, Prof. Pandey is Executive Vice President of the Indian Water Resources Society (IWRS) and Chairman, Institution of Engineers (Roorkee Local Centre). He has trained about 1000 + senior-level scientists/engineers under 37 various capacity-building programs from India and Afro-Asian Countries. He has also served on several National Committees of the Government of India. He is the recipient of several national and international fellowships/awards and honors. He is Fellow member of (1) Institution of Engineers (India); (2) Indian Association of Hydrologists (IAH); and (3) IWRS.

Prof. Vijay P. Singh is Distinguished Professor, Regents Professor, and Inaugural Holder of the Caroline and William N. Lehrer Distinguished Chair in Water Engineering at Texas A&M University, USA. He received his B.Sc. and Tech., M.S., Ph.D. and D.Sc. degrees, all in engineering. He is a registered professional engineer, a registered professional hydrologist, and Honorary Diplomate of ASCE-AAWRE. He is Distinguished Member of ASCE, Honorary Distinguished Member of IWRA, Distinguished Fellow of AGGS, and Honorary Member of AWRA, and Fellow of EWRI-ASCE, ASCE, IAH, ISAE, IWRS, and IASWC. He is a member of National Academy of Engineering (NAE), an academician of Georgia Fazisi Academy, a member of European Academy of Science and Arts, and a fellow/member of 12 other international science/engineering academies. He has published extensively in the areas of hydrology, irrigation engineering, hydraulics, groundwater, water quality, water resources, entropy theory, copula theory, and climate change impacts with more than 1530 refereed journal articles; 40 books; 92 edited reference books, as well as the *Handbook of Applied Hydrology* and *Encyclopedia of Snow, Ice and Glaciers*; 130 book chapters; and 330 conference papers. He has received four honorary doctorates from Italy and Canada, and more than 110 national and international awards, including Norman Medal (twice), Chow Award, Torrens Award, Arid Lands Hydraulic Engineering Award, and Outstanding Projects and Leader (OPAL) Award, Lifetime Achievement Award, Best paper Awards (twice), and Roy C. Tipton Award, all given by American Society of Civil Engineers; Hancor Award and Lalit and Aruna Verma Award for Global excellence, both given by American Society of Agricultural and Biological Engineers; Merriam Improved Irrigation Award, given by International Commission on Irrigation and Drainage; Linsley Award, Wetzel Award, and Founders Award, given by American Institute of Hydrology; Crystal Drop Award and Chow Memorial Award, given by International Water Resources Association; Outstanding Scientist Award given by Sigma Xi; Distinguished Scientist Award, given by Chinese Academy of Science; Distinguished Professor Award, given by Mexican Academy of Science; Prof. Gajendra Singh Gold Medal for Education, given by Indian Society of Agricultural Engineers; to name but a few. He has served as President of the American Institute of Hydrology (AIH); President of American Academy of Water Resources Engineers; President of International Association for Water, Environment, Energy, and Society; and Chair of Watershed Council of

American Society of Civil Engineers. He has served as the editor-in-chief of five journals and two book series and has served/serves on editorial boards of more than 40 journals and three book series. He has given more than 175 keynote lectures and 300 invited seminars all over the world and has organized more than 25 international conferences and his Google Scholar citations are more than 95,550; h-index of 128; and I10-index of 1151.

Professor Singh has long been engaged in philanthropic activities. He founded G. B. School in 1994 in his native village Naglavishnu in District Agra, Uttar Pradesh, India, in memory of his parents. The school imparts quality education to children in rural Agra. He has been bearing most of the expenditures involved in the operation of school, including the cost of building construction, staff salaries, furniture, maintenance, management of the school, and so on. The school now has four campuses: (1) Primary School (Grade 1–5), (2) Inter College (Grade 6–12), (3) Degree College (B.A. and B.Sc.); and (4) Industrial Training Center (Fitter and Electronics).

List of Figures

List of Tables

Chapter 1
An Integrated View of Irrigation and Agriculture

Abstract This chapter introduces a systemic view of irrigation and irrigated agriculture which aids a systematic understanding of the processes of irrigation and irrigated agriculture. Irrigated agriculture requires natural inputs (land, soil, rainwater, and climate), resource inputs (irrigation water, chemicals, fertilizers, etc.), and management inputs (human skill, material, energy, and equipment) to achieve biophysical objectives and contribute to a higher-level societal goals. The process of irrigated agriculture is visualized as (i) a project involving five stages (planning, design, installation or implementation, operation and maintenance, and monitoring and feedback); (ii) a planned system of management actions, implementation tools, and institutional arrangements; and (iii) a set of linked subsystems (water control, agronomic, social, and economic). Physical and managerial linkages between the subsystems are illustrated using schematic diagrams. Coordination among these subsystems is essential to achieve specific social and biophysical objectives (optimum agricultural production under prevailing constraints). Irrigation management is just one component of integrated water resources management (IWRM). Concepts, principles, and interlinkages in IWRM are explained. Differences in institutional and socioeconomic characteristics affect the success of an integrated water management strategy between developing and developed countries. The most important challenge that developing countries face in setting up successful IWRM is the dominance of irrigation and agricultural sectors. Implications of the dominance of the agriculture sector in the adoption of IWRM are discussed.

1.1 Introduction

Agricultural production ought to be thought of as the outcome of favorable interaction among various inputs, such as seeds, soil, water, weather, fertilizer, pesticides, labor, and energy. This interaction occurs through the application of technology, such as water control and delivery, and provision of facilities, such as roads, credit, and markets. These various inputs and facilities can be visualized as interactive components of the irrigated agriculture system, aiming at optimum agricultural production under prevailing constraints. The control and delivery of water through a network

U. C. Chaube et al., *Canal Irrigation Systems in India*, Water Science and Technology Library 126, https://doi.org/10.1007/978-3-031-42812-8_1

of canals, outlets, watercourses, and field channels are the most important components in an irrigated agriculture system and are often treated as a separate system. An irrigated agriculture system can be decomposed into several subsystems, such as a water control system, agronomic subsystem, social subsystem, and economic subsystem. Often these subsystems have functioned independently. Coordination among these subsystems is essential to achieve specified objectives and higher-level goals (Skogerboe 1990).

A wide range of issues arise in managing irrigated agriculture for achieving specified objectives and contributing to higher-level goals. Management is the process of operating a system (and its sub-system and components) to achieve specified objectives and higher-level goals. The steps involved in the management process are:

1. Specify the objectives and higher-level goals of irrigated agriculture;
2. Prepare plans to meet objectives;
3. Assign priorities to achieve objectives;
4. Set targets for meeting each objective;
5. Determine the activities necessary to meet targets;
6. Monitor activities for feedback in management; and
7. Operate the system and its components according to the management process.

Based on the review of the literature (Skogerboe 1990; Murray-Rust 1992; Smith 1970), and the author's experience, a systemic view of irrigation and irrigated agriculture is presented in this chapter to have a systematic understanding of the process of irrigation and irrigated agriculture.

1.2 Irrigation as a Project

In India, irrigation development activities are carried out in a project mode. An irrigation scheme is often referred to as an irrigation project. An irrigation project passes through the stages of planning, design, installation, operation, and maintenance sequentially, as shown in Fig. 1.1.

Planning begins with the perception of a problem, such as a deficit in the supply of food grains or the need to improve the socio-economic condition of an area, which leads to a decision to prepare an irrigation project. A feasibility report is then prepared, which provides a basis for a decision to take up the project. Then, detailed designs are prepared.

The first stage of implementation often takes many years and often involves substantial capital expenditure. This is followed by the operation and maintenance stage. Planning and implementation should proceed in tandem so that information gained in the implementation is used to further improve planning activities or projects. Unfortunately, this is not done, partly because an authorized agency must approve plans (project report) before a project can be authorized. Most of India's irrigation

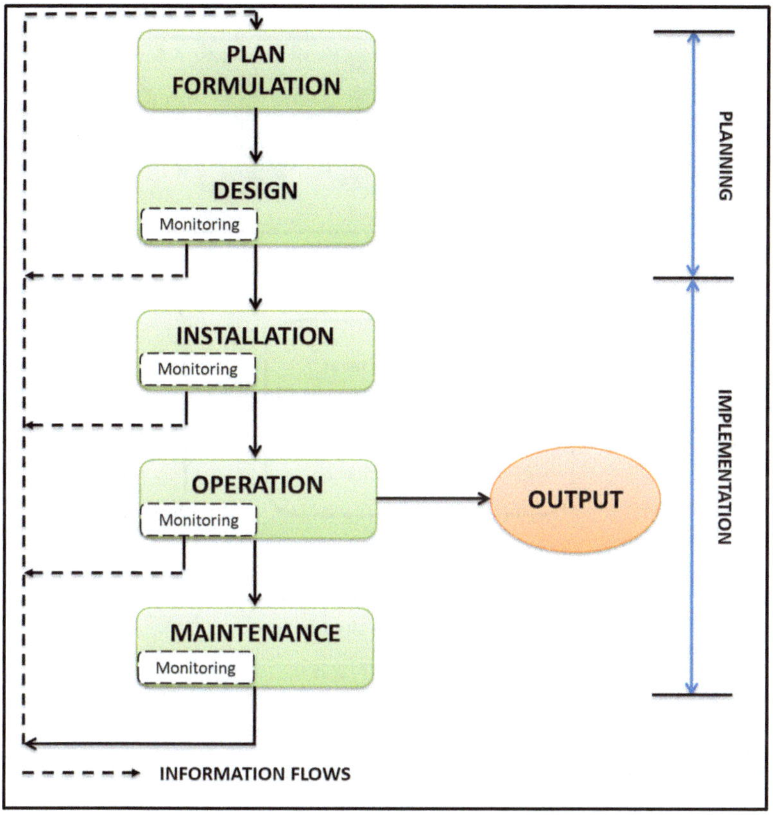

Fig. 1.1 Five stages of an irrigation project

projects are sponsored by government agencies, and the projects have been mainly aimed at social welfare.

Monitoring must start at the design stage and should continue in all stages so that information flows can be utilized to make appropriate revisions.

1.3 Irrigated Agriculture as a Planned System

Figure 1.2 depicts irrigated agriculture in an area as a planned system. Inputs (except climate) have related costs, and outputs have related benefits. There are onsite effects (changes in waterlogging, soil salinity, environmental health, and changes in water yielding capacity) and off-site effects (changes in the time pattern of streamflow and groundwater flow; channel degradation; water quality) on the natural system.

The information depicted in Fig. 1.2 can be used to depict a system in the (i) planning; (ii) design; (iii) construction or (iv) operation stages.

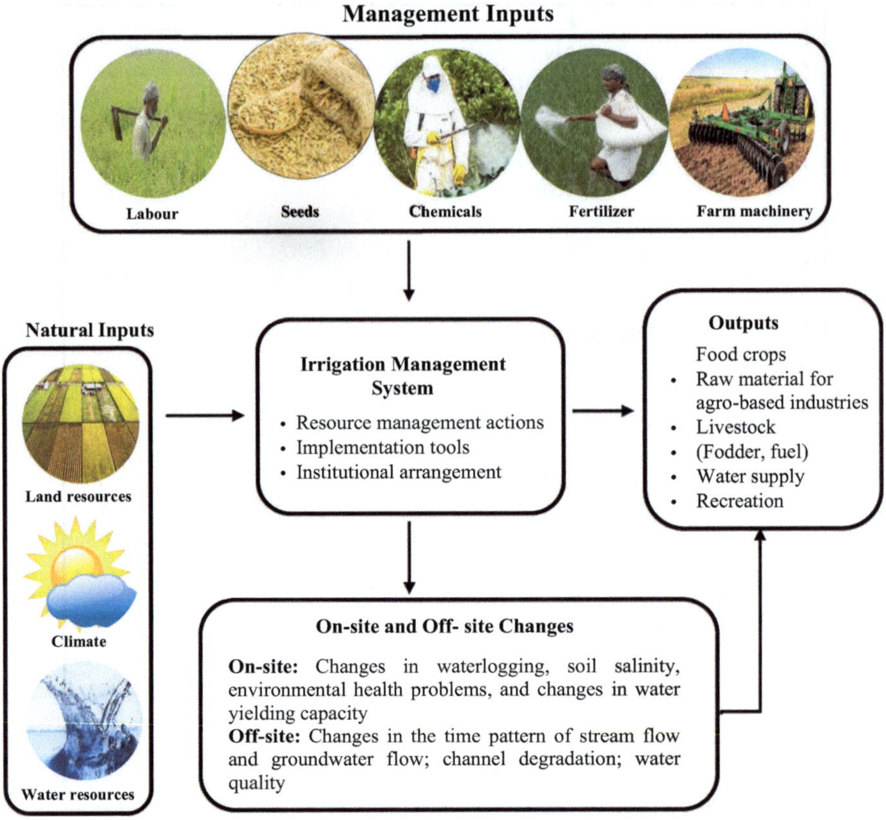

Fig. 1.2 Generalized system for management of irrigated agriculture in physical terms

1.4 Irrigated Agriculture as a Set of Linked Subsystems

There are four major subsystems: (i) water control, (ii) agronomic, (iii) social, and (iv) economic, as shown in Table 1.1. Each of the subsystems has specific functions to be performed. Each subsystem has distinctive elements (structural and non-structural). It is necessary to understand that these subsystems are not entirely independent of each other. Resource utilization and management practices of these subsystems would need coordination to improve the overall efficiency of the irrigated agriculture system.

Table 1.1 System functions and major elements

Sl. no.	Sub-system	Major functions	Major elements
1	Water control		
(a)	Water delivery subsystem	To convey an adequate amount of water to the crop area	Conveyance network above and below outlets
(b)	Water application subsystem	To distribute water over the crop area uniformly and at the proper time to meet crop water requirement and leaching water requirements, while satisfying leaching and erosion control standards	Stream size (outlet discharge), field topography, soil infiltration rate, and method of irrigation
(c)	Water use subsystem	Use water for crop growth, maintain acceptable salinity levels, maintain appropriate environment (soil-air) temperature, ensure adequate nutrients, and provide appropriate soil conditions	Water quantity and quality, soil type, nutrient availability, and evapotranspiration
(d)	Water removal subsystem	Provide necessary surface drainage, maintain given salinity levels, and improve the workability of land	Leaching requirement, evapotranspiration rate, drainage facilities, and soil type/ subsoil type
2	Agronomic (Cropping) subsystem	Manage physical and biological resources to produce food, fiber, and specialty crops to ensure the long-term productivity of crops	Plants; climate; temperature; water; topography; physical, biological, and chemical aspects of soil; nutrient supply; insect control; management practices
3	Social subsystem	Achieve individual and social goals through participatory management at the tertiary level	Institutional facilities, Water User Associations, rules (norms, laws, etc.),
4	Economic subsystem	Allocate appropriate resources, maximize production, and optimize the decision-making process	Land, labour, capital, markets, risk/uncertainty, cost/benefits, and consumption

1.5 Water Control System and Its Subsystems

The water control system/subsystem is also referred to as the physical irrigation system and is presented in Fig. 1.3. It consists of (a) water delivery, (b) water application, (c) water use, and (d) water removal subsystems.

1.5.1 Water Delivery Sub-System

The water delivery subsystem covers water conveyance from the water supply source through the main canal and distributary canals to canal outlets, and from there to crop fields through farms and field channels. Depending upon the irrigation system's

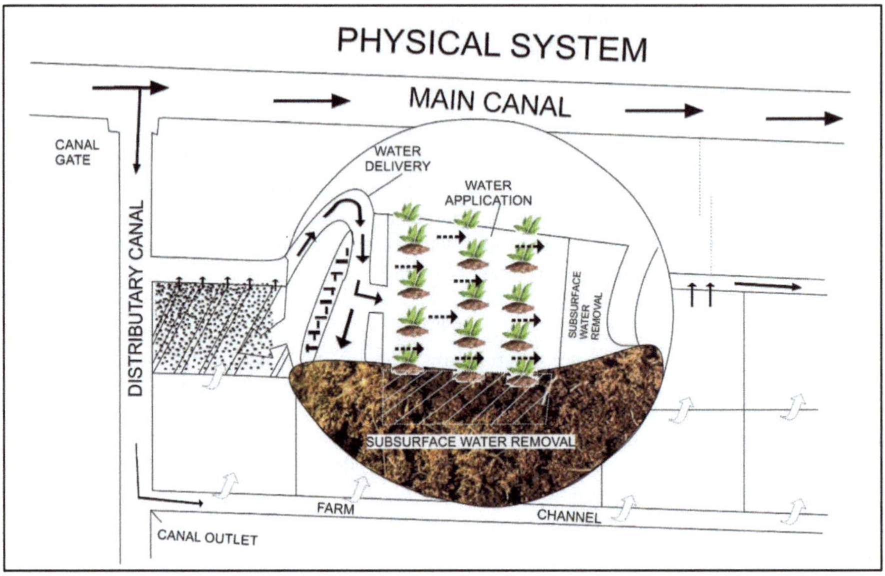

Fig. 1.3 Water control system (delivery, application, use, and removal subsystems)

boundary, the water supply source may be a well, a storage reservoir, or a canal. It may be operated by a private, public, or governmental organization. The water delivery sub-system (Fig. 1.4) typically consists of a water conveyance network above outlets and a farm conveyance network below outlets. The main conveyance network (main, branch, distributary, and minor canals) is frequently managed by a government or other organizations. It is often referred to as the main delivery subsystem. Farmers usually manage the water conveyance network below the canal outlets (farm channels, field channels, farm roads, field outlets, and field drains). It is often known as an on-farm subsystem.

The water delivery sub-system serves the water application subsystem, which, in turn, supplies water to the water use subsystem. A major design variable of the water application system is the design water application rate for a particular method of irrigation. This specifies the desired flow rate that is necessary to properly irrigate a field. Thus, the water delivery sub-system's primary function is to supply this design flow rate to the field.

The purpose of the water delivery sub-system is to convey water from the supply source to the crop field:

- at a constant regulated rate,
- at the proper elevation with seepage controlled,
- without erosion or sedimentation in the distribution channels, and
- with safety.

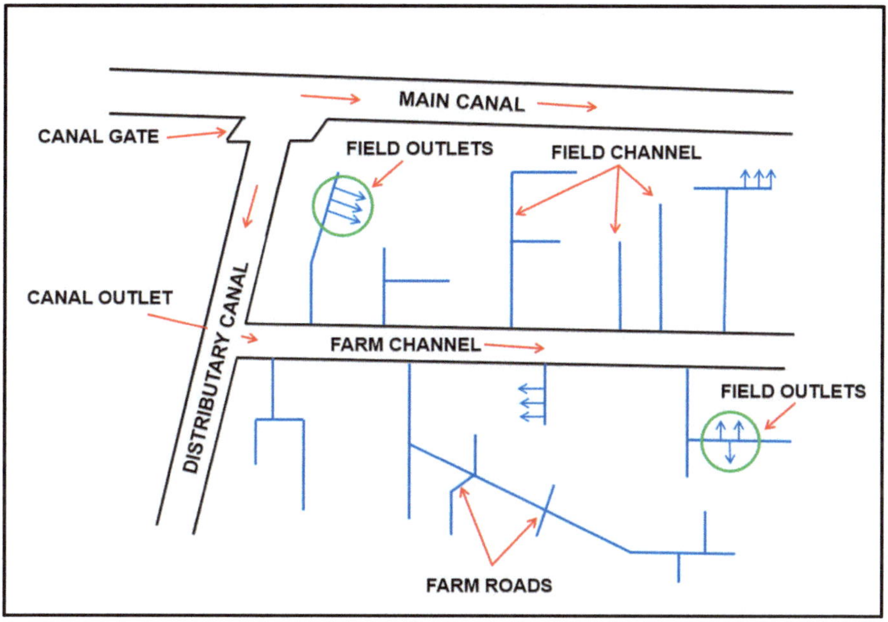

Fig. 1.4 Water delivery sub-system

These functions are performed by a delivery system, based upon the physical and management factors. The factors that influence water delivery are discussed below.

Discharge: The rate at which water is supplied to the field is regulated according to the following requirements:

- The total quantity of water to be supplied,
- To meet peak demand,
- Constant flow for an appropriate time of application, and
- To provide dependable flow.

These are the key factors that establish the capacity of a delivery system.

Cross-section, hydraulic radius, and roughness of the channel: An appropriate channel cross-section must be provided to maintain the head and deliver water at a proper elevation. The cross-section also must be provided for the design flow rate defined by the water application sub-system. The hydraulic radius should be minimum for the design flow to minimize the cut and fill volume of earthwork associated with channel construction and to minimize the cost of construction. The roughness of a channel must be carefully selected for the design to conform to the design cross-section.

Slope: The design slope is important to maintain the minimum cross-section to reduce the cost of channel construction and to ensure that sedimentation or erosion does not occur in the channel.

Seepage rate: The effect of seepage rate on the depth, channel storage, and operational losses of the delivery sub-system should be evaluated and explicitly included in the design. Realistic assumptions concerning the system maintenance should also be considered in adopting a seepage rate and included as part of the system design parameters.

1.5.2 Water Application Sub-System

The water application subsystem supplies water to the water use subsystem by distributing water over the field surface.

Water application must provide water for the functions of the water use subsystem as well as fulfill the following functions:

1. distribute the desired amount of water with the designed uniformity;
2. satisfy erosion control standards;
3. provide necessary surface drainage; and
4. be economically appropriate and socially acceptable to the management abilities of the farmer.

The farmer manages the water application subsystem by operating it to meet both water use and the functional objectives of water application. In the process, the following three basic management questions must be answered:

1. How do we irrigate?
2. When do we irrigate?
3. How much water do we apply?
4. For how long do we apply water?

1.5.3 Water Use Sub-System

The water use subsystem accepts water from the application subsystem and transmits water through the soil for storage or deep percolation. Plants transport water through the plant structure to the leaves where it is transpired. From the soil, surface water evaporates. Thus, evapotranspiration occurs from the field. The excess water flows through the root zone as deep percolation, and it is input to the drainage or water removal subsystem.

The water use subsystem has the following functions:

1. To meet water requirements for crop growth (quantity and quality);
2. To control the level of soil salinity;
3. To maintain appropriate environmental (soil and air) temperature;
4. To provide nutrients; and
5. To maintain proper soil condition.

To prevent excessive crop stress, it is necessary to supply an adequate quantity of irrigation water at the appropriate time. Crop stress is a function of several factors, including; (i) crop and stage of growth, (ii) soil, (iii) climate, (iv) irrigation system characteristics, and (v) economics.

1.5.4 Water Removal Subsystem

The water removal subsystem (drainage subsystem) aims to remove and dispose of surplus surface and subsurface waters from crop fields to improve agricultural operations and productivity. The objective of drainage is to provide an environment for plants that will result in the optimal production of crops. A crop field may receive water from precipitation, seepage from ponds, irrigation from canals, seepage from adjacent aquifers, floodwaters from streams, and the application of water to the field for special purposes, such as salinity control. In irrigated areas, natural drainage is usually inadequate; therefore, a water removal subsystem (drainage subsystem) is needed to supplement natural drainage.

The water removal subsystem has the following primary functions:

- To provide proper root aeration;
- To control salinity levels within the soil profile; and
- To improve the workability of land.

1.6 Agronomic Subsystem (Irrigated Cropping Subsystem)

This subsystem consists of the elements required for producing a set of crops and depicts the inter-relationship between the set of crops, water, and the environment. The function of the agronomic subsystem is to produce food, fiber, and other organic products at optimum levels with the desired quality and to ensure long-term productivity. Implied in the definition and function of the agronomic subsystem is the interdependency among crops, water, man (the manager), and the natural environment (the input).

1.6.1 Classifying Irrigated Cropping Sub-Systems

The system approach to an irrigated cropping subsystem recognizes that we are dealing with a complex system involving several crop varieties and various crop and water management methods. To reduce this complexity, irrigated cropping subsystems are classified according to the pattern of cropping and rotational patterns as shown in Fig. 1.5. A brief description is given in what follows.

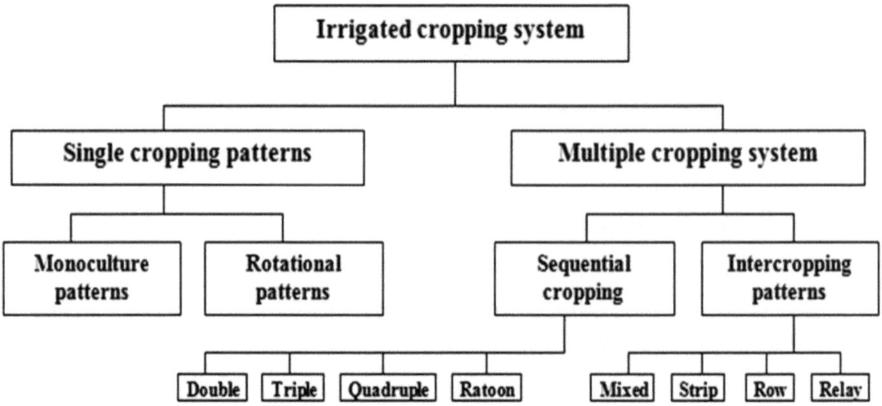

Fig. 1.5 Classification of the irrigated cropping system

1. **Single Cropping Patterns**

   Single cropping patterns are subdivided into monocultural and rotational patterns. A monocultural pattern is characterized by the growth of the same crop on a field year after year as opposed to an alternation of two or more crops in multiples of a yearly rotational pattern. Monocultural and rotational patterns are most often characteristic of temperate zone climates, but they are also found in tropical or arid zones on irrigation systems incapable of supplying crop water requirements on a year-round basis.

2. **Multiple Cropping Patterns**

   Multiple cropping patterns are subdivided into sequential cropping patterns and intercropping patterns. Sequential cropping patterns are multiple cropping patterns. One crop is planted immediately after the harvest of the previous crop. Specific sequential cropping patterns that may be identified in an irrigation system are double, triple, quadruple, and ratoon cropping. Double, triple, and quadruple refer to the number of crops grown sequentially in a year; ratoon cropping refers to the cultivation of crops grown after harvest from the same rootstock.

3. **Intercropping Pattern**

   The intercropping pattern is distinguished from sequential cropping patterns in that two or more crops are grown simultaneously in the same field. Mixed intercropping patterns have no distinct row arrangement, while intercropping patterns maintain distinct row arrangements.

   Strip intercropping patterns feature the growth of two or more crops in strips wide enough to permit independent cultivation but narrow enough for the crops to interact agronomically. A second crop is usually planted after the first crop has reached its reproductive stage.

1.6.2 Plant Environment

Data concerned with climate, soils, pest infestation, and farmer's management practices are useful for evaluating the irrigated cropping subsystems. While a great deal of this information is provided by soil survey reports, climatic records, and previous research in the area, it is necessary to gather specific data on soils, irrigation water, crops, pests, and management practices followed by the farmers in the study area. Some important data used to describe the irrigated cropping systems are summarized in the following:

1. **Climate:** Climate exerts a major influence on soils, natural vegetation, and types of crops that are grown in a given area. Climatic parameters that are considered important in evaluating the irrigation cropping system are: solar radiation; temperature; precipitation; relative humidity; climatic extremes; and wind.
2. **Soils:** Analysis of soils in crop fields requires that each soil is described vertically (with depth) and horizontally. The morphological characteristics observed with depth are used to identify specific soils. Changes in these characteristics in the horizontal direction distinguish one soil from another. Some of the soil parameters considered important in the analysis are:

   topography; soil horizon; soil depth; soil texture; soil structure; bulk density; soil moisture regimes; infiltration and permeability; organic water content; soil mineral nutrient status; salinity; sodicity; and specialized soil problem

3. **Biological Constraints to Crop Production:** The identification of biological pests and their damage potential is another aspect of analysis important to understand the irrigated cropping system. Biological pests may be subdivided into two categories:

   (1) Those that affect the irrigated crops; and
   (2) Those that physically may affect the farmers and their families.

   The first category includes animals, insects, weeds, and plant density. The background information concerning the identification of the most serious agricultural pests in the study area is usually obtained from local, district, or state-level agricultural agencies. In addition, field surveys are used to identify pests and the damage caused by them.

1.6.3 Management Practices of Farmers

The management practices of farmers are critical to the understanding of the biological potential of the irrigated cropping system. Ideally, the agronomist would prefer to observe the farmers' management practices throughout the cropping season. Some of the more critical aspects of the farmers' management practices are:

(i) Land preparation and tillage operations,

(ii) Irrigation method,
(iii) Managing soil fertility,
(iv) Managing seedbed,
(v) Cropping practices,
(vi) Pest management, and
(vii) Special management procedures.

1.7 Socio-Economic System

(i) The primary functions of the socio-economic system/subsystem are to examine the social and organizational aspects of the operating system, and how the operational efficiency of the irrigation system influences the farmers in respect of:
(ii) irrigation management,
(iii) crop productivity, and
(iv) the living standard of the farmers.

1.7.1 Objectives of Socio-Economic Study

The objectives of the socio-economic study are to:

(a) Examine the social and organizational aspects of water availability, equity, dependability, and distribution;
(b) Examine the farmer's values and perception, level of knowledge, and decision-making processes related to irrigation and its management, use of technologies, and crop selection;
(c) Identify the level and use of selected institutional services and their significance to farmers in improving farm management performance;
(d) Study the input and outputs of important crops grown and resource use efficiency;
(e) Identify the problem areas and potential constraints, including the validity of the system; and
(f) Estimate the level of resource allocation, yield, and average income for various crops using the farm budget.

1.7.2 Economic Subsystem

The economic subsystem is concerned with the productivity and allocation of resources. It impacts all other systems as well as the ultimate decision-making process adopted by the farmer. Productivity resources, also known as factors of production, usually are grouped into four main categories:

(i) Natural resources,
(ii) Labor,
(iii) Capital, and
(iv) Management.

The land is made productive as a result of human effort in cultivating, fertilizing, irrigating, and draining.

The part on the technical and socio-economic constraints includes a discussion of the following:

(i) Role of economics in farming and as part of a diagnostic analysis;
(ii) Define the difference between economic and physical efficiencies;
(iii) Explore various farm management activities;
(iv) Demonstrate some of the economic principles utilized in decision-making at the farm level; and
(v) Examine the economic implications of long-term investment briefly.

1.7.2.1 Economics of Farming

The contribution of economics to the farmer's decision-making process lies in the estimation of costs and returns. The word cost is used here in a broad sense, meaning not only the expenditure of money but also the sacrifice of leisure, food, or anything else valued by the farmer's family. The farmer's objective is to achieve the maximization of his well-being. The benefits of farm output increase well-being. The value of income over cost is called profit. Before we can compare benefits and costs, we need technical information on the physical relationships between input and land, labour and capital, and the expected outputs for each alternative open to the farmer. The allocation of farm resources also needs to be assessed in an environment of social and governmental regulations as well as changing market conditions. Therefore, when we investigate a particular farm problem, we need the following diverse information:

1. The availability of resources, both on and off the farm, such as land, labour, water, fertilizer, credit, and equipment.
2. The technical information from various disciplines, including agricultural engineering and agronomy.
3. Market price information for both inputs and products.
4. The governmental and social organizational constraints limit the farmers' production choices.

1.7.2.2 Economics and Diagnostic Analysis

The role of economics in the diagnostic analysis is to delineate economic problems, identify linkages between economic and other constraints, and determine the losses associated with technical, institutional, social, and economic conditions.

Irrigation engineers will measure the problem of inadequate water supply to part of the command area. Economic analysis is evaluated, which includes yield data, supplementary inputs, cropping pattern, and cropping intensity; and the overall impact of inadequate water supply on output and farm income and assessment.

Furthermore, in searching for a solution to inadequate water supply, an irrigation engineer may look at the possibilities for increasing the water allotment to the command area, improving the command area's distribution, minimizing conveyance and seepage losses, and supplying groundwater.

1.7.2.3 Economic Efficiency and Physical Efficiency

Economic efficiency is concerned with the optimal allocation of resources among alternative uses and the optimal combination of inputs in a production activity to maximize profit. Economic efficiency consists of both technical efficiency and allocation efficiency.

Physical efficiency can be defined as achieving the maximum level of output from the given amount of input, or conversely, achieving a certain level of output with the least possible use of input. Output can be a physical good or a service, and input can be a physical input or knowledge.

The objective of irrigation engineering is to improve the system's physical efficiency to improve crop production. Such an objective would include minimum water loss in the delivery and application, uniformity of application, and minimization of erosion. On the other hand, the agronomist concerns himself with factors affecting yield maximization on a sustained basis.

The maximum physical efficiency of an irrigation system or the maximum yield per acre coupled with the improvement in the farmer's welfare, however, can be achieved only, if all resources were free, in other words, input and practices that might achieve the maximum physical efficiency might be undesirable or inefficient from the economic point of view.

1.7.3 Social Organizational Subsystem

Defects in water delivery, water application, water use, and water removal are often associated with problems in the irrigation system's social organization, including constraints on farmers and official decision-making.

1.7.3.1 Water Users Association

With education and technology transfer, engineering principles can be adopted by farmers. Still, the major problem is finding ways to seek the involvement of farmers more effectively in the operation and maintenance. A water users association has become necessary to perform the following tasks:

(a) Allocation of water to users within a scheme;
(b) Operational control of water;
(c) Modernization of the watercourse;
(d) Devising the assessment methods and then rates and collecting the assessment;
(e) Administration of budget for operation and maintenance; and
(f) Liaison with government administration (at the village, district level, or province level).

1.7.3.2 Farmer's Irrigation Behavior

A diagnostic analysis of an irrigation system should be carried out to study the factors which influence the farmer's behavior. By directly studying the farmer's behavior patterns and decision-making, a richer and more detailed picture emerges as to how an irrigation system operates and the possible constraints on that system.

1.8 Irrigation: Component of Integrated Water Resources Management

1.8.1 Meaning of Integrated Water Resource Management

Integrated Water Resources Management (IWRM) has been defined by the Technical Advisory Committee of the Global Water Partnership as "a process which promotes the coordinated development and management of water, land and related resources, to maximize the resultant economic and social welfare in an equitable manner without compromising the sustainability of vital ecosystems."

IWRM is a framework for planning, organizing, and controlling water systems to balance all relevant views and goals of stakeholders (Mitchell 1990; Grigg 1996). The IWRM concept embodies the integration of water resources development and management (Fig. 1.6):

- across local, regional, and national levels;
- across sectors, such as energy, transport, agriculture, etc.;
- across multiple uses (hydropower, irrigation, municipal water supply);
- with the environment; and—with the people.

Water Resources: It is available for use and susceptible to human interventions, but its availability is limited. Water can be surface water or groundwater. It is characterized by both quantity and quality.

Development and Management: Cover assessment, planning, implementation, operation and maintenance, monitoring, and control. They include the management

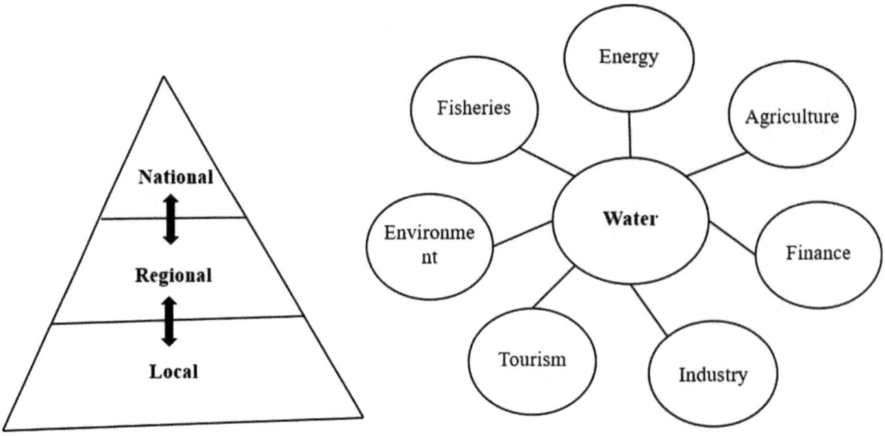

Fig. 1.6 Integration across sectors and multiple uses and with the environment

of water as a resource, management of water supply, and management of water demand for various purposes.

Multiple Purposes: Water resources have been developed and managed all over the world to serve various purposes, such as irrigation, hydropower, municipal and industrial water supply, wastewater and water quality services, flood control, navigation, recreation, and water for the environment, fish, and wildlife. In the Indian subcontinent, irrigation water supply has been the most significant purpose of river valley projects. As the practice of water resources management evolved, the term "multipurpose" (or "multi-objective") water resources development (or management) was used to refer to projects with more than one purpose.

Integrated: means the development and management of water resources as regards both the use and protection and considering all sectors and institutions which use and affect water resources (cross-sectoral integration).

1.8.2 Interlinkages and Components of IWRM

The physical, institutional, economic, and social linkages of water are depicted in Fig. 1.7. IWRM begins with the term "water resources management" itself, which uses structural measures and nonstructural measures to control natural and human-made water resources systems for irrigation and other uses. Water-control facilities and environmental elements work together in water resources systems to achieve water management purposes.

Structural components used in ***human-made systems*** control water flow and quality and include conveyance systems (channels, canals, and pipes), diversion structures, dams and storage facilities, treatment plants, pumping stations, and hydroelectric plants, wells, and appurtenances.

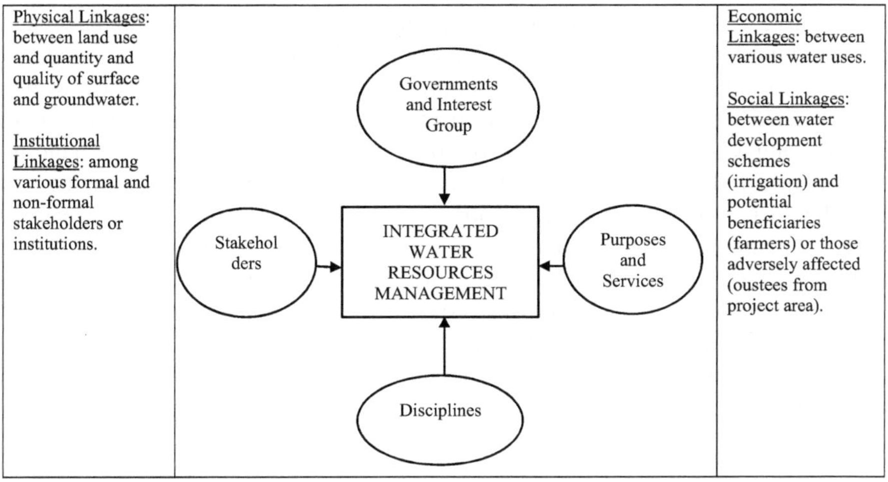

Fig. 1.7 Physical, Institutional, Economic, and Social Linkages of Water

Elements of ***natural water resources systems*** include the atmosphere, watersheds, stream channels, wetlands, floodplains, aquifers, lakes, estuaries, seas, and the ocean. Examples of ***non-structural components***, which do not require constructed facilities, are pricing schedules, zoning, incentives, public relations, regulatory programs, and insurance.

1.8.3 IWRM Principles

At the International Conference on Water and the Environment (ICWE), held in Dublin, Ireland in 1991, the following principles were recommended to guide IWRM:

Principle 1 "Ecological": Freshwater is a finite and vulnerable resource, essential to sustain life, development, and the environment. Recognizing the catchment area or river basin as the most appropriate unit for WRM, Principle 1 calls for coordination across the range of human activities that use and affect water in a given river basin. IWRM approaches to incorporate this principle into its emphasis on integration between all concerned water sectors.

Principle 2 "Institutional": Water development and management should be based on a participatory approach, involving users, planners, and policy-makers at all levels. This participatory approach is to raise awareness of water issues among policy-makers and the general public. It advocates increased accountability of management institutions and full consultation and involvement of users in the planning and implementation of water projects. The capacity of certain disadvantaged groups may need to be enhanced through training and targeted pro-poor development policies for full participation.

Principle 3 "Gender": Women play a central part in the provision, management, and safeguarding of water. Worldwide, women play a key role in the collection of water for domestic—and often agricultural—use, but in many societies, women are excluded from water management decisions. IWRM includes an emphasis on empowering women in its focus on participatory management and capacity building.

Principle 4 "Instrument": Water has an economic value in all its competing uses and should be recognized as an economic good. Known as the "instrument principle", the approach emphasizes economic and financial sustainability. The human right to access clean water and sanitation at an affordable price must be recognized, but the scarcity of water demands that economic perspectives should not be ignored. Managing water as an economic good is also a key to achieving the financial sustainability of water service provision, by making sure that water is priced at levels that ensure full cost recovery.

1.8.4 Challenge for IWRM in Developing Countries

Most examples of successful initiatives for IWRM are from developed countries. There are certain aspects that policymakers must keep in mind when designing IWRM policy for developing countries versus developed countries. Differences in institutional and socioeconomic characteristics affect the success of an integrated water management strategy between developing and developed countries. The following are the main challenges that developing countries face in setting up successful IWRM frameworks (ADB 2007).

Lack of coordination across Relevant Line Departments: Regulatory conditions are most often not conducive to establishing organizations that require a cross-disciplinary and integrated approach to resource management. Reductions in discretionary authority on the part of existing management agencies may be difficult because of their unwillingness to relinquish control.

Dominance of Agricultural Sector: Irrigation accounts for almost 70% of water use in developing countries. Irrigation accounts for as little as 3.3% in some developed countries (ADB 2007). The dominance of agriculture sector has several implications; with a large proportion of the (mostly small-scale farmers) population dependent on irrigation, the government finds it hard to regulate resource use across several users.

In addition, farmers have a significant voice in the political system, and accordingly diluting the powers of the irrigation department would amount to political suicide in several developing countries. In parallel, the irrigation departments in these countries are often large bureaucracies with an army of irrigation specialists. Hence, relinquishing power and resources to other departments is hard. However, the very concept of IWRM requires a more balanced cross-sectoral management structure.

Conditions imposed by Donor Agencies: Projects that fund domestic water supply focus on drinking water, while industry focuses on hydropower. With the clear sectoral delineation of funds for water projects, some of them are tied to the conditionality of implementing the water sector reform. The system does not allow for cross-sectoral cooperation and leads to competition, which is not conducive to IWRM.

IWRM, as practiced, in developed countries allocates water through regulation and incentive mechanisms such as water pricing. Water pricing will increase the efficiency of water use and will allow for better maintenance of the water-related infrastructure. However, in some countries, putting a price on water may not be possible because the informal nature of water use may make monitoring and enforcement difficult. A large proportion of the population in developing countries does not have access to the most basic facilities of drinking water and sanitation, and putting a price on the water for these sections (although in several cases they end up paying private water providers [tankers] for their daily quota) will not be possible.

In developing countries, financial, infrastructure, and human capacities to fulfill the responsibilities of government are often lacking. This is especially important in the case of monitoring water use and water quality.

1.8.5 Risks in Hierarchical Integration

The hierarchical integration of irrigation systems into local-regional-national economic systems is shown in Fig. 1.8. Interlinkages occur through various inputs and outputs. Key inputs and outputs among the hierarchical systems are also indicated. However, there are limitations and risks in such integration:

Getting mired in complexity and not making good use of specialist expertise;

a) Overlooking negative impacts on the environment and other sectors;
b) Inefficient use of resources—natural and financial;
c) Lack of coordination across relevant line departments of the government;
d) Dominance of the agriculture sector in the economy; and
e) A very large number of stakeholders having different self-interest.

Each state or department needs to decide where integration makes sense, based on its social, political, legal, financial, and hydrological situation.

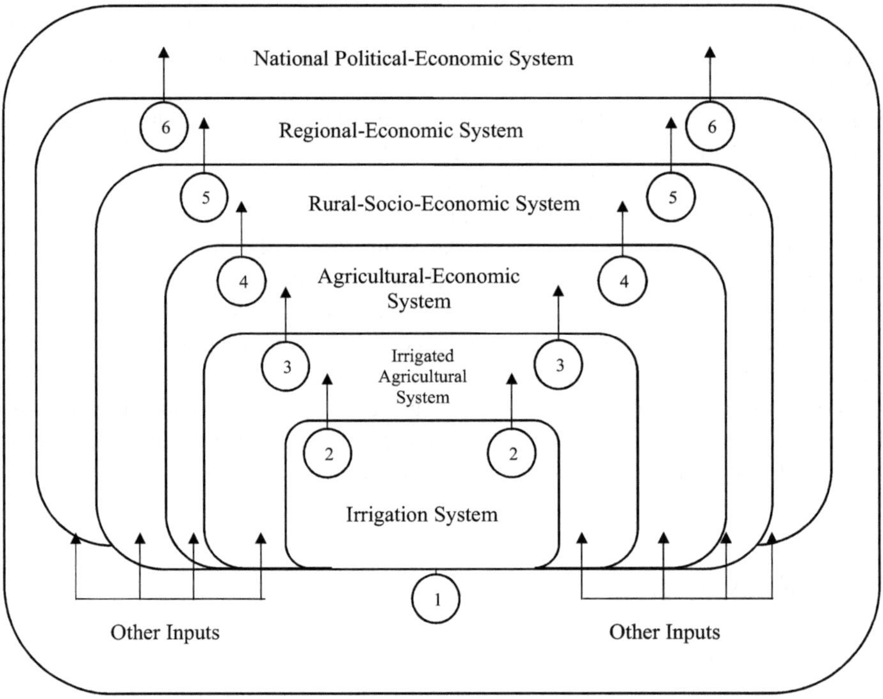

Key to Inputs/ Outputs:
1. Operation of Irrigation Facilities, 2. Supply of Water to Crops, 3. Agricultural Production, 4. Income in Rural Sector 5. Rural Socio-Economic Development, 6. Development of regional economy

Fig. 1.8 Hierarchical integration of irrigation system into higher-level systems

Questions

1. List structural components (physical facilities) of a water control subsystem.
2. Draw a figure to depict physical facilities for transferring water from a river to an agricultural field.
3. Besides increasing crop production, what other social aims/goals are served by an irrigation project?
4. Explain in brief the following terms:

   (i) Natural inputs, (ii) management inputs, (iii) irrigated agriculture system, (iv) self-reliance, (v) integration (vi) tasks, (vii) system elements, (viii) management activities, and (x) implementation tools.
5. Explain the importance of monitoring and feedback information during the operation and maintenance stage of an irrigation project.
6. Explain the need for taking a system view of irrigated agriculture?
7. Draw a figure and explain irrigated agriculture management as a planned system.

8. Discuss functions and elements of the water control subsystem. Explain interlinkages in the sub-system.
9. Discuss functions and elements of the socio-economic subsystem. Explain interlinkages in the sub-system.
10. Draw a figure and discuss the method of classifying the cropping subsystem.
11. In which of the five stages of an irrigation project involvement of farmers is most needed?
12. Planning and implementation can and should proceed in tandem, Elaborate on this statement.
13. How is an economic subsystem different from a social and organizational subsystem?
14. Explain with examples, the meaning, concept and principles of integrated water resource management.
15. A river valley project serves the purposes of irrigation, hydropower and flood control. Discuss various interlinkages with examples.
16. Why implementation of IWRM in India has not been as successful as in developed countries?
17. Search the literature and prepare a note on successful implementation of IWRM in any developed country.
18. Explain hierarchical interlinkages between irrigation system and other sectoral economic systems.

References

ADB (2007) Institutional options for improving water management in India—the potential role of river basin organization, a publication of Asian Development Bank, New Delhi, India (Publications stock No 110507)

Grigg NS (1996) Water resources management: principles, regulations, and cases. McGraw-Hill, New York

Mitchell B (1990) Integrated water management. In: Integrated water management: international experiences and perspectives. Bel-haven Press, London, UK

Murray-Rust DH (1992) Strategic and operational management in irrigation systems. In: FAO. Regional Office for Asia and the Pacific (RAPA); Japan International Cooperation Agency (JICA). Irrigation-water management for sustainable agricultural development: report of the Expert Consultation of the Asian Network on Irrigation-Water Management, Bangkok, Thailand, 25–28 August 1992. Bangkok, Thailand: Japan International Cooperation Agency (JICA); FAO. RAPA, pp 31–44 (RAPA Publication 1992/24)

Skogerboe GV (1990) Development of the irrigation M&O learning process in irrigation and drainage systems, vol 4. Kluwer Academic Publishers, Netherlands, pp 151–169

Smith RA (1970) Management structure for irrigation. In: Proceedings of the journal of irrigation and drainage division on ASCE, December 1970, pp 475–488

Reading Material

Gideon P, Skogerboe GV (1979) Evaluation and improvement of irrigation system. Colorado State University, Report 49A, USA

Government of India (1979) Manual on irrigation and water management. Ministry of Agriculture & Irrigation, New Delhi

Lowdermilk MK et al (1983) Diagnostic analysis of irrigation system, volume 1—concept and methodology. Colorado State University, USA

WAPCOS (1989) Handbook for improving irrigation system maintenance practices. Technical Report No. 19-A, Power Consultancy Services (India) Ltd., January 1989, New Delhi

Chapter 2
Irrigation Management in India: Problems and Issues

Abstract In an agriculture-based economy such as in India, the purpose of irrigated agriculture is not only to improve agricultural production but also to serve higher-level societal goals-most importantly self-reliance in food crop production and improving the socio-economic condition of backward regions/groups in the country. This chapter deals with problems and issues related to irrigation implementation and management in India. Though the content is India-centric, it is equally relevant to other regions/countries in South Asia as socio-economic and physical conditions are similar. Water and land resources are discussed. Much of the agricultural land in India is privately owned in the form of fragmented small holdings. Most of the irrigation schemes in India are financially sponsored and owned by the state governments. The government follows an irrigation policy to provide irrigation water to at least one major crop in a year and a large number of scattered land holdings as is possible. A large number of major, medium and small irrigation schemes have been developed in the country and a large irrigation potential has been created. Command Area Development works (mainly on-farm development works) have been taken up to rehabilitate projects and bridge the gap between potential created and potential utilized.. The need and scope for improving hardware (canal carrying capacity, regulators, escapes, unregulated fixed outlets, canal lining, and poor maintenance) and software components of irrigation systems mainly relating to water conveyance and water distribution in the field are explained. Hardware and software components of an irrigation system are schematically depicted in Fig. 2.1 The pricing of irrigation water has been critically examined. The need for the rationalization of irrigation water charges is discussed. Issues related to irrigation administration, tube well irrigation, and participatory irrigation management, etc. are critically examined. The meaning and scope of rehabilitation and modernization activities are also explained.

U. C. Chaube et al., *Canal Irrigation Systems in India*, Water Science and Technology Library 126, https://doi.org/10.1007/978-3-031-42812-8_2

2.1 General

The primary objective of any irrigated agriculture project is to improve crop production. In addition, in developing countries with dominance of rural economies such as in India, improved agricultural production serves a combination of the following higher-level goals:

(i) Self-reliance in food crop production;
(ii) Import substitution;
(iii) Increase in export of agriculture-related commodities;
(iv) Improving the socio-economic condition of backward regions/groups;
(v) Transmigration and resettlement of population from high population density areas to low population density areas;
(vi) Self-employment in rural areas; and
(vii) Growth of agro-based industries.

Based on the cultivable command area (CCA) of irrigation schemes, irrigation projects are termed as

1. Major irrigation projects- CCA > 10,000 ha;
2. Medium irrigation projects-CCA > 2000 ha and < 10,000 ha; and
3. Minor irrigation projects-CCA < 2000 ha.

Major and medium irrigation projects make use of surface water resources, whereas minor projects may have surface water or groundwater as the source. Minor irrigation works have great importance (dug wells, tube-wells, tanks, etc.) in the Indian irrigation sector. The total area benefitting from minor irrigation works has historically always exceeded the corresponding area covered by major and medium projects. Irrigated agriculture in India has made the country self-reliant in food production. Yet, many irrigation projects are suffering from inadequacies of management, as a result of which the utilization of irrigation potential has been rather unsatisfactory.

An irrigation project consists of water conveyance facilities above outlets (main conveyance system) and water distribution facilities in the outlet command area (on-farm system). Figure 2.1 depicts the hardware and software components of the main conveyance system and on-farm system. Problems and issues in irrigation management as related to these hardware and software components are discussed in this chapter.

2.2 Water Resource and Agricultural Land Resource

India has 16% of the world's population and its geographical area is only 2.4% of the world's total area. The country has only 4% of the total available fresh water on earth. The livelihood of about 70% of the population depends directly on agriculture

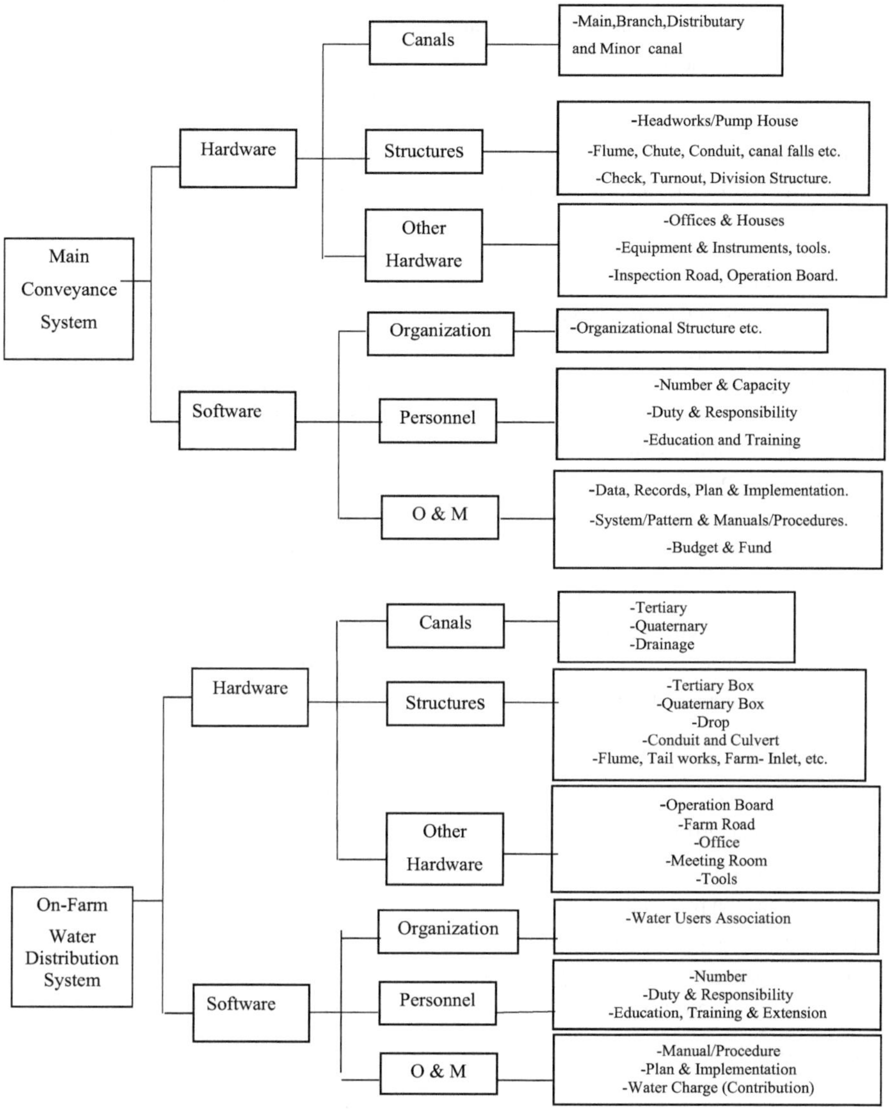

Fig. 2.1 Hardware and software components of irrigation system

or agro-industry 30% of our GNP is derived from agriculture which, in turn, depends on land and water resources. Land and water resources facts of India are shown in Table 2.1 (CWC 2020).

Table 2.1 Water resources and agricultural land in India

Geographical area and location	328.7 M ha Latitude; 80 4′ N–370 6′ N Longitude: 680 7′ E–970 25′ E
Population (2011)	1210.19 Million
Major River Basin (catchment area more than 20,000 km^2)	12 nos. having total catchment area 253 M ha
Medium river basin (catchment area between 2000 and 20,000 km^2)	46 nos. having total catchments area 25 M ha
Total navigable length of important rivers	14,464 km
Rainfall, utilizable water, storage potential	
Rainfall variation	100 mm in Western most regions to 11,000 mm in Eastern most region
Average annual rainfall (1985–2015)	1105 mm (3880 BCM)
Annual rainfall (2018)	1074 mm
Mean annual natural run-off	1999.2 BCM
Total utilisable water	1122 BCM
Estimated utilisable surface water potential	690 BCM
Net ground water availability (2013)	411 BCM
Live storage created	253 BCM
Additional live storage of dams under construction/ consideration	155 BCM
Agricultural land and irrigation potential	
Total cultivable land	182.2 M ha
Gross sown area (2014–15)	198.4 M ha
Net sown area (2014–15)	140.1 M ha
Ultimate irrigation potential	140 M ha
Ultimate irrigation potential from surface water	76 M ha
Ultimate Irrigation Potential from ground water	64 M ha
Net irrigated area (2014–15)	68.4 M ha

Source CWC (2020)

2.2.1 Water Resources

Central Water Commission (CWC 2019) reports that the mean annual precipitation volume over the country is about 3880 Billion Cubic Meters (BCM). The average annual water availability, after evaporation, is assessed at 1999.20 BCM. Due to geological and other factors, the utilizable water is limited to 690 BCM of surface water and 432 BCM of groundwater per year. The water potential utilized is around 699 BCM, comprising 450 BCM of surface water and 249 BCM of groundwater. The total requirement of the country for different uses for high-demand scenarios for the years 2025 and 2050 has been assessed as 843 BCM and 1180 BCM, respectively.

The average annual per capita water availability for the years 2001 and 2011 was assessed at 1816 cubic meters and 1545 cubic meters, respectively. The average annual per capita water availability is expected to decline to 1340 cubic meters and 1140 cubic meters in the years 2025 and 2050, respectively. The annual per-capita water availability of less than 1700 cubic meters is considered as water-stressed condition, whereas annual per-capita water availability below 1000 cubic meters is considered as a water scarcity condition.

Agricultural Land Resources

India has a geographical area of 328.7 million hectares (m ha) out of which the arable land is 182.2 million hectares (m ha). The net sown area in 2014–15 was about 140.1 m ha (76.89%). The population of the country was 1210.855 million as per the 2011 census the second largest in the world. The estimated population in the year 2021 is 1393.41 million.

Thus, agricultural land supports 9.95 persons per ha of the net sown area, a high but by no means an unsustainable figure. Much of the land in India is flat and culturable; the Indo-Gangetic plains and the coastal belts are some of the most fertile areas in the world, well-favored by water and sunshine, the other essential ingredients of productivity.

The dominant pattern of land ownership in India is small, privately owned farms. The pernicious system of hereditary intermediaries or zamindars evolved during the chaos of the last days of the Moghul Empire and institutionalized by the British has been largely abolished. The pattern of land holdings in the country is shown in Table 2.2.

Even the small holdings of the farmers are generally fragmented into disconnected plots. This pattern presents problems in the development of an efficient water distribution system and other services, such as farm roads, drains, etc., but many countries of South-East and East Asia, such as Taiwan, South Korea, and Japan, have developed highly productive agriculture under similar patterns of land ownership.

Table 2.2 Pattern of landholding

Plot holdings	Percentage of households	Percentage of land
Below 2 ha	49	24
2–10 ha	21	53
More than 10 ha	3	23

2.3 Irrigation Policy and Potential

2.3.1 Irrigation Policy

In India, the development of irrigation and its management has been traditionally based on the concept of social welfare. Irrigation projects are designed to provide irrigation water to at least one major crop in a year over as large a number of land holdings as is technoeconomically possible. Irrigation projects aim at the removal of poverty, self-reliance in food crop production, and improvement in the quality of life particularly in backward regions and weaker groups of society.

It has been estimated that almost 80% of additional food production in India is contributed by irrigated agriculture. Increased agricultural production has diminished rural poverty in several ways. It has kept food prices in check enabling poor families to meet their nutritional requirements.

Besides increased productivity, irrigation development has had an important influence on the livelihood of rural people. It is observed that after the introduction of canal irrigation, there has been a significant increase in the number of days worked by agricultural labor and the wages also increase. Improved security against impoverishment has resulted in better health, better education, better shelter, and more time for the care of children.

2.3.2 Irrigation Potential

Created Irrigation Potential: It refers to the total gross area that can be irrigated annually by the water made available by the completion of the distribution network and other related works up to the end of the watercourse or the last point in the water delivery system.

Utilized Irrigation Potential: It refers to the total gross area irrigated during the year under consideration.

Table 2.3 shows the total irrigation potential created and utilized up to the year 2012. As seen from the table, the potential utilized is significantly less than the potential created in the plan periods. The delay in the construction of distribution networks, diversion of water for domestic/industrial water supply, change in cropping pattern resulting in the adoption of crops with higher water consumption, etc. are some of the reasons for the gap in potential created and utilized.

Table 2.3 Irrigation potential created and utilized during plan periods (in M ha)

Plan	Created potential					Utilized potential				
	Major and medium	Minor			Total	Major and medium	Minor			Total
		S.W	G.W	Total			S.W	G.W	Total	
Up to 1951 (pre-plan)	9.70	6.40	6.50	12.90	22.6	9.70	6.40	6.50	**12.90**	22.60
Cumulative up to to 2012 (XI the plan)	45.34	N.A	N.A	63.57	108.91	34.66	N.A	N.A	**52.73**	87.39

Source Government of India (2011), 'Report of the Working Group on Major and Medium Irrigation and Command Area Development for The XII Five Year Plan (2012–2017)', 'Ministry of Water Resources, Government of India, New Delhi

2.4 Command Area Development

Centrally sponsored Command Area Development (CAD) Programme was launched in 1974–75. In the year 2004, water use management was also included in the CAD Programme and renamed as Command Area Development and Water Management (CAD & WM) Programme. The program is under implementation as a sub-component of Har Khet Ko Pani (HKKP) component of Pradhan Mantri Krishi Sinchayee Yojna (PMKSY)—from 2015 to 16 onwards. The ongoing CAD & WM program were restricted to the implementation of CAD works of 99 prioritized AIBP projects during 2016–17 to December 2019.

Coverage: At the time of inception in 1974–75, it covered 60 selected major and medium projects with a cultivable command area (CCA) of 15 million hectares. In the year 1998, there were 203 projects under the program with a CCA of 211 million hectares. As of March 2014, there were 150 ongoing projects under CADA with a CCA of 16.3 M ha. But during 2016–17 through December 2019, only 99 prioritized AIBP projects were taken under PMKSY. As of December 2019, the CAD works had been completed in respect of 9 projects, and the remaining 90 CAD & WM projects under PMKSY have yet to be completed.

Table 2.4 shows CCA covered under CAD and under CAD & WM program during plan periods.

Table 2.4 Physical and financial achievement during the given plan periods (in M ha)

Sl. no.	Plan period	CCA covered (in Million ha)
1	Up to VIII plan (ie upto 1997)	13.952
2	IX plan (1997–2002)	1.801
3	X plan (2002–2007)	2.314
4	XI plan (2007–2012)	2.08
5	XII plan	
(a)	Up to 2015–16	1.419
(b)	2016–17 to 2019–20 (up to Dec. 2019)	1.472
	Total	23.038

Source CWC (2022)
Note Data for the period from 2012–13 to 2019–20 are under revision/compilation as per CWC Report of the year 2022 "Water Resources at a Glance 2022"

2.5 Problems Related to Canal Design, Operation and Maintenance

Figure 2.2 schematically depicts the hardware (canal and related structures) of a surface irrigation scheme. Problems related to hardware and software components of Irrigation schemes in India have been discussed by several authors (Singh 1985; Chaube and Varshney 1992).

2.5.1 *Channel Capacity*

In the past, the approach to capacity determination has been based on an adopted 'duty' often irrationally determined. The scientific procedure for estimating crop evapotranspiration and other requirements over ten-day periods and adopting the peak water requirement for channel design is now generally accepted. Design cropping patterns and design intensity of irrigation are based on planning policy and socio-economic decisions at the regional level. Many of the existing projects have been designed for 30–40% annual irrigation intensity. New systems will have to be designed for 100–180% annual irrigation intensity with conjunctive use of surface water and groundwater wherever possible. Rotational water delivery and conjunctive use of surface water and groundwater can be utilized to reduce the design capacity of irrigation channels. Crops should also be so planned that the canal capacity is fairly uniformly utilized over the year. The cropping patterns adopted by the farmers are quite different from the design cropping patterns in several of the project areas.

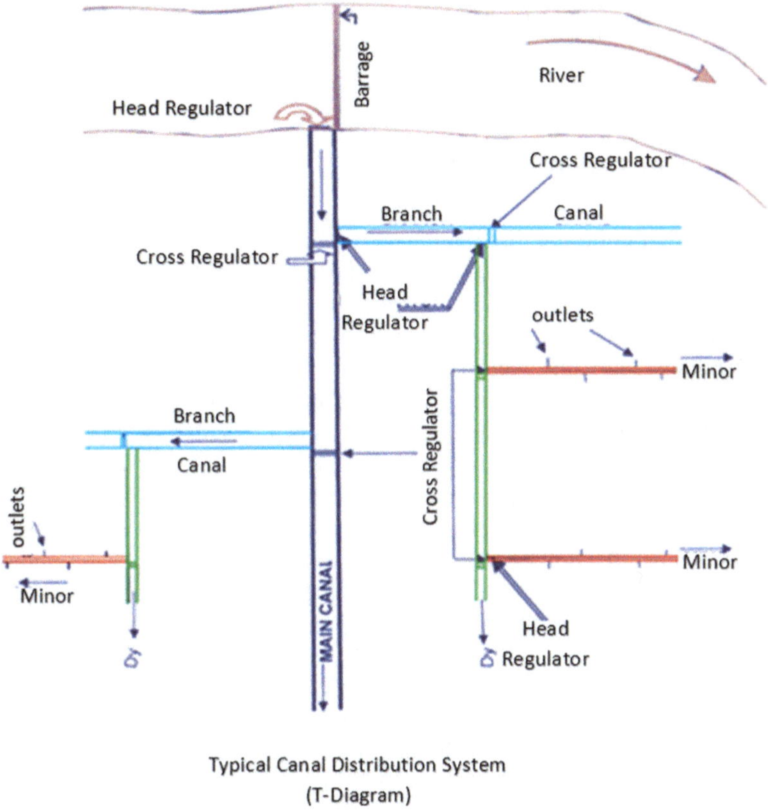

Fig. 2.2 Layout of canal and related structures

2.5.2 *Inadequacy of Regulators and Escapes*

Due to the scarcity of financial resources, an adequate number of regulators and escapes have not been provided in the distribution network to regulate the distribution of water according to the adopted management policy.

2.5.3 *Unregulated Fixed Ventage Outlets*

The outlet is crucial in so far as the water distribution to farmers is controlled at the outlets. It is also an administrative divide between the irrigation department at its upper end and groups of farmers at the lower end. The Command Area Development Authority, where established, generally takes over works beyond the outlet. Most of the projects have unregulated, fixed ventage outlets, which are not allowed to be closed except by decision at a very senior level and in practice are seldom closed. For supply aiming to meet the variable demand, and arguable demand, regulated outlets become necessary to adjust the flow by the requirement.

2.5.4 Issue of Canal Lining

Many of the Irrigation water distribution networks in India are not lined. Canal lining involves large initial investments which could not be possible due to the scarcity of financial resources. The seepage losses from unlined canals are generally high, possibly 40% or more. In unconsolidated alluviums, such as in the states of Uttar Pradesh, Punjab, Haryana, etc., the intensive development of well irrigation has stabilized or is tending to stabilize the water tables. The seepage from canals in such a situation may not be termed as a loss—it is a recharge of groundwater from which the farmer may draw water at his convenience and as needed. It does involve the expenditure of energy, but the manufacture of lining materials, such as cement, bricks, etc. also uses a large block of energy. Economic analysis need to be carried out to find the cost of a unit of water saved by lining and a unit of water obtained by pumping and should compare the two. Even if the pumped water is a little costlier, it may turn out to be more profitable in terms of productivity due to timely and efficient application.

If the lining is to be done, the most cost-effective is the lining of watercourses. Large-scale watercourse lining has been carried out in Haryana and Punjab. This work has not yet been taken up in UP. Studies have shown that even simple measures, such as cleaning and compaction of unlined watercourses, yield better returns on investment.

2.5.5 Operation

Generally, the main canal and branches run continuously, while the distributaries and minors are operated intermittently such as in the Upper Ganga Canal (UGC) system in Uttar Pradesh. Assessment of demand is often subjective and approximate. A water distribution plan, based on cropping patterns, net irrigation requirements, and water availability, should be prepared at the beginning of each season and updated every fortnightly. Field data on cropping patterns, crop growth stage, soil moisture at the beginning of the irrigation period, and climatic parameters need to be utilized in preparing and revising the water distribution plan.

2.5.6 *Maintenance*

The expenditure on the operation and maintenance activities, including repairs, replacement, etc., is met from the non-plan funds. A major portion of maintenance provision is spent on the staff salary payments, with the result that very little is left for the actual maintenance of work and it remains unattended and consequently, the system deteriorates.

The irrigation water rates, as per the National Water Policy, are required to meet at least full operation and maintenance costs. It is desirable that they also bear part of the interest in capital expenditure. However, due to low water rates and low recovery of water rates, the actual revenue received is much less than the working expenses of irrigation systems. As a result of this situation, the irrigation systems are not maintained properly. The tertiary level system, part of which is to be maintained by farmers, is at times facing obliteration due to non-maintenance.

2.5.7 *Night Irrigation*

Night irrigation has lower efficiency but cannot be avoided in non-storage schemes. It can also not be avoided on large and long canals due to the large time required for filling and emptying (example-Upper Ganga Canal in Uttar Pradesh). But it can be minimized on short-length canals from small tanks. During the night the flow could be minimized and increased again during the day. Such a system has successfully experimented on Morna Project in Maharashtra.

2.6 Water Distribution Below Outlets

Water distribution below the outlet aims to provide water in proper quantity and at the right time to the crops. Rotational delivery and field application and field drainage are important components (Fig. 2.3).

The vantage of the unregulated outlet is related to a nominal discharge, though actual flows under non-modular (submerged flow) conditions would vary considerably. Most of the canals in Uttar Pradesh are not suited to semi-modules due to the small available working heads. If direct outlets on branches or main canals and just upstream of falls and regulators, which draw a disproportionate share of discharge, are converted to sub-proportional semi-modules, the task of the irrigation engineer in attempting to secure an equitable distribution of water will become that much easier (Singh 1985).

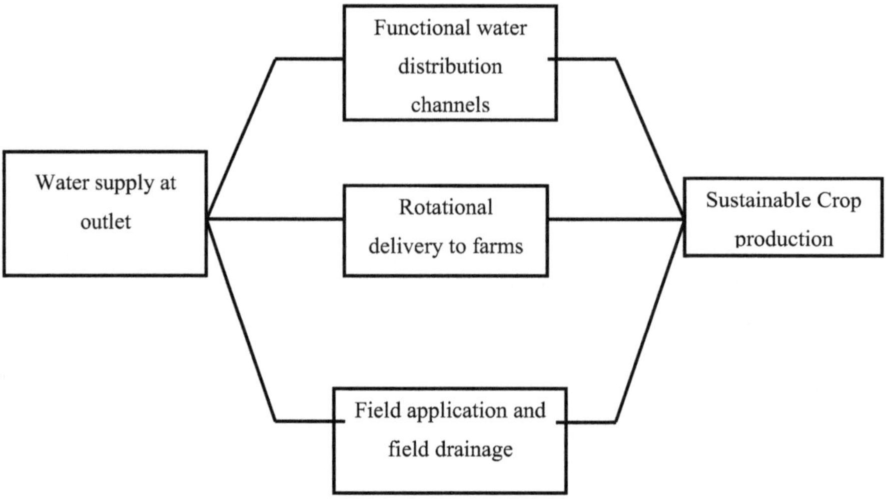

Fig. 2.3 Water distribution, field application, and drainage in outlet command area

2.6.1 Rotational System

'Warabandi' (rotational delivery) allows for the delivery of water to all farmers within an outlet command according to their predetermined entitlement. 'Wara' means turn and 'andi' means fixation; warabandi means the fixation of turn. The entire flow of the outlet is allowed to the farmer for his allotted time during the rotation period. In northern states, the time allotment is based on the area of the farm without consideration of seepage losses in the watercourse, crop type, and crop irrigation requirements. In Maharashtra and southern states, the farmer has to seek approval for irrigation of specific areas of specific crops, and on allotment gets a proportionate time. This is known as the 'Shejpali' system. He is supposed to use the water for those crops only for which it is allotted. Rotation is feasible and is practiced in both systems. The Northern Warabandi is simpler than the southern Shejpali system.

The Warabandi system has been found in practice to be efficient, acceptable to the farmers and hence minimizes disputes such as in Upper Ganga Canal Command in the western parts of Uttar Pradesh.

2.6.2 Equity and Timeliness of Supplies

Equity, as related to water delivery, may be defined as the fair share of water to users at different locations in the command area. It is one of the main objectives of irrigation management. What is usually meant by equitable distribution is the allocation of water through time rationing according to the proportional area held by the farmers in the outlet command. But, in a strict sense, it is a proportional distribution of water and not equitable distribution of water. There is an implied objective in the distribution aspect of irrigation water which is normally lost sight of. This is related to the welfare and social justice objective of water between various classes of landholders, namely, large, medium, and small farmers, and as between different system reaches, namely, head, middle, and tail-enders. The prevalent practice of water distribution and even the ideal implementation of Warabandi considered presently to be a panacea of all ills of equity in the irrigation water supply do not address themselves to this basic question of equity in a complex sense.

All farmers must have equal security or timeliness of supplies. They feel secure and less risky if the timeliness of supplies is maintained at the outlet. Mismanagement of water is most visible in the farmers' fields, while it is less visible in the main system. But the fact remains that reliable, equitable, and timely irrigation water supply is a precondition for good water management by the farmers below the outlets. The gap between demand and supply is always present, demand being generally more than supply. Effective water management plans, therefore, focus on the equitable and orderly distribution of water among the farmers well in time, safeguarding against any crop stress.

There is considerable head end-tail end conflict, with head-reach farmers taking more water than needed, thereby constraining water supply to tail-end farmers. The net result of inefficient water management is reduced productivity and income for all recipients along the distribution chain.

2.6.3 Need to Improve Field Application Efficiency

Most of the cereal crops in the country are irrigated by the check-basin method. For rice fields, free flooding is used. The border strip method is being adopted by a few farmers in recent years. Row crops are generally irrigated by the furrow method. All these are surface irrigation methods. With proper land preparation and care during irrigation, reasonably high field application efficiencies can be obtained by these methods. The stream sizes are generally small, there is little runoff from the fields. However, percolation below the root zone does occur particularly in the early stages of crop growth when the root zone has a shallow depth and this needs to be corrected.

2.7 Tubewell Irrigation

In India, there are two systems of tubewells, those owned and operated by the state and others by private farmers. The latter is primarily for their use, though they may sell water to their neighbors when there is surplus water in excess of their needs. Irrigation from open wells by animal or human power has sharply declined. Rather surprisingly there has been a little trend towards cooperative ownership. Also, private tube wells installed for the sole or main purpose of selling water as practiced in Bangladesh are not popular in India.

2.7.1 *Performance of State Tubewells*

Performance studies of state tubewells reveal that their potential is not fully utilized. One of the reasons for low performance is the erratic and inadequate power supply. Another reason is poor maintenance. The cost of irrigation from state tubewells is also much higher than that from private tubewells. This suggests the need to encourage private development of groundwater in the future and the turnover of state tubewells to water user associations. However, it would be necessary to protect the interest of small farmers who can neither afford a tube well on their own nor buy tubewell water at a higher price. Depending on the local situation, alternatives could be cooperative or joint ownership, or private ownership with state subsidy and loan, with the obligation to sell water at fixed rates for a certain number of years.

2.7.2 *Electricity Subsidy*

The general trend in most states is to charge a flat rate per horsepower per month. At this rate, the resource cost of supplying power for pumps is three to four times the price paid by the farmers and involves a heavy subsidy to farmers with energized pumps. Besides this, the flat rate encourages large-scale misuse of power for purposes other than pumping. Due to weak administration, metered supply has, in any case, become impractical, and the flat rate has the consolation of having a definite positive impact on agricultural production.

2.8 Underpricing of Water

The underpricing of water adversely affects the availability of resources for the management of irrigation systems. Inadequate allocation for maintenance and repairs acts as a direct consequence of the poor financial position of the states and is responsible for the low, possibly deteriorating quality of service. The National Water Policy Statement of 1987 specifically states:

> The Water Rates should be such as to convey the scarcity value of the resource to the users and to foster the motivation for the economy in water use. They should be adequate to cover the Maintenance and Operation Charges and a part of the fixed costs of irrigation works. Efforts should be made to reach this ideal over a period of time while ensuring assured and timely supplies of irrigation water. The Water Rates for Surface Water and Ground Water should be rationalized with due regard to the interest of small and marginal farmers.

Similar sentiments have been expressed by the National Water Policy Statement of 2002 and 2012.

The National Irrigation Commission (1972) had stressed levying water charges on a crop basis considering (i) the adequacy and dependability of water supply, (ii) the need for common policy among the neighboring states, (iii) the water requirement of crops, and (iv) the revision of water rates in every five years.

Planning Commission of the Government of India have also been recommending from time to time various principles for fixing irrigation rates (GOI 1992). The 14th Finance Commission inter-alia recommended that (i) All States, irrespective of whether Water Regulatory Authorities (WRAs) is in place or not, consider full volumetric measurement of the use of irrigation water; and (ii) States which have not set up WRAs, consider setting up a statutory WRA so that the pricing of water for domestic, irrigation and other uses can be determined independently and judiciously. Further, WRAs already established should be made fully functional.

2.8.1 Wide Variations in Water Rate Structures Across States

There are wide variations in water rate structures across states and the rate per unit volume of water consumed varies greatly across crops. The rates vary widely for the same crop in the same state, depending on the irrigation season, type of system, etc. A multiplicity of principles is followed in fixing water rates, such as the recovery of the cost of water, the capacity of irrigation to pay, based on gross earnings or net benefit of irrigation, the water requirement of crops, sources of water supply, and its assurance, classification of land linked with cess, betterment levy, etc.

The presently existing system of water charges in the States/Union Territories (UTs) has been detailed in the report of Central Water Commission (CWC 2017). The CWC report provides State-wise crop-specific water rates for some of the important crops (viz. paddy, wheat, sugarcane, cotton, oilseeds, and pulses) in respect of flow as well as lift irrigation. Apart from this, an analysis in respect of capital expenditure and working expenditure along with gross receipts over the period 2000–01 to 2013–14 in major and medium projects has been provided. Finally, it gives details on the gap in revenue assessed and realized for the same period for all States and Union Territories. The following tables have been taken from CWC (2017). Table 2.5 presents the overall range of water rates State-wise in respect of flow vis-a-vis lift irrigation and the dates when the rates were last revised.

Similarly, State-wise crop-specific water rates in respect of flow irrigation and State wise crop-specific water rates in respect of lift irrigation are given in the CWC report (CWC 2017).

Table 2.5 State/UT-wise water rates for flow and lift irrigation (unit Rs/hectare)

State/UT	Flow irrigation range		Lift Irrigation range		Date since applicable
	Max.	Min.	Max.	Min.	
1	2	3	4	5	6
Andhra Pradesh Arunachal Pradesh Assam Bihar Chhattisgarh Delhi Goa Gujarat Haryana Himachal Pradesh Jammu and Kashmir Jharkhand Karnataka Kerala Madhya Pradesh Maharashtra Manipur Meghalaya Mizoram Nagaland Orissa Punjab Rajasthan Sikkim Tamil Nadu Tripura Uttarakhand Uttar Pradesh West Bengal A and N Islands Chandigarh* Dadra and Nagar Haveli Daman and Diu Lakshadweep Puducherry	864.50	148.20	NA		01-07-1996
	No water rates				29-12-2008
	751.00	150.00	751.00	150.00	30-03-2000
	370.50	74.10	NA		Nov-2011
	741.00	123.50	741.00	123.50	15-06-1999
	148.20	34.03	148.20	33.35	2009
	360.00	72.00	720.00	144.00	01-04-2013
	300.00	160.00	100.00	53.33	01-01-2007
	197.60	24.70	98.80	12.35	27-07-2000
	49.92	49.92	99.81	99.81	01-04-2015
	298.87	121.03	2998.58	298.87	01-04-2015
	370.50	74.10	370.50	74.10	26-11-2001
	988.40	37.00	1976.80	74.00	13-07-2000
	99.00	37.00	148.50	93.00	18-09-1974
	960.00	50.00	960.00	50.00	31-12-2005
	6297.00	119.00	5405.00	20.00	01-07-2003
	602.00	184.00	602.00	184.00	24-08-2013
	No water rates				
	No water rates				
	No water rates				
	930.00	60.00	NA		05-04-2002
	123.50	123.50	123.50	123.50	12-11-2014
	286.52	29.64	573.04	14.82	24-05-1999
	250.00	10.00	NA		2002
	61.78	2.77	NA		06-11-1987
	312.50	312.50	312.50	312.50	01-10-2003
	474.00	30.00	237.00	15.00	18-09-1995
	474.00	30.00	237.00	15.00	18-09-1995
	123.50	37.06	2015.52	251.94	01-07-2003
	No water rates				
	NA				
	830.00	110.00	275.00	75.00	29-01-1996
	286.00	286.00	286.00	286.00	2007

(continued)

Table 2.5 (continued)

State/UT	Flow irrigation range		Lift Irrigation range		Date since applicable
	Max.	Min.	Max.	Min.	
	No water rates				
	NA				

* In rural areas of Chandigarh, the water rates for irrigation purposes is Rs. 23 per hour with effect from 01.01.2010. *NA* not available

2.9 Participatory Irrigation Management

It has now been recognized that unless farmers are progressively involved, in an organized way, in the operation, management, and maintenance of irrigation systems, the objective of increased utilization and production from irrigation commands cannot be realized and sustained in the long run. Many shortcomings of the present irrigation management could be reduced by effectively involving farmers in the irrigation management.

In India nearly 60% of landholdings receiving irrigation water are less than one hectare in size, belonging to marginal farmers. Irrigation department dealing with a very large number of farmers tends to administer water as per the departmental procedure rather than to manage it to the optimum satisfaction of the farmers.

A large number of discussions, seminars, workshops, etc. have been held on the desirability, scope, and purpose of water user associations (WUAs). The most important features of the consensus that have emerged in this regard are:

- The farmers need to organize themselves into society or co-operative society which would be the water users associations (WUAs);
- The irrigation departments should ensure bulk supplies to the water users association at distributary or minor level rather than dealing with a large number of individual farmers.
- Necessary changes required for nurturing such farmers' organizations need to be made in the Irrigation Act.

A large number of irrigation water users' co-operatives/societies have already been established in various states of the country. Irrigation Acts have been enacted at the state level to legalize the formation of water users associations.

2.10 Land Degradation Due to Irrigation

The development of irrigation facilities by itself is not responsible for drainage-related problems. Lack of adequate drainage provision, improper water management, seepage from canals, obstruction to natural drainage on account of various developmental activities, absence of a realistic operational plan especially during the initial stages of construction, inadequate support for maintenance, etc. have gradually led to the rise in groundwater level resulting in water-logging and soil salinity/alkalinity problem in some of the project command areas.

According to an estimate made recently an area of 2.46 M ha is affected by waterlogging and 3.06 M ha by alkalinity in irrigation commands (IWRS 1998).

2.11 Training

Lack of training of the irrigation and agricultural staff at all levels has been considered as one of the reasons for inefficient management. A large program for the training of staff and farmers has been taken up. For this purpose, Water and Land Management Institutes (WALMI) and Irrigation Management Training Institutes (IMTI) have been established in 11 major States and also in the northeastern region. As a result of training, the staff learns management techniques, and the management of irrigation systems is improving. They also study the problem of important commands through an action research program which requires a detailed study by diagnostic analysis of existing irrigated areas. The remedies and interventions which do not require much investment are suggested.

2.12 Rehabilitation and Modernization

Concepts and policies related to the rehabilitation and modernization of irrigation schemes are briefly explained below (Chaube 1985; Varshney 1992).

2.12.1 Meaning of Rehabilitation and Modernization

Maintenance is the process of keeping the irrigation, drainage, and other infrastructural facilities in good repair and working order, fulfilling the intentions for which they were originally designed. It would also involve improvements of a relatively minor nature, which could be performed during the normal process of maintenance.

Rehabilitation is the process of renovating an existing project, whose performance is failing to meet its design targets. Rehabilitation covers improvements (a) to the physical infrastructure, (b) to management and institutional aspects, and (c) to policy measures influencing the overall project,

Modernization is the process of updating and modernizing an existing project (which otherwise is meeting its design targets), to meet enhanced technical, social, or economic objectives. The definitions indicate that rehabilitation, as well as modernization, include structural and non-structural measures.

2.12.2 Need for Rehabilitation

Activities under the Command Area Development Program discussed earlier in this chapter aim at the rehabilitation of the project.

An irrigation scheme can only function properly if design, construction, operation, and maintenance are adequate. A default in any of these phases may require rehabilitation works. Inadequate operation and maintenance is perhaps the most frequent cause of the need for rehabilitation, especially in developing countries. In particular, insufficient maintenance may render an irrigation scheme completely obsolete and unable to meet project objectives.

Top-down Versus Bottom-up Approach: A top-down approach has the indisputable advantage of speeding up the process of decision-making as well as design and construction. However, the top-down approach tends to emphasize the technological component and disregard the human factor. Moreover, large schemes fit better in the investment criteria of funding agencies.

A bottom-up approach can take into account the historical and human factors, such as the average cultural standards of the rural population, its readiness in learning new techniques, the collectivism existing in the rural community, and especially the endurance of sometimes very old traditions. However, a pure bottom-up approach can afford satisfactory results only in projects composed of a multitude of small-scale systems. It cannot easily face the problems of large irrigation schemes, including design, construction, supervision, and operation and maintenance aspects of major structures and canal systems.

2.12.3 Need and Scope of Modernization

The objective of the modernization of canal irrigation is to bring a significant increase in crop production per unit of available water and land at an economic cost. This requires improvements of the following significant components.

Canal Lining: High-yielding varieties of crops need an assured water supply. India has limited water resources and multisectoral water demand is increasing. Therefore, the conservation of water by the lining of the conveyance system is necessary. It is necessary not to confine lining only up to the canal and distributary system, but to extend it in the canal commands in blocks of up to 8 ha, to contain seepage losses.

Conjunctive Use: Canal irrigation increases groundwater recharge in the irrigated area. Conjunctive use of surface water and groundwater can be made as follows:

(1) Irrigation of pockets exclusively with groundwater in a canal command especially where the terrain is uneven.
(2) Augmentation of canal water by putting tubewells along the canal.
(3) Conjunctive use of groundwater during the period of low canal supply or canal closures.

Modernization of Structures: Modernization of canal structures viz. Head Regulators, Cross Regulator, Cross Drainage, Works, falls, Bridges, Escapes, etc. will direct the following benefits in the canal system:

1. Capability to pass designed discharge at every point in the canal system.
2. Replacement of old obsolete plank-controlled regulators by steel gated regulators will ensure better operational efficiency and saving of water.
3. Development of the capability to remove sediment in the canal system to maintain the uniform capacity of big as well as small channels in the monsoon or rainy season.
4. Introduction of water measuring devices will ensure timely and equitable distribution of supplies and will be useful in the evaluation of distribution efficiencies.
5. It would be possible to adopt a user-oriented canal operation policy.

Remodeling and Construction of Additional Escapes: It is necessary to provide adequate numbers of escapes in the major canal in suitable locations to enable canals to be run with full discharge even during the monsoon season without any danger of a breach in the case of suddenly no demand due to rainfall. The escaping capacity on a major canal system has got to be at least 75% of the head discharge of a canal.

Improvement of Drainage in the Command: The drainage system in the canal command has to be reshaped with the introduction of a large number of artificial drains, construction of link drains, and improving the capacity of natural drainage. It will also be necessary to construct carriers, as well as links, drains up to natural drainage to effectively control canal seepage.

Improvement of Tele-communication on Canal Systems: Communication is an important factor for the efficient running of the canal system. The communication system needs modernization. The Internet communication system is a prerequisite to improving canal system operation and management efficiencies.

Engineering Infrastructure: The existing status of engineering infrastructure, such as communication systems, inspection houses, residential and non-residential buildings, vehicle facilities, etc., are not adequate to meet the present-day requirement of effective management and operational efficiency. These have to be adequate.

On-Farm Development Works

Some of the essential features of the modernization of O.F.D. works are indicated in table below:

Item	Feature
The lining of water courses	Lining up to 8-hectare blocks in the command can result in the saving of water up to 35% in the field. The cost of the lining can be balanced by the value of water saved
Land leveling and farm drainage	Should form part of irrigation project implementation
Field application	Sprinkler and drip irrigation should be introduced wherever possible to reduce the wastage of water
Outlets	Adjustable proportionate Module (A.P.M.) should be adopted
Credit and marketing facilities	Facilities should be provided in cooperative sectors (banking facility, storage facility, market (mandies)

Crop Planning: The trend in cropping patterns indicates that farmers switch over to cash crops under irrigated agriculture. Crop planning must be realistic, reflecting the aspirations of farmers as well as a strategy to bring required changes in cropping patterns. Accurate knowledge of crop behavior and crop water requirements is necessary.

Economic Viability: Modernization projects must envisage achieving yields 2 to 3 times the yields under average irrigated conditions at present. The cost per unit of additional water saved or provided must be less than the cost per unit of irrigation water for a new project.

Staff: Staff of the irrigation department should be capable of monitoring the construction program, particularly for OFD works. Much of the success of a modernization project depends upon the management efficiency during the operation stage. In many projects, there is overstaffing (to take care of unemployment). A major portion of the O&M budget is usually spent on the payment of the salary of staff.

2.13 Features of Irrigation Administration

The present system of administration of the canal systems in India has the following features:

- Operation and maintenance are controlled by Civil Engineers who are being trained by Water and Land Management Institutes in the field of irrigation.
- The system is administered and revenue-oriented.

- The water rights of the farmers are not recognized. If a canal system fails to deliver the water in time or delivers inadequate quantity, the most that the farmer can claim is remission in water charges.
- The responsibility of the canal administration ends at the outlet. The distribution within the outlet command is the responsibility of farmers.

Command Area Development (CADA) program was launched to increase the productivity of major canal system commands. These were supposed to improve not only the irrigation facilities but all other needed agricultural inputs. There have been wide variations in their structure, scope, and performance. There have been conflicts of jurisdiction with the Irrigation Department on the one hand and other agencies dealing with agricultural inputs on the other.

Canal systems are subject to physical and social entropy which means that unless efficient management is done, the system is subject to physical and administrative decay. For various reasons, this has been happening at increasingly faster rates (Singh, 1985), an eminent Irrigation Engineer, stated.

> With all the deficiencies of the system and authoritarian attitudes, there used to be a continuous pressure on the canal staff to deliver the water to the tails, to increase irrigation under the jurisdiction, and to improve the duties of water entrusted to them. ...No system, however well it may be devised, can perform well unless the men who work it are devoted to and take pride in their jobs-no system is so bad in which a determined individual cannot make his mark.
>
> The old system of the officers being required to stay out, meet the farmers, and hear and remove their grievances on the spot must be revived. ...Above all, the prestige of the operation and maintenance wing must be restored. This is the wing that ultimately renders service to the farmers and is the end objective of all design and construction work.

Questions

1. Explain in brief the following terms.

   (i) major, medium, minor irrigation projects; (ii) land holding; (iii) irrigation potential; (iv) on-farm development; (v) water rate(price or charge); (vi) night irrigation; (vii) rotational delivery; and (viii) equity in supply.
2. Write brief notes on the following:

   (a) Agricultural land resource and constraints
   (b) Land degradation due to irrigation.

3. Write brief notes on the following:

(a) Surface and groundwater resources and constraints
(b) Irrigation policy in India
(c) Performance of state tube well.

4. Discuss the need, scope, and performance of the command area development program in India.
5. Critically examine the design parameters of the canal system (capacity, annual irrigation intensity, control points, outlet, and night irrigation).
6. Critically examine the performance of rotational delivery.
7. What are the issues related to tube well irrigation?
8. Discuss the need to revise the existing water rates.
9. Explain features of irrigation administration in India.
10. Discuss the need and scope for the participation of farmers in irrigation management.
11. Explain the meaning and scope of rehabilitation and modernization activities.
12. Elaborate the Command Area Development Program mainly aimed at rehabilitation of irrigation projects.

References

CWC (2017) Pricing of water in public systems in India: information system organization. Water Planning and Projects Wing, Central Water Commission, New Delhi

CWC (2019) Reassessment of water availability in basins using space inputs. Central Water Commission, Government of India, New Delhi

CWC (2020) Annual report of central water commission for 2019–20. Government of India, New Delhi

Chaube UC (1985) Search for appropriate water use management technology, Institution of Engineers (India). J Agric Eng Div

IWRS (1998) Modernization of canal irrigation. In: Proceedings of the Indian Water Resources Society, WRDTC, University of Roorkee, Roorkee 247667 (U.P.)

Singh B (1985) Irrigation management in India. In: Proceedings of the seminar on irrigation water management, institution of engineers (India), Roorkee Local Centre, Roorkee

Report of the Committee on Pricing of Irrigation Water, Planning Commission (1992) Government of India, New Delhi

Varshney RS (1992) Rehabilitation and modernization of irrigation projects—concepts and policies. Write up in Spl. Course 661 on Modernization of Canal Irrigation, Continuing Education Department, University of Roorkee, Roorkee, (U.P.) March/April 1992

Reading Material

Central Board of irrigation and Power (1987) Methodology for Evaluation of Irrigation and CAD Projects, New Delhi

Seckler D (1985) The new era of irrigation management in India. J Indian Water Resour Soc

Chapter 3
Irrigation in Indian Subcontinent: A Brief History and Some Lessons

Abstract This chapter describes the practice of irrigation and its management in the Indian subcontinent since ancient times. Emphasis is not to prove the existence of several works of eminence but to draw some lessons therefrom. Items such as profession and diffusion of agriculture, agricultural land classification, geographic variation in rainfall, irrigation management, lift irrigation, and irrigation from anicuts and storage type works are discussed and compared with modern techniques. Some of the ancient guidelines for irrigation management are still relevant. On the subject of dams and canals, evidence from south India and Sri Lanka offers special material. Tropical monsoon hydrology over the Indian subcontinent necessitates the storage of monsoon runoff for irrigated agriculture and other purposes. Therefore, storage dams have been constructed and maintained in the Indian subcontinent throughout history. The ancient Sudarshan dam (300 B.C.–450 A.D.) had a long life, as soil erosion in the catchment was much less than soil erosion rates being observed nowadays. Comparison of irrigation administration in the Mauryan period (322–185 B.C.) with present-day practice shows that irrigation charges were much higher in the Mauryan period than the charges being paid by farmers at present. The salary structure of government employees during the Mauryan period is also compared with the present structure. Privatization, taxation, and severe penalties point to relatively efficient irrigation systems in the Mauryan period. *During the period of Arab, Mughal, and Sikh rulers* (800–1840 A.D.), several canal works were constructed mainly for water supply to gardens, hunting grounds, and forts. These were diversion canals on perennial rivers such as Yamuna, Ravi, Chenab, Sutlej, etc. Examples of canal works (Western and Eastern Yamuna canals, Ravi canal, Hasli canal, inundation canals from Chenab) have been provided. Irrigation works during the British period in the present geographic region of India consisted of (i) improvements in the existing works namely Western Yamuna Canal System, Eastern Yamuna Canal System, and Cauveri Delta System (ii). New works such as the Upper Bari Doab canal, Ganges Canal, and Godavari Delta System, and (iii) taking up of a large number of storage schemes (earthen and masonry dams). Salient features of masonry dams constructed during the 19th Century A.D. and up to 1947 A.D. have been compiled. Irrigation works in India during the British period consisted of (i) improvements in the existing works namely Western Yamuna Canal System, Eastern Yamuna Canal System, and Cauveri Delta System; (ii). New works such as the Upper Bari Doab canal, Ganges

U. C. Chaube et al., *Canal Irrigation Systems in India*, Water Science and Technology Library 126, https://doi.org/10.1007/978-3-031-42812-8_3

Canal, and Godavari Delta System; and (iii) taking up a large number of storage schemes (earthen and masonry dams). Masonry dams constructed during the 19th Century A.D. and up to 1947 A.D. are listed in tabular form. *At the time of independence* (after partition), the net irrigated area in India and Pakistan (including Bangladesh) was 19.4 Mha and 8.8 Mha, respectively. The irrigation works which remained with India, barring some of the old works in Uttar Pradesh and the deltas of the south, were mostly protective and meant more to ward off famine than to produce significant yields. *Since independence from the British rule in 1947 A.D.*, the development of irrigation and its management has been based on the concept of social welfare. Irrigation projects are designed to provide irrigation water to at least one major crop in a year over as large an area as possible and serve a variety of social goals. A large number of major, medium and minor irrigation schemes have been implemented in the country. Irrigation development and problems and issues related with irrigation management in the post-independence period have been discussed in Chap. 2.

3.1 Introduction

Literature on Indian history is generally overladen with details of kings and their feuds and wars. Hence, it is not surprising that historical documents on kings and their kingdoms do not contain detailed information on the science and technology of agriculture, irrigation, and management of irrigated agriculture systems. Randhawa (1980) had done pioneering work by providing a synthesis of agriculture in ancient India. Srinivasan (1970) had shown that irrigation was widely practiced in ancient India. Brohier (1934) had provided detailed information on ancient irrigation works in south India and Sri Lanka. At a period when Europe was in the rudest and most primitive state, irrigation in the Indian subcontinent had transformed the area into plentiful prosperity. Extensive works of irrigation were constructed with an immense amount of labour, engineering skill, and patronage of benevolent kings.

The objective of this chapter is to synthesize the historical information (from 3300 B.C. up to 1947 A.D. i.e. over a period of more than 5000 years) on various technical and managerial aspects of irrigation in Indian subcontinent and draw some lessons. Scattered information on irrigation and agriculture is available for the following historical time periods (Table 3.1).

Table 3.1 Time periods of history

Identification name of the period	Historical time period	Region for which some information is available
Neolithic and chalcolithic period	3300–1500 B.C.	Indus valley—Mohanjodaro, Harappa
Early and later Vedic (Aryan) period	1500–300 B.C.	Eastern Punjab, North Rajasthan, Kathiwar
Buddhist period (pre-Mauryan period)	600 B.C.	North, East India
Mauryan empire period	300–232 B.C.	Indian subcontinent (excluding southern part of India)
Post Mauryan period (Gupta period)	200 B.C.–300 A.D.	Northern India
Chola period (Kingdoms in south India and Sri Lanka)	First century A.D. to 1000 A.D.	South India, Sri Lanka (Ceylon)
Medieval period-Arab, Mughal, and Sikh Rule	1100–1800 A.D.	Indus basin, Yamuna river
British colonial rule period	1800–1947 A.D.	All over India, Pakistan, Bangladesh
Post-independence period	1947 A.D. upto present time	All over present day India

3.2 Background

3.2.1 Original Evidence

Original evidence on the development of agriculture and irrigation in the Indian subcontinent from ancient time to about 1000 A.D. are indicated in Table 3.2. Only significant developments are identified in this table, based on the information available in the literature. This evidence is in the form of archaeological findings, inscriptions on rocks and other historical monuments, as well as ancient literature. Evidence on the progress of agriculture & irrigation is scattered. However, when such evidence is put together in chronological order and viewed in terms of concerned regions, a synthesized scenario spanning over a long period and a large area emerges. Most of this evidence, except for works in Sri Lanka, has been discussed by several investigators in a historical context only.

Brohier (1934) had given a detailed account of ancient irrigation works in Sri Lanka, based upon field investigations of these works. These works in Sri Lanka were constructed under the benevolent patronage of Sinhalese kings who were descendants from the royal families of kings (Kalinga, Chola) from India. Further, there was a technology transfer from Sri Lanka to lands as far as Kashmir.

Table 3.2 Evidences on Ancient Irrigation and Agriculture

S. no.	Period	Item concerning irrigated agriculture	Region	The original source of information
1	Neolithic and Chalcolithic 2295–1300 B.C.	The earliest evidence of irrigation in India	Ghod river in Maharashtra	Discovery of large mud embankment on the stone foundation
2	Neolithic 2000 B.C.	Origin of rice in India	West-Bengal, Orissa, Bihar	Archeological excavations at Chirand (Bihar), Pandu, Rajar and Dhibi sites (W.B.)
3	Pre-Harappan Neolithic	Stone-built dams (Gabar Bunds)	Baluchistan	Discovery of a series of dams
4	Chalcolithic 3000–1700 B.C.	The invention of the plough, wheeled cart, irrigated farming (reservoir, canal)	Valleys of Indus, Tigris and Euphrates	Excavation at Ur
5	–do–	The decline of civilization due to the salinity of irrigated fields	–do–	–
6	Chalcolithic 2300–1600 B.C.	Harappan learn use of ploughs and bullock-carts, Granaries, payment in kind, use of cotton	Sind and North West India	Excavations at Harappa and Mohenjo-Daro (Pakistan), Lothal (Gujrat-India)
7	Early Vedic 1500–1000 B.C.	Pastoralism of early Vedic Aryans, agriculture profession attached with social stigma	Punjab, North-West India	Rigveda (Literature)
8	Later Vedic 1000–800 B.C.	Adoption of wheat, cotton, and rice cultivation	U.P., Bihar	Yajurveda, Satapatha, Brahmana, Taittirya Samhita (Literature)
9	Buddhist 600 B.C.	Removal of social stigma from agriculture, ring wells, use of iron, and cattle	North India	Jataks and Suttra (literature), Excavations at Ropar
10	Mauryan 322–232 B.C.	Guidelines for the management of agriculture and irrigation	The northern part of the Indian subcontinent	Arthashastra by Kautilya, Kamajataka (literature), Junagarh Rock Inscription

(continued)

Table 3.2 (continued)

S. no.	Period	Item concerning irrigated agriculture	Region	The original source of information
11	Ashoka 274–234 B.C.	Arboriculture and horticulture as state policy	Indian subcontinent	Inscription on stone pillars
12	Sathavahanas 100 B.C.–200 A.D.	Brahmins as pioneers of progressive agriculture in South	Andhra Pradesh	
13	300 B.C.	Diffusion of irrigated rice cultivation from Orisa	Coastal Andhra Pradesh, Tamil Nadu	
14	Kushan 78 A.D.	Brick wells, 32 cm × 20 cm × 8 cm brick size	Punjab, Ganga Valley	
15	100–200 A.D.	Deccanese learn the use of brick and ring well	Deccan	
16	47 B.C.–459 A.D.	Irrigation works dams, canals, interbasin transfer schemes	Sri Lanka	Writing of Medhatithi, Parashara, Kashyapa Mahavansa (Sri Lanka), field investigation by R. L. Brohier
17	190 A.D.	160 km long flood protection embankment, irrigation canals	Along Cauvery River	Mahavamsa (Sri Lanka)
18	300–600 A.D.	Tanks, Importance of irrigation	Tamil Nadu	Sangham literature
19	300–500 A.D.	Soil classification and land use	Region of Gupta	Kamasutra by Vatsyayana, Brahit Samhita by Varahmihir, Amarakosh by Amarsinha
20	985–1013 A.D.	Anicut, Irrigation canals	Tamil Nadu	Sangham literature
21	700–1200 A.D.	Chain tanks and maintenance of the tank	Andhra Pradesh	The inscription, Viswakarma Vastusastram
22	Gupta period 900–1100 A.D.	Guidelines for soil–water-crop management	Tamil Nadu	Inscription-Nandivarman Pallava

3.2.2 Agriculture: Profession and Diffusion

While the profession of agriculture was extolled for producing surplus food in Vedic age, the profession was regarded as fit only for the illiterate and those devoid of wisdom.

References in the Vedic literature indicate that the cultivators possessed a fair knowledge of agriculture. However, in contrast to the Vedic age, agriculture as a profession was not associated with either social prestige or social stigma in the Buddhist period (600 B.C.). In both Jatakas and Sutra, Brahmins are frequently found pursuing tillage but no reflection is passed upon them for so doing.

The Brahmins (100 B.C.–200 A.D.) acted as pioneers of progressive agriculture. The rulers in Deccan land (Satavahanes) made land grants to Brahmins, who came from the Ganga basin as they possessed knowledge of distant markets, organization of village settlements, and trade. The Brahmins claimed and generally received an exemption from all taxes and loans on the low-interest rate. The practice of irrigated rice cultivation diffused from adjoining areas of Orissa to the coastal areas of Andhra Pradesh and Tamil Nadu (300 B.C.)

Poets of the Sangama literature (300–600 A.D.) emphasized the dignity of labour. Kashyapa (900–1100 A.D.) realized the importance of agriculture and advocated its practice by the rulers, their advisers, and officials so that they could realize the difficulties which the farmer faced: "Both *bipeds and quadrupeds on the face of the earth would face misery if there were no cultivation*", says Kashyapa.

In the present-day India, the proportion of nonagricultural communities has considerably increased due to industrialization. This coupled with the need for self-reliance has necessitated the production of surplus food in large quantities and yet in the management of irrigated agriculture system, the farmer has often been considered ignorant, non-cooperative, devoid of wisdom mainly due to illiteracy. A literate person is reluctant to take up agriculture as a profession even now.

Lesson: Agriculture should not be associated with any social stigma (cast, illiteracy). In ancient India, Brahmins were frequently found pursuing tillage but no reflection was passed upon them for so doing. Farmers' formal education though is useful but not necessary for the adoption and success of farming as a profession.

3.2.3 Annual Rainfall and Its Geographic Variation: Then and Now

Drona was the unit of rainfall measurement in the Mauryan period. Kautilya's Arthashastra (Kangle 1963) mentions the average annual rainfall in the Indian subcontinent as below:

Sixteen dronas are the amount of rain in forest areas, one-and-a-half times of that in wetlands, where sowing conforms with the nature of region; thirteen-and-half dronas in the Ashamakas, twenty-three in the Avantis, unlimited in the Aparantas

Table 3.3 Average annual rainfall and its geographic variation—then and now

Region as in Mauryan period	Corresponding present-day region	Rainfall observed 2400 years ago (Mauryan period)	Rainfall as observed now
Lands where sowing is made with the help of camel	Deserts, Rajasthan	Unpredictable in time	< 40 cm and high coefficient of variability
Avantis	Western M.P.	13.5 Drona (67.5 cm)	40–60 cm
Forest area	Central Indian forests	16 Drona (80 cm)	60–100 cm
Ashamaka	Godavari basin	23 Drona (115 cm)	100–200 cm
Aparanta	Western coast/ Himalayan ranges	Unlimited	> 200 cm

and snowy regions and unpredictable as to time in lands where sowing was done with the help of camels.

Equating one drona to two inches or five centimeters (Mate 2006), the geographic variation of rainfall 2400 years back and now has been compared as shown in Table 3.3. A general correspondence is observed between geographic variations in rainfall then and now.

3.3 Lift Irrigation

The oldest evidence of the existence of water-lifting devices is provided by the excavations at Mohanjo-Daro and Harappa. The techniques of raising water by machinery were practiced in Sri Lanka at least as early as 19 B.C. King Bhatikabhayo is said to have raised the water of the Abayo tank in this manner (Brohier 1934). Literary and epigraphic evidence indicates the use of various types of water–lifting devices, which could be arranged under three broadheads.

1. *Intermittent or discontinuous water supply from stream and canals*

The basket bag or bucket moved by pulley wheels and the water drawn from wells by animal power has been indicated in the literature and inscriptions. Pulley wheels of stone or wood are mentioned in Rigveda harnessing animals for drawing water from wells was practiced at least as early as 500 B.C.

2. *Semi-mechanical devices*

Counterweights and the weight of the human body were used in the balanced bucket mechanism. Water was raised from wells using a bucket mechanism in which buckets were tied by rope to one end of a long wooden pole, working about a fulcrum near the other end that carried a heavyweight.

3. *Continuous water supply by water-lifting machinery*

A fully mechanized water hoist moved by a water wheel was put into practical application as early as the second millennium B.C. The notion that the device was introduced into India from Persia is not tenable. Srinivasan (1970) has discussed in detail the origin of this mechanism in India.

The water lifting mechanism indicated in categories 2 and 3 above i.e. the bucket and levers and the wheel of pots are still in use in many parts of India, particularly in arid and water scarce areas.

3.4 Dams in North India

A large number of dams had been and have continued to be constructed in the Indian subcontinent. The storage of rainwater is necessitated due to the prevalence of tropical monsoon hydrology over the region. The region experiences plenty of rainfall during the monsoon season (June/July to September/October) and very little rain in the remaining period of the year, necessitating storage of surplus water of the monsoon season for use during the non-monsoon period. Regarding dams in ancient India, scattered information about different periods in history is available in various documents. Ancient religious texts (such as Veda) abundantly highlight the importance of creating surface water storage. Since independence from the British rule in 1947 A.D., several large dams have been constructed in India.

3.4.1 Dam Construction Technology in Mauryan Period

Kautilya in Athasastra (322 B.C.) mentions that the tanks and dams should be constructed only in the center of a province or village. According to Kautilya, the following are the 12 technical considerations and by following these considerations, an excellent tank is easily attainable on (this) earth.

Essential Requirements

(i) A king (State) endowed with righteousness, rich, happy, and desirous of (acquiring) the permanent wealth of fame.
(ii) Experts in the science of hydrology (Pathas-Sastra).
(iii) The straight and long stones should be available and the quarry should be as close as possible to the site of the intended structure.
(iv) A gang of men (craftsmen, masons, skilled in the construction of such a structure) should be available.

Site Requirements

(v) The bed should be extensive and deep.

(vi) Two extremes (srimga) pointing away from fruit (giving) land (phalasthira) outside.
(vii) The river should contain sweet water.
(viii) The dam should be located at least three yojanas downstream from its source.
(ix) The site should have strong hills on either side of the river so that the dam of a compact stone wall (not too long but firm) could be pinned against them.
(x) The ground should be firm i.e. hard clay (not sandy, not porous).
(xi) The neighboring fields should be fruit-bearing (i.e. fertile) level and large enough to absorb the water stored in the lake. A stone bund should be erected and sluices should be
(xii) Provided in the wall to carry water to fields through channels.

Six possible defects according to Kautilya are:

(i) water oozing from the dam, (ii) saline soil, (iii) situation at the boundary of two kingdoms, (iv) elevated mounds (karma) in the middle (of the tank) bed, (v) scanty supply of water and extensive stretch of land (to be irrigated), and (vi) scanty ground.

The Sudarshan Dam and Lake: Sudarshana dam was constructed in the third century BC in the Girnar (Junagarh) area of Gujarat. The dam had a base width of 100 feet across river Suvarnrekha. Supplementary channels/conduits were later added during the reign of Emperor Ashoka.

Lesson: In contrast to small size tanks during the ancient period, India after independence embarked on the construction of large size multipurpose dams. Several of these projects are now being criticized for high cost, long gestation period, and problems of poor management. It is now being realized that small irrigation schemes can be better planned and managed and their performance can be monitored more accurately as compared to large projects.

3.4.2 Dams in Sanchi, Vidisha Area (Central Part of India)

Shaw and Sutcliffe (2003) carried out field investigations of 16 ancient dams/embankments (c. second–first century B.C.) located in the vicinity of Vidisha and Sanchi towns in the state of Madhya Pradesh, India. Comparison of reservoir volumes with estimated inflows suggests that their design was based on hydrological understanding. The salient features of some of the dams are given in Table 3.4. Dam height varied from 3.6 to 6 m. The study of Shaw and Sutcliffe (2003) showed very rough estimates of the size of dams which need to be verified.

The work of Shaw and Sutcliffe (2003) on ancient dams in the Sanchi area is significant, as it not only discusses the distribution and form of the dams but also examines the socioeconomic and religious infrastructure that underlay the construction and management of these dams. The dams are located in close vicinity to the early historic complexes of Buddhist stupas or monasteries. Some reservoirs have spillways, while others have deep trench on one side.

Table 3.4 Salient features of some dams in the Vidisha region

Name	Location	Phase	Height (m)	Length (m)	Reservoir area (km^2)	Catchment area (km^2)	Reservoir volume MCM
Chandna	N23° 24′; E77° 49′	2nd–first century B.C.	6	450	0.25	1.29	0.75
Devrajpur	N23° 27′; E77° 52′	fifth century A.D.	4	1100	1.52	13.49	3.036
Dhakna	N23° 27′; E77° 42′	2nd–first century B.C.	4.5	630	0.40	4.11	0.893
Ferozpur	N23° 29′; E77° 40′	Fifth century A.D.	4	500	1	17	2
Morel Kala	N23° 24′; E77° 51′	Second century AD	6	350	0.88	12.83	2.625
Morel Khurd	N23° 26′; E77° 49′	2nd–first century BC	3.6	1100	0.6	2	1.08

3.5 Dams and Canals in South India and Sri Lanka

3.5.1 Salient Features of Irrigation Works in South India and Sri Lanka

On the subject of dams and canals, South India and Sri Lanka offer special material, because the failure of monsoon, the uneven rainfall, scarcity, and excess of water in non-perennial rivers made the people devise methods for storing and diverting available runoff for irrigation. Table 3.5 gives salient features of some of the important tanks and embankments.

Historically, it was in the reign of Chola king Karikala (100 A.D.) that important irrigation works had been constructed with the help of 12,000 captives from Sri Lanka, a 160-km long floor protection embankment along River Chavery was constructed in his reign. Several tanks and canals taking off from River Kavery were also constructed by king Karikala and his successors.

Table 3.5 Salient features of irrigation works in South India and Sri Lanka

Sl. no.	Name of irrigation project	Year	Place/river	Technical details
1	Flood protection embankment	190 A.D.	Kavery	160 km long, constructed with help of 12,000 captives brought from Sri Lanka
2	Srirangam Anicut	–	Kavery	329 m long, 12–18 m broad
3	Nagarjun Konda canal	400 A.D.	Nagarjun Konda	Width 50 feet, depth 6 feet, embankment made of limestone mixed with kankar, stones
4	Pratap Bukkaraya Mandala channel	1338 A.D.	Pennar River	The water of the Pennar river diverted to the Siruvera tank at Penugonda
5	Citramegha Tank	900 A.D.	Chinglaput (Tamil Nadu)	Tank bund rests on the bases of two hills
6	Paramewara Takaka (Tank)	700 A.D.	–	One-fifth of irrigation land for public purposes
7	Sudarshana lake	322 A.D.	Junagarh	Lake with conduits suffered a breach in 200 B.C. due to heavy floods
8	Kavery pak lake	900 A.D.	–	6.4 km long bund
9	Bahur tank	985–1100 A.D.	Pondicherry Arikesarimanglam	
10	Porumamilla tank	1369 A.D.	South India	The inscription mentions 12 merits of the tank and its complete details
11	Sulekera tank		Karnataka	64 km in circumference
12	Palar system of tank		Karnataka	1000 tanks in the valley, the last one being the Ramasagar tank
13	Wannakannam Canal	47 B.C.	Sri Lanka	A large canal work
14	Pandawewa tank	505 B.C.	Sri Lanka	Earthen bund height 24 ft., Length is about 1.5 mile. Reservoir area about 1100 acres
15	Parakrama Samudha	1250 A.D.	Ellahera (Sri Lanka)	80–90 ft. high embankments
16	Minipe Yodi Ela	459 A.D. (Sri Lanka)	River Mahaweli	Cut stones placed at the right angle to channel; 6 ft. high retaining wall in cut sections of the canal. Embankment 37 ft. high and 6 ft. wide

3.5.2 Anicuts Across River Kaveri

The River Kaveri and its tributaries support an extensive system of irrigation using channels drawn from small height dams, called the anicuts (in Kannada language: *ane katte* or dam, dyke, or embankment). Tradition attributes to the Chola kings the construction of the famous anicuts across the River Chavery in the Tanjore District (Table 3.5). The canals offtaking from the anicuts meander over the adjoining tracts of country on each bank, following the sinuosity of the ground, the total length running in Karnataka being upward of 1915 km.

Example: Kallanai Dam/Grand Anicut, India

Kallanai Dam, also known as Grand Anicut, is the fourth oldest dam in the world, still in use. It still serves the people of Tamil Nadu, India. The dam was constructed by King Karikala Chola of the Chola Dynasty in the second century A.D. The dam is located on the River Kaveri, approximately 16 km east of the city of Tiruchirappalli. Improvements were made to the dam in the nineteenth century by Arthur Cottons, a British general and irrigation engineer.

It is perhaps the oldest hydraulic structure built on permeable foundation (sandy bed) in the world which is still functioning; the dam provides water for irrigating 400,000 ha of land along the Kaveri Delta Region.

Dimensions: The structure measures 329 m in length, 12.2–18.3 m in width and 4.57–5.49 m in height.

Construction material: Large cyclopean granite stones would have been brought and dumped across the flowing stream and continuously replenished, as these boulders sank in the sandy riverbed, until they got embedded in the clayey layer below. The structure was thus raised up to the crest level as seen today.

The anicut, as seen today, consists of a core of rough stones in clay covered with facing of rough stone in mortar. A portion of the crest was built with a curved top and the rest with a series of steps, the foot of the solid dam being protected by a rough stone apron.

3.5.3 Chain Tanks

The topography of Telangana and Karnataka is well suited for the construction of storage reservoirs. Telangana, where many of the ancient irrigation tanks were located is known as "The Land of Thousand Tanks". A special feature of the tank in these tracts is their construction in series by bunding the same river at several sites forming a chain of reservoirs. The surplus water escaping from one tank supplies water to the lower down and so on. In addition to surplus water, the return flow from the irrigation commands under the upper tank also flows into the lower tank. Owing to the porous nature of the soil and the sloping terrain, often large quantities of water are drained into the lower reservoir in this manner. There are more than 38,080 tanks in the Karnataka region alone. The largest tank is Sulekera. It is 64 km in circumference. Palar system is one example of chain tanks. It has 1000 tanks in the valley, the last one being the Ramasagar tank. The village assembly and especially the Tank Supervision Committee (eri-variyam) looked after the maintenance of the irrigation work of a village, by repairing breaches and dams, removing silt, and regulating the distribution of water supply.

3.5.4 Dams and Canals in Sri Lanka

Henry Parker and Brohier (Parker 1889, 1909; Brohier 1934, 1979) had given a detailed account of ancient dams and irrigation canals in Sri Lanka. A project report (available on internet) on The *Ancient Technology on Dam Construction and Irrigation in Sri Lanka* has been prepared by Chandana Rohana Withanachchi of University of Sri Lanka. (https://www.researchgate.net/publication/281937136_Technology_and_techniques_applied_in_ancient_Sri_Lanka_in_constructing_dams). This report provides some details on dam technology (planning, construction material and construction methodology as viewed by an Archeologist).

Several researchers are of the view that in the Pre-Christian time, Sri Lanka had attained the knowledge of controlling the river water. Extensive works of irrigation constructed with an immense amount of labour, skill, and science had transformed arid plains into areas of plentiful prosperity at a period when agriculture in Europe was in the rudest and most primitive state. The salient features of some of the ancient irrigation works in Sri Lanka are given in Table 3.4.

Two different systems were adopted in Sri Lanka for conserving runoff during the seasonal rains of the two monsoons. According to one, the natural and effective plan of storing water in the upper reaches of a valley was reported to the other system that was based on much more scientific and ambitious methods of inter-basin transfer. This was affected by the construction of massive causeways and anicuts across large rivers and diverting water through excavated channels which conveyed it sometimes many miles, over the apparently flat country and impounded the water eventually in large reservoirs or a chain of reservoirs. Sluices were provided in the dams and canals to regulate irrigation water supplies to the rice fields.

Another branch of engineering which had unquestionably attained a very high degree of perfection in Sri Lanka was that of surveying and leveling. Most of the irrigation schemes in Sri Lanka are confined to tracts of land which appear to be flat. Yet, evidence indicates that long channels were traced on the gradient. It required the use of precise instruments of the modern age. The ingenuity displayed in the layout of these works is baffling.

Lesson 1: The main advantage of the series of tanks on the same river is that the benefits of irrigation are distributed over the entire watershed. The return flow from the irrigation commands under the upper tank also flows into the lower tank and thus available for utilization. In a big storage tank, the catchment area does not derive any benefit and the entire irrigation is downstream of the tank.

Lesson 2: In contrast to these small-size chain tanks, emphasis in modern India has been on the construction of large-size multipurpose dams in head reaches since independence from colonial rule. Several of these projects have been criticized for high cost, long gestation period, and problems of poor management. Without denying the need as well as advantages of multipurpose projects, it is now being realized that small irrigation schemes can be better planned and managed and their performance can be monitored more accurately as compared to large projects.

Lesson 3: Technology of inter-basin transfer of water, high degree of perfection in surveying and leveling, and knowledge of controlling river water in Sri Lanka and south India should make us proud of our engineering skills.

3.6 Long Life of Ancient Dams (Tanks)

The useful life of a dam depends on the erosion of soil from the catchment which gets deposited in the live storage and dead storage zones of the reservoir. Unlike present times (the 21th century AD), in the earlier periods of history, human pressure on forest lands was not so much as to cause excessive soil erosion in the catchments of dams.

Example of Sudarsan Lake and Dam: Chaube et al. (2020) analyzed technical aspects of the ancient Sudarshan dam and the lake. Figure 3.1 shows a cross section of the valley at possible site of Sudarshan dam and Fig. 3.2 shows a contour map of the mountains and Sudarshan Lake vicinity.

Sudarshana dam was constructed in the third century BC in the Girnar (Junagarh) area of Gujarat. The dam had a base width of 100 feet across River Suvarnrekha. Supplementary channels/conduits were latter added during the reign of Emperor Ashoka. The dam was overtopped and failed due to a flood on margasirsa krsnapaksha pratipada samvat 72 (19 November, A.D. 150). The breach was 420 cubits (~192 m) long and 75 cubits (~34.5 m) deep. A cubit is an ancient unit of length based on the forearm length from the tip of the middle finger to the bottom of the elbow (usually equal to about 18 in. or 46 cm). The dam was repaired and made three times more durable under the supervision of Suvisakha (Governor of Rudradaman). With renovations and repairs, the dam continued to exist over the ensuing period, as is attested by a Junagadh inscription of Skandagupta pertaining to 455–456 A.D.

Long Life of Sudarshan Reservoir

The useful life of several present-day dams in India has significantly reduced due to accelerated soil erosion in the catchments (CWC 2015). The useful life of some

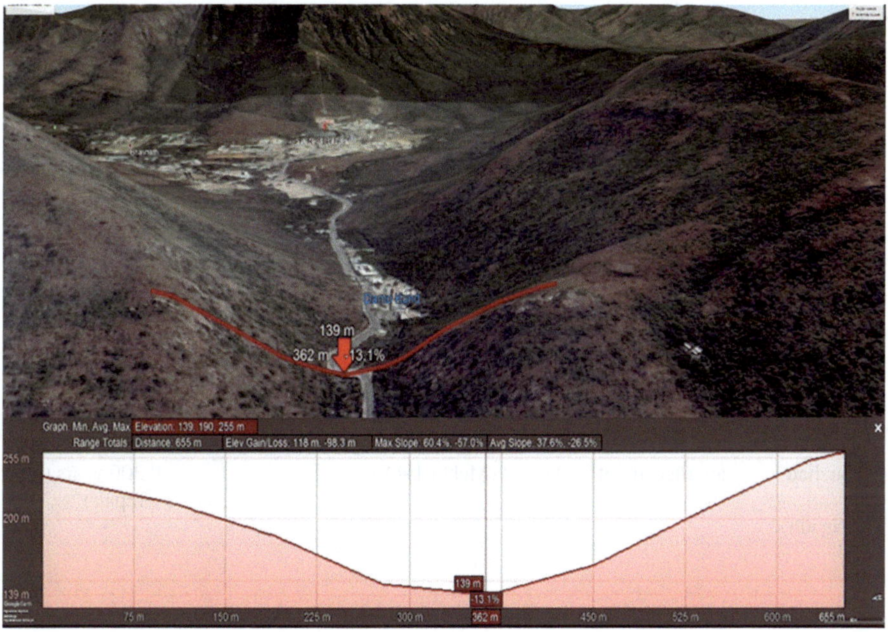

Fig. 3.1 Cross section of the valley at possible site of Sudarshan dam based on google map

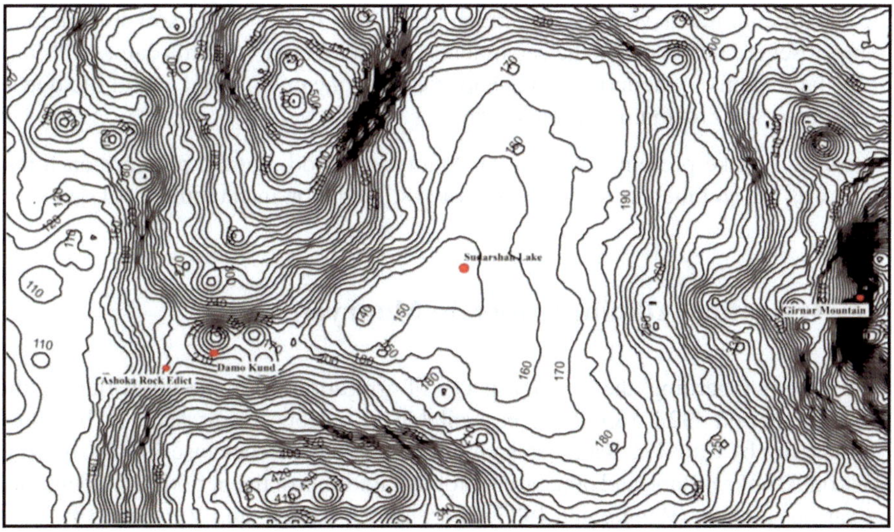

Fig. 3.2 Contour map of the mountains and Sudarshan lake vicinity

Table 3.6 Useful reservoir life of existing dams in the vicinity of ancient Sudarshan dam

Name of dam/ (district)	Catchment area km^2	Period of sediment observation (years)	Change in dead storage capacity (M.C.M.)	Av. Annual rate of depletion of dead storage (MCM/year)	Expected period for the depletion of dead storage (years)
Bhadar (P)/ (Panchmahal district)	407	1983–2009 (26 years)	5.044	0.194	35.15
Moj/(Rajkot district)	440	1956–1999 (43 years)	2.91	0.0667	43.0
Venu II/ (Rajkot district)	751	1989–1999 (10 years)	1.33	1.132	28.64
The ancient Sudarshan dam (Junagadh)	The first dam lasted for about 450 years; repairs gave it another life of 300 years (Mehta 1968)				450 years + 300 years (after repair)

present-day dams located in the vicinity of the ancient Sudarshan dam is less than 50 years as compared to the long life (750 years) of the Sudarshan dam (Table 3.6).

Lesson 1: Such a long life of the dam and the reservoir shows that soil erosion from the catchment must have been very much less at that time than the soil erosion rates being observed nowadays in the catchments of Bhakra dam, Matatila dam, etc. Many of the large dams in modern India are approaching the end of their useful life due to accelerated soil erosion in the catchments.

Lesson 2: Unlike present times, in the Mauryan period, human pressure on forest lands was not so much as to cause excessive soil erosion in catchments of dams.

3.7 Irrigation Administration in Mauryan Period

Information regarding the fiscal aspects of irrigation administration during the Mauryan period (322–185 B.C.) is available in various forms such as literature (Arthasastra by Kautilya, Kamajataka literature), Junagarh rock inscriptions, inscriptions on Ashoka pillars, and other archeological findings of the period. Commentaries by various authors (Rangarajan 1987; Sharma 1995; Zubin and Vylder 2014; Srinivasan 1970) on the Arthasastra and other literature provide useful information on the subject of irrigation and its management. Chaube (2019) synthesized the available information on irrigation administration during the Mauryan period (322–185 B.C.) and compared it with the practice in present times in India to draw some lessons.

Fig. 3.3 Pana (weight 3.5 g) equivalence to Indian rupee in the year 2016

Equivalence of Pana in Rupee Currency

In the Arthashastra it is mentioned that "*an annual salary of 60 Panas is equal to one adhaka of grain per day and one adhaka is enough for four meals for one Arya male*" (Rangarajan, 1987). Mulla et al. (2014) considered individuals in urban areas earning less than Rs. 10,314.20 per annum to be "below the poverty line". Based on this definition of the poverty line, Mulla et al. (2014) equated an annual salary of 60 Panas to Rs. 10,314.20 per annum and suggested that the value of one pana in Kautilya's period is equal to Rs. 171.90 in the present time.

In the study by Chaube (2019), the valuation of Pana into Rupee was made. *Adhaka* is the Sanskrit name of the weight unit corresponding to 2.56 kg (Sharma 1998). The annual grain requirement of one person is 934.4 kg grain (=365 × 2.56 kg grain per day) which he could buy from his salary of 60 Pana in a year. In other words, one Pana could buy 15.573 kg of grain. In the year 2016 (year of implementation of 7th Pay Commission), the minimum support price for wheat grain was Rs. 1525 per quintal (GOI 2015). It means 15.573 kg grain could be bought from Rs. 337.5. Therefore, one Pana may be equated to Rs. 237.5 in the year 2016 (Fig. 3.3). Similarly, in the year 2019, the minimum support price for wheat grain is Rs. 1840 per quintal (GOI 2018). Therefore, one pana may be equated to Rs. 286.5 in the year 2019.

Pay Differential in Government of Mauryan Period and Now

The Maurya Empire was extremely efficient in irrigation administration but officials at a higher level were paid disproportionately higher than their relative output simply under their position in the hierarchy. Organizations that have relatively large pay differentials are said to be hierarchical while organizations that have relatively small pay differentials are said to be egalitarian (Mulla et al. 2014). The extent of hierarchy in a particular organizational structure can be measured by the ratio of the wages of the highest-paid member in the hierarchy to that of the least paid member of the hierarchy.

Table 3.9 shows the maximum and minimum salaries of government employees and the ratio of the salaries in the Mauryan period compared with those in the present time. The five different years of present time correspond to the years of pay revision by Central Pay Commission (CPC) in India (Table 3.7). The ratio of maximum and a minimum salary of Government employees was 400 in the Mauryan period

Table 3.7 Monthly salary of high-level officers and skilled labor and ratio of salaries

Time in history	Unit	Monthly salary of highest government employee (councilor/bureaucrat)	Monthly salary of skilled labour	Ratio of salary
400 B.C.	Pana	4000	10	1:400
1947 A.D. (Ist CPC*)	Rupee	2000	55	1:36.4
1957 A.D. (IInd CPC)	Rupee	3000	80	1:37.5
1972 A.D. (IIIrd CPC)	Rupee	3500	196	1:17.9
1996 A.D. (Vth CPC)	Rupee	26,000	2500	1:10.2
2016 A.D. (VIIth CPC)	Rupee	182,200	18,000	1:10.12

Note * Central Pay Commission

suggesting an extremely hierarchical organization. On the other hand, this ratio is 10.1 in 2016 A.D., indicating the highly egalitarian structure of society in present times. Further, the ratio has been gradually declining from 36.4 in 1947 to 10.1 in 2016.

The government of India has made massive investments in the irrigation sector since independence in 1947. But crop production has been much below optimal level even after the implementation of command area development program in several irrigation projects. Egalitarian wage structure could be one of the reasons for inefficiency in the irrigation sector (Chaube 2019).

Lesson: High pay differentials and the severe penalties for violation of irrigation rules during the Mauryan period suggest that the Mauryan state could not be defrauded by the people or its officials. This is in striking contrast to the taxation and other regimes prevalent at present in India.

3.8 Salary of Irrigation Staff–Then and Now

A vast discrepancy prevailed in the payment of salaries to labour, craftsman, and engineers as compared to the highest salaried job. The officiating priest, King's Guru, and the councilors received the highest salary of 4000 Pana per month, whereas King's physician and Chief Engineer received 167 Pana per month only. Table 3.8 gives a comparison of the salaries paid in 400 B.C. and now in present times i.e. after 2400 years. Labour for digging irrigation canals got very low salaries in the Mauryan period as compared to present-day salary. The salary paid to labor in 400 B.C. was almost the same as was paid to the lowest Indian labour by the British East India

Table 3.8 Monthly salary-then and now

Time in history	Unit	Highest salary of councilor/ bureaucrat	Salary of chief engineer	Salary of carpenter, craftsman	Salary ratio Chief Engr.: Highest	Salary ratio Craftsman: Chief Engr.
400 B.C.	Pana	4000	167	10	1:24	1:17
1996 A.D.[@]	Rupee	26,000	20,000	2550	1:1.3	1:8
2016 A.D.[$]	Rupee	182,200	144,200	18,000	1:1.3	1:8

Note [@] Vth Pay Commission GOI; [$] VIIth Pay Commission GOI

Company in 1800 A.D. Engineers, craftsmen, and labor receive much higher salaries now (in terms of equivalent wheat grain) than in the Mauryan period.

Lesson: The extent of inequality in the pay structure as measured by the salary ratio has important implications for organizational outcomes such as individual effort, individual performance, risk-taking, employee turnover, and organizational performance.

3.9 Ownership of Waterworks and Water Tax in the Mauryan Period

Conditions for the ownership of waterworks, water tax, and its exemption are compared in Table 3.9.

Irrigation charge was considered as an important item of revenue in the Arthasastra which provides a long list of taxes as sources of revenue. Arthashashtra mentions: '*The king shall bestow on cultivators only such favour and remission as will tend to swell the treasury and shall avoid such favour which deplete it*'.

Lesson: Irrigation charges were much higher in the Mauryan period as compared to the charges being paid by farmers at present. Present-day water rates are neither based on project cost nor farm benefits. As such several irrigation projects have an actual benefit–cost ratio lower than one.

3.10 Penalty for Violation of Rules

At present, there are various irrigation Acts and Codes of practice to regulate irrigation practices in different parts of India. For example, "The Northern India Canal and Drainage Act VIII" was enacted in 1873 for the regulation of irrigation practice in the United Provinces (now Uttar Pradesh, Punjab, and Haryana). However, the penalty for violation of irrigation rules is not so severe as it was in the Mauryan

Table 3.9 Ownership of water works, water tax and its exemption

As practiced in the Mauryan period	As practiced now in India
Ownership	
Irrigation works such as embankments and tanks could be privately owned and the owner was free to sell or mortgage them	Tanks are mainly owned by village-level societies. Private tanks also exist
The ownership of the tank lapsed if the tank had not been in use for a period of five years, except in case of distress	No such condition exists at present
Anyone leasing, hiring, sharing, or accepting waterworks as a pledge, with the right to use them, was required to keep them in good condition	There are a large number of government-owned tanks which need proper maintenance but no rule for transfer of ownership
Owners may give water to others in return for a share of the produce grown in the field, parks, or gardens	Water at the head of the minor canal is given to a cooperative society and society is responsible to collect the water charges
In the absence of the owner, either charitable individuals or the people of the village acting together shall maintain waterworks	Gram Panchayats/Pani panchayats (Water User Associations) have been formed for the maintenance of irrigation works at minor canal levels
Water tax/rate	
Tax for water use from waterworks built by the king: (a) One-fifth of the product if water is manually transported (b) One-fourth of the product if water is carried by bullocks (c) One-third of the produce if water is lifted by mechanism into channels	Irrigation development in India is based on the concept of social welfare. Irrigation charges are nominal. Not adequate to meet even maintenance expenditure. Due to socio-political reasons, the collection of irrigation charges from the farmers is rather difficult
Tax for water use from natural reservoirs: One-fourth of produce when the field is irrigated from rivers, lakes, tanks, and springs	No tax is levied. The use of natural sources such as a river, lake, spring is not controlled by the government
Exemption from payment of tax	
Five years exemption for building or renovating new tanks and embankments	There are several government-sponsored schemes providing incentives to the rural population in building/renovating new tanks, rejuvenation, and maintenance of tanks
Four years exemption for renovating ruined or abandoned waterworks	
Three years for clearing waterworks over-grown with weeds {3.9.33}	

period. For example, delayed payment of irrigation water charges by farmers and sometimes even nonpayment are very common in north India. Furthermore, irrigation development in India is based on the concept of social welfare and it is not viewed as a source of revenue.

Quarrels among farmers concerning irrigation facilities had existed during the Mauryan period and such quarrels exist in the present times as well. Arthashastra had laid down strict rules for water use and penalties to settle conflicts about priority water use. The penalties prescribed during the Mauryan period are given in Table 3.10. Arthashastra mentions three levels of standard fine for various types of crimes and violations of rules (Rangarajan 1987).

(a) Lowest SP: 48–96 Panas; (b) Middle SP: 200–500 Panas; and (c) Highest SP: 500–1000 Panas. It is also mentioned in Arthashastra that Magistrates shall determine whether to levy the highest, middle, or lower standard penalty (SP) taking into account the person sentenced, the nature of the offense, the motive, and its gravity, the circumstances prevailing, time, place, and consequences, while maintaining a balance between the interests of the king (state) and the offender. Thus, flexibility was provided to the magistrates to decide the level of penalty and fix amount from within the range of decided levels of penalty.

Lesson: Large pay differentials as discussed earlier and the severe penalties for the violation of rules suggest that it was most unlikely that the Mauryan state could be defrauded by the people or its officials. This is in striking contrast to the taxation and other regimes prevalent at present in India.

Table 3.10 Penalty for violation of irrigation rules

Cause	Penalty
Causing damage to another's ploughed or sown field by letting water overflow from a reservoir, channel, or field	Compensation according to damage
Causing damage to gardens, parks and embankments	Double the damage
The natural flow from a higher tank is prevented to fill the lower tank which has been in use for at least three years	Lowest SP & emptying of the higher tank
Failure to maintain an irrigation facility	Double the loss caused by failure
Letting water from a dam out of turn, obstructing the flow of water to a user with a right to it	6 Panas
Obstructing or diverting a customary watercourse	72 Pana[@] (lowest SP)
Building a well or a dam on someone else's land	72 Pana[@] (lowest SP)
The person selling or inducing someone to sell waterworks	350 Pana[#] (middle SP)
Witnesses to the transaction (for not preventing it) of selling or mortgaging a charitable waterworks	750 Pana* (highest SP)
Breaking a dam having water in the reservoir	Drowning in the same place
Breaking a dam having no water in the reservoir	750 Pana* (highest SP)
Breaking a dam that is abandoned or it is in ruins	350 Pana[#] (middle SP)

[*] Average of the highest standard penalty range (SP), [#] Average of range for the middle standard penalty (SP); [@] Average of range for the lowest standard penalty (SP)

3.11 Irrigation Works in Medieval Period (800–1840 A.D.)

During the periods of Arab, Mughal, and Sikh rulers (800–1840 A.D.), several canal works were constructed mainly for water supply to gardens, hunting grounds, and forts. These were diversion canals on perennial rivers, such as Yamuna, Ravi, Chenab, Sutlej, etc. Examples of some canal works are given below.

The Western Yamuna Canal: Canal was built by Feroz Shah around 1355 A.D. The main purpose was to transport water to Emperor's hunting lodge at Hissar (Haryana) rather than to irrigate agricultural land. In fixing the alignment, the advantage was taken of channels having favorable slope and direction towards Hissar. Thus, the resulting work was a linked series of drains rather than canals. About 1568 A.D., Akbar renovated the canal and used it for irrigation in the Hissar district.

The Eastern Yamuna Canal: It was first constructed during the reign of Muhammad Shah (1719–1748 A.D.). Rohilla chiefs partially restored it in about 1780 A.D. and water was brought up to Saharanpur (Uttar Pradesh).

Ravi Canal: Jahangir (1605–1623 A.D.) built an 80 km long canal to take water to his fort and hunting ground (and garden) near Sheikhpura.

Hasli canal: Alimardan Khan a celebrated engineer of Shahjahan built another canal (Hasli canal) in 1693 A.D. This canal carried Ravi water to Shalimar garden in Lahore. Hasli canal was 177 km long and had a discharge capacity of 14 cumecs. During the Sikh rule (1763–1849 A.D.), a branch of this canal (208 km, capacity 4.25 cumec) was constructed to carry water to the Golden Temple at Amritsar. During the British period, this canal was replaced by the Bari Doab canal.

The Khanwali inundation canal from Sutlej River: The canal is said to have been built by one of the ministers of Akbar. The head of the canal was choked by sand during its operation. In 1813 A.D., Raja Kharak Singh got the sand cleared.

Inundation canals from Chenab: A system of thirteen inundation canals taking off from the left bank of the Chenab River were constructed by a Pathan ruler of Multan and Shujabad. Later their heads were amalgamated resulting in four canals. The total discharge of these was around 143 cumecs.

3.12 Irrigation Works in Indus Basin During British Period

The Indus River has been regulated by various institutions through artificial irrigation structures such as reservoirs and barrages constructed on the main river since the mid of the eighteenth century. The salient features of some of the canals are given in Table 3.11. Central Board of Irrigation and Power, Government of India publication (CBIP 1992) provides details of irrigation works in Indus basin.

Table 3.11 Salient features of some diversion canals in the Indus basin

Name	Location	Salient features
Upper Bari Doab canal	Ravi River in Punjab	1000 m long masonry weir, 520 km long main canal and branches, 2500 km long distributaries
Begari canal	Indus River, Sind province, Baluchistan	224 m^3/s, length of the distribution network is 390 km
Fuleli canal	Indus River Sind province	Natural flood channel, 80 m^3/s, length of the distribution network is 1600 km
Ghar canal	Indus River near Sukkur	Natural flood channel, 280 m^3/s
Sukkur canal	Indus River near Sukkur	Old canals linked together and enlarged length of the distribution network is 208 km, irrigation area is 40,000 ha
Eastern Nara system	Indus River near Rohri	Originally 266 cumecs, later enlarged to 304 cumecs, 419 km network
Sirhind canal	Sutlej near Ropar above the junction with Beas	720 m long weir, Budki and Siswan super passages, Daher, Haron syphons
Lower Swat canal	Swat River in northwest Frontier province	20 cumec capacity,20 cross drainage works including Nawadand aqueduct (17 spans of 5.5 m each)
Sindhani canal (1882 A.D.)	Ravi River in Multan	51 cumec capacity,739 km network
Lower Chenab canal (1892 A.D.)	Chenab at Khanki	303 cumec capacity, 3589 km network of canals
Kabul canal	Kabul River, northwest Frontier province	Restoration of old Moghul work, irrigation in Peshawar and Naushera districts

3.13 Irrigation Works in India During the British Period

3.13.1 Some Important Irrigation Works

Table 3.12 shows some important irrigation works completed during the British period. In the nineteenth century, efforts were first directed to improve the existing works, namely Western Yamuna Canal System, Eastern Yamuna Canal System, and Cauveri Delta System. These three systems provided the first lessons in India in hydraulic engineering on a large scale. Dyas in Punjab, Cautley in the United Provinces, and Cotton in Madras were engaged in the canal improvement works in their provinces. Each carried the lessons they learned on their systems to take up new

Table 3.12 Some important irrigation works completed during British period

State	Important irrigation works	Year of completion	Irrigation benefit area (lakh hectare
Andhra Pradesh	Godavari delta system	1890	5.58
	Krishna delta system	1898	4.42
Bihar	Sone canal system	1874	3.47
Haryana	Western Yamuna canal system	1820	4.31
Punjab	Upper Bari Doab canal	1859	3.35
	Sirhind canal	1873	6.0
Rajasthan	Gang canal	1927	3.04
Tamil Nadu	Kaveri delta system	1889	5.05
Uttar Pradesh	Upper Ganga canal system	1856	6.99
	Lower Ganga canal system	1880	5.28
	Eastern Yamuna canal system	1830	1.91
	Sarda canal system	1926	6.12

works: Upper Bari Doab canal, Ganges Canal, and Godavari Delta System. Some important aspects of irrigation works during the British period are:

(i) Taking up of a large number of storage schemes (mainly earthen and masonry dams), and
(ii) Application of design engineering and construction technology.

It is not possible to enlist all the works in this chapter. The reading material given at the end of this chapter may be referred to for detailed information.

3.13.2 *Upper Ganga Canal During British Period*

River Ganga is a source for large irrigation potential throughout its length of more than 2000 km for the vast agriculture community in Ganga basin in India. In old days Ganga–Yamuna doab which forms the most fertile agriculture land in the country had been facing recurring severe droughts and famines. The well recorded severe famine in the doab land in 1837 had specially drawn the attention of engineers to protect this agricultural land against recurring droughts. The construction of Upper Ganga Canal was conceived and constructed by Proby T. Cautley during the period 1840–1854.

Headworks of UGC

In the beginning, at the close of monsoon season every year, the Ganga River water used to be diverted into a supply channel by constructing a temporary bund of timber crates filled with boulders. Another bund was constructed on the left bank of supply channel opposite. Har-ki-pauri to close the spill between Belwala and Laljiwala islands.

This bund was replaced by pucca weir called Hardwar dam in the year 1876. Originally it consisted of 5 bays of 53.65 m (176 ft.) each with 110 drop gates. After 1894 Gohna flood which caused major damage to Hardwar dam, an additional bay of 53.65 m was added on its right side in the year 1897. The Hardwar dam was capable of passing a flood discharge of 1700 cumecs. Permanent headworks consisting of Bhimgoda barrage, undersluices, canal head regulator and river training works were constructed from 1971 to 1992.

Landmarks

Date of commissioning of UGC	April 8, 1854
Date of start of canal irrigation	May 01, 1855
First remodelling done in year	1868
Osrabandi system started in year	1880
Remodelling & increase in capacity at the head from 191.16 to 240.72 cumec	1932
Increase in capacity from 240.72 to 297.36 cumecs	1951–52
Construction of new headworks i.e. Bhimgoda Barrage	1978–84

The Canal

Initially, the Ganga Canal was designed to carry 6,750 cusecs (191 cumecs). The supply of irrigation was restricted from 30 to 45 percent of the commanded area. Subsequently the canal was remodeled and it is now capable of passing 10,500 cusecs (297.20 cumecs). The system comprises 570 miles (912 km) of the main canal and branches, 3,560 miles (5696 km) of distributaries and minors. The area statistics are as below:

1. Gross commanded area: 50.60 lacs acres (20.48 lac ha)
2. Culturable commanded area: 39.34 lacs acres (15.92 lac ha)
3. Area irrigable at present: 19.12 lacs acres (7.74 lac ha)
4. Area irrigated in Kharif: 8.38 lacs acres (3.39 lac ha)
5. Area irrigated in Rabi: 7.70 lacs acres (3.12 lac ha)

The UGC system has undergone modernization in the past few decades. The modernization scheme is discussed in Chap. 14.

3.13.3 Masonry Dams in the British Period

Several gravity dams were constructed during the British period mainly for the protection from drought and famine. Table 3.13 provides salient features of some of these dams. The dams are listed in the historical sequence of construction. Approximate geographic locations of the dams are shown in terms of the stream and nearby town or district. Height, crest length, the volume of construction material, and storage capacity are also indicated, based on the literature. Compared to dams in ancient and medieval periods, dams of higher height and storage capacity were constructed during the British period.

Dams in India are classified as small, intermediate and large on the basis of height (from bed of river to full reservoir level and gross storage capacity (CWC 2001). Table 3.14 shows the classification criteria.

Based on this criterion, 16 large dams, 6 intermediate dams and 3 small masonry dams were constructed during the 19th Century A.D. and up to 1947 A.D.

3.13.4 World Heritage Irrigation Structures (WHIS) of the British Period

International Commission on Irrigation and Drainage (ICID) awards the World Heritage Irrigation Structures (WHIS) awards for Irrigation Structures that are more than 100 years old and are still functioning. It includes both old operational irrigation structures as well as those having archival value. Table 3.15 shows the irrigation works of the British period in India that have been declared as World Heritage Irrigation Structure.

Dhukwan Weir Sukwan-Dhukwan Weir: The weir is constructed with granite stone masonry with a heating of cement concrete. The weir has been provided with 383 no falling shutter gates of size 10 ft. × 8 ft. above the crest level 890 ft., making full reservoir level 898 ft. The weir is also known as Sukwan-Dhukwan weir connecting Sukwan village in MP and Dhukwan village in UP through a passage tunnel inside the weir body. The water from the reservoir is released into the river for utilization at Parichha weir through two upper sluices. The sluice gates are operated from the sluice tower located in the middle of the weir. The weir was designed to pass a maximum flood of 6,52,000 cusecs with a water column height 12.75 ft. above the crest, RL 902.75 ft. The afflux bund with top-level kept at 907 ft. was constructed with black cotton earth placed upstream of the core wall with side slope 1:3 and downstream slope with common earth.

At the time of independence, the net irrigated area of India under British rule which included Bangladesh and Pakistan was 28.2 Mha. After partition, the net irrigated area in India and Pakistan was 19.4 Mha and 8.8 Mha, respectively.

Table 3.13 Masonry dams constructed during the 19th century A.D. and up to 1947 A.D.

S. no.	Name of dams	Year of completion	Location stream/town or district	Height above river bed (M)	Length of crest (M)	Volume content of dam (1000 cum)	Gross capacity (M cum)
1	Khadakwasla	1879	Mutha river	31.20	1470.00	290.00	87.50
2	Muchkundi	1884	Kisina basin	18.30	158.50	18.35	17.70
3	Paricha	1885	Betwa/Jhansi	16.10	1174.40	–	91.41
4	Bhatool	1892	Mehekari/ Ahemadnagar	15.20	837.30	–	1.14
5	Periyar	1897	Periyar/ Madurai	48.16	378.20	141.44	443.49
6	Kodayar (Pechipara)	1906	Kodayar/ Madras	30.20	425.50	107.30	226.43
7	Vanivislas Sagar	1907	Vedavati/ Chitaldurg	43.20	405.40	2166.00	849.60
8	Dhukwan	1909	Betwa/Jhansi	14.20	1172.20	–	106.44
9	Lamura	1910	Dhasan/ Jhansi	14.00	542.20	–	35.96
10	Chankpur	1911	Girnakalwan	30.80	458.00	–	43.70
11	Darna	1912	Darna/Nasik	25.00	1630.00	183.00	249.00
12	Pahari	1913	Dhasan/ Jhansi	13.90	580.00	–	79.34
13	Gangao	1915	Ken/Satna	13.10	801.30	77.80	99.79
14	Tigra	1917	Sank/Tigra	24.10	1181.00	284.76	121.41
15	Ghagar	1917	Ghaga/ Mirzapur	20.40	695.80	5801.00	152.50
16	Pocharam	1922	Aylair/ Hyderabad	14.90	640.00	–	51.50
17	Borina upper dam	1923	Jabalpur	21.00	527.60	72.93	1.67
18	Wilson	1926	Pravara/ Maharashtra	82.30	507.00	336.00	315.74
19	Chandia Nala	1927	Chandia/ Sagar	25.30	329.50	19.64	5.86
20	Bhatghar	1927	Yalwandi/ Maharashtra	51.20	1625.00	580.00	682.00
21	Nizam Sagar	1932	Manjira/ Hyderabad	35.20	2286.00	868.00	454.80
22	Krishna Raj Sagar	1932	Kaveri/ Mysore	39.60	2621.30	840.00	1368.70
23	Wyra	1933	–	18.59	1167.84	114.52	56.29

(continued)

Table 3.13 (continued)

S. no.	Name of dams	Year of completion	Location stream/town or district	Height above river bed (M)	Length of crest (M)	Volume content of dam (1000 cum)	Gross capacity (M cum)
24	Mettur	1934	Kaveri/ Madras	53.60	1615.40	1545.90	2708.80
25	Thippaya Palem	1938	Pallavagu/ Kurnool	12.20	129.20	9.18	14.00

Table 3.14 Size classification

Category	Gross storage (MCM)	Height (m)
Small	Between 0.5 and 10 MCM	Between 7.5 and 12 m
Intermediate	Between 10 and 60 MCM	Between 12 and 30 m
Large	Greater than 60 MCM	Greater than 30 m

Source Central Water Commission, Ministry of Water Resources, GOI (CWC 2001)

Table 3.15 World heritage irrigation structures (WHIS) of British period

State	Irrigation work	River/basin	Year of construction	Size	Irrigation area
Telangana	Tank (Pedda Cheru	Manair/ Godavari	1897	1.8 km long tank bund	900 acre
Telangana	Sadarmatt anicut	Godavari	1891–92	437.4 + 23.8 m	13,100 acre
Andhra Pradesh	Kurnool-Cuddapah canal	Tugabhadra/ Krishna	1871	Interconnects Pennar and Tungabhadra river	2.6 lakh acre
Uttar Pradesh	Dhukwan weir	Betwa river/ Yamuna	1909	3845 ft. lonk, 50 ft. high, 890 ft. at crest	62000acre original/ 601927 acre present

Source http://www.incid.cwc.gov.in/whis.html

Major canal systems, including the Sutlej and Indus systems, fell to Pakistan's share. East Bengal, now Bangladesh, which comprises the fertile Ganga–Brahmaputra delta region, also went to Pakistan. The irrigation works which remained with India, barring some of the old works in Uttar Pradesh and the deltas of the south, were mostly protective and meant more to ward off famine than to produce significant yields.

3.14 Irrigation in the Post-independence Period

Agriculture and irrigation have continued to receive a very high priority in India's Five Year Plans and lessons learned from famines have made India self-reliant in food production. Irrigation development in the post-independence period has the following broader social objectives besides food crop production. The development of irrigation and its management is based on the concept of social welfare. Most of the irrigation development has taken place through government-sponsored and managed irrigation works.

Irrigation Policy: Irrigation projects are designed to provide irrigation water to at least one major crop in a year over as large an area as possible.

The broader societal objectives are

(i) removal of poverty,
(ii) self-reliance in food crop production,
(iii) improvement in the quality of life particularly in backward regions and marginal groups of society,
(iv) import substitution and increase in export of agriculture-related commodities,
(v) draught area protection,
(vi) self-employment in rural areas, and
(vii) growth of agro-based industries.

Agriculture, as an industry, has made tremendous progress due to the practice of irrigated agriculture, use of fertilizer, pesticides, and high-yielding variety seeds. A large number of major, medium and minor irrigation schemes have been implemented in the country. More details on irrigation in the post-independence period are given in Chap. 2.

A wide range of problems and issues have been encountered in managing irrigated agriculture for achieving the specified objectives and making contributions to higher-level goals stated above.

Questions

1. What lessons could be learnt from irrigation works in ancient period?
2. Write a note on geographic variation in rainfall in ancient time
3. Explain the following terms:

   (i) Persian wheel, (ii) social stigma, (iii) Drona, (iv) anicut, (v) size of dam.

4. Write short notes on (a) long life of Sudarshan dam, (b) Egalitarian wage structure, and (c) penalty for violation of irrigation rules during the Mauryan period.
5. What was the purpose of water conveyance canals during the Arab and Mughal periods?

6. Discuss some important diversion canals in the Indus basin during the British period.
7. Discuss some important diversion canals in South India during the British period
8. Discuss historical evolution of construction technology.
9. Write a note on the type and size of irrigation works in ancient Sri Lanka.
10. What was the main reason for constructing the upper Ganga Canal system
11. Describe the modifications in headworks of Upper Ganga Canal.

References

Brohier RL (1934) Ancient irrigation works in Ceylon, Part I and Part II. Govt. Publications Bureau, Colombo, Sri Lanka

CBIP, Central Board of Irrigation and Power (Pub. No. 230) (1992) History of irrigation in Indus basin

Chaube UC et al (1997) Lessons from ancient works of irrigation and agriculture in Indian subcontinent (Up to 1000 A.D.). J Indian Water Resour Soc 17(3)

Chaube UC et al (2019) Fiscal aspects of irrigation administration in Mauryan period: a comparative study with the present. J IWRS 39(3)

Chaube UC et al (2020) Some technical aspects of the ancient Sudarshan dam in India. ICID News Bulletin, 3–4 quarter, 2020, International Commission on Irrigation and Drainage, New Delhi

CWC (2001) Manual on estimation of design flood. Central Water Commission, New Delhi

CWC (2015) Compendium on silting of reservoirs in India. Central Water Commission, Government of India, New Delhi

Dass M (2006) Water systems at Udaigiri: a search for its meaning, in the book traditional water management systems of India. In: Chakravarty KK et al (eds) Indira Gandhi Rashtriya Manav Sangrahalaya, Bhopal-462013. ISBN-81-7305-315-4

GOI (2015) Notification on minimum support price by ministry of agriculture and farmers welfare. Government of India, 16 Nov 2015

GOI (2018) Notification on minimum support price by ministry of agriculture and farmers welfare. Government of India, 10 October 2018

Govt. of India, Public Works Department (1922) Triennial review of irrigation in India, 1918–1921

Govt. of U.P. (1988) Modernization of upper Ganga Canal. Brochure brought out by UGC Modernization Circle (WB) Roorkee, 1 Dec 1988

Gupta PL (2013) Coins. National Book Trust India, p 21

ICID (2001) Book "historical dams". In: Fahlbusch H (ed) International commission on irrigation and drainage (ICID), New Delhi

Indian Irrigation Commission Report, (1901–1903)

Jha HK (1991) Irrigation in ancient and early medieval period in Indian subcontinent, (up to 1200 A.D.) (Special Problem report). WRDTC, University of Roorkee, Roorkee, India

Kangle RP (1993) Kautilya Arthashastra. University of Bombay, Bombay

Mate MS, Chakravarty KK et al (2006) Traditional water management systems in India. Indira Gandhi Rashtriya Manav Sangrahalaya, Bhopal

Mehta RN (1968) Sudarsana Lake. J Orient Inst M.S. Univ Baroda XVIII(1 and 2):20–28

Mulla ZR, De Vylder G (2014) Wages in the Indian bureaucracy: Can Kautilya's Arthashastra provide an answer. Great Lakes Herald 8(2):16–39

Parker H (1889) Irrigation in the North Western provinces report on the proposed Deduru Oya Project, Vol III. Colombo, Government Printer

Parker H, 1909 (1981) Ancient Ceylon. New Delhi, Asian Educational Services.

Randhawa MS (1980) History of agriculture in India, Volume I. Beginning to 12th century, Published by Indian Council of Agriculture Research, New Delhi
Rangarajan LN (1987) Kautilya—the Arthashastra. Penguin Books, New Delhi, pp 231–233
Roman U (1994) Historical development of irrigation structures in India. M.E. Dissertation, WRDTC, University of Roorkee, Roorkee, January 1994
Sarma KK (1962) Anicut Canal. J Indian Hist
Shaw J, Sutcliffe J (2003) Ancient dams, settlement archaeology and Buddhist propagation in central India: the hydrological background. Hydrol Sci J 48(2):277–291. https://doi.org/10.1623/hysj.48.2.277.44695
Sharma RS (1995) Perspectives in social and economic history of early India, (chapter XI-Irrigation in North). Munshi Ram Manohar Lal Publishers Ltd., New Delhi
Srinivasan TM (1970a) A brief account of the ancient irrigation engineering system prevalent in South India. India J Hist Sci 2(5)
Srinivasan TM (1970b) Water-lifting devices in ancient India: their origin and mechanisms (from the earliest times to C.A.D. 1000). Indian J Hist Sci 2(5)
Thomas M (2018) 2019 Standard catalog of world coins, 2001—date. Krause Publications, Wisconsin, USA
Thomas A (2019) Indus river basin: water security and sustainability. Elsevier/Academic Press. ISBN: 9780128127827

Reading Material

Central Board of Irrigation and Power (Pub. No. 76) (1965) Development of irrigation in India
Central Board of Irrigation and Power (Pub. No. 138) (1979) Design and construction features of selected dams in India
Central Board of Irrigation and Power (Pub. No. 148) (1981a) Barrages in India
Central Board of Irrigation and Power (Pub. No. 149) (1981b) Design and construction features of selected Barrages in India
Central Board of Irrigation and Power (Pub. No. 138, Vol. II) (1983) Design and construction features of selected dams in India
Central Board of Irrigation and Power (Pub. No. 185) (1987a) History of Cauvery Mettur project
Central Board of Irrigation and Power (Pub. No. 197, Vol. No. I) (1987b) Large dams in India
Govt. of U.P., "Upper Ganga Modernization Project" I&P (World Bank) Govt. of Uttar Pradesh, Deptt. of Irrigation, Roorkee (1991).
Sharma PV (1998) Essentials of ayurveda: text and translation of Sodasangahrdayam. Motilal Banarsidass Publications, New Delhi
Tracy S, Thomas M (2018) 2019 Standard catalog of world coins, 1901–2000. Krause Publications, Wisconsin, USA
WAPCOS (1990) Manual on O&M of upper Ganga canal. Water and Power Consultancy Service, New Delhi

Chapter 4
Irrigation Administration

Abstract The irrigation system is viewed in terms of the social structure of managers and farmers. This chapter explains the conventional system of irrigation administration, its deficiencies, and possible reforms. The functions of work-charged staff and the technical and financial responsibilities of the engineering staff are discussed. Some typical norms for the deployment of work-charged staff of maintenance are presented. The existing organizational structure for irrigation administration is illustrated with an example, and the existing organizational structure is critically examined. The existing structures are not based on integrated management of land, water, and other inputs. The system is administration and revenue-oriented. The organizational arrangements for water control, water distribution, and on-farm development differ from State to State. There are multiple agencies for water control, water distribution, and on-farm development works. Several factors are responsible for the underutilization of irrigation potential and the low productivity of irrigated agriculture, including job functions and the organizational structure for irrigation. Irrigation managers are senior staff, almost all of whom have engineering degrees. Their incentives include convenience and amenities, promotion and career, status, income, the avoidance of stress, and professional satisfaction. Their motivation can be powerfully affected by the political control of postings. Corruption of canal irrigation systems has five adverse effects: costs to farmers, especially the poorer and weaker; bad physical work in maintenance; bad canal management, indiscipline of field staff; and managers demoralized and distracted from their proper work. Possible reforms in irrigation administration include vigilance, political reform, discipline, separate operations and maintenance cadres; rights and information for farmers; accountability and incentives for managers; and enhanced professionalism. Political reform implies less control or no control over transfers by politicians In India, the irrigation department usually has four wings—investigation, design, construction, and O&M. A separate O&M wing can weaken the transfer trade and the control of politicians. The reasons are briefly explained. As a case study, the organizational structure of a maintenance division (headed by an Executive Engineer) in the Sagar district of Madhya Pradesh state in India is discussed. The number of junior engineers is 42 against the sanctioned strength of 16. The number of watchmen on work assigned and daily wage basis is 42. The salary component of the budget allocation to a field division is disproportionately high, leaving a lesser budget for actual maintenance

U. C. Chaube et al., *Canal Irrigation Systems in India*, Water Science and Technology Library 126, https://doi.org/10.1007/978-3-031-42812-8_4

work. Based on farmers interviews and field investigations, suggestions are made for improving staff performance.

4.1 Features of Conventional Canal Administration

The irrigation administration is mainly by Civil Engineers who may receive in-service training at Water and Land Management Institutes in the irrigation field. The system is administered (instead of managed) and revenue-oriented (instead of physical performance-oriented). Table 4.1 shows characteristic features of irrigation administration in India.

The water rights of farmers are still not fully recognized. If a canal system fails to deliver the water in time or inadequate quantity, the most that the farmer could claim is remission in water charges. The responsibility of the canal administration ends at the outlet. The distribution within the outlet command is the responsibility of farmers.

Canal systems are subject to physical and social entropy, which means that unless efficient management is done, the system is subject to physical and administrative decay. For various reasons, this has been happening at an increasingly faster rate.

The main objective of this chapter is to understand the conventional system of irrigation administration in terms of staffing patterns, staff functions, and the financial responsibilities of staff.

Table 4.1 Features of conventional canal administration

Aspect of administration	Characterizing features
Focus of administration	More emphasis on financial performance(annual budgetting, timely expenditure) and supply-oriented operation of conveyance system
Physical facilities under the control	Canals and regulation structures upto outlets
Organizational control	Mainly by civil engineers
Staff	A large contingent of technical, non-technical, and work-charged staff
Staff duties	Mixing up duties related to planning, construction, operation, and maintenance of the irrigation, drainage, and flood control facilities. The distribution of duties is based on the command area jurisdiction
Important item of expenditure	More expenditure on staff salary; relatively much less on physical maintenance

4.2 Organization Structure

The conventional administration of irrigated agriculture in India is not based on the integrated management of land, water, and other inputs to achieve physical, social, and economic objectives. Responsibilities are bifurcated between the irrigation department and the agriculture department from top to bottom (ministerial secretary level to field staff). Even irrigation organization structures differ from state to state and even from project to project within a state. A general organizational pattern of an irrigation department is depicted in Fig. 4.1. Whereas most of the states have a separate department dealing with irrigation, in Tamil Nadu it is part of the public works department. In West Bengal, it is part of the irrigation and waterways department. Organization chart of the Water Resources Department of Gujarat is shown in Fig. 4.2.

At the ministry level, there is an administrative head known as the Secretary of Irrigation (usually from the Indian Administrative Service Cadre and not from the Engineering Service Cadre) and a technical head known as Engineer-in-Chief (belonging to the state-level Engineering Service Cadre). At the project (scheme) level, there are chief engineer (project head), superintending engineer (100,100–500,100 ha), executive engineer (40,000–1,00,000 ha), sub-divisional engineer (10,000–25,000 ha), and section officer (2500–6000 ha). The project staff is usually classified as:

(a) Regular (long-term employment) and work charged (ad-hoc employment),
(b) Gazetted (officer) and non-gazetted (office staff), and
(c) Technical and non-technical.

Table 4.2 shows multiple agencies involved in water control, water distribution, and on-farm development works in some states in India. An example of the organizational hierarchy of the operation and maintenance staff for the Nagarjunsagar project (Telangana State of India) is given below in Table 4.3.

4.3 Functions of Work-Charged Staff

4.3.1 Functions of Work Charged Staff

The work-charged staff is employed on an ad-hoc/contract basis for a short period for particular works. After completion of one work, their services are generally utilized on other works. However, on many projects, work-charged staff have been serving for several years. The normal maintenance program needs the services of various type of work-charged staff as shown in Table 4.4. The table shows the type of the work- charged staff, literacy requirement, and functions/services provided by them.

Besides these, gardeners, sweepers, cooks, mechanics, pipefitters, electricians, and masons, etc., are also needed on a work-charged basis.

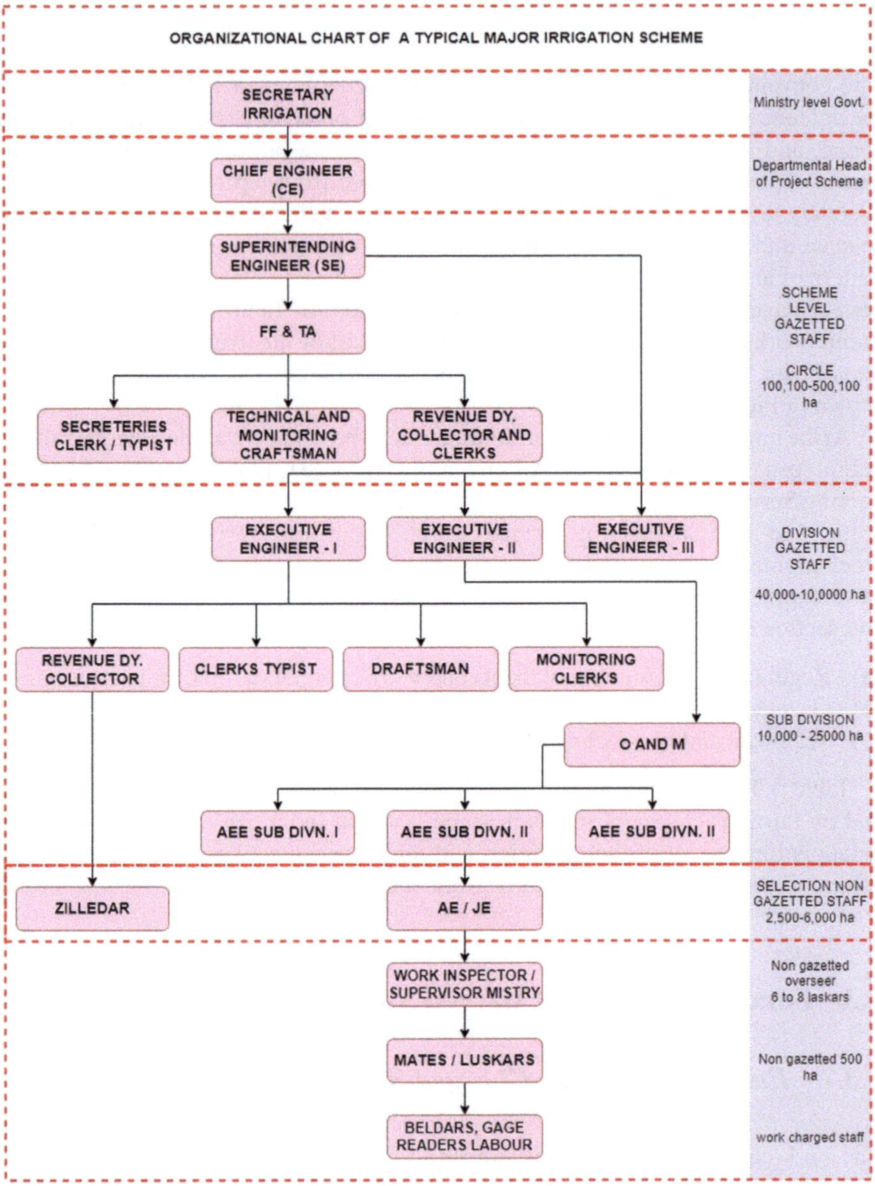

Fig. 4.1 General organizational pattern of an irrigation department

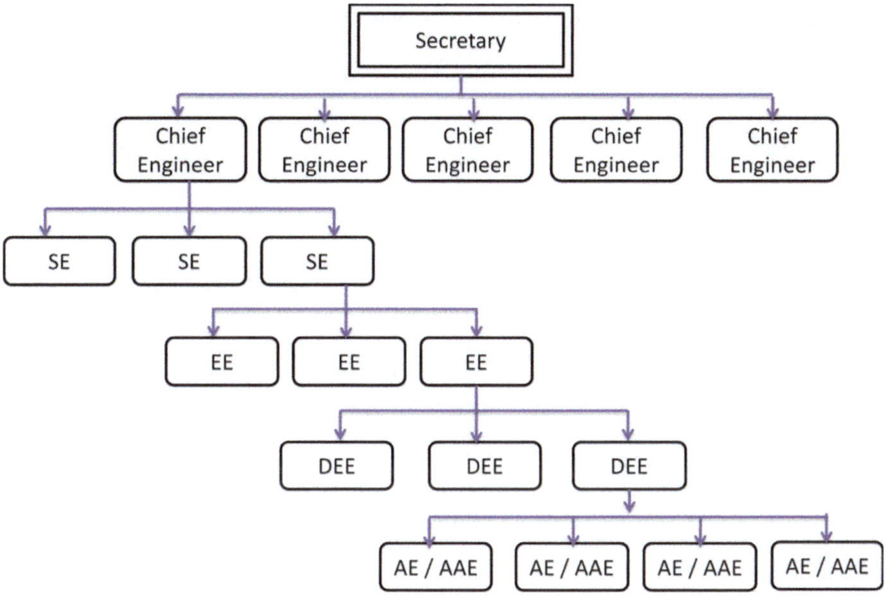

Fig. 4.2 Organisation chart of water resources department Gujarat

4.3.2 Norms for Deployment of Work-Charged Staff

Norms for deploying work-charged staff for the maintenance of irrigation works are indicated in Table 4.5.

4.4 Technical Responsibilities of Engineering Staff at Field Level

Engineering staff at the field level consists of Junior Engineer (JE), Assistant Engineer (AE) and Executive Engineer (EE). JE may be a diploma holder or a degree holder. Assistant Engineer and Executive Engineers are usually degree holders in Civil Engineering, but diploma holders with sufficient experience could also become AE and EE. Technical responsibilities of field engineering staff are stated in Table 4.6.

Junior Engineer

Within a maintenance unit, he is the key person on whom a good standard of maintenance depends. Junior Engineer is concerned with periodic supervision of items that may need maintenance, their surveys, preparation, and estimation of works to be done, supervision of work, and measurements of the actual work done at the site for making payments.

Table 4.2 Multiple agencies for water control, distribution, and on-farm development

State	Water control structure	Water distribution upto outlet	Water distribution below outlet	On farm development
Andhra Pradesh	ID	ID	Farmer	CADA
Bihar	ID	ID	Farmers	AD
Gujarat	ID	CADA	Farmer/CADA/ ID	AD
Maharashtra	Management Wing of ID	Management Wing of ID	Farmer/ID	ID
Tamil Nadu	1. Anicut—PWD (irrigation)	PWD (Irrigation)	Farmer	AD
	2. Small works—Revenue Department	PWD (irrigation)	Farmer	AD
	3. Reservoirs—ID and Revenue Department	PWD (Irrigation)	Farmer	AD
Uttar Pradesh	Irrigation Department	ID	ID	ID&AD
West Bengal	Irrigation and Waterways Department	Irrigation and Waterways Department	Farmer/ Agriculture Department/ CADA	AD

Note ID Irrigation Department, *CADA* Command Area Development Authority; *AD* Agriculture Department; *PWD* Public Works Department, *OFD* On-Farm Development
Source Report of CAD and WM Division, Ministry of Irrigation, Government of India, December 1982

Table 4.3 Organization hierarchy of O&M Staff for Nagarjun Sagar project

Level	Position
i. Projects	Chief Engineer for Nagarjun Sagar for the right command (4.8 lakh ha) Chief Engineer for Nagarjun Sagar for the left command (3.2 lakh ha)
ii. Circle	Superintending Engineer (Circle): In-charge of about 2–2.5 lakh ha
iii. Division	Executive Engineer: A circle normally has 4 divisions, each headed by an Executive Engineer for 0.8–1.0 lakh ha
iv. Sub-division	Deputy Exec Engineer: A division has four subdivisions, each with a deputing executive engineer and operates over 20,000–25,000 ha
v. Section	Junior Engineer: There are usually 4 sections, each headed by a Junior Engineer to look after around 5000–6000 ha. This forms the main functional unit with 15–20 minors and 200–250 outlets or chaks
vi. Working units	a. Work Inspector: In a section, there are 2–3 work inspectors. He oversees 6–8 last field functionaries, called laskars. Sometimes he is also directly in charge of a minor/distributory b. Laskar is the last working unit, in charge of about 500 ha, consisting of 20–25 outlets

Table 4.4 Type and functions of work-charged staff

Designation of work charged staff	Literacy	Functions of the work-charged staff
1. Beldar (labor)	No literacy requirement	(a) Jungle clearance, removal of floating weeds, plantation, and pruning of trees; (c) Operation of regulation gates; (d) Watering of service road, filling of rain cuts and patrolling of channels; and (e) Assisting in survey work and repair of structures
2. Chowkidar (watchman)	No literacy requirement	Watch and ward of the inspection bungalow, site stores, office complex, staff colonies
3. Mate/Munshi (supervisor of beldars)	High School pass	(a) Marking the attendance of the labor, distributing and supervising the work (b) Measuring and recording the progress of work and reporting to the supervisor or junior engineer; (c) Keeping watch and ward on the canal and other properties and reporting to his seniors about any causality, accident, damage, or encroachment upon property; (d) Keeping account of tools and plant and materials, trees (e) Assisting the junior engineer in taking measurements of works and surveys
4. Gauge reader	High School pass	Gauge reading at regular intervals at the head and tail of channels, at the head outlet sluice of the dam, and all the off-taking points on the distributaries and minors Maintain a register of gauges and communicate gauge readings to Engineer in charge in a prescribed gauge slip every 3 h/6 h/daily as instructed

(continued)

Table 4.4 (continued)

Designation of work charged staff	Literacy	Functions of the work-charged staff
5. Signaller (telephone/ wireless communication operator)	High School with training in telegraph service	They are essential links for communicating the canal messages regarding gauges, increasing or decreasing the canal discharges, etc. to higher authorities
6. Mistri/supervisor	High School with experience as a mate	A Mistri/Supervisor organizes and supervises about 5 to 10 mates as per the type of work and is directly responsible to the Junior Engineer. He is promoted from the post of the mate Collects the daily progress of work from mates and submits to Assistant Engineer Maintains the discharge at the head outlet and in the channels as per directions from the executive engineer by adequately regulating the gate openings He also maintains site stores and tools and plant and field equipment under Junior Engineer's charge He assists the Junior Engineer in surveys and measurement of works

He is in charge of all stores, materials, and equipment at the site. Also, he maintains accounts, both their receipts and issues, which he submits to his Assistant Engineer every month. He also maintains the muster/acquaintance rolls of the daily-waged or work-charged labourers under his charge. At the close of the month, he submits the rolls to the Assistant Engineer with the work done by that laborer for pass and payment orders.

Assistant Engineer

He may be a direct recruit if a degree holder through the Public Service Commission, but some posts are also filled by promotion from Junior Engineer's position. The percentages and eligibility for direct recruitment and promotion vary from state to state.

Table 4.5 Typical norms for deployment of work-charged staff of maintenance

S. no.	Particular	Staff-category	Number	Remark
A-minor irrigation works (CCA up to 2000 ha)				
1	Tanks up to 50 Mcft	Beldar	1	For the tanks and canals
2	Tanks above 50 Mcft	Beldar	1	For tank
B-medium (CCA from 2000 to 10,000 ha) and major irrigation work (CCA above 10,000 ha)				
1	Medium tanks	Beldar	4	For dam and appurtenant work and day and night watch
2	Major tanks	Beldar	6 to 8	For dam and appurtenant work and day and night watch
C-canal and distribution system				
1	General maintenance of canals	Mate	1	For 10 beldars
		Beldar	1	For 10 km up to 25 cusec discharge
		Beldar	1	For 8 km from 25 to 100 cusec
		Beldar	1	For 5 km, from 100 to 500 cusec
		Beldar	1	For 3 km, from 500 to 1000 cusec
		Beldar	1	For 2 km, beyond 1000 cusec discharge
2	Regulation of canals	Gauge Reader	1	For discharge upto 50 cusecs
		Gauge Reader and Beldar	1 + 1	For discharge from 50 to 100 cusecs

(continued)

Table 4.5 (continued)

S. no.	Particular	Staff-category	Number	Remark
		Gauge Reader and Beldar	2 + 2	For discharge from 100 to 500 cusecs
		Gauge Reader and Beldar	3 + 3	For discharge above 500 cusecs
*D-**telegraph/telephone** or **wireless station***				
1	At divisional headquarter	Signaller/Telegraph/ Telephone operator& Beldar	2 + 2	Per instrument
2	At subdivisional level	Signaller/Telegraph/ Telephone operator& Beldar	1 + 1	Per instrument
3	At other places	Signaller/Telegraph/ Telephone operator	1	Per instrument
*E-**inspection hut** and rest houses*				
1	Inspection hut with no catering	Watchman	1	
2	Class II rest house (only tea is provided)	Watchman &Beldar	1 + 1	
3	Class, I rest house	Watchman (2), Gardener (1), Beldar(2), Sweeper (1), Cook (1), Supervisor (1)		
F-irrigation colonies				
	Colonies up to 25 houses	Watchman	1	With or without office complex
	Colonies from 25 to 50 houses	Watchman (1), Beldar (2), Sweeper (1), Artisian (1)		With or without office complex
	Colonies from 50 to 200 houses	Watchman (2) Beldar (4), Sweeper (2), Gardener (1), Mistri (1), Meson(1), Artisian(1), Electrician (1)		With or without office complex

(continued)

Table 4.5 (continued)

S. no.	Particular	Staff-category	Number	Remark
G-plantation nurseries				
	Nurseries providing minimum 5000 plants per year	Gardener and Beldar	1 + 2	

Table 4.6 Technical responsibilities of engineering staff in the field

Designation and educational qualification	Areal jurisdiction	Technical responsibilities
Junior Engineer (JE)	One section: 2500–3500 ha	(a) Periodic inspection of the system (dam, canals, roads, building etc.) for the command area under his charge and preparation of maintenance plan/estimates for his area (b) Submission of the maintenance plan to the Assistant Engineer for arranging approval (c) Organizing the maintenance work to be done either with machines or manually (d) Ensure proper technical standards while doing the maintenance work (e) Ensure adequate safety measures during maintenance works; and (f) Measurements of the maintenance work completed

(continued)

Table 4.6 (continued)

Designation and educational qualification	Areal jurisdiction	Technical responsibilities
Assistant Engineer (AE) (sub division officer)	4–5 JEs:10,000–15,000 ha	(a) Evaluate annually the maintenance work needed (b) Plan the maintenance work in order of priority to achieve the maximum utilization of the available budget (c) Prepare technical and economic specifications of the work to be undertaken by the contract (d) Arrange timely sanction of the maintenance estimates from the Executive Engineer (e) Order and account for material and tools and plants (f) Supervise the upkeep of equipment in his charge (g) Issue instructions to the subordinates regarding the maintenance work to be done; and (h) Pass payment to laborers and contractors
Executive Engineer (DE) Divisional Engineer	3–4 Assistant Engineers 40,000–50,000 ha	(a) Sanction the maintenance estimates as per budget provisions (b) Decide the method of implementation for the maintenance program, whether departmentally or contractually (c) Get the maintenance of the system done as per the program and make payment for the work done (d) Check some percentage of the work done quantitatively and qualitatively (e) Make payment to the labourers (skilled and unskilled) employed on maintenance (f) Plan proper regulation and distribution of water among the farmers as per the availability of water and convening water distribution committees (g) Listen to the farmer's petition and give judgments as per provisions in the Irrigation Act and rules of the State (h) In some states, he is also in charge of assessment and collection of water charges and depositing them in the treasury; and (i) He is to function as per directions of the Superintending Engineer or Chief Engineer from time to time and is responsible to them

Executive Engineer

An executive Engineer is promoted from among Assistant Engineers based on seniority cum merit and generally has an experience record of 10–15 years as an Assistant Engineer before promotion. An Executive Engineer is the head of all maintenance activities in the division and is competent to sanction all maintenance estimates up to the budget provisions.

4.5 Financial Responsibilities of Engineering Staff

Engineering staff for operation and maintenance consists of Junior Engineer, Assistant Engineer, Executive Engineer, Superintending Engineer, and Chief Engineer in order of increasing hierarchy. At the field level, Executive Engineer has the authority to make O&M expenditures according to budget provisions. The process of O&M cost estimation and financial approval is depicted in Table 4.7.

Table 4.7 Process of O&M cost estimation and financial approval

Work item	Responsibility
Prepare L-section, X-section, and other maintenance/repair work drawings	Design Engineer
Work out quantities of various items of work	Junior Engineer
Obtain recently approved unit rates from PWD/irrigation department	Junior Engineer
Prepare detailed cost estimates based on necessary plans	Junior Engineer
Prepare abstract cost estimates for different items of work	Junior Engineer
Review and checking the estimates for quantities, unit costs, and total costs and submit to Executive Engineer	Assistant Engineer
Second review, checking, and approval of the estimates within the allocated budget for his division	Executive Engineer
Prioritization of maintenance works for execution and conveying his sanction to each Assistant Engineer for execution during financial year	Executive Engineer
If additional funds are required during that financial year, EE makes a complete case and requests the Superintending Engineer to arrange for additional funds	Executive Engineer
Preparing a complete case with his recommendations to the Chief Engineer for additional funds or requesting an appropriation of budgets from one division to the other, if possible	Superintending Engineer
a. Recommending additional budget from the State Government or b. Arranging for re-appropriation from one division to the other after reviewing the overall budget needs of each division; and c. Ensuring with the State Government that the budget to be provided for maintenance is quite adequate	Chief Engineer

4.6 Conditions and Incentives Affecting Irrigation Managers

Whether it is improving scheduling, reducing losses at night, instituting joint management with farmers, or other interventions, the managers hold the initiative. They are not just part of the system; for purposes of reform, they are the key. Understanding the environment they work in, their motivation, and their behavior are, thus, crucial in any search for performance improvement. The discussion in this section is based on Chambers (1988).

Staff managing canal irrigation systems in South Asia can be divided into engineers and field staff. Engineers with engineering degrees are normally liable to transfer away from one project to another or from one part of a very large system to another. Field staff do not have engineering degrees and normally remain in the same project or the same part of a very large project for the duration of their service.

Conditions and incentives for engineer-managers vary by person, stage of career, stage of family development, post, period in time, and project. The incentives, disincentives, satisfaction, and dissatisfaction of irrigation engineers can be analyzed below.

4.6.1 Convenience and Amenity

Most professionals prefer to live in towns and be not too distant from their areas of origin. Many irrigation postings are to areas regarded as remote and inconvenient, far from good amenities. In such postings, they may find themselves spending more on education, sending their children to boarding school instead of the good-day schools they could have found in most large towns. Remote postings are also far from sources of information and seats of power.

4.6.2 Career

There is a normal stepwise progression of promotion, with neither reward for exceptional diligence nor penalty for moderate delinquency, and with ceilings at certain levels which it can be difficult to pass. Though adverse entries in confidential reports can hinder promotion, that is unusual. For most officers, accelerated promotion is neither a great award nor is deferred promotion a serious penalty.

4.6.3 Status

An operation and maintenance (O&M) engineer may be sought after by big farmers, contractors, and politicians, especially when water is scarce. There are psychic rewards in having people waiting outside the office and behaving deferentially. O & M probably confers more status of this sort than most other irrigation engineering activities.

4.6.4 Income, Stress, and Professional Satisfaction

At a formal level, salaries have been subject to long-term declines in real terms. Pay differential among salaries of engineers and those of skilled/unskilled field workers has greatly reduced since the time of independence. Stress is the other side of the status coin, dealing with farmers' and politicians' complaints and pressures. A further obvious point is that engineers' training is for design and construction, not for canal management, for which they have in the past received virtually no training at all. Not surprisingly, a preference is expressed for work in design and construction.

The nexus of relations which are broadly described by the term 'corruption' appears to influence the behaviour of many engineers. Corruption affects the behaviour of managers, and therefore it is part of the system and its environment. To ignore it could be not just unscientific but gravely misleading since prescriptions would not be based on reality. Like any other part of the system, it must be considered. It is in that spirit that it is analyzed here.

4.6.5 Political Control of Bureaucracy

Transfers of officials are at the core of political control of the bureaucracy in India. For obvious reasons of convenience, amenity, and professional satisfaction, officials prefer particular posts. Normally there is a three-year term for any post. But in practice, turnover could be faster. The reasons include; transfers accepted because they carry promotion, the transferee immediately negotiating for retransfer carrying the promotion with him or her; and unpopular and punitive postings from which the transferee seeks early release. Transfers are indeed a powerful instrument for punishment and patronage. Nor is there any mystery about the control of transfers by politicians. The Indian press and social media provide ample evidence of the prevalence and effects of transfers as a means of political patronage and control. Threatened and actual transfers, as a means of control, punishment, and reward, are thus a recognized phenomenon in India.

4.6.6 Effects of Corruption

Corruption in canal irrigation has the following five adverse effects (Chambers 1988).

(i) **Costs to farmers**: The first adverse effect is the cost to farmers in cash and kind. Estimates of per-acre costs vary widely. For example, farmers on a project in Gujarat paid 20 to 50 rupees each time they needed water, rising to as high as 100 rupees if they needed water badly. In South Indian systems, it appears that only some groups of farmers pay, mainly tailenders and encroachers: that being so, the per acre average will be much lower than the figures cited for those who do pay. In systems like Mahi-Kadana where there are water indents, payments may be per watering rather than per acre since each watering requires a bureaucratic procedure.

To have to negotiate and pay for each watering involves time, uncertainty, and financial cost, which are likely to hurt and worry smaller farmers more than large. The richer farmers seem to have less problems getting water because of their position in the village and ability to give more gifts. The poor face more hassle because they have to adjust around the needs of the rich even when they make illegal payments.

(ii) **Bad physical Work**: Evidence of corruption in construction is reflected in the poor quality of irrigation works in many of the command areas of India. Quality control suffers when the quality controllers, the EEs and AEs, profit from poor quality and when contractors, engineers, and politicians are anyway all deep in the plot together.

(iii) **Bad Canal Management**: An adequate, timely predictable, and hassle-free water supply is a precondition for the adoption by farmers of high-yielding practices and a test of good canal management. But, in stark contrast, it pays canal managers and their staff to make water inadequate, untimely, unpredictable, and difficult to obtain.

(iv) **Indiscipline of field staff**: A fourth harmful and linked phenomenon is indiscipline and intransigence among field and clerical staff. Junior staff demand and require their share in the bribe. Indiscipline has many aspects. In many projects, work-charged staff (not pensionable and not on the establishment) make unions and may be reluctant to work.

(v) **Demoralization and distraction**: Engineers caught up in corrupt systems are not necessarily happy and fulfilled. On the one hand, they may have opportunities to accumulate large sums of money quickly; but on the other, it is risky, stressful, and unprofessional. Many engineers find the pressures to which they are exposed from all sides (especially on O&M jobs) very trying and find the behaviour needed to stay in the post in varying degrees distasteful. Morale in the Irrigation Department is certainly low.

Even if engineers want to manage their systems well, within the constraints of these various pressures, they face a serious problem. The management of flows of large amounts of money, negotiations with politicians, farmers, and staff, and the supply or non-supply of water must surely drain their energies and demand skill, ingenuity, and guile. Fudging figures takes time and demands calculation with care. Not much time may be left over, at least during some periods of the year, for managing canals.

It would be hard to imagine a stronger distraction from the formal task of managing an irrigation system in a productive, equitable, and stable manner. Managers' antennae have to be sensitive to signals from precisely the wrong people—the powerful and those with money—rather than the poor and deprived unless they, too, are able and prepared to pay. Possibilities of being double-crossed and cheated are legion. Managers are like double agents: with a formal task to be performed in public with sparing use of time and energy so that an informal, covert, and demanding set of activities can have priority.

Not all engineers are corrupt. No doubt many fight it, some bravely, some prudently. No doubt, there are many variations by region and by project. But, if something like the system described is widespread, it is a massive obstacle to improved system management. With short and insecure tenure in the post and a need to raise large sums quickly, a manager has little incentive to improve the main—system scheduling or irrigation at night or farmer participation.

4.7 Possible Reforms in Irrigation Administration

Reforms should change the conditions and incentives of managers so that they will wish and be able to manage canal irrigation better. Possible reform proposals are: vigilance, political reform, and discipline separate O and M cadres; rights and information for farmers; managers' incentives and accountability; and enhanced professionalism (Chambers 1988).

Vigilance: Vigilance, meaning investigation and inspection by outside bodies, is the most common reform adopted in a department. Vigilance Cells have now been created within irrigation departments.

Political reform: Political reform implies less control or no control over transfers by politicians.

Discipline: Another attractive option, much advocated, is 'discipline'. Often this is applied to irrigators. Discipline within Irrigation Departments touches directly on the motivation and performance of managers.

4.7.1 Separate O&M Cadres

In India, the irrigation department usually has four wings—investigation, design, construction, and O&M. Some characteristics of these four wings, as perceived by an engineer (manager) are shown in Table 4.8. Many variations can be expected within and between States in India, and between South Asian countries. In Sri Lanka, for example, design is more attractive because it is likely to mean residence in Colombo and also gives useful experience for a subsequent overseas job.

O&M posts vary widely from, e.g., a head-reach post with much-unauthorized irrigation, which provides engineers with a reliable and high take, to a tail-reach post where water is uncertain, farmers are known to be violent and payments unreliable.

A separate O&M wing can weaken the transfer trade and the control of politicians. The reasons are that there would be:

(i) More engineers in O&M who resist the transfer trade. Under the current system, selection for O&M exercises a bias towards those able and willing to purchase their posts and against those who are not. If initial recruitment to the new cadre attracted a representative cross-section of staff, it would offset these biases and include more engineers inclined to resist buying their posts
(ii) O&M would be seen more as a long-term career with disincentives for risky exploitation.
(iii) Encouragement to O&M engineers to opt out of the transfer trade. Being a member of the cadre, he would have to be posted at another place but only in the O&M cadre. That post would then lose its market value in the transfer trade and become another bridgehead of integrity.
(iv) A fostering of professionalism and esprit de corps, and of collective professional resistance to unethical practices.

Some states have been slow to set up separate O&M cadres.

Table 4.8 Perception of an assistant engineer on the investigation, design, construction and O&M activity

	Investigation	Design	Construction	O&M
University degree prepares engineers for the professional task	Yes	Yes	Yes	No
Opportunities for making money	Low	Rarely	High	Medium to low
Status and power	Medium	Low	High	Low
Professional interest of work	Medium	Medium	High	Low
Political hassle	Low	Low	Medium	High
Used as punishment postings for recalcitrant (honest) officers	Yes	No	Rarely	Yes
Competition for postings	Medium	Low	High	Medium

4.7.2 Rights and Information

Farmers on canal irrigation systems have varying senses of the right to water. In practice, as we have seen, they can be prepared to pay to ensure a supply to their outlets and chaks. Where CADAs exist and are confined below outlets, they, like farmers, have an interest in more adequate, timely, predictable, and hassle-free water supplies over large areas, including tailends; in short, in tighter scheduling and delivery. To this end, CADAs can exert pressure on Irrigation Departments. Moreover, by introducing warabandi' they stimulate the sense of the rights to water, which makes farmers more demanding.

Farmers can complain to administrators and political leaders. They can invite them to come and see for themselves. Several villages, a distributary, or some tail-end minors, can combine to exert pressure. They can demand the removal by transfer of any engineer who fails to serve them.

Government regulations can demand operational plans for water scheduling and publication of those plans. For their part, farmers can demand more information and become more adept at finding it out. Under the dual pressure of orders from above and farmers' questions from below, engineers and the politicians they pay may find management by rumor and muddle more and more difficult, embarrassing, and ultimately untenable.

On a system with farmers' representatives at minor, distributary, and other levels, and regular meetings between staff and farmers, management may become less tense for engineers.

4.7.3 Enhanced Professionalism

Irrigation engineers receive professional training for design, construction, and maintenance but scarcely any at all for operation. Construction is the professional ideal, involving mathematical precision, the management of materials, and the prestige of making big new things, especially dams. Nehru shared these values in his much-repeated remark, 'These irrigation project sites are the places of pilgrimage for me.' Few can fail to wonder at the great works of civil engineering. Constructions, and even designs, leave a permanent, visible outcome, a mark in history. A design or construction engineer can leave this world content, knowing the work he leaves for posterity.

The prestige of construction serves to debase the standing of O&M. Maintenance is a poor relation to construction, a constant struggle against entropy. In most operations, there is no precision. Flowing water is hard to measure, even 5–10% gauges are often badly calibrated. Water leaks through the cracked lining, or is stolen, or flows as unknown about the system. Distribution is often little supervised. Benefits are not counted. More distasteful and disorienting work would be hard to imagine for a civil engineer trained in the precise management of physical things.

Sample Survey of Perception of Project Engineers: Jayaraman and Jayaraman (1981b) have reported the findings of a sample survey of views indicated by irrigation engineers on Mahi—Kadana and Panam Projects, Gujarat. Respondents were asked to rank eight statements on a five-point (strongly agree-agree-do not know-disagree-strongly disagree) scale. The questionnaire was mailed, and the response was 81%. Table 4.9 shows the results.

Important conclusions of this survey are

- Water management is a monotonous activity, whereas construction and design offer a greater variety of experiences.
- Construction and design give greater job satisfaction than water management.
- There is a constant fear in water management that the dissatisfied local politicians may successfully attempt to transfer the personnel to a distant place.
- Construction and design do not require those difficult public relations with irrigators which are generally required in water management.

In sum, O&M was not only not interesting professionally; it also involves bothersome relations with politicians and farmers

These professional gaps are also found in education and training. Even now, there is no textbook or field practice manual on canal management or any of these gap subjects. There is no accepted methodology for determining optimal scheduling in South Asian conditions. There is no manual on combating the transfer trade or other pressures for corruption. The Water and Land Management Institute (WALMIs), in various states, have started field learning experiences for farmers and irrigation staff.

4.7.4 Other Measures

To enhance the professional status of O&M requires many complementary actions. In addition to the reforms suggested above, other measures include:

(i) the testing, development, and dissemination of gap methodologies;
(ii) textbooks on canal irrigation management, and curriculum changes to include training based on them;
(iii) a journal of Canal Irrigation Management which would specialize in articles on methodologies and reports of practical experience by engineer-managers;
(iv) professional training both in forms appealing to engineers, such as computer simulation games, and involving learning from farmers;
(v) annual awards for good system management; and

Table 4.9 Views indicated by irrigation engineers on Mahi–Kadana and Panam projects, Gujarat

Ranking of agreement	Statement	Scored responses (N = 289)				
		Strongly agree	Agree	Do not know	Disagree	Strongly disagree
1	Water management is a monotonous activity, whereas construction and design offers a great variety of experiences	80	169	10	27	3
2	Construction and design gives greater job satisfaction than water management	73	164	15	31	6
3	There is a constant fear in water management that the dissatisfied local politicians may successfully attempt to have the personnel transferred out to a distant place	63	119	71	31	5
4	Construction and design does not require those difficult public relations with irrigators which are generally required in water management	39	179	13	53	5
5	Construction and design offers a high degree of independence of action and it does not require liaison with other departments	38	167	14	55	15

(continued)

Table 4.9 (continued)

Ranking of agreement	Statement	Scored responses (N = 289)				
		Strongly agree	Agree	Do not know	Disagree	Strongly disagree
6	Construction and design is for 'hard' applied science people, whereas water management is for 'soft' applied science people	43	130	29	68	19
7	Water management needs multidisciplinary skills is difficult	13	123	55	81	17
8	Construction and design offers greater promotional opportunities	28	105	38	98	25

Source Jayaraman and Jayaraman (1981b, 288)

(vi) support from professional bodies such as the International Commission on Irrigation and Drainage, the Indian National Commission on Irrigation and Drainage, and India's Central Board for Irrigation and Power. They might be especially influential in helping members of the profession to fight corruption. Each could appoint a panel and invite anonymous case studies from practicing engineers, stressing tactics for resisting pressures for unprofessional behaviour.

4.8 Case Study on Evaluation of Irrigation Administration

The Operation and Maintenance (O&M) structure for four minor (tank) irrigation projects in the Sagar district of Madhya Pradesh state in India is discussed here (Nissanka 2007).

4.8.1 Staffing Pattern

All four tank irrigation projects are under the administrative and financial control of Chief Engineer Dhasan Ken basin of the Water Resources Department of the Government of Madhya Pradesh. Earlier, Chief Engineer had 2–4 field circles, each headed by a Superintending Engineer, but now circle offices have been abolished, and five field divisions are now under the direct control of the Chief Engineer. The four projects are under the jurisdiction of two divisions, as shown below:

Water Resources Division I: Mahuakheda and Khairana tanks
Water Resources Division II: Maheri and Hinauta Kharmau tanks

A division has 3–4 subdivisions, each headed by an Assistant Engineer. A subdivision has the technical and non-technical supporting staff employed as regular/temporary/work-charged or on a daily wage basis. The strength of regular staff in Water Resources Division II is shown in Table 4.10. Besides, a large contingent of staff on a work-charged basis and daily wage basis is employed in the division, as shown in Table 4.11.

4.8.2 *Evaluation of Staff Strength and Staff Performance*

Staff: A large contingent of technical and non-technical staff has been employed on a regular/temporary/work-charged and daily wage basis. The number of junior engineers is 42 against the sanctioned strength of 16. The number of watchmen on work assigned and daily wage basis is 42. The salary component of the budget allocation to a field division is disproportionately high, leaving a lesser budget for actual maintenance work. A typical norm for deploying work-charged staff for the maintenance of minor irrigation works is given below.

Tank up to 50 Mcft: one Watchman (Beldar) for both tank and canal
Tank above 50 Mcft: one Watchman for the tank and one Watchman for canal

Besides, a one-gauge reader for canal regulation for discharge up to 50 cusecs is also posted.

It is expected that with the transfer of O&M to the Water User Association, the staff strength of a field Division of Water Resources Division (WRD) can be significantly reduced, particularly the number of Junior Engineers and Watchmen.

Staff Performance

The poor maintenance of works was not entirely due to the non-availability of an adequate O&M grant. The staff's substandard functional performance, particularly of the Beldar, Gauge Reader, and Junior Engineer in discharging their duties, was also responsible for poor maintenance, as was evident from the observed deficiencies in the projects' physical status.

During the farmers' interview, it was stated that Junior Engineers had made very few field visits to project works. Belders did not seem to be fully aware of their duties and responsibilities. At one of the project sites, Beldar was also found to be involved in unauthorized tank bed cultivation. At another project site, Beldar seemed to be reluctant to report the unauthorized activities of some farmers.

Table 4.10 Regular staff in division II

	Executive Engineer	Assistant Engineer	Accounts officer	Draftman	Junior Engineer	Amin	Asst. Grade II	Asst. Grade III	Irrigation inspector	Nahar Samahata	Steno	Peon
Sanctioned	1	6	1	6	16	18	10	15	1	1	4	13
Existing	1	7	1	3	42	16	11	15	–	–	1	7

Table 4.11 Staff on work charge/daily wage basis in division II

	Driver	Time keeper	Pump operator	Gauge reader	Helper	Water man	Date runner	Watchman	Khalasi	Sweeper	Typist	Supervisor
Work charge	4	7	3	1	4	1	8	13	1	1	–	–
Daily wage	1	–	–	–	4	–	1	29	–	–	3	1

Suggestions

1. Executive Engineer's frequent inspection visits are necessary to keep the field staff alert and vigilant in the discharge of their duties.
2. The benefit of the government service being provided to the Junior Engineer (promotion, salary) should be based on the confidential reports containing Water User Associations' performance assessment.
3. Accountability of the field subdivisions for O&M work should be strictly reinforced.
4. The number of Junior Engineers and Beldars should be reduced.

4.8.3 Capabilities and Need for Capacity Building

The sustained productivity of an irrigation project depends on implementing the agency's capacity, the ability of technical personnel, and the use of improved/modern practices by farmers. This calls for effective human resource development regularly to achieve production goals and safeguard public investment.

Construction works of the four projects have been carried out on a contract basis under the supervision of technically qualified engineers of the Water Resources Department. This is a normal practice all over the country. The engineers (Assistant Engineer, Executive Engineer, and Chief Engineer) are graduates of the civil engineering branch. The Junior Engineers are diploma holders in civil engineering or mechanical engineering. As discussed earlier, each division has many Junior Engineers responsible for directly supervising construction work at the site. The technical capability of the Water Resources Department's engineering staff is satisfactory regarding conventional construction supervision. However, there is scope for improvement in their capability in project management. A review of project documents and the discussions held with the project engineers indicate that an improvement in capabilities is not only possible but also necessary.

Engineers' attitude is that of governance and administration. Managerial skills need to be improved. The following suggestions are made for capacity building:

1. Refresher and specialist training courses should be organized in (i) construction management, (ii) monitoring and quality control, (iii) management of operation and maintenance, and (iv) on-farm development works.
2. Departmental engineers will be required to train the staff of Water Users Associations in operation and maintenance, i.e., an engineer will have to perform the role of trainer. Thus, training of trainers is necessary.
3. Frequent inspection of works by senior engineers and recording inspection notes would help junior engineers in learning by following instructions.
4. Capacity-building programs should further improve the technical knowledge of junior and middle-level engineers. For this purpose, incentives need to be provided to the engineering staff for the up-gradation of their skills through graduate and postgraduate degree/diploma/training certificate programs.

5. Contractors may not have an engineering background/experience, which is also responsible for low construction quality. Screening of contractors should be done based on engineering capacity and past performance.
6. Subcontracting by the main contractor to petty contractors should be based on certain minimum qualifications.
7. Junior and Assistant Engineers need to have sufficient knowledge of command area development.
8. The introduction of pre-service training of junior engineers and assistant engineers in irrigation development and management is necessary. This will effectively reduce the time taken by fresh engineers to learn by experience.

4.8.4 Coordination with Other Departments

Formal interdisciplinary coordination at the project level does not exist for the cluster of these four projects. However, at the block and district levels, government agencies exist to deal with activities such as (i) land revenue, (ii) financing, (iii) agriculture development, (iv) forest conservation, (v) cooperative banking, and (vi) agro-industry. The agencies representing disciplines other than irrigation enter the decision-making process mainly at the field level, i.e., below the outlet locations. Coordination with other departments is made for specific works, such as coordination with the Revenue Department for the acquisition of private land and the formation of water users associations, coordination with the Forest Department for the acquisition of forest lands, coordination with the Agriculture Department for the adoption of cropping pattern, and coordination with financing agencies such as NABARD for arranging the bank loan.

Questions

1. Explain in brief the following terms:

   (i) system of personnel, (ii) hierarchy of staff, (iii) work charged staff, (iv) annual work plan, (v) delegation of authority, (vi) fiscal policy, (viii) irrigation administration.
2. Why democratic spirit should prevail in the organization structure for irrigation management?
3. Discuss the deficiencies in the existing organization structures for irrigation management.
4. Identify the role of Beldar, Mistri, and gage reader in the maintenance of an irrigation project.
5. Discuss technical responsibilities of Junior Engineer and Assistant Engineer.
6. Discuss financial responsibilities at the level of Executive Engineer, Superintending Engineer, and Chief Engineer.

7. Suggest possible reforms in irrigation administration.
8. Explain the following terms:

   (i) Demonstration farm, (ii) infrastructure, (iii) extension work, (iv) cadre of professionals, (v) job incentives, and (vi) reforms in irrigation administration.
9. Write a note on the adverse effects of corruption.
10. What are the job incentives for irrigation managers?
11. What are advantages of having separate staff cadre for operation and maintenance?
12. Why do irrigation engineers prefer posting in construction wing?
13. Write a note on political control on job posting and job transfer.
14. What lessons can be learned from the case study on the evaluation of the O&M organization structure?

References

Chambers R (1988) Managing canal irrigation: practical analysis from South Asia. Cambridge University Press, Cambridge [England]

Jayaraman P, Jayaraman TK (1981b) Attitudes of the irrigation bureaucracy in India to scientific water management tasks in irrigated agriculture: a case study from Gujarat State. Zeitschrift Fur Auslandische Landwirtschaft, Quarterly Journal of Agriculture, Berlin Technical University

Nissanka N (2007) O&M aspects of small tank irrigation projects—some case studies. M.Tech Dissertation, supervised by Prof U C Chaube, IIT Roorkee, Roorkee

Report of the High-Level Committee on Organizational Set-up of Command Area Development Programme in Major and Medium Irrigation Projects and Creation of a Water Management and Land Development wing in State Irrigation Departments, Govt. of India, Ministry of Irrigation (CAD & W. Division) New Delhi, December 1982

Report of the Sub-group on Management Information System for Command Area Development Programme Published by Ministry of Irrigation, Govt. of India, New Delhi, 1982

Smith RA (1970) Management structure for irrigation. J Irrigat Drain Div Proc ASCE, 475–488

WAPCOS (1989) Handbook for improving irrigation system maintenance practices prepared by Louis Berger International Incorporation, Water and Power Consultancy Services (India) Limited, New Delhi 1989.

World Bank (1981a) Comparative study of the management and organization of irrigation projects. World Bank Staff Working Paper No. 458

Reading Material

Central Board of Irrigation and Power (1987) Methodology for evaluation of irrigation and CAD projects, New Delhi

Indian Water Resources Society (1998) Five decades of water resources development in India. Theme Paper for Water Resources Day 1998. Indian Water Resources Society at WRDTC, University of Roorkee, Roorkee, 247667 (U.P.), India

Siriwardana Nihal KD (2007) Financial and economic evaluation of tank irrigation projects. M. Tech Dessertation, supervised by Prof U C Chaube, IIT Roorkee, Roorkee

Chapter 5
Organisational Structure for Management of Irrigated Agriculture

Abstract The objectives of this chapter are to (i) understand the importance of the carefully planned transition from construction to operation and maintenance (O&M) stage; (ii) understand possible shortcomings in the existing irrigation administration; (iii) understand the need and scope of command area development works; and (iv) critically examine the organizational structure for command area development works and learn possible improvements. Organizational structures for the management of irrigated agriculture need to be based on integrated management of land, water, and other inputs. Possible shortcomings in the O&M organization are discussed. The scope of activities under the Command Area Development (CAD) is explained. On-farm development (OFD) works constitute the most important ingredient of CAD programs; thus, the focus is rehabilitation. The organization structures for CAD at the central government level are explained. A critical review of the organizational structures is made, and possible improvements in the organizational structures at the state level and project levels are discussed. Lierature available on the websites of the Ministry of Water Resources (MOWR), Government of India, Planning Commission of Government of India, and other official agencies (such as i. www.Jalshakti-dowr.gov.in; ii. www.mowr.gov.in; iii. www.dapl.karnataka.gov.in; iv. www.wrmin.nic.in) has been synthesized and relevant information extracted for the preparation of this chapter.

5.1 Transfer to Operation and Maintenance Stage

Operational readiness would involve carefully planned physical trials and model simulations. Sometimes, even before the completion of the entire construction work, it may be possible to start the irrigation and drainage service in part of the project command area. This should be based on sound engineering judgment rather than on short-term economic gains or petty political considerations. The completed portions may be transferred to another division of the construction agency for operation and maintenance purposes. Also, farmers need to be trained in the methods of irrigation and the practice of irrigated agriculture. Therefore, sometimes a period of 3–5 years

U. C. Chaube et al., *Canal Irrigation Systems in India*, Water Science and Technology Library 126, https://doi.org/10.1007/978-3-031-42812-8_5

is involved in the transition from rainfed farming to irrigated farming (irrigation transition) and from traditional cultivation practices to improved agricultural practices (agriculture transition).

Policies, rules, and regulations for the physical system and system of trained personnel need to be ready well in advance so that the transition from construction to full operation and maintenance stage takes place smoothly and within a short period of, say, 3–4 years (Fig. 5.1).

Objective of CAD

The objective of CAD is to enhance water use efficiency and production and productivity of crops per unit of land and water for improving socio-economic condition of farmers. On-farm development and beneficiaries' participation in farm water management form essential components of the CAD programme. The programme envisages integrating all activities relating to irrigated agriculture in a coordinated manner with the multi-disciplinary team under the CAD Authority (GOI 1982; CWC 2019).

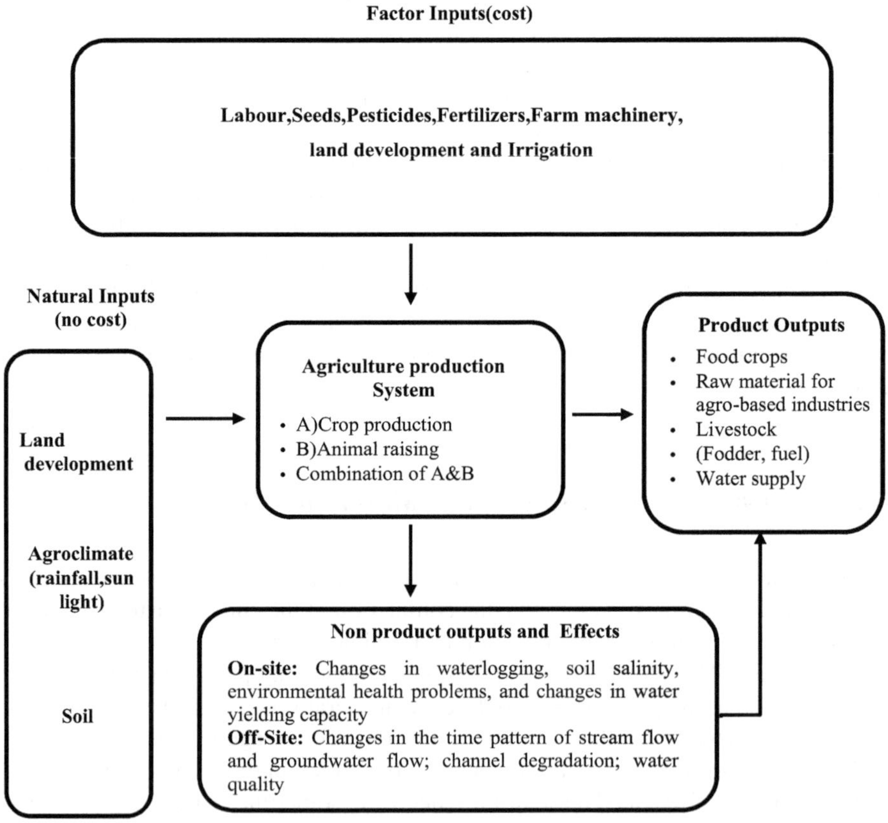

Fig. 5.1 Inputs and outputs of an agriculture production system

Scope of Command Area Development

The Command Area Development (CAD) programme covers the following activities (GOI 1982):

1. Modernization, maintenance, and efficient operation of irrigation system.
2. Development and maintenance of the main drainage system beyond the farmer's block.
3. Development of field channels and field drains and field access roads within the command of each outlet.
4. Land leveling on an outlet command basis for the type of irrigation that is to be given.
5. Consolidation of holdings and redrawing of field boundaries on an outlet command basis.
6. Enforcement of proper rotational delivery system for equitable, timely, and adequate water supply to individual fields.
7. Development of groundwater (tube wells and bore wells) and its conjunctive use.
8. Adoption and enforcement of suitable cropping patterns.
9. Supply of inputs (seeds, fertilizers, and pesticides) and services including credit.
10. Development of mandies, regulated markets, processing facilities, warehouses, and link roads.
11. Special programmes for small farmers, marginal farmers, and agricultural labourers in the project area.
12. Diversification of agriculture and development of activities, such as animal husbandry, farm forestry, poultry, pisciculture, etc.
13. Soil conservation and afforestation, where necessary.
14. Extension programme and farmers training through T&V system.

On-Farm Development Works

On-farm development (OFD) works constitute the most important ingredient of the CAD programmes. The OFD works consist mainly of the following activities:

- Construction of field channels, field drains, and farm roads.
- Construction of water control structures (siphons, silting tanks, diversion boxes, tail boxes, field outlets, and culverts).
- Land leveling, land shaping, and bunding.
- Consolidation of land holdings/realignment of field boundaries.

The earlier emphasis on bringing canal water to 40–60 ha blocks at the government cost has been changed to 5–8 ha blocks. This changed emphasis has extraordinarily enlarged the extent of OFD works required to be done in irrigation projects' commands. This enormous and yet urgent task of on-farm development calls for the active involvement of beneficiary farmers for various reasons. Farmers need a countervailing power and voice to assure that their needs are met by those who should be more accountable to farmers.

5.2 CAD Program Revised as CADWM Program

The CAD Programme has been termed as Command Area Development and Water Management (CADWM) in the XII Five Year Plan and it has been implemented pari-passu with Accelerated Irrigation Benefits Programme (AIBP) during the XII Five Year Plan. Since the year 2015–16, the CAD programme is under implemented as a sub-component of *Har Khet Ko Pani* meaning water for every field component of *Pradhan Mantri Krishi Sinchayee Yojna* meaning Prime Minister's Plan for Agriculture and Irrigation (PMKSY) (www.Jalshakti-dowr.gov.in).

Programme Components

(a) **Structural Intervention**: includes survey, planning, design, and execution of:

   (i) On-Farm Development (OFD) works;
   (ii) Construction of field, intermediate and link drains; Correction of system deficiencies; and
   (iii) Reclamation of waterlogged areas.

(b) **Non-Structural Intervention**: includes activities directed at strengthening of PIM:

   (i) One-time Functional Grant to the registered Water Users' Associations (WUAs);
   (ii) One-time Infrastructure Grant to the registered WUAs;
   (iii) Training, demonstrations, and adaptive trials with respect to water use efficiency increased productivity and sustainable irrigation under a participatory environment.

To promote water use efficiency in irrigation, financial assistance is provided to the States for the development of infrastructure for micro-irrigation to facilitate the use of sprinkler/drip irrigation as an alternative to the construction of field channels. At least 10% CCA of each project is to be covered under micro-irrigation. Micro-irrigation infrastructure includes components of the sump, pump, HDPE pipelines, and pertinent devices needed for bringing efficiency in water conveyance and field applications (through sprinklers, rain guns, pivots, etc.). In the case of micro-irrigation, other components, such as land leveling, drainage works, etc., would be reduced, or entirely discarded; enabling certain cost-saving which is expected to offset the higher cost of micro-irrigation infrastructure. The devices, such as sprinklers/rain guns/drip sets, etc., needed to be installed by individual farmers below farm outlets are not part of the micro-irrigation infrastructure. Farmers are expected to bear the cost of such devices or avail of subsidies available in the extant schemes of the Ministry of Agriculture (CWC 2019).

Table 5.1 Funds under PMKSY (HKKP) for the CAD component

S. no.	Activities eligible for funding	Cost sharing ratio
(a)	All activities of structural interventions	50:50 (centre: state)
(b)	All activities of non-structural interventions excluding functional grant to WUAs	60:40 (centre: state)
(c)	Functional grant to registered WUAs	45:45:10 (centre: state: farmer)
(d)	Incremental establishment cost	50:50 (centre: state)

Source CWC (2019)

5.3 Coverage of Projects Under CAD and CADWM

At the time of inception in 1974–75, CADA covered 60 selected major and medium projects with a cultivable command area (CCA) of 15 Mha. In the year 1998, there were 203 projects under the programme with a CCA of 211 Mha. As of March 2014, there were 150 ongoing projects under CADA with a CCA of 16.3 Mha. The ongoing CAD&WM programme has now been restricted to the implementation of CAD works of 99 prioritized AIBP projects during 2016–17 to December 2019. Out of these 99 prioritized AIBP projects under *Pradhan Mantri Krishi Sinchayee Yojna* PMKSY, the CAD works have been completed in respect of 9 projects, and the remaining 90 CAD&WM projects under PMKSY are yet to be completed as of December 2019 (www.Jalshakti-dowr.gov.in, CWC 2019).

Funds under PMKSY (HKKP) for the CAD component are being provided to the State Governments as per Cost Sharing Ratios (which are applied to the Ceiling Costs) as Table 5.1.

For the eight North Eastern States and the three Himalayan States/UTs of Himachal Pradesh, Jammu and Kashmir, and Uttarakhand, the cost-sharing norms for 'All activities of non-structural interventions except Functional Grant to WUAs' is 75:25 (Centre: State) as against 60:40 norm applicable for other states.

The physical and financial achievement of the projects during the given plan periods is summarized in Table 5.2.

5.4 Examples of CAD Organisational Structures

Presently different states in India have different types of organizational structures to plan, implement, and monitor the CAD programs, depending upon an emphasis on particular items of CAD. Some of the existing models are briefly explained Table 5.2.

Table 5.2 Physical and financial achievement of the projects

Sl. no.	Plan period	Financial	Physical
		Central assistance released (in Rs. crores)	CCA covered (in Mha)
1	Upto VIII Plan	1680.738	13.952
2	IX Plan	751.656	1.801
3	X Plan	818.568	2.314
4	XI Plan	1957.264	2.08
5	XII Plan		
(a)	Up to 2015–16	1887.876	1.419
(b)	2016–17 to 2019–20 (up to Dec. 2019)	2544.25	1.472
	Total	9640.352	23.038

Source CADWM Wing, Ministry of Water Resources, Government of India, New Delhi (www.Jalshakti-dowr.gov.in)

5.4.1 At Central Government Level

At the Central Government level till 1980–81, the CAD Programme was under the purview of the Ministry of Agriculture and Rural Reconstruction. However, from 1980 to 81 onwards, the CAD staff was transferred to the Ministry of Irrigation (Now Ministry of Jal Shakti). The strength of the CAD Wing at the Central level has been considerably increased. A Chief Engineer (CAD) with 2 Joint commissioners, 9 Deputy commissioners, and 2 specialists now man the Water Management and CAD Division at the Central level. The CAD Programme has been termed as Command Area Development and Water Management (CADWM) in the XII Five Year Plan. The CAD wing is now known as the CADWM wing in the Ministry of Jal Shakti (www.Jalshakti-dowr.gov.in).

5.4.2 At State Government Level

The organization structures in only some of the states are explained below to illustrate the difference.

Bihar

At the state level: The CAD program is within Agriculture Department. One senior IAS, Officer known as Commissioner (Special Agriculture Programme) is in-charge of CAD programms. He is assisted by one Special Secretary (Senior IAS) at the headquarters. There is a technical cell at the headquarters headed by one Superintending Engineer (Agricultural Engineer) for project planning, monitoring, and evaluation.

At project level: Seven projects with a CCA of 2517 thousand ha are covered under the CAD Programme. Authorities have been established as autonomous bodies known as Agencies under an Act called Bihar Agricultural and Rural Area Development Agency Act. These CAD agencies are headed by administrators from various disciplines, such as Agriculture, Agricultural Engineering, Irrigation, and Indian Administrative Service. The CAD Agencies are under the administrative control of the Agriculture Department. The administrators for Sone and Gandak projects are known as Area Development Commissioners-cum-Chairman CADA, whereas, for other projects concerned Divisional Commissioners (IAS) function as Area Development Commissioners. On the agency level, there is one Board that has a strength of twenty-five members. The Board meeting is chaired by the Area Development Commissioner. The Board meets once every three months.

Each administrator is assisted by the following staff at the headquarters:

(a) Secretary to CADA: an additional collector rank officer from Bihar Administrative Service.
(b) Financial Adviser-cum Chief Accounts Officer.
(c) Superintending Engineer (Irrigation Agricultural Engineering).
(d) Joint Director of Agriculture.
(e) Deputy Director Soil Conservation.
(f) Deputy Director Ground Water.
(g) Deputy Director Statistics.

For the construction of field channels, contour survey, mapping, and planning, etc., field engineering units headed by Executive Engineer rank officers with supporting staff drafted from Irrigation/Agricultural Engineering have been placed in charge of canal distributaries and minors.

Gujarat

Thirty-six projects with a CCA of 1007 thousand ha are covered under the CAD programme. At the state level, there is a Coordination Committee headed by the Minister of Irrigation to look after water utilization in all its aspects. The concerned Secretaries to the State Government and the Area Development Commissioners are the Members of the Committee. The Command Area Development Authority comprises the concerned Heads of Departments and other organizations, such as State Cooperative Banks, Land Development Banks, and Panchayats. They are responsible for carrying out the integrated programme in the entire area and supervising its implementation. The Area Development Commissioner is the Chairman of the Authority and is also its executive authority.

The distribution of water is regulated by the Gujarat Irrigation Act. The maintenance and operation of canals up to the outlet is the responsibility of the State Government which has been entrusted to the Command Area Development Authority. The maintenance of channels beyond the outlet is the responsibility of the concerned cultivators enforceable through the provision of the Irrigation Act.

On-farm development for improving the efficiency of water use is carried out through the divisions of the soil conservation and irrigation department under the control of the authority. Training of farmers, carrying out adaptive trials, and training of functionaries are carried out by separate divisions under the control of the authority. The authority is also responsible for agricultural marketing, and livestock development as part of the integrated development of the area. The road development work is entrusted to Panchayats for categories of roads and state highways as roads; major district roads and state highways are managed by the Building and Communication Department whose Chief Engineer is a member of the authority: agricultural extension is the direct responsibility of the Director of Agriculture working through the Joint Director of Agriculture and the District Agricultural Officers. The Joint Director of Agriculture has been placed under the operational control of the Area Development Commissioner. The Director of Agriculture is a member of the 'authority'. The Cooperative Banks are responsible for the credit input for agricultural operations. They are members of the 'Authority'.

Karnataka

Fifteen projects with a CCA of 2078 thousand ha are covered under the CAD programme. The task of operation and maintenance of major projects, including the distribution system up to the outlet rests with the Irrigation Department. Beyond the outlet point, i.e. for field channels having a capacity of 1 cusec and below, the jurisdiction of CADA begins. The overall control of CADA rests with Development Commissioner at the state level (www.dapl.karnataka.gov.in).

The Command Area Development Authority (CADA) envisages amongst other things the construction of field channels, efficient maintenance of field channels, and field drains. on-farm development (OFD) works, conjunctive use of surface and groundwater, and regulation of cropping patterns.

CADA's are functioning as independent statutory bodies. Each CADA is operating not only as an effective link between several development authorities/agencies but acts on its own equipped with adequate legislative powers and plays an effective role in synchronizing the various development activities in the command Areas.

The membership of each authority consists of a Chairman, the Administrator, the Secretary to Government in the Finance Department, the Vice-Chancellor of the University of Agricultural Sciences, the Deputy commissioners of the concerned Revenue Districts, the Director of Agriculture, the Chief Engineer of concerned projects, the Registrar of Co-operative Societies, the Director of Town Planning, one person nominated by the State Government to represent banks and financial institutions, and ten persons nominated by the State Government of whom one shall be a small farmer, one shall be a person belonging to scheduled caste or scheduled tribe, one shall be an Agricultural laborer, and one shall be a rural artisan.

Figure 5.2 shows the CADA organization chart in Karnataka. The Administrator is an officer not below the rank of Secretary to the Government. In Karnataka, at present, the CADAs are headed by IAS Officers, the Chief Engineer, the Chief Conservator of Forests, and the Director of Agriculture. Each Administrator is assisted by three specialists designated as Land Development Officers. They are a Joint Director of Agriculture and Joint Registrar of Cooperative Societies and a Superintending Engineer. These specialists in turn have their field formations. For accounting matters, the

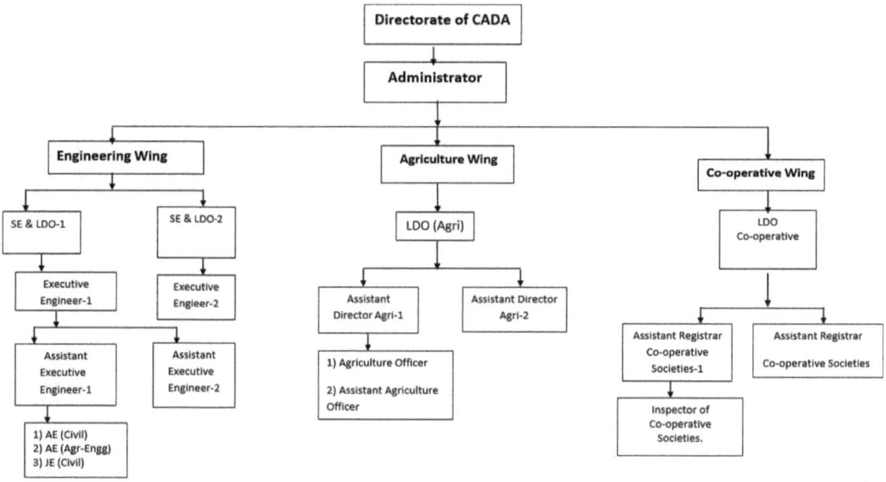

Fig. 5.2 Proposed organisation structure of CADA, Karnataka. *Source* www.dapl.karnataka.gov.in

Administrator has a Chief Accounts Officer. On the Administrative and Development side, the Administrator is assisted by a Deputy Administrator who also functions as the Secretary to the Authority. The Deputy Administrator is drawn either from the administrative services or from the Development Department.

Maharashtra

Twenty-one projects with a CCA of 1481 thousand ha are covered under the CAD programme. CADAs are under the Irrigation Department and form part of the management wing. Each CADA is headed by an Administrator, of the rank of Superintending Engineer. He has under him one division for irrigation management headed by Executive Engineers. These divisions look after the maintenance of the canal system, planning of irrigation. programmes, giving water sanctions, preparations of demand statements, and recovery of water charges accordingly.

There are one or more divisions under each CADA for planning and construction of community items (field channel system and allied works) and land leveling etc. These divisions are manned by Divisional Soil Conservation Officers or Executive Engineers with 5 sub-divisions each under them. For assistance in the detailed design of the field channel system, technical cells, headed by Executive Engineers have been set up under the CADAs. The agricultural extension was earlier managed through 'Irrigation Units' working under the guidance of the Deputy Director of Agriculture working under the Officer of each Administrator. These units have now been converted to the Training and Visit System of Agricultural Extension and as such these have been transferred to the administrative control of the Agriculture Department.

There is a separate Monitoring and Evaluation Cell at the Ministry level under a Superintending Engineer and Deputy Secretary and Administrator's level under the control of the Irrigation Department for Monitoring and Evaluation of World Bank-aided projects. For compiling statistics, there is a unit headed by the Joint Director of Statistics at the Mantralaya Level. Suitable statistical staff is also attached to the offices of Administrators.

The Government has also set up a Water and Land Management Institute at Aurangabad for providing training in land development and water management to Officers and staff of the Irrigation Department in the state. The Institute has been registered as a Society, with a governing council chaired by the Secretary of Irrigation. The Institute has five faculties in soils, agronomy, irrigation, engineering and civil engineering, and basic sciences. The first regular course commenced on 1st October 1980.

Rajasthan

Four projects with CCA of 973 thousand ha are covered under the CAD programme. The Government of Rajasthan has created a separate CAD and Water Utilization Department headed by a Special Secretary under the Ministry of Agriculture. The Secretary also looks after Ground Water Development, Drought Prone Area Programmes, and Dairy Development. He is guided by a Coordination Committee for CAD at the state level. The Committee is chaired by the Chief Minister and includes Ministers, Secretaries, and other senior officers of all departments concerned. The CADA is headed by an area Development Commissioner who is assisted by an Additional Area Development Commissioner. The CAD authority has under its direct control the irrigation and on-farm development works division, the agricultural division, the planning, and coordination division, and the town planning and mandi development division which take up on-farm development and allied activities at the field level.

The organizational chart of the Rajasthan Water Resources Department and Chambal Command Area Authority is given in Figs. 5.3 and 5.4.

West Bengal

Four projects with a CCA of 1882 thousand ha are covered under the CAD programme. Administratively, the CAD programme comes under the purview of the Agriculture Department of the State and is overseen by Secretary to the Government in the Agriculture Department. The CAD authorities have been set up under the chairmanship of the Divisional Commissioner. The project is headed by the Project Administrator who is of the level of Superintending Engineer or Joint Director level of Agriculture. The CAD authorities do not have their field staff and are entirely dependent on the various departments involved in the program.

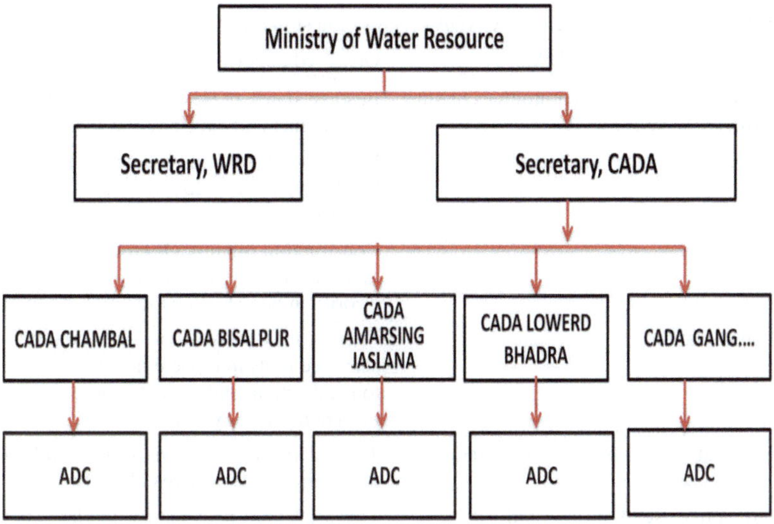

Fig. 5.3 Organisation chart of water resources department Rajasthan

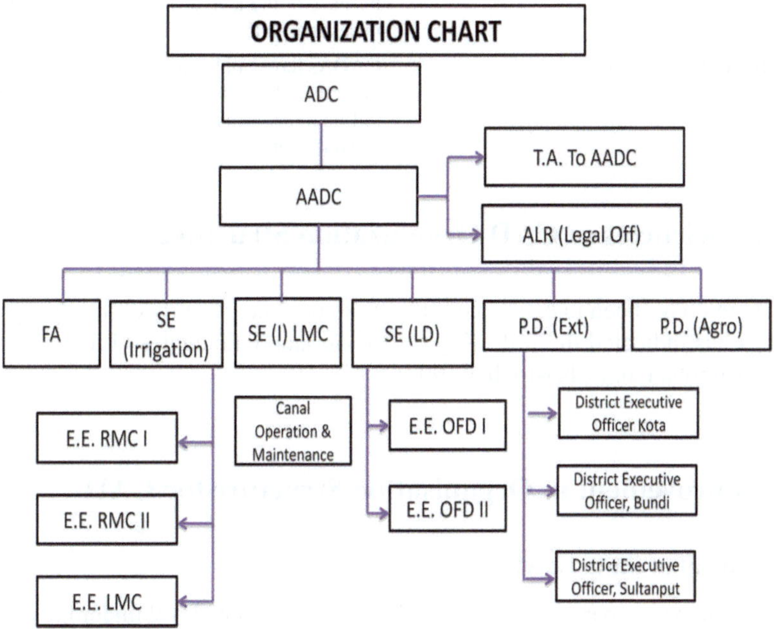

Fig. 5.4 Organization chart of Chambal CADA

Table 5.3 Observed deficiencies in CADA organisation structure

Organizational aspect of CADA	Observed deficiency
Policy guidelines by CAD committees or state water utilization cells	In some states, these committees are more or less defunct and have not met for years
Coordination with other agencies	Lack of coordination with agencies responsible for the supply of seeds, fertilizers, credit, and extension work
Priority of work	Construction (mainly OFD works) has the highest priority. O&M activities have low priority in budget allocation and expenditure
Land consolidation	Farmers have resisted the consolidation of their scattered plots into one unit. They have resorted to legal action, and the CAD programs have been delayed for years
Staff on deputation from other departments	These deputed persons may not be fully loyal to the aim and objectives of CADA
The conjunctive use of surface and groundwater	This aspect has been paid secondary attention in almost all CADAs
Rotational water delivery	It could not be introduced in several projects because of the lack of field channels and necessary staff to enforce the same
Equitable water distribution to tail-enders	CADAs have not been able to achieve the objective of equitable distribution up to the tail-enders because of inadequate organizational structure at the field level

5.5 Deficiencies in CAD Organization Structure

The CADAs have been in existence for quite some time. A study of their programme activities has indicated the following organizational deficiencies (Table 5.3) (CBIP 1987; Sivamohan and Christopher 1992).

5.6 Improvement in Organisation Structure for CAD

Improvement at State Level

There should be a separate department for CAD at the state level under the charge of a Secretary of Government. He should promote and coordinate actions by concerned departments such as Irrigation, Agriculture, Public Works, Cooperatives, etc. To assist the Secretary-in-Charge of CAD, a necessary administrative set-up for monitoring and evaluation cells representing different disciplines should be created. The Secretary of CAD should also have a cell for organizing internet-based management information systems (CBIP 1987; IWRS 1998).

Improvements at Project Level

As far as practicable, there should be a separate CADA for each major project having more than 200,000 ha culturable commanded area (CCA) and 10,000 ha per year as the rate of growth for potential utilization. However, in very large projects, such as the Sarda Sahayak Canal command in Uttar Pradesh, there could be more than one CADA. Since the CADA administrator is required to coordinate the works of different disciplines, he should be a senior officer of the rank of Commissioner/ Secretary to the Government/Chief Engineer/Director of Agriculture, etc.

Coordination Between CADA and Other Departments

The success of the CAD programme depends on how cooperation could be secured amongst the institutions dealing with various disciplines. The administrative and organizational arrangements for agricultural extension under the T&V system for the command areas should be so devised that they could be effectively controlled and directed by the CADAs and that live linkages are established so that all the Development Departments function in unison. A similar organizational arrangement should be devised wherever necessary in relation to the functionaries of Irrigation, Energy, cooperative, Forests, and other Departments wherever they are not directly functioning under the CADAs.

Administrative Control of CADA Staff

For these purposes, the concerned staff should be under the direct administrative control of the CADA and draw their pay and other emoluments from the CADA. However, the respective departments' technical guidance and supervision would be provided from which the staff is drawn on deputation. To ensure that talented staff is provided to CADA by respective departments, it should be ensured by the State Governments that necessary deputation reserve is provided in the concerned departments and the CAD Secretary/Administrator should have a choice in the selection of staff and the staff should not be withdrawn/replaced except with the approval of the CADA administrator. Unless proper incentives are provided, it may not be possible to attract competent staff.

Questions

1. Discuss the need and scope of planning for a smooth transition from construction to the Operation and Maintenance stage.
2. Discuss the deficiencies in the existing organization structures for irrigation management.
3. Explain the following terms:

   (i) Demonstration farm, (ii) Infrastructure, (iii) Extension work, (iv) Cadre of professionals, (v) Multi-disciplinary activities, and (vi) Management information system.

4. Discuss the factors responsible for the underutilization of created irrigation potential and low productivity of irrigated agriculture.
5. Suggest improvements in the existing command area organization structure.
6. What are the essential ingredients of the CAD organization?
7. Compare and comment on the CAD organization structures in Andhra Pradesh, Maharashtra.
8. Compare and comment on the CAD organization structures in Rajasthan and Uttar Pradesh.
9. What could be the reasons for including water management in the command area development program?
10. The focus of CAD program has been on On-Farm Development works. Why?

References

CBIP (1987) Methodology for evaluation of irrigation and CAD projects. Central Board of Irrigation and Power, New Delhi

CWC (2019) Compilation of statistics of ongoing major and medium projects. Water Related Statistics Directorate Information System Organisation Water Planning and Projects Wing Central Water Commission, June 2019

GOI (1982) Report of the high-level committee on organizational set-up of command area development programme in major and medium irrigation projects and creation of a water management and land development wing in state irrigation departments, Government of India. Ministry of Irrigation (CAD&W. Division) New Delhi, December 1982

IWRS (1998) Five decades of water resources development in India. Theme Paper for Water Resources Day 1998. Indian Water Resources Society at WRDTC, University of Roorkee, Roorkee, India

Report of the Sub-group on Management Information System for Command Area Development Programme Published by Ministry of Irrigation, Govt. of India, New Delhi, 1982

Sivamohan MVK, Scott CA (1992) The command area development program in India—a policy perspective. Overseas Development Institute, U.K., October 1992

Websites i. www.Jalshakti-dowr.gov.in; ii. www.mowr.gov.in; iii. www.dapl.karnataka.gov.in; iv. www.wrmin.nic.in

Reading Material

Roush FM (1955) Operation and maintenance of irrigation systems. Paper 623, Jl. Of Irrigation and Drainage Division, Proceeding, ASCE, pp 1–8

Smith RA (1970) Management structure for irrigation. J Irrig Drain Div Proc ASCE, 475–488

WAPCOS India (1989) Handbook for improving irrigation system maintenance practices prepared by Louis Berger International Incorporation. Water and Porer Consultancy Services (India) Limited, New Delhi

World Bank (1981) Comparative study of the management and organization of irrigation projects. World Bank Staff Working Paper No. 458, May 1981

Chapter 6
Farmers' Participation

Abstract The objectives of this chapter are to: appreciate the necessity of involving farmers in O&M of irrigation works, and investigate the scope and forms of existing farmers' organizations. The role of NGOs in encouraging farmers' participation is highlighted and illustrated with an example. Experience in several developing countries suggests that irrigation systems implemented on behalf of farmers and sponsored by farmer's cooperative agencies have been successful. Non-Government Organizations (NGOs) and other voluntary agencies act as a catalyst in organizing the farmers, as illustrated through an example of Shri Datta Water Management Society (canal irrigation). Cooperative irrigation societies, such as Mohini Cooperative Society (Kakrapar project—Gujarat), have also had limited success. Benefits and issues involved in transferring (turnover) the ownership to farmers' association and associated responsibilities for managing tertiary irrigation schemes are discussed. Examples of turnover in India and Indonesia are provided. About 400 small irrigation systems (< 500 ha each) have been transferred to WUAs in Bali Island of Indonesia. The turnover has greatly improved the performance of these irrigation systems. Social, technical, legal, financial, and administrative issues in the turnover process are discussed. There could be several conflict interfaces that need to be taken care of. A brief explanation of these conflict interfaces is given. WUA is only one of the many social groups to which a farmer belongs. The preferential affinity of farmers to other social groups compared to that of WUA happens to be a big hindrance to the functioning of WUAs. Legal acts have been enacted or amended by several states to resolve various issues and to provide legal authority for WUAs. Status of the enactment of legal Acts is explained. The content and scope of the Tamil Nadu Farmers' Management of Irrigation Systems Act, 2000 (Tamil Nadu Act 7 of 2001) is discussed as an illustration of legal acts enacted by state governments. This legal act covers the formation and functioning of WUAs, offenses, and penalties, settlement of disputes, and elaborates formation of canal distributary committees. This will help increase the awareness of legal acts regarding WUAs. A field survey based study on the implementation of WUAs in four tank irrigation projects in the Sagar district of Madhya Pradesh is discussed. Literature available on the websites of the Ministry of Water Resources, Government of India, Planning Commission of Government of India, and other official agencies (NABARD-National Bank for Agriculture and Rural Development, Government of Tamil Nadu, Government of Odisha,

U. C. Chaube et al., *Canal Irrigation Systems in India*, Water Science and Technology Library 126, https://doi.org/10.1007/978-3-031-42812-8_6

etc.) has been studied and the material taken for the preparation of this chapter. For example: (a) www.wrd.tn.gov.in; (b) www.mowr.gov.in; (c) www.dowrodisha.gov.in; (d) www.wrmin.nic.in; and (e) panchayat.gov.in etc.

6.1 Need for Farmers' Participation in Irrigation Management

Farmers' involvement/participation in irrigation management (PIM) means the farmers play an active role in implementing, operating, maintaining, and evaluating irrigation projects and programs. The involvement of farmers in the irrigation system's management has been recommended in the National Water Policy documents.

National Water Policy 1987: "*Efforts should be made to involve farmers progressively in various aspects of management of irrigation systems, particularly in water distribution and collection of water rates. The assistance of voluntary agencies should be enlisted in educating the farmers in efficient water-use and water management.*"

National Water Policy 2002: "*Management of the water resources for diverse uses should incorporate a participatory approach: by involving not only the various governmental agencies but also the users' and other stakeholders, effectively and decisively, in various aspects of planning, design, development, and management of the water resources schemes. Necessary legal and institutional changes should be made at various levels for the purpose of duly ensuring women's appropriate role. Water Users' Association and local bodies such as municipalities and Gram-Panchayats should particularly be involved in the operation, maintenance, and management of water infrastructures/facilities at appropriate levels progressively, to eventually transfer the management of such facilities to the user groups/local bodies.*"

The desirability of involving farmers in water management has been increasingly advocated over the past several years in the plan documents and various forums. Behind the advocacy of associations lie two assumptions, viz. it would reduce the cost of operation and maintenance, and it would raise the effectiveness of the system with the help of local situational knowledge and reduce bureaucratic intervention from above. Experience in other developing countries suggests that irrigation systems implemented on behalf of farmers and sponsored by farmer's cooperative agencies have been successful.

6.2 Farmers Involvement at Different Levels

The legal framework in India provides for the creation of farmers organizations at different levels, such as:

(a) Water Users' Association (WUA): covering a group of outlets or a minor.
(b) Distributary Committee: comprising presidents of five or more WUAs.
(c) Project Committee: presidents of the distributary committees in the project area will constitute this committee's general body.

6.2.1 *Example 1: Andhra Pradesh*

The organization structure, composition, and functions of three-tier WUAs in irrigation projects in Andhra Pradesh are depicted in Table 6.1.

Table 6.1 Example—three-tier WUAs in Andhra Pradesh

Level	Name of farmers organization	Composition and functions
At the primary level for a group of outlets	Water User Associations (WUA), also called Pani Panchayats	Area of a WUA to be divided into territorial constitutions (TC) to give adequate representation to all farmers in head, middle and tail reach All landholders and tenants within the notified area are members with voting rights All other water users are members without voting rights
At distributary level	Distributory Committee (DC)	Presidents of 5 or more WUAs constitute a Distributory Committee All WUAs in its jurisdiction are its members Look after medium drains and distributories Resolve disputes of WUAs
At project level	Project Committee (PC)	A part or whole of the project to have a project committee The presidents of the DC and WUA are its members in respect of project committees of major and medium projects only Resolve disputes between DCs and WUAs

6.2.2 Example 2: Odissa

In Odissa, farmers involvement at various levels is depicted in flow chart Fig. 6.1. There are executive committees of Pani Panchayat at the outlet level. President, secretary, and treasurer of Pani Panchayat are members of the general body at the distributary and project levels (Odissa 2016).

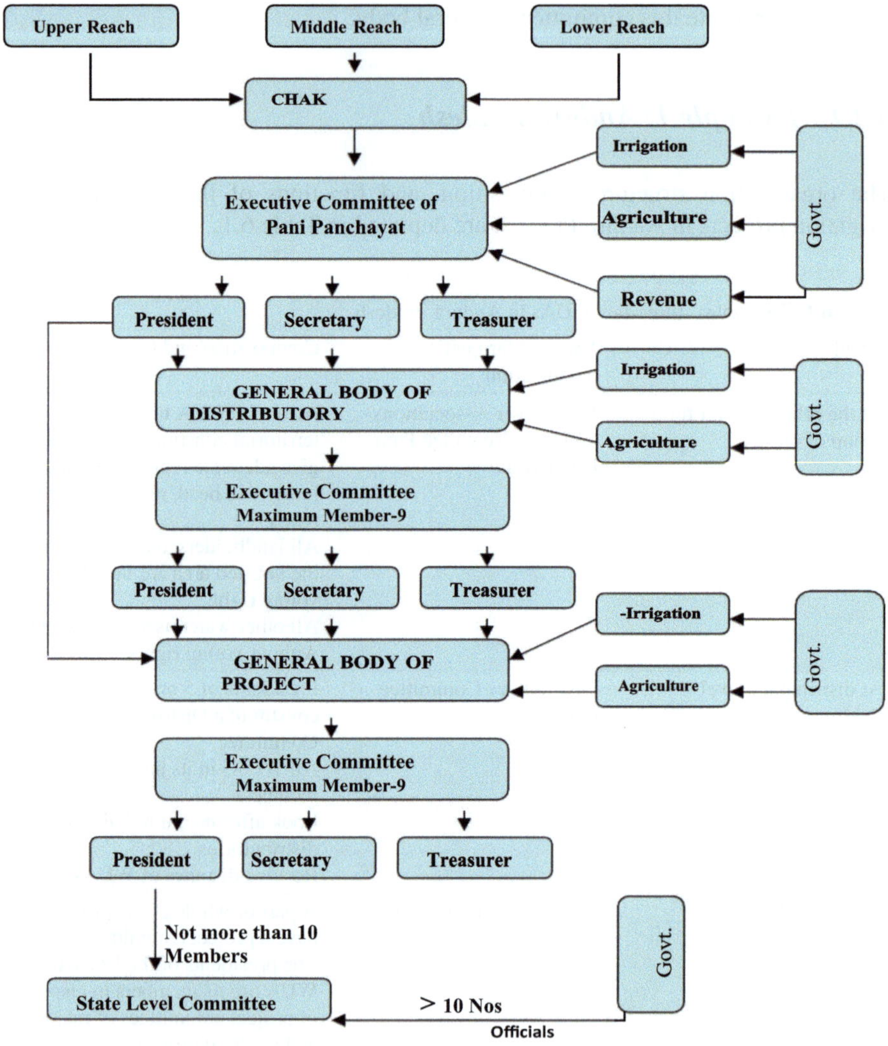

Fig. 6.1 Farmers Involvement at Various Levels 2016 (Source Odissa)

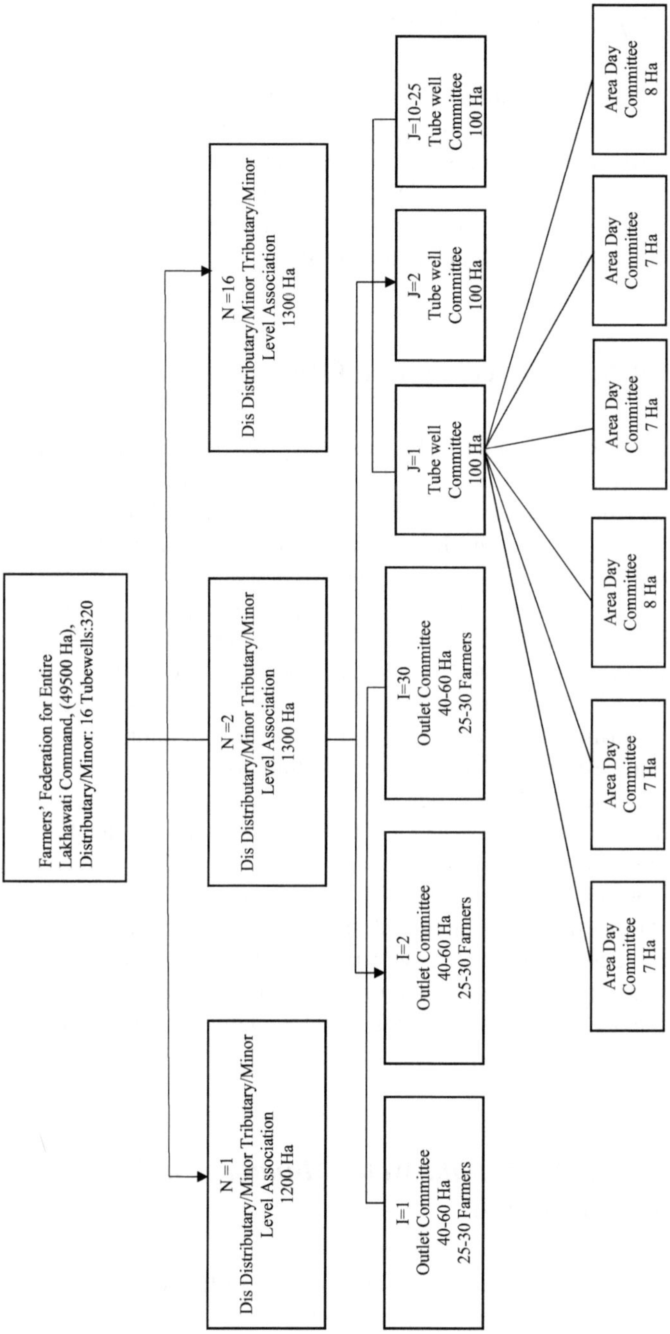

Fig. 6.2 Farmers associations at various levels in Lakhauti Canal command area (Uttar Pradesh)

6.2.3 *Example 3: Uttar Pradesh*

Lakhauti Branch canal command is situated in the western part of Uttar Pradesh. The CCA is 49,500 ha. There are 16 distributaries and minors of this branch canal and 320 tube wells. The farmers association is at a different level in the command area of the Lakhauti branch canal in the western part of Uttar Pradesh. There are outlet committees and tubewell committees, distributary/minor canal level WUAs, and Farmers Federation and branch canal level.

6.3 Role of Non-Governmental Organisations for Farmers

Non-Government Organizations (NGO's) have played an active role in forming farmers associations in the country. There are several success stories of farmers' involvement, such as pipe outlet committees in Andhra Pradesh, water cooperatives in Gujarat, the Phad system in Maharashtra, and the Staddar System in Bihar.

Datye and Patil (1987) narrated the success story of the Mohini Water Cooperative Society in the command of Kakrapar project in Gujarat state. The society was registered in September 1978 in a village Mohini, about 25 km from Surat city. It was registered under the State Cooperative Society Act of 1961. Good leadership was available, and water assurance was given to shareholders. Water charges were paid to the government by society, and the government supplies water to society on a volumetric basis at a mutually agreed volumetric rate. The society charges the shareholders for the supply of water on a crop area basis.

Datye and Patil (1987) and WAPCOS (1991) discussed the role of several farmers associations in maintaining irrigation facilities. Case studies covering the type of organization, water rates, maintenance, income to the farmers' association, and advantages due to the involvement of farmers association in irrigation management have been described. These case studies illustrate the collective involvement of farmers in canal irrigation, lift irrigation, tank water use, and field drainage. As an illustration, the Shri Datta Water Management Society (canal irrigation) is discussed below.

6.3.1 *Example: Shri Datta Water Management Society (Canal Irrigation)*

This has been one of the most all-around successful farmers' organizations. It has most of the idealized features and advantages.

Location: Chanda Village, Nevasa Taluk, Ahmednagar, District, Maharashtra.

Type of Organisation: The Shri Datta Co-operative Water Management Society is in the command area of Minor Canal No. 7 of the Mula Command. There are 200 farmers having 361 ha of agricultural land.

Water Rates: Before the society was formed, many farmers did not pay their water fees, which were supposed to be paid through ID officers, but they would give them a gratuity to get their water again the next year. In nearby villages, half the water fees had not been paid.

Now the Society collects the water fees in advance. Farmers willingly pay per ha Rs. 50 per kharif, Rs. 75 for rabi, and Rs.150 for hot weather, plus a Society service charge of Rs. 25 per ha. There are no overdues. This is still far cheaper than groundwater, which farmers also use.

The Government Irrigation Department has negotiated with the Society the volumetric water supply rates for kharif, rabi, and hot weather. Water quota from rabi carried over to hot weather costs 30% less, and water stored by the Society is paid for by the rate of the season in which it is used.

Maintenance: Irrigation Department was spending unreasonable amounts to maintain this minor, though it is only 2 km long. Before the Society was formed, part of the maintenance money was not properly utilized. After handing over the maintenance work to the Society, there has been less cost to the State, and the workloads of the Patkari (Canal Inspector) and Junior Engineer are reduced. Field channels are maintained by farmers.

Income: The Society collects water charges from farmers on a per-hectare basis, regardless of the crop, which covers its personnel and office expenses. It rents a peasant office room.

Advantages:

1. The state saves on maintenance money and duties of engineers and field functionaries.
2. The farmers get all water dues on time.
3. The inefficient old procedure of recording farmer applications for water by season and crop is avoided.
4. Since water is paid for in proportion to use, it is not wasted, and more land is irrigated. The threat of waterlogging has subsided.
5. Water comes more or less at specified times so that agriculture activities can be properly planned.
6. Cropping intensity and crop variety are greater, and specialized crops requiring cooperation can be grown.
7. Farmers save wages by posting a man in the field to wait one or two days when water might come.
8. Society controls patkaris and guards who must now work in the interest of farmers and so are flexible in giving water turns, if necessary.
9. Conflicts are resolved.
10. The environment is improved through tree planting.

6.4 Turnover (Transfer) of Irrigation Facilities to Water Users Associations

Turnover is the process of transferring ownership and associated responsibilities (partly or fully) of an asset from an existing agency to another agency, which is legally recognized and acceptable to individuals having a common interest in the asset. There are two categories of turnover; (i) from construction agency to operation agency, managed within the irrigation department, and (ii) ownership transfer from Government irrigation department to Water User Association.

6.4.1 Operational Readiness

Operational readiness is necessary in both categories of the turn over process, as depicted in Fig. 6.3. In India, land holdings are generally small and fragmented. Individual ownership of canals or tubewells is economically unviable. Therefore, collective ownership and management are critical to the success of turnover. Several social, technical, legal, and administrative issues are involved in the turnover process.

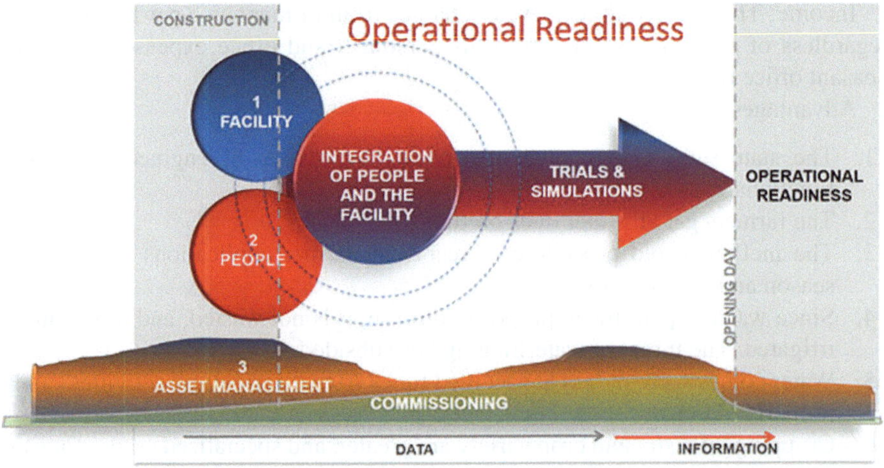

Fig. 6.3 Operational readiness for transfer from construction to operation stage

6.4.2 *Example of Transfer of Minor System Management in Odissa (India)*

In line with Irrigation Act and Rules, the WUAs will be involved in the O&M management of water infrastructure and facilities. The process of management transfer involves many activities to be taken up in sequential order. An inventory gathers information on the physical condition of the irrigation scheme and existing management. Socio-technical profile and institutional profile are prepared. Design and construction improvements are identified. An example of the process of management transfer is given below.

The Gohira Irrigation Scheme is located in the Brahmani River basin and comes under the northwestern plateau zone (agro-climatic zone) of Odissa state in India. The Department of Water Resources (DoWR) of the Government of Odissa has prepared an O&M Manual (Odissa 2016). The manual explains the procedure and process for management transfer agreements between Government Irrigation Department and *Pani Panchayats* (WUAs).

The process for management transfer agreements (MTA) is depicted in Table 6.2. Usually, the management handover agreements expire after a set number of years and need to be renewed periodically.

6.4.3 *Example: Turnover of Small Irrigation Schemes in Indonesia*

WUA is only one of the many social groups to which a farmer belongs. In some of the projects in India, the affinity of farmers to other social groups (religion, cast, subcast etc.) has been found to be stronger than that to WUA. Unlike in India, farmers in Indonesia have demonstrated a stronger affinity to WUA than other social groups.

Water users associations (WUA) have been formed for tubewells in Madura Project Area, East Java (Indonesia) (Chaube et al. 1996) have carried out a detailed analysis of the issues involved in the transfer of tubewells in Indonesia. Under the Madura Project, much effort was devoted to formulating practical guidelines for training procedures, monitoring, and evaluating the physical and financial performance of WUAs. Economic and social issues were found to have a direct bearing on the performance of WUAs. Sociological studies were used to plan farmers' participation and provide WUA training in financial accounting and water management. The user experience was gained on the Madura Project in developing suitable procedures for WUA establishment, training, and monitoring.

Table 6.2 Process for Renewal of Management Transfer Agreements (MTA)

Step No	Name of the activity plan	Details of the activities	Main responsibility	Supporting responsibility	Supporting responsibility
Step-1	Preparation of the schedule, for signing of the management transfer agreement (MTA), between project Manager, Gohira with each of the 19 PPs separately	Deputy Manager will prepare scheduled for signing MTA with 19 PPs in consultation with Manager and, AE, JE of Gohira	Deputy Project Manager Gohira	Project Manager, A.E., J.E	DD(PIM/O&M) under Addl. Dir. O&M to monitor implementation of program
Step-2	Preparation of Documents of assets, about canal and its structures, to be handed over to PP at the time of signing of MTA	Prepare the following documents for the signing of MTA (1) Map showing the area of operation of PP, (2) Inventory of canals (3) list of structures to be handed over to pp (4) List of any other assets to be handed over (Each AE, JE will prepare pp wise Chak wise area of operation map of their jurisdiction)	Project Manager	Deputy project Manager, AE, JE	DD PIM/O&M under Addl. Dir. O&M is to monitor the progress of documentation, with respect to schedule in Step-1

(continued)

Table 6.2 (continued)

Step No	Name of the activity plan	Details of the activities	Main responsibility	Supporting responsibility	Supporting responsibility
Step-3	Preparation of Documents of assets, about CADA system (WC & FC and structures)	(2)CADA is to prepare List of Chak wise, pp wise F C & W C Inventory and hand over it to SIO Manager, Gohira	EE, CADA working under Addl. Dir. CADA of PD PMU	CADA Unit of PD PMU	DD PIM/O&M under Addl. Dir. PMU to coordinate and monitor the program
Step-4	Compilation of irrigation and CADA documentation	Project Manager has to Compile (list1 + list 2) and to hand over the same to PP at the time of signing of the MTA	Project Manager	Deputy Project Manager, AE, JE	DD PIM/O&M under Addl. Dir. PMU to coordinate and monitor the program
Step-5	Arrange public ceremony for signing of MTA with PP	As per the guide lines Deputy SIO Manager has to arrange ceremony in the PP area for signing of MTA with PP, publicly	Project Manager	Deputy Project Manager	DD PIM/O&M under Addl. Dir. PMU to coordinate and monitor the program
Step-6	Complete signing of MTA with 19 PPs as per schedule	Arrange public ceremony as per scheduled step-1 for signing of MTA with 19 PPs	Project Manager	Deputy Project Manager AE/ JE	DD PIM/O&M under Addl. Dir. PMU to coordinate and monitor the program

Note *PP* Pani Panchayat (= WUA)
Source Odissa (2016)

6.5 Issues Relating to Turnover (Transfer)

Issues involved in turnover (transfer) have been discussed by Chaube et al. (1996) and Bruns and Dwi (1992).

6.5.1 Economic and Financial Issues

The following issues are important in the context of the turnover process:

(i) What is the potential benefit to farmer agencies participating in the turnover? (water users association and Government)
(ii) To what extent can the investment cost be recovered from WUA?
(iii) Should O&M costs be subsidized by Government? If so, for how long and how much?

The turnover of an irrigation facility will succeed only if it becomes economically viable. For example, the financial performance of state tubewells in Uttar Pradesh has not been satisfactory due to the underutilization of the potential of tubewells, the existing low water rate structure, and the non-realistion of charges.

The revenue of WUAs can be increased by better utilization of irrigation potential, increasing water rates, and timely collection of the same. But the significant increase in water rates would be a disincentive for the farmers to become a member of WUA, particularly for those farmers who can get the water in adequate quantity at a cheaper rate from other sources, such as owning a well or purchasing from a private tubewell owned by others.

6.5.2 Role of Voluntary Agencies

Voluntary agencies can play an important role as a catalyst in motivating user farmers to form WUA. The catalyst generally acts as a channel of effective communication between the government department and user farmers. It can integrate the viewpoints of both these groups into an action plan for the collective good. These field workers are given different nomenclature in different countries and locations, e.g., *community organizers, change agents, motivators, institutional workers,* etc. However, they all perform similar functions.

If voluntary agencies are not available, the Irrigation Department may identify Government officials who have the right background and aptitude to function as community organizers. It is also possible to use farmer leaders as honorary community organizers. In the Philippines, the Turnover Program is increasingly using farmer leaders called Farmer Irrigation Organizers to motivate and organizing farmers. They may not be appointed as employees on a full-time basis, but their actual expenditure incurred for the turnover program and honorarium may be reimbursed.

6.5.3 Guidance After Turnover

The Government agency must continue to provide guidance and assistance (technical and financial) to farmers after turnover. Some of these are:

(i) Continuous technical guidance and financial assistance for special repairs;
(ii) Monitor WUA activities, particularly about the interests of small and marginal farmers and religious minorities;
(iii) Guide and improve the management of funds by WUAs. Proper accounting is necessary;
(iv) Replication is possible only if WUA can demonstrate its capability to manage tubewells. The government will, therefore, need to pay continuous attention in the beginning years; and
(v) Even if a cluster of tubewells is turned over to WUA, the government may still hold control rights, authority, and groundwater. The government will have to control the annual withdrawal of groundwater to mitigate environmental hazards.

6.5.4 Transfer of Assets or Only Management

The authority and responsibility of WUA may not match if irrigation facilities continue to belong to the State Irrigation Department. In the turnover program for tubewells, it is better to transfer the ownership of assets so that farmers have some sense of responsibility for the operation and maintenance and authority to manage tube-wells. Also, the assets of tubewells cannot become a guarantee to borrow money from the bank if tubewells continue to belong to Government. However, even after transferring the assets to the farmers, the government should continue to monitor performance and assume financial responsibility for a major repair.

6.5.5 Staff Adjustment After Turnover

The government irrigation agencies need to be restructured and re-organized, and necessary legal changes are required to be made to promote participatory management of irrigation systems and help in the promotion, formation, and sustainability of WUAs for self-management and turnover. After the turnover, the partnership of the state Irrigation Department and WUA would need a transformation of the Irrigation Department from the Irrigation Officers' usual role of the regulator to the new role of a supporter in technical and administrative matters along with financial support.

6.5.6 Turn Over Only After Rehabilitation

Only those irrigation systems can be handed over, which are working satisfactorily and have a sufficient service life. The WUA can work better if the irrigation system is free from technical defects and the water/electric supply is reliable. The watercourses and field channels should carry the required discharge, and each member's field should be connected. A joint inspection should be done to demonstrate the proper functioning of the system. The preference for rehabilitation should be given to those systems where WUAs are willing to accept the management responsibility. The best involvement of WUA in the rehabilitation of new work is when it takes on responsibility for the execution of works. If the WUA cannot take on such responsibility, it may be offered to the NGO, which may support such WUA. If the NGO cannot take on the responsibility, the Irrigation Department may carry it out.

The performance evaluation of public tubewells in the Chhutmalpur area shows that breakdowns (mechanical) are frequent. Framers need to be convinced that with proper care of assets, such breakdowns can be reduced, if not eliminated.

Also, the success of the turnover of tube wells will depend on a reliable electric supply. Dedicated electric supply lines and penalties to be paid by the government for inadequate supply of electricity can help restore farmers' confidence in turnover.

6.5.7 Social Heterogeneity—A Big Hindrance

For the successful implementation of the turnover program, it is necessary to understand the existing groups, their objectives, and group psychology. The attitudes and behavior of farmers are greatly influenced by the functioning of the existing groups they traditionally belong to.

Every farmer participates in a variety of groups. Some of the groups, such as panchayat and caste groups, are formally organized groups. The implementation agency must identify these groups and plan the strategy of its approach to the influence of these groups in the promotion of turnover (Chaube 2006).

Religion is a powerful social institution that has a direct bearing on the effectiveness of turnover. If all the farmers belong to the same religion, turnover becomes easier.

For example, in Indonesia, two forms of traditional organizations, viz. Desa-adat (concerned with social aspect) and Subak (concerned with agricultural prosperity), have been successfully used in irrigation management. In Indonesia, Subak organizations have rules based on religious sanctions. Farmers do not violate rules for fear of God. Shrines (temples) are used as a meeting place for Subak, and meetings are arranged on auspicious days (Chaube et al. 1996). In India, religious festivals and meals form an important part of village life. If farmers in tubewell command belong to a different religion, which is generally the case, it is challenging to implement the Subak model.

In India, most of the farmers belong to the Hindu or Muslim religion. Within the Hindu religion, farmers have a strong attachment to caste groups. Unlike Islam and Christianity, Hinduism is not an institutionalized religion. Within Hinduism, there are different sects, each having its own God/Goddess. Further, caste division has made several groups based on sect, caste, and subcaste. Therefore, even if all the farmers belong to the Hindu religion, differences emerge.

The objectives of water user associations and farmers' common interest in sharing irrigation water transcend family and caste groups' interests. But family and caste groups are so strong that it is difficult to accept the higher-order hierarchy of groups like water user associations. Therefore, the objectives of water user association get defeated whenever in conflict with family or caste group objectives.

6.6 Conflict Interfaces

Legal Acts are enacted to resolve various conflicts that arise at several interfaces. Therefore, in framing legal acts, it is first necessary to recognize multiple interfaces and the type of conflicts. A farmer belongs to several social groups. There can be many conflict interfaces, such as groups, religious groups, and government agencies. For example, the state tube well corporation may have conflicts with individual farmers and the electricity supply agency. Some conflict interfaces are schematically depicted in Fig. 6.4 and discussed below.

Water Users Association-Individual Farmer

Water Users Association (WUA) will be involved in the management of tube wells and tertiary systems and the distribution of water at the micro-level. WUA may have conflicts with the individual farmer(s) in the supply of surface water and groundwater, O&M of a watercourse and field channels, collection of irrigation charges for the use of surface water and groundwater, protection of irrigation works, and illegal irrigation and cropping patterns adopted by farmers. Some of these are discussed below.

WUA-Project Authority

The project authority (a project management agency) is responsible for the operation and maintenance of irrigation systems and water delivery up to the distributary and minor levels, while O&M of the tertiary canal system and the tube wells within the canal command will be the responsibility of WUAs. WUAs may have conflicts with the project authority regarding water supply, O&M of watercourse and field channels, O&M of tube wells, collection of irrigation water charges, and flow measurement.

WUA—Electricity Department

The Electricity Department provides electricity to WUAs to operate pumps and tube wells in lift irrigation areas. Conflicts may arise between WUA and the electricity supply department due to unreliable electricity supply, theft of power, and fixation of electricity charges. With dedicated electricity supply lines, the chances of power interruption can be greatly reduced. The theft of power is very common, which results in heavy revenue loss for the electricity department. Checking the theft of

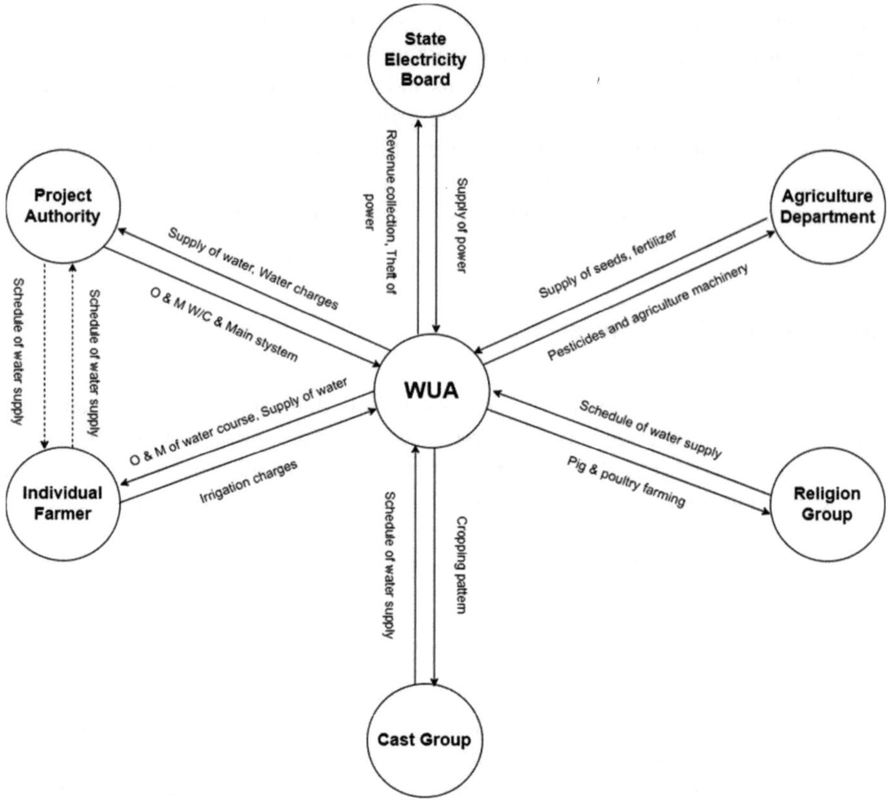

Fig. 6.4 Conflict interfaces between WUA and other agencies

power will be the responsibility of WUA. Under the written contract, the electricity supply department will have the right to claim for any loss due to the theft of power.

Project Authority-Private Tube Well Owners

Conflicts between individual farmers and project authority may arise due to the indiscriminate use of groundwater, spacing, and depth of private tube wells, which could result in groundwater depletion, besides the inequitable distribution of groundwater. To avoid this, the project authority should declare the whole project command as a notified area. In such a case, any person desiring to sink a well in the area will have to apply to the project authority to grant a permit for the purpose. As far as existing users in the area are concerned, they will have to apply to the authority for a grant of a certificate. The holder of agricultural land in which a well is situated will not use groundwater other than for agriculture or drinking and will not waste water for any reason. WUA can help project authority to regulate the pumping or use of groundwater, especially through private tube wells.

WUA—Agriculture Department

Conflicts may arise regarding the time of supply or quality of agricultural input. Therefore, the collection of seeds, pesticides, and fertilizer should be done under the written contract with WUA. In the case of untimely supply and unsatisfactory output or adverse effect of agricultural input on crops, WUA can claim compensation. Similarly, in the supply of machinery on a hire basis, if the department fails to provide machinery in time or bad order, WUA can claim compensation for any loss due to delay or machinery disorder. On the other hand, if there is any loss or damage to machinery in the custody of WUA, WUA will have to pay for the cost of repair or the cost of the machinery.

WUA—Religious Group/Cast Group

An individual farmer may have a stronger affinity to their religious or cast groups than his/her affinity to WUA as a member, which could be detrimental to the performance of WUA. For example, if a Hindu farmer is asked to receive water on Holi, Deepawali and Ramnavmi, he will object to it, similarly, a Muslim farmer will object to taking water on Eid and, also from 12 noon to 2.00 PM, on each Friday.

Conclusions on Conflict Interfaces

Generally, farmers are not highly educated and may not be fully aware of irrigation Acts; therefore, resolving conflicts through legal acts could be a long-drawn process. Existing penalties for irrigation offenses, such as damage to irrigation works and unauthorized irrigation, need to be rationalized. A fine of, say Rs. 100.0 or so can easily be paid by the farmers if they can do unauthorized irrigation. Moreover, there is no provision for social sanctions against offenders. Also, irrigation officers are not equipped with sufficient powers to exercise their duties, especially in settlement of disputes and requisition of land for the construction of watercourses. There is no Groundwater Act to control and regulate its development, specially through private tube wells.

To control and regulate groundwater development through private tube wells, the irrigation service area should be notified, and the tube well owner should first get a certificate from the project authority for using groundwater.

6.7 Enactment of Legal Acts

6.7.1 Enactment/Amendment of Irrigation Act in India

There has been an increased consciousness in states about the need to actively involve farmers in irrigation management. Table 6.3 shows legal acts enacted by various states for the involvement of farmers in irrigation management.

Table 6.3 Enactment/amendment of irrigation act in India

Sl. No	Name of state	Enactment/amendment of irrigation act
1	Andhra Pradesh	Enacted "Andhra Pradesh Farmers" Management of Irrigation Systems Act, March, 1997"
2	Assam	The Assam Irrigation Water Users Act 2004
3	Bihar	"The Bihar Irrigation, Flood Management, and Drainage Rules, 2003" under the Bihar irrigation Act, 1997
4	Chhattisgarh	Enacted "Chhattisgarh Sinchai Prabandhan Me Krishkon Ki Bhagidari Adhiniyam, 2006"
5	Goa	Enacted "Goa Command Area Development Act 1997 (Goa Act 27 of 1997)"
6	Gujarat	Gujarat Water Users Participation Management Act, 2007
7	Karnataka	Promulgated an Ordinance on 7th June 2000 for amendment of the existing Karnataka Irrigation Act 1957
8	Kerala	Enacted "The Kerala Irrigation and Water Conservation Act 2003"
9	Madhya Pradesh	Enacted "Madhya Pradesh Sinchai Prabandhan Me Krishkon Ki Bhagidari Adhiniyam, 1999" during September 1999
10	Maharashtra	"The Maharashtra Management of Irrigation Systems by Farmers Act, 2005"
11	Nagaland	Nagaland Farmers Participation in Management of Irrigation Systems Act, 2013
12	Orissa	Enacted "The Orissa Pani Panchayat Act, 2002"
13	Rajasthan	Passed the "Rajasthan Sinchai Pranali Ke Prabandh Me Krishkon Ki Sahabhagita Adhiniyam, 2000"
14	Sikkim	"Sikkim Irrigation Water Tax 2002" and "Sikkim Irrigation Water Tax (Amendment) Act 2008"
15	Tamil Nadu	Enacted the "Tamil Nadu Farmers" Management of Irrigation Systems Act, 2000"
16	Uttar Pradesh	Enacted the "Uttar Pradesh Irrigation Management Act, 2009"

6.7.2 *Progress of Formation of WUAs*

By the end of 31.3.2018, 87,401 WUAs had been formed in the country. Table 6.4 shows details of State-wise Water User Associations (WUAs) formed and the area covered up to the end of the financial year 2017–18.

6.7.3 *Example of a WUA: Irrigation Panchayats (Madhya Pradesh)*

The government enacted "Madhya Pradesh Sinchai Prabandhan Me Krishakon Ki Bhagidari Adhiniyam, 1999" during the year 1999.

Table 6.4 Details of state-wise WUA formed and area covered

Sl. No	Name of state	Number of WUAs formed 31.03.2016	Area covered (Thousand hectare) 31.03.2016	Number of WUAs formed 2016–17	Number of WUAs formed 2017–18
1	2	3	4	5	6
1	Andhra Pradesh	10,884	4179.25	0	0
2	Arunachal Pradesh	43	10.97	0	0
3	Assam	847	95.02	25	35
4	Bihar	80	209.47	–	10
5	Chattisgarh	1324	1244.56	0	6
6	Goa	84	9.54	0	0
7	Gujarat	8278	662.99	606	445
8	Harayana	8490	1616.27	0	0
9	Himachal Pradesh	1173	140.56	0	0
10	Jammu & Kashmir	383	32.79	0	0
11	Jharkhand	0[a]	0.00	0	25
12	Karnataka	2787	1418.66	5	391
13	Kerala	4398	191.22	0	0
14	Madhya Pradesh	2062	1999.64	19	0
15	Maharashtra	2959	1156.22	457	331
16	Manipur	69	29.40	18	20
17	Meghalaya	159	20.17	0	0
18	Mizoram	390	18.23	0	0
19	Nagaland	24	3.44	0	0
20	Odisha	20,794	1757.71	64	34
21	Punjab	4845	610.29	0	42
22	Rajasthan	1994	1144.45	56	33
23	Sikkim	0	0.00	0	0
24	Tamil Nadu	1910	935.66	0	0
25	Telengana	0[a]	0.00	0	0
26	Tripura	0	0.00	0	0
27	Uttar Pradesh	802	318.69	0	0
28	Uttrakhand	0[a]	0.00	0	0

(continued)

Table 6.4 (continued)

Sl. No	Name of state	Number of WUAs formed 31.03.2016	Area covered (Thousand hectare) 31.03.2016	Number of WUAs formed 2016–17	Number of WUAs formed 2017–18
1	2	3	4	5	6
29	West Bengal	10,000	37.00	0	0
Total	**84,779**	**17,842.21**	**1250**	**1372**	

Source Command Area Development and Water Management Wing Data on website www.mowr.gov.in

[a] Details of WUAs, if any, are included in the list of WUAs of parent States viz. Bihar, Andhra Pradesh and Uttar Pradesh, respectively

According to section 61(i) and section 62 (ii) of the Irrigation Act of Madhya Pradesh, an Irrigation Panchayat should be established for every village or Chak and can also be established for a group of villages in the command area of a canal at the discretion of the Collector. The irrigation panchayat will consist of a Sarpanch and two or more members elected by the land's permanent holders and occupants from among themselves. The functions of the irrigation panchayat are:

To assist the Irrigation Department in (i) preventing the encroachment on canal-lands and to avoid damage to irrigation canals; (ii) construction of water courses; (iii) keeping a record of irrigation; and (iv) collecting water charges and arranging for the repair of watercourses.

The Irrigation Panchayats have effectively helped in the assessment of irrigation and collection of water revenue. They are also helpful in distributing water where the irrigation Panchayat is influential in commanding village farmers' respect and enforcing discipline. But some of the irrigation panchayats have not been able to work effectively for want of undisputed leadership and command over local farmers. Another important reason is that there are no financial resources to systematically control and distribute irrigation water from each outlet.

6.7.4 Bundelkhand Jal Saheli Manch-An Informal Water Committee

Title: BUNDELKHAND JAL SAHELI MANCH (WATER FRIENDS)

Location: Chhatarpur,

Area: Villages in Bundelkhand (Madhya Pradesh and Uttar Pradesh).

Jal Saheli (Water Friends): Sirkoo, a 39-year-old woman in Bundelkhand, Uttar Pradesh, walked 8 km daily to fetch water. As a woman, it was her responsibility to ensure the household's water availability. This put additional stress on her already depleted health and time-until she decided to tackle the issue head on. Three years ago, she and a few other women came together to form an informal water committee or '**Paani Panchayat**' to work on water issues which is what affected them the most. Their agenda was simple—ensure water availability for all through the creation and conservation of water resources in their villages, so that water was available as a basic right. With the help of a local organisation called the Parmarth Samaj Sevi Sansthan, they began to take steps in this direction. The 'Pani Panchayat Sanghatan' nominated two women as 'Jal Saheli's' or 'water friends'. They now meet, discuss and decide on how to tackle local water-related problems, have a say in where a new handpump should be constructed, how to revive a dying 'talaab' or village pond, and also where check dams are needed for better irrigation. All this work is done through the village panchayat and at the block level.

Outcome: Nearly 500 Sahelis are distributed across 7 districts of Madhya Pradesh and Uttar Pradesh. And they do not stop just at meetings but ensure that their voice reaches the state-level officials. They even wrote a letter to the Chief Minister of Uttar Pradesh for the renovation of water bodies in dire need of repair. Sirkoo recalls," Rainfall had decreased, and there was always a problem of clean water. So, the water became our priority. They have built check dams with government allocations and 'shramdan' or voluntary contributions by the community. The 'panchayat' considers the village Water Security Plan (WSP) or the 'Jal Suraksha Karya Yojana' prepared by them before any major decisions on water. Dressed in blue saris, they flaunt their 'water' purpose to change the world or, in their own words, 'Badlenge Zamana'.

6.8 Tamil Nadu Farmers Management of Irrigation Systems Act, 2000

The **Tamil Nadu Act 7** (2001) was published in Part IV-Sect. 6.2 of the Tamil Nadu Government Gazette Extraordinary, dated the 5th March 2001. This was last updated 22nd May 2019 [tn453]. For details of this legal Act; refer to the following internet link of the Government of Tamil Nadu.

http://www.wrd.tn.gov.in/Docu/TNFMISRules.pdf and https://www.indiacode.nic.in/handle/123456789/13163?view_type=browse&sam_handle=123456789/2507

6.8.1 Content of the Act

It has the following seven chapters and notifications

Chapter I	Preliminary
Chapter II	Farmers Organisations
Chapter III	Functions of Farmers Organisation
Chapter IV	Funds of Farmers Organisation
Chapter V	Offences and Penalties
Chapter VI	Settlement of Disputes
Chapter VII	Miscellaneous

Notifications: The notifications regarding the appointment of competent authorities, delineation of area of committees etc. are stated.

6.8.2 Functions of the Water Users Association

The act provides for farmers' participation in managing irrigation systems and for matters connected therewith or incidental thereto. The act provides for delineating the Water Users Association area on a hydraulic basis, which may be administratively viable. Every Water Users Associations area will be divided into territorial constituencies, which will not be less than four, but will not be more than ten, as may be prescribed. The Water Users Association will perform the following functions, namely:

- to prepare and implement an operational plan and a Rotational Water Supply for each irrigation season, consistent with the operational plan prepared by the Distributory Committee and the Project Committee, and based upon the entitlement, area, soil, and cropping pattern as approved by the managing committee or Distributory Committee, or as the case may be, of the Project Committee;
- to prepare a plan for the maintenance of the irrigation system in the area of its operation at the end of each crop season and carry out the maintenance works of both the distributory system, water courses, and field drains in its area of operation with the funds of the Water Users Association, from time to time.
- to regulate the use of water among the various sluices under its area of operation according to the rotational water supply;
- to promote the economy through the use of water allocated;
- to assist the authorities of the Revenue Department of the Government in the preparation of demand and collection of water charges;
- to maintain a register of water users, as published by the Revenue Department of the Government;
- to prepare and maintain an inventory of the irrigation system within the area of operation;
- to monitor the flow of water for irrigation;

- to resolve the disputes, if any, between the members of the Water Users Association in its area of operation;
- to raise resources;
- to maintain accounts;
- to cause an annual audit of its accounts;
- to assist in the conduct of elections to the managing committee;
- to maintain such other records as may be prescribed;
- to abide by the decisions of the Distributory Committee and Project Committee;
- to conduct general body meetings in such a manner as may be prescribed;
- to encourage avenues for plantation on canal and tank poramboke, and to protect and maintain such plantations;
- to conduct regular water budgeting and also to conduct a periodical social audit, as may be prescribed; and
- to remove the encroachments on canal, drains, and tank poramboke in the area of jurisdiction of the Water Users Association.

6.8.3 Distribution Committees

The Appendix of the Act (see **Section 6.6 of the Act and rule 6**) provides information in the tabular form (Table 6.5) regarding Distributary committees (and related WUAs, area, taluq, and district) for Aiyar Reservoir Irrigation System, Tirumoorthy Reservoir Irrigation System, and Palar Anicut Irrigation System.

Aiyar Reservoir Irrigation System: The Aiyar Reservoir Irrigation System has the following four Distributary committees:

1. Pollachi Kalvoi Distributory Committee No. 1 (covers 6 WUAs in Coimbatore district)
2. Pollachi Kalvoi Distributory Committee No. 2 (covers 5 WUAs in Coimbatore district)
3. Vettaikaaranpudur Distributory Committee (covers 3 WUAs in Coimbatore district)
4. Aliyar Uoottu Kalvoi and Sethumadai Kalvoi Distributory Committee (covers 2 WUAs in Coimbatore district)

Tirumoorthy Reservoir Irrigation System: The Tirumoorthy Reservoir Irrigation System has the following nine Distributary committees.

Table 6.5 Information regarding distributary committees

SI. No	Name of the distributory committee	Location of the off-take sluice	Name of the water users association	Name of the taluk	Name of the district
(1)	(2)	(3)	(4)	(5)	(6)

1. Udumalai Kalvoi Distributory Committee (covers 25 WUAs in Coimbatore and Erode districts)
2. Parambikulam Pira-thana Kalvoi Distribu-tory Committee No. 1 (covers 12 WUAs in Coimbatore district)
3. Parambikulam Prirathana Kalvoi Distributory Committee No. 2 (covers 12 WUAs in Coimbatore district)
4. Parambikulam Pirathana Kalvoi Distribu-tory Committee No. 3 (covers 10 WUAs in Coimbatore district)
5. Parambikulam Pira-thana Kalvoi Distribu-tory Committee No. 4 (covers 12 WUAs in Coimbatore and Erode districts)
6. Parambikulam Pira-thana Kalvoi Distribu-tory Committee No. 5 (covers 16 WUAs in Coimbatore and Erode districts)
7. Parambikulam Pira-thana Kalvoi Distribu-tory Committee No. 6 (covers 13 WUAs in Coimbatore and Erode districts)
8. Parambikulam Pirathana Kalvoi Distributory Committee No. 7 (covers 13 WUAs in Coimbatore district)
9. Parambikulam Pira-thana Kalvoi Distribu-tory Committee No. 8 (covers 21 WUAs in Erode district)

Form IV

(see Sect. 6.6 of the Act and rule 6)

Palar Anicut Irrigation System: The Palar Anicut Irrigation System has the following twelve distributary committees.

1. Palar Anicut Distributory Committee—1 (covers 5 WUAs in Vellor district)
2. Palar Anicut Distributory Committee—2 (covers 4 WUAs in Vellor district)
3. Palar Anicut Distribu-tory Committee—3 (covers 5 WUAs in Vellor district)
4. Palar Anicut Distribu-tory Committee—4 (covers 7 WUAs in Vellor district)
5. Palar Anicut Distribu-tory Committee—5 (covers 9 WUAs in Vellor district)
6. Palar Anicut Distribu-tory Committee—6 (covers 8 WUAs in Vellor district)
7. Palar Anicut Distribu-tory Committee—7 (covers 8 WUAs in Vellor & Kanchipuram district)
8. Palar Anicut Distribu-tory Committee—8 (covers 5 WUAs in Vellor district)
9. Palar Anicut Distribu-tory Committee—9 (covers 7 WUAs in Kanchipuram district)
10. Palar Anicut Distribu-tory Committee—10 covers 4 WUAs in Kanchipuram district)
11. Palar Anicut Distribu-tory Committee—11 (covers 9 WUAs in Kanchipuram & Thiruvallur districts)
12. Palar Anicut Distribu-tory Committee—12 (covers 7 WUAs in Vellor and Arcot districts)

6.9 Farmers Participation-A Field Survey-Based Study

This is a field survey-based case study of four tank irrigation projects in the Sagar district of Madhya Pradesh. The study was carried out by one of the authors as part of a sponsored research project at IIT Roorkee (Chaube 2006) and Nissanka (2007). The four projects did not have a separate WUA for each project; WUAs were formed for different groups of tank irrigation projects in the vicinity. For example, the Khairana Irrigation Project is under WUA-Vijayapura. The WUA chairman is from the village Bichchiya (outside command of Khairana project). The Mahuakheda Project is under WUA Ratauna, which covers Ratauna, Mahuakheda, Baigwar Padrai, and Madaiyagaur. The Mahuakheda is not yet represented in the WUA because dissatisfied farmers did not participate in the election process. Sarpanch of Mahuakheda has again received notification for filing of nominations. The WUA of Hinauta Kharmau has six members representing different villages. During the interview, the WUA Chairman expressed unhappiness over the physical status of the canal work.

Recommendations Based on the Case Study:

Some important lessons from this case study are:

(a) It will be useful to involve the farmers in planning the layout of minor canals and water courses. This will instill a high level of commitment and a sense of ownership. Farmers' participation need not be delayed until the completion of the balance of works and the formation of WUAs.
(b) Implement social water allocation (water allocation preference to small and marginal farmers).
(c) Do not allow large/powerful/politically influential farmers to influence the layout of the distribution network to improve water supplies to their land.
(d) Motivate the farmers to show stronger affinity and respect for WUA as a social group compared to other social groups (such as religion/caste, village, and political affiliation) to which they belong. Religious leaders, and local OS can serve as catalysts.

Questions

1. Explain in brief the following terms: (i) participatory irrigation management, (ii) water tariff on volumetric basis, (iii) irrigation panchayat, (iv) rights, (v) functions, (vi) responsibilities, (vii) turnover, (viii) voluntary agency, (ix) social homogeneity, (x) social group, and (xi) rehabilitation.
2. Explain with examples the need for farmer's participation in operation and maintenance of conveyance network below outlet.
3. Describe alternative organisations of farmers.
4. Critically examine the feasibility of involving farmers in planning and design of an irrigation project.

5. Describe the success stories of farmer associations in improving irrigation management.
6. Discuss the role of farmers associations at the outlet level, minor level, and project level.
7. Discuss the need for legal support for the rights of WUAs.
8. Explain the delegation of powers to WUAs at the outlet level.
9. Explain the process of turnover of an irrigation system.
10. Explain economic and financial issues relating to turnover.
11. Explain the role of voluntary agencies and village leaders in encouraging farmers participation.
12. Why rehabilitation of irrigation system is necessary before turnover.
13. Explain in brief, the following terms. (i) Water charge, (ii) cooperative society, (iii) service charge, (iv) shramdan, (v) subsidy, (vi) replicability, (vii) unregistered society, (viii) pattadar.
14. For any one project in your district, compare water rates for various crops as charged from the farmers by the irrigation department/farmers organizations.
15. Compare the performance of the farmer's organizations in state of Gujarat.
16. Write a note on financial soundness of the farmer's organizations.
17. Discuss various problems/constraints in proper functioning of farmers organizations.
18. How enactment of legal Acts has helped in participation of farmers in irrigation management.
19. Explain salient features of Tamil Nadu farmers' management of irrigation systems act, 2000.
20. What is the purpose of the canal distributary committee?

References

Bruns B, Atmanto SD (1992) "Issues in the Turn over Programme in Indonesia", Network paper 10. Overseas Development Institute, London

Chaube et al (1996) "Issues in turnover of state tubewell", by Chaube UC, Chawla AS, Suli Yanti in Sixth National Water Convention, National Water Development Agency Bhopal, 4–6 Jan 1996

Chaube (2006) Report on evaluation of rural infrastructure (irrigation) projects in Sagar District of Madhya Pradesh; Sponsored Research by National Bank for Agriculture and Rural Development (NABARD), Mumbai, India

Datye KR, Patil RK (1987) "Farmer managed irrigation systems" Centre for Applied System Analysis in Development, Mumbai, India

Internet source: (a) www.wrd.tn.gov.in; (b)www.mowr.gov.in; (c) www.dowrodisha.gov.in; (d) www.wrmin.nic.in; (e) panchayat.gov.in etc

Nissanka N (2007) O&M aspects of small tank irrigation projects—some case studies, M.Tech Dessertation, supervised by Prof U C Chaube, IIT Roorkee, Roorkee (2007)

Odissa (2016) Gohira operation and maintenance manual, Volume 1 Main Report, Project Management Unit, Odissa Department of Water Resources

Tamil Nadu Act 7(2001) Tamil Nadu Government Gazette Extraordinary, dated the 5th March 2001. Internet Link: http://www.wrd.tn.gov.in/Docu/TNFMISRules.pdf and https://www.indiacode.nic.in/handle/123456789/13163?view_type=browse&sam_handle=123456789/2507
WAPCOS (1991) Farmer's Organization for Irrigation: Profile Series Tech. Report No. 47 of Water & Power Consultancy Services (India) & Louis Berger Intl. & Inc. Unpublished Report of Indo-US Water Resources Management and Training Project, March 1991

Chapter 7
Operation of Dams and Barrages

Abstract This chapter covers reservoir water balance and operation of gates of dam and barrage for canal water supply and for the safety of structures. The components of the dam structure and surface reservoir are explained. Reservoir operation is based on the concept of water balance, which accounts for various inflows, outflows, and resulting changes in water storage. A reservoir operation study is carried out to plan seasonal water allocations for irrigation and other purposes. Parameters to evaluate the reservoir performance in meeting irrigation demand are discussed and illustrated with an example. The procedure for the operation of a dam covers (a) regulation of gates as per requirements, (b) keeping records and online transmission, and (c) ensuring the safety of the dam and the public. Guidelines for normal and emergency operations are explained. The concept of still pond regulation and semi-still pond regulation to avoid sediment entry into the canal is explained. The operation of silt excluders and silt ejectors is described. Shoal formation near the barrage structure, both upstream and downstream, needs to be avoided. River surveys should be conducted regularly to keep a watch on river behavior. The operation and Maintenance (O&M) Manual for Bhoothathankettu Barrage in Kerala state of India is reviewed. The manual includes operation, inspection, maintenance, and repair of barrage components, and replacement of equipment and appurtenant structures, as required. A brief description of some important items is given in the chapter to illustrate the procedure for operating a barrage.

7.1 Introduction

Tropical monsoon hydrology in the Indian subcontinent necessitates the construction of dams to store river runoff during the monsoon season. Dams in India are classified as small, intermediate, and large based on the height of the dam (from the river bed to full reservoir level) and gross storage capacity (CWC 2001). Table 7.1 shows the classification criteria.

The primary purpose of a weir or a barrage is to create a pond and raise the pond water level to feed the off-taking canal. There is no substantial storage of river water.

U. C. Chaube et al., *Canal Irrigation Systems in India*, Water Science and Technology Library 126, https://doi.org/10.1007/978-3-031-42812-8_7

Table 7.1 Size classification

Category	Gross storage (MCM)	Height (metres)
Small	Between 0.5 MCM and 10 MCM	Between 7.5 and 12 m
Intermediate	Between 10 and 60 MCM	Between 12 and 30 m
Large	Greater than 60 MCM	Greater than 30 m

Source Central Water Commission, Ministry of Water Resources, GOI (CWC 2001)

Table 7.2 Distinguishing features of barrage and weir

Barrage	Low set crest	Ponding by gate	High gates	Less afflux	Bridge possible	Costly	Less silting in u/s
Weir	Higher level crest	Ponding by the raised crest and by a shutter	Approx 2 m high shutters	Excessive afflux	Bridge not possible	Relatively cheap	Silting in u/s due to raised crest

It is a type of low-height diversion dam. The distinguishing features of the barrage and weir are depicted in Table 7.2.

The Central Water Commission of the Government of India has provided guidelines for preparing a manual on the operation and maintenance of dams (CWC 2018). The Bureau of Indian Standards has provided guidelines for the operation and maintenance of barrages and weirs BIS (2012). This chapter deals with reservoir operation and regulation of gates of dam and barrage.

7.2 Components of Dams, Reservoirs, and Barrages

7.2.1 Components of Dams and Reservoirs

The storage capacity of a reservoir is conceptually divided into a number of zones, based on the useful purposes that a reservoir is required to serve. Figure 7.1 depicts various storage zones of reservoir and dam components. The dead storage zone is the bottom-most zone of a reservoir. Major storage space is occupied by the conservation zone. If the reservoir is operated to control floods, then the flood control storage is provided above the conservation zone, followed by the surcharge storage.

Top of Dam—Top of the non-overflow section of a dam.

Surcharge storage—The space in a reservoir between the controlled retention water level (Full Reservoir Level) and the maximum water level. Flood surcharge flows over the spillway until the controlled retention water level is reached.

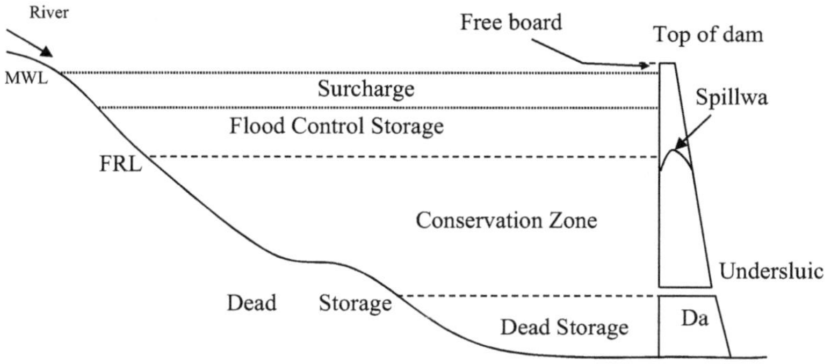

Fig. 7.1 Components of dam and reservoir

Freeboard—The vertical distance between a stated reservoir level and the top of a dam. Normal freeboard is the vertical distance between the Full Reservoir Level (FRL) and the top of the Dam. The minimum freeboard is the vertical distance between the Maximum Water Level (MWL) and the top of the Dam.

Full Reservoir Level (FRL)/Normal water level—For a reservoir with an un-gated spillway it is the spillway crest level. For a reservoir, whose outflow is controlled wholly or partly by movable gates, siphons, or other means, it is the maximum level to which water can be stored under normal operating conditions, exclusive of any provision for flood surcharge.

Intake—Any structure in a reservoir, Dam, or river through which water can be drawn into an aqueduct.

Low-level outlet (bottom outlet)—An opening at a low level from a reservoir generally used for emptying or for scouring sediment and sometimes for irrigation.

Maximum water level (MWL)—The maximum water level, including flood surcharge, the dam is designed to withstand.

Minimum operating level—The lowest level to which the reservoir is drawn down under normal operating conditions.

Outlet gate—A gate controlling the outflow of water from a reservoir.

Primary Spillway (Principal Spillway)—The principal or first-used spillway during flood flows.

Reservoir area—The surface area of a reservoir when filled to controlled retention level.

Reservoir routing—The computation by which the interrelated effects of the inflow hydrograph, reservoir storage, and discharge from the reservoir are evaluated.

Reservoir surface—The surface of a reservoir at any level.

Sill—(a) A submerged structure across a river to control the water level upstream (b). The crest of a spillway, (c) A horizontal gate seating made of wood, stone, concrete, or metal at the invert of any opening or gap in a structure, hence the expressions *gate sill* and *stop log sill.*

Stilling Basin—A basin constructed to dissipate the energy of fast-flowing water, e.g., from a spillway or bottom outlet, and to protect the riverbed from erosion.

Stop logs—Large logs or timber or steel beams placed on top of each other with their ends held in guides on each side of a channel or conduit, providing a cheaper or easily handled temporary closure than a bulkhead gate.

Tail water Level—The level of water in the tailrace at the nearest free surface to the turbine or in the discharge channel immediately downstream of the dam.

Toe of Dam—The junction of the downstream face of a dam with the ground surface, referred to as the *downstream toe.* For an embankment dam, the junction of the upstream face with the ground surface is called the *upstream toe.*

Top of Dam—The elevation of the uppermost surface of a dam, usually a road or walkway, excluding any parapet wall, railings, etc.

Top Width—The width of a dam at the level of the top of the dam.

Trash rack—A screen located at an intake to prevent the ingress of debris.

Weir—(a) A low dam or wall built across a stream to raise the upstream water level, called a *fixed-crest weir* when uncontrolled. (b) A structure built across a stream or channel for measuring flow, is sometimes called a weir.

Dead Storage—Dead storage (Fig. 7.2) is provided in a reservoir for the deposition of sediment entering the reservoir. Dead storage is the zone where sediments mainly settle (coarse sediment settles in the live storage zone).

Bottom outlets are provided above the sediment deposition level.

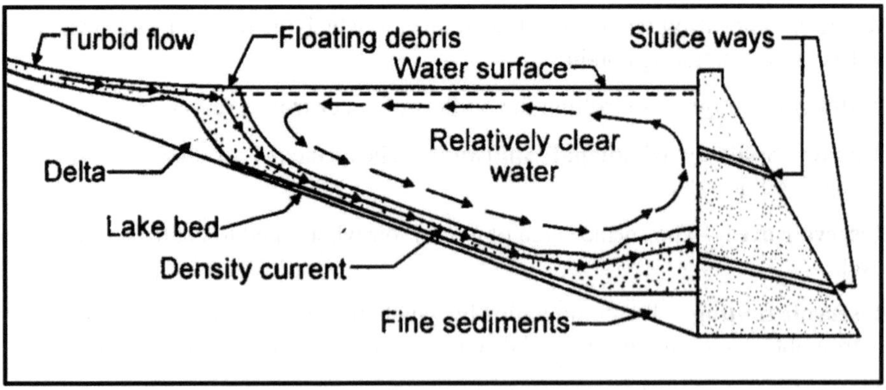

Fig. 7.2 Sediment entry and deposition in reservoir

Conservation (Live) Storage—Reservoir storage between the dead storage level and full reservoir level (FRL). This storage is used for meeting the demand for water for various purposes.

7.2.2 Components of Barrages

The main components of a barrage are:

Guide bund,
Afflux bund,
Main barrage,
Undersluices,
Divide wall,
Fish ladder, and
Head regulator.

Figure 7.3 shows the arrangement of these components in a barrage.

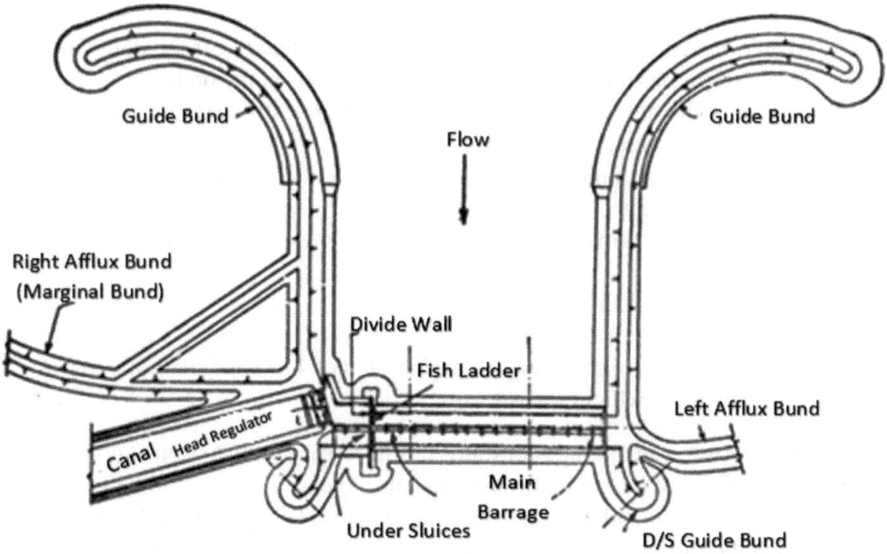

Fig. 7.3 Main components of a barrage

7.3 Reservoir Operation (Simulation) and Reliability

7.3.1 Reservoir Operation Table

It is essentially a search technique used to examine and evaluate the performance of a reservoir for meeting the specified release pattern and to satisfy the conditional operation rules. Simulation study can be done either for a selected critical period or for an entire period of observed or synthetic record. It can be done either for planning a reservoir or for evaluating the performance of an existing reservoir. The storage capacity/yield obtained from the mass curve or the sequent peak method provides a preliminary design for the reservoir. These methods do not consider gains and losses in the reservoir (due to rainfall, evaporation). A simulation study is usually done for monthly flows accounting for evaporation loss, rainfall over reservoir, head available for power generation, and operation rules. However, if the reservoir is small, the sequence of flow within the month may become important, and weekly or daily data may be required. Table 7.3 depicts a reservoir operation study.

Col 1 Usually a month. If the reservoir is small or demand variation within a month is significant, then weekly or daily time intervals can be considered.

Col 2 The rain gauge station should be near the dam site.

Col 3 Pan evaporimeter should be at the ground level and near the dam site.

Col 4 Long-term observed flows or synthesized flows at the dam site.

Col 5 Equal to the d/s release requirement or the unobstructed flow in the period, whichever is less.

Col 6 Reservoir area at the end of previous Period $\times$ RF (col. 2). Part of it goes into bank storage.

Col 7 Col (11) of the previous month $\times$ col. (3) of the current month $\times$ adjustment coefficient.

Col 8 As specified for the purpose.

Col 9 If col 10 values is $> S_{max}$.

Col 10 Col 10 values for the previous period $+$ col 4′ $-$ col 5 $+$ col 6 $-$ col 1 $-$ col 8 or S_{max}, whichever is less.

While performing an operational study, it is necessary to start from a month with a full reservoir. The first month in the analysis must, therefore, be a month in which the flow is in excess of demand. It is also necessary to complete the full cycle of operation and end with the full reservoir. If the inflow series is available for a sufficiently long period, say 30–40 years, a simulation study may be started with any suitable initial storage value. After a simulation study for the entire period of record, reliability is checked.

Table 7.3 Reservoir operation table

Period (m) Year (t) (1)	RF (mm) (2)	Evap. (mm) (3)	Inflow (MCM) (4)	DS water right (MCM) (5)	Gain due to RF (MCM) (6)	Loss due to evap. (MCM) (7)	Release requ't (MCM) (8)	Spill (MCM) (9)	End Storage (MCM) (10)	Reservoir Elevation (M) (11)	Reservoir Area (M^2) (12)
								Value S_{max}		From storage Elevation curve	From elevation area curve

7.3.2 Reliability of Water Supply from Reservoir

The following parameters can be used to estimate the reliability of physical performance for the quantity of water or energy. Let the target function of time be q(t) and the supply function of time be y(t).

(a)
$$Occurence\text{-}based\, reliability = \frac{no.of\ yrs.in\, which\, ensured\, supply\, of\, water\ (energy)\, is\, not\, limited}{Total\, no.of\ yrs.}$$

$$P_0 = \frac{n-m}{n}$$

The number of failure years (m) does not reflect the duration and depth of failure nor the amount of deficit water or energy. Po does not affect the percentage of economic losses out of total production.

(b)
$$Time\, base\, reliability = \frac{duration\, water\ (energy)\, supply\, without\, any\, break\, downs}{Total\, duration}$$

$$P_t = \frac{\sum \Delta t}{T}\ for\ \Delta t\ such\, that\, release\, in\ \Delta t\ is > target\ g\ for\ \Delta t$$

T is the total period of operation.

(P_E) is a better index than (P_o), but (P_t) also does not reflect the failure depth.

(c)
$$Quantity\, based\, reliability\ P_d = 1 - \frac{\int_{y<q}(q-y)dt}{\int_0^T qdt}$$

$$\text{i.e.}\frac{supplied\, water(energy)}{demand}$$

.

If each unit of deficit water or energy leads to the same economic loss then P_d accurately reflects the economic consequences

(d)
$$Total\, deficit\, of\, water\ (energy) = total\, amount\, of\, water\ (energy) in\ \text{m}^3\ \text{(KWh)}\ that\, is\, lacking\, during\, low\, flow\, periods\, to\, ensure\, the\, volume\, of\, water\ (energy)\, that\, is\, to\, be\, supplied.$$

$$= \int\limits_{y<q} (q-y)dt$$

$$Relative\,deficit\,of\,water\,(energy) = \frac{Total\,deficit\,of\,water\,(energy)}{Demanded\,quantity\,of\,water\,(energy)}$$

$$during\,whole\,period$$

(e)

$$Mean\,annual\,deficit\,of\,water\,(energy) = \frac{total\,deficit\,of\,water\,(energy)}{total\,no\,of\,years} = \frac{\int_{y<q}(q-y)dt}{n}$$

.

7.3.3 Standards of Design Reliability (Po Design)

It is given as a share (%) of years in which water is supplied without any failure. The following standards are normally followed:

Po design (%).

(a) Drinking water supply 95–99
(b) Irrigation (for the growing season) 90–95
(c) Livestock farming 95–91
(d) Hydropower production (reliability can be

decreased for HP if it is part of a large power.
system with base load thermal stations) 80–95.

(e) Navigation 90–95
(f) Maintained minimum discharge in streams 95–99

Over year release control must have higher design reliability than seasonal control.

Relationship Between Reliability Indices

Relation between time-based reliability (P_t) and occurrence based reliability (P_o).

Reservoirs that hardly regulate their discharge $P_t > P_o$.

Reservoirs with long-term regulation values: P_t and P_o are close.

Relation between occurrences-based reliability of a real design low flow year P_o and the design duration-based reliability of normal supply P_t des:

$$P_o = 1 - \frac{1 - P_t des}{\mu}$$

$$\mu = \frac{total\,duration\,of\,shortage\,period}{total\,duration}$$

where for a given change in P_o, (ΔP_o) corresponding to the change of P_t des by ΔP_t *des* is given by $\Delta P_o = \frac{1}{\mu}\Delta P_t des$

Example: Given $\mu = 0.2$

P_t design (%)	95	96	91	98	99	100
P_o (%)	15	80	85	90	95	100

A small change in P_t design brings a large change in P_o.

7.3.4 Example of Reliability Parameters Computation

Water balance of a reservoir was simulated for 20 years at monthly time intervals to meet the uniform demand of 15 MCM per month. Following are the deficits in meeting the demand.

Years	2007–08		2011–12		2016–17		2019–2020
Month	July	May	Nov	Apr	Dec	June	Sept
Deficit (MCM)	2.1	1.0	2.5	1.8	0.01	1.0	2.0

Assuming that 10% deficit is tolerable, estimate the time-based and quantity-based reliability parameters for the Rabi season (November to May), Kharif season (Jun to October), and on an annual basis.

What would be quantity-based reliability parameters for the Rabi season (November to May), Kharif season (Jun to October), and annually if no deficit is tolerated?

Solution:

Season	Number of months in a year	Number of months in 20 years	Seasonal water demand in a year (MCM)	Total demand in 20 years (MCM)	Number of failure months in 20 years
Rabi season	7	140	$7 \times 15 = 105$	2100	2
Kharif season	5	100	$5 \times 15 = 75$	1500	2
Annual	12	240	$12 \times 15 = 180$	3600	4

Years	2007–08		2011–12		2016–17		2019–2020
Month	July	May	Nov	Apr	Dec	June	Sept
Deficit (MCM)	2.1	1.0	2.5	1.8	0.01	1.0	2.0
Tolerable deficit (MCM)	1.5	1.5	1.5	1.5	1.5	1.5	1.5
Net deficit (MCM)	0.6	NIL	1.0	0.3	NIL	NIL	0.5
Season	Kharif	Rabi	Rabi	Rabi	Rabi	Kharif	Kharif

Season	Time based reliability (%)	Total deficit quantity (MCM)	Quantity based reliability (%)
Rabi season	(1–2/140) × 100 = 98.6	1.0 + 0.3 = 1.3	(2100–1.3) × 100/2100 = 99.94
Kharif season	(1–2/100) × 100 = 98.0	0.6 + 0.5 = 1.1	(1500–1.1) × 100/1500 = 99.93
Annual	(1–4/240) × 100 = 98.3	2.4	(3600–2.4) × 100/3600 = 99.93

If no deficit is tolerable, then the total deficit in the Rabi season is 5.31 MCM, the total deficit in the Kharif season is 5.1 MCM, and on an annual basis, the total deficit is 10.41 MCM.

So, the quantity-based reliability will be 99.74% in Rabi and 99.66% in Kharif, and 99.71% annually if no deficit is tolerable.

Tutorial exercise: 20 years simulation of reservoir operation in monthly time steps was carried out to meet the constant demand of 22.5 MCM per month. Deficits, as shown below, were found. 10% deficit is tolerable:

8th Year 2.5 MCM (May), 4.8 MCM (June).
11th Year 1 MCM (November); 1.5 (April), 14 MCM (June).
12th Year 4 MCM (July).
15th Year 2 MCM (August), 1 MCM (November).
20th Year 1 MCM (October).

Work out various reliability parameters on annual basis and in Kharif season (June to October).

7.4 Dam (Gates) Operation

Operating guidelines are needed for both normal operations and emergency operations. Central Water Commission has provided detailed guidelines for the operation and maintenance of dams (CWC 2018).

7.4.1 Normal Operations

The normal or day-to-day operating procedure consists of (a) instructions for operating control mechanisms; (b) instructions for operating the reservoir in accordance with reservoir operation rule curve, and (c) general instructions for the safe operation of the dam and appurtenances.

The following aspects also need to be included:

- Releases are to be made for various purposes, and releases are to be made as per various agreements, if any.
- Rule curves.
- Inflow forecasting.
- Flood release procedure.
- Limitation on reservoir drawdown rate to prevent failure of u/s slope.

7.4.2 Emergency Operations

Instructions for operating spillway gates are based on flood inflows and flood situations downstream. The emergency action plan helps in minimizing losses of life and property. The Central Water Commission has brought out *'Guidelines for Developing Emergency Action Plans for Dams'* (CWC 2016). The Bureau of Indian Standards Code I.S. 7323—Operation of Reservoirs—Guidelines should be used to develop rule curves/reservoir regulation plans, including those for emergency operations.

7.5 Operation and Regulation of Barrage Gates

A barrage has gates for controlling flow across almost the entire river section, with the crest levels being close to the riverbed. Hence the operation of gates affects the flow pattern in the upstream and in downstream. River morphology is also affected. Further, the undersluice bays have to be operated in such a way that there is not significant entry of silt into the off-taking canal. In general, the operation of barrage gates should ensure the following features:

(a) The required pond level is maintained during non-monsoon flows and falling flood periods.
(b) Non-monsoon flows remain near the under-sluice bays so that the canal(s) feeding through the head regulator(s) is not affected.
(c) A fairly uniform distribution of discharge along the width of the barrage is obtained as far as practicable.
(d) The flow parallel to the barrage axis, both on the upstream and the downstream of the barrage, is avoided at all times, as far as practicable.
(e) The risk of deep scour and shoal formations in the vicinity of the barrage, both upstream and downstream, is minimized as far as possible.
(f) If shoal formation has taken place, the gates in front of the shoal should be opened more to accentuate the flushing of sediment with the forward flow wherever sufficient discharge is available. Alternatively, the sudden and simultaneous opening of the required number of gates in front of the shoal could be tried.
(g) To evolve the operation of gates to exclude maximum silt/debris deposits on the upstream side and also to minimize the entry of the same in canals/channels.

(h) Hydraulic jumps should not be allowed to form beyond the toe of the downstream glacis in any case.
(i) A relatively high intensity of flow is avoided in the deep scour zones, if formed.
(j) If a shoal has formed on either upstream or downstream or both sides of the barrage, it is washed out and kept away from the barrage, as far as practicable.
(k) The gate operation schedule should also consider constraints regarding the rates of lowering/raising of ponds. It should also consider the safe rate of filling of the canals.
(l) Constant and regular supply in canals/channels, even during fluctuations in discharges from powerhouses located upstream.
(m) The approach channel should be trained so that the tendency of the river to meander near the barrage or outflanking the barrage can be arrested.

7.6 Exclusion of Sediment Entry into Canal

There are, generally, two common types of layouts for sediment excluders. The first type has the mouths of tunnels that are laid staggered and is sometimes called the Khanki-type, the name of which is derived from one of the projects where it has been used. The other type of excluder has all the tunnel mouths aligned at an angle in the plan, which is called the Trimmu-type excluder. Details of the sediment excluders are given in CBIP (1985)

7.6.1 Requirement of Sediment Excluders

Sediment excluders are required to be provided in the river pockets of barrages when the following characteristics predominate:

- When there is high ponding upstream of the barrage to meet the canal discharge requirement;
- When the river/tributary brings in a sediment load of the order of 1500 ppm and above and contains a significant percentage of coarse and medium-sized sediment;
- If the river is in the aggrading stage or wherever the formation of bed bars/shoals is noticed due to unfavorable approach conditions;
- Bed building stage of the river may occur due to barrage obstruction to flow as well as improper regulation of barrage gates;
- Due to adverse flow curvature upstream of the barrage head regulator, most of the sediment load in high river stages may likely settle in front of the head regulator and may enter its way into the canal;
- In spite of the suitable location of the head regulator, river training measures for arriving at the favourable curvature of flow, providing divide walls for separating pockets from barrage bays and suitable gate regulation of barrage undersluice bays for sand exclusion, a large quantity of coarse material may find its way into

the pocket. In such cases, for efficient working of the canal, silt excluders are required to be provided; and

- Exclusion of gravels and boulders could be achieved by providing a barrage crest at river bed level and ponding operation only during non-flood season. Similarly, features such as frontal intake, provision of river sluice, undersluices with low crests, provision of divide-walls their orientation and lengths, and barrage gate regulation are some of the features depending upon sediment load and hydrograph characteristics of the river as well as cut-off available and pond level to be maintained. Generally, gravels and boulders are excluded by keeping the crest of spillway bays low and suspended load and bed load is removed through special silt-excluding structures.

The sediment-excluding devices are not required to be provided in river pockets under the following circumstances:

- If sediment trap like flood control dams exist in the upper reaches of the river and there are no major tributaries in between the dam and the barrage downstream;
- If the sediment load coming in the river reaches upstream of the barrage is only wash load;
- Wherever proper regulation of the undersluices is possible so that the incoming sediment load in the pocket could be flushed completely; and
- If the sill of the canal head regulator is 4 m or higher above the floor level of the pocket, the decision on the omission may be taken after the assessment on the basis of model studies.

7.6.2 Still Pond or Semi Still Pond Regulation

In order to create most favourable conditions for sediment exclusion from the canal, still-pond regulation should be resorted to, as far as possible. In still pond operation, all the gates of the undersluice bays are kept closed so as to limit the discharge flowing into the pocket to be equal to the canal withdrawal. As the undersluice bays are kept closed, the low velocity in the pocket causes the sediment to settle down and relatively clear water enters the canal. However, the pocket gets silted up in this process after sometime.

However, in locations where canals cannot be closed for flushing, semi-still pond regulation may be adopted (as in the case of power channels). In the semi-still-pond operation, the gates of the canal head regulator are not closed for flushing of the silt deposited in the pocket. The gates of the undersluice bays are kept partially open to the minimum necessary so that the bed material in the pocket could be passed downstream. For details, reference may be made to BIS Code of Practice IS 7349 (BIS 2012).

7.6.3 Silt Excluders and Silt Ejectors Operation

Excluders, where provided, should be kept open and while doing so, the limitations imposed by the safety of the structure should be kept in view. The required intensity of discharge q, which may be sufficient to flush the deposited silt in the pocket, can be calculated from the Lacey/Blench scour equation. The required intensity of discharge thus calculated may be generated by suitable gate openings when sufficient head is available in the pond. Under no circumstances should the under-sluice gates be allowed to be overtopped. Silt ejectors in the canal should be operated as much as possible so that the chances of heavy siltation in the canal posing a problem of flushing due to its compaction are minimized.

During monsoon month, it is important keep a constant watch over the sediment entering the headworks, the portion thereof ejected by the extractor if any, and the sediment deposition taking place in the canal and to ensure that sediment deposition only to the extent that can be washed out early in the fair weather before the full demand develops is allowed.

7.7 River Behaviour Observations

Gate operation of barrage structure should be attempted in such a manner that the shoal formation in the vicinity of barrage structure both upstream and downstream is avoided. The Following discussion is based on Bureau of Indian Standards Code of Practice IS 7349 "Barrages And Weirs—Operation And Maintenance—Guidelines" (BIS 2012).

In order to keep a close watch on the river behaviour and bed configuration both upstream and downstream of the barrage, river surveys should be conducted regularly, once before floods and another after floods. For major structures, powerful launches fitted with echo-sounders or any other state of art instrument is desirable to measure cross-sections even during medium flood stage. The survey should be conducted over a stretch of the river close to the barrage at least up to the end of guide bank both on the upstream and downstream. The bed levels should be determined at close intervals of at least 10 m. Depending upon the bed configuration, the pattern of gate regulation should be modified suitably to ensure safety and better hydraulic performance of the barrage.

If a study of the survey data indicates that shoal formation has occurred on the upstream and/or downstream of the barrage, the shoal should be removed by dredging by the use of suitable dredgers to the extent possible so that satisfactory flow conditions are established and also desired capacity is restored.

Satellite imageries obtainable from the National Remote Sensing Agency may be helpful in identifying the variations of bank lines, flow patterns, formation of submerged shoals, etc., in the upstream pond from year to year. Studies with satellite imageries may be made and remedial measures for improving the river behaviour and flow pattern may be taken up.

7.8 Case Study of Bhoothathankettu Barrage (Kerala)

This case study illustrates the procedure for the operation and regulation of barrage gates.

7.8.1 The Bhoothathankettu Barrage

The barrage constructed across the River Periyar at Bhoothathankettu is the head work of Periyar Valley Irrigation Project (PVIP). This barrage is located in the Ernakulam district of Kerala state. Bhoothathankettu barrage is a masonry structure with 15 No of vertically operated shutters. The Government of Kerala has prepared an Operation and Maintenance (O&M) Manual for Bhoothathankettu Barrage (Kerala 2020). The manual includes procedures for carrying out reservoir and gates operations, inspection, maintenance and repair of barrage components, replacement of equipment and appurtenant structures, as required. This discussion is based on the O&M manual of the Bhoothathankettu barrage.

7.8.2 Collection and Reporting of Barrage and Reservoir Data

The following information is recorded and documented for the record.

- Reservoir: water surface elevation, inflow, spill over spillway; downstream release; release for irrigation and other purposes
- Weather-related data
- Instrumentation data
- Water quality

Table 7.4 is proforma for recording reservoir data.

7.8.3 Elevation-Area-Capacity Relation

The elevation-area capacity data are given in Table 7.5. The FRL and the original maximum water level are + 34.95 m and the crest level is + 24.59 m/ + 25.81 m. The gross capacity of the barrage is 0.60 TMC. The reservoir capacity may reduce with time because of reservoir sedimentation. Bathymetric survey on a periodic (every 10 years) basis is to be considered.

Table 7.4 Example Proforma for recording reservoir (Pond) data

For the Month of________									
Date	Time	Water level(m)	Inflow in cumec	Out Flow in cumec					Reservoir Capacity in Million cubic metre (MCM)
				Gates (15 nos. Vertical shutters)	Irrigation canal	...		Total outflow	

Table 7.5 Elevation—area—capacity table

Elevation (m)	Water spread area (× 1000 m^2)	Cumulative capacity (in million cubic meters)
23	20	0.093
24	50	0.136
25	150	0.27
26	360	0.577
27	810	1.248
28	1550	2.625
29	2720	5.216
30	3800	9.484
31	4780	15.416
32	5460	23.2
33	5740	32.208
34	5780	41.61
34.2	5800	43.491
34.95	6100	44.35

7.8.4 Normal Operation Procedures

The Bhoothathankettu barrage operation plan contains instructions for operating the barrage and reservoir during routine (normal) and emergency conditions. The operating procedures for normal operations are given below, including operating criteria for the reservoir, barrage and irrigation outlets for canal system.

All the shutters are kept open from the start of monsoon to the November/ December month of every year. On the first week of December every year, the shutters are closed so that the reservoir starts filling. About one week is required for obtaining full reservoir level. At other times, a minimum discharge of 20 cumec is supplied to meet the drinking purposes and saltwater intrusion. The shutters are opened from centre to sides. As the storage capacity of this barrage is very low, considering the inflow, the rule curve is not applicable. This reservoir acts as a balancing reservoir only. During floods, flood level reports are submitted to District Collector as well as Department Authorities. The local citizens on the downstream of the barrage are intimated by revenue authorities.

7.8.5 Operating Instructions for Gates of Barrage

For maintaining river discharge of 20 cumec, one or more shutters are opened in a random manner such that the total discharge comes to 20 cumec. The calculated discharge of shutter number 1, 6, 10, 15 for an opening of 10 cm is 10.66 cumec. The discharge for opening of 10 cm is 8.75 cumec. The same for shutter number 7, 8 and 9 is 7.95 cumec. If the water level rises to FRL, the opening of shutters is increased to maintain the water level below FRL. As the inflow is mainly from the discharge of Hydro Electric Project such as Idamalayar and Lower Periyar Power houses, the inflow increases during peak hours of power production. Table 7.6 provides a checklist of actions to be undertaken before the operation of gates.

Table 7.6 Checklist of actions to be undertaken before operation of gates

Description	Actions to be carried out
Seal clamp	The seal clamp fixing should be such as to ensure that the side seals are flexible enough
Guide roller and wheels of gate	The smooth movement and alignment of guide rollers and wheels of gate should be checked
Limit switch	Limit switches shall be checked for satisfactory operation (function) to stop the hoist manually when the gate reaches the fully opened or fully closed position
Nut and bolts	All nuts, bolts and screws shall be checked for wear, tear and tightness
Pins	Gate pin, connecting pin and all other pins should be in proper position and ensure that the nuts and split pins are tight
Oil level	To ensure that the proper gear oil of SAE140 is used and should be filled to the desired level in the worm reducer. The oil should not be allowed to be mixed up with water for which necessary precaution has to be ensured during operation and maintenance of gates
Control panel	Control panel shall be checked for its suitability for the purpose envisaged
Starter switches	Starter Switches shall be checked for their working condition

7.8.6 Operating Instructions for Vertical Lift Gates (Barrage/ Irrigation Sluice Gate)

The gates are operated electrically or manually by the hoisting mechanism. The hoists are capable of fully opening or fully closing the gates, holding them fully opened or in any intermediate position. Each hoist can be individually operated from the control unit located on the operating platform. Each control panel contains the necessary main line fuse, switch, reversing contacts relay, overload relays, WRC-Fuses, indicating lamps etc.

For every gate there is a gate position indicator in the control unit stating the exact position in terms of opening. There is also a provision for operating the hoist manually in the case of power failure.

The following precautions are to be taken before dry testing of gates:

1. Checking of all critical dimensions and proper seating of gate over embedded parts is to be ensured. Record of readings may be maintained for future reference.
2. Worm reducer, plumber blocks, gate wheels, and gear wheels should be lubricated.
3. Removal of temporary supports if any.
4. Checking of weld between horizontal girders, tees and horizontal girders, cross girders and hoist bridge girders, final welding of lifting bracket etc., are to be welded, if left over. A comprehensive check list is to be maintained.
5. Checking tightness of bolts between the end box with horizontal girders, lock plate bolts of wheel assembly pin, chain lock pin bolts and other bolts, if used, are to be tightened properly if loose.
6. The gates are to be inspected thoroughly for projections, and temporary supports coming in the way of gate movement and excess concrete is to be removed if any.
7. All the rubber seals are to be made wet before lowering the gate preferably with water to reduce heat generated between seal and seal seat and ensure fixing of all bolts and mouldings of site joints.
8. Light test may be conducted for checking the gap between seal and seal seat. All weld tests are to be conducted.

Do's and Don'ts for Operation and Maintenance of Gates are indicated in Table 7.7

Table 7.7 Do's and don'ts for operation and maintenance of gates

Do's	Don'ts
1. Authorized personnel should be allowed near control panel for operation of gates	1. Unauthorized persons should not be allowed to operate gates
2. Technically qualified or trained operators should be allowed to operate gates	2. Unqualified persons technically should not be allowed to operate gates
3. Operate the gates only when there is required power supply as per the design is available	3. Not to operate gates during low voltage period
4. Adjust the brakes when the gate is lowered fully and rest on sill	4. Not to adjust the brakes when gate is not dogged
5. Attend maintenance during pre-monsoon season	5. Maintenance works should not be attended during rains
6. Use proper tools for attending repairs/adjustments	6. Do not use improper tools for attending repairs/adjustments
7. Conduct dry test before putting into operation	7. Do not operate gates without dry testing
8. Maintain cleanliness of hoist platform	8. Not to keep any slippery material on hoist walkway
9. Insulate damaged electrical wiring, which are exposed to atmosphere	9. Do not allow any persons to operate gate if Bare wires are seen
10. Protect hoist gear box and motors from rains	10. Not to keep gear box cover open after daily maintenance
11. Ensure chain//steel wire rope tightness on either side equally before operating gate	11. Not to operate gate when the chain/steel wire rope is slacked
12. Safety precautions should be taken during maintenance works	12. Not to wear loose dresses during operation of gates
13. Cut off power supply after operation of gates	13. Do not allow power supply to motors when it is not at all required
14. Check connections and functioning of limit switches before operation	14. Advised not to operate gate if limit switches are not functioning
15. Ensure before operation that no foreign materials fall in the gear teeth	15. Not to switch on hoist motors if foreign material found in between teeth of gear wheels
16. Check tightness of Plummer block, drive unit, line shaft, coupling bolts	16. Do not operate gate if any bolts of Plummer block and coupling bolts are found loose
17. Ensure no foreign particles stuck up in between roller and roller track/wall plate and rubber seals etc	17. Not advised to operate gates if foreign material found in between rollers and roller track and wall plates and rubber seals etc
18. Maintain oil as per level indicator in the gear box	18. Not to operate gates with low oil level in gear box
19. Check tightness of chain pin in the lifting plate before operation	19. Not to operate gate if the chain pin and clamps are found loose
20. Operate gate only when grooves are clean without any obstructions	20. Do not operate gate if there is any projection, in the grooves which obstruct movement of gates

Questions

1. Explain the difference in the following terms
   (i) Valley storage, surcharge storage, bank storage
   (ii) Dead storage level, sill level, minimum operating level
   (iii) Full reservoir level, spillway crest level, maximum water level.
2. How long term simulation would provide more reliable estimate of storage or yield to meet seasonal irrigation water demand
3. Discuss various indices of reliability for an irrigation reservoir.
4. How the sediment entry into canal can be avoided or reduced
5. Why the shoal formation in the vicinity of barrage should be avoided
6. Describe purpose of various gates in a dam and in a barrage.
7. What aspects of operation need to be covered in preparation of O&M manual of a barrage

References

BIS Code I.S 7323—Operation of reservoirs—Guide-lines, Bureau of Indian Standards, Manak Bhawan, New Delhi
BIS (2012) IS 7349: 2012, Barrages and weirs—operation and maintenance—guidelines, Bureau of Indian Standards, Manak Bhavan, 9 Bahadur Shah Zafar Marg, New Delhi
CBIP (1985) Manual of barrages, published by Central Board of Irrigation and Power, New Delhi
CWC (2001) Manual on estimation of design flood. Central Water Commission, New Delhi (Mar 2001)
CWC (2016) Guidelines for developing emergency action plans for dams (Doc No. CDSO_GUD_DS_01_ v2.0). Central Water Commission, Government of India
CWC (2018) Guidelines for preparing operation and maintenance manual for dams (Doc.No.CDSO_GUD_DS_03_v1.0) Central Water Commission, Government of India (Jan 2018)
Kerala (2020) Operation and maintenance (O&M) manual for Bhoothathankettu Barrage, Government of Kerala. www.irrigation.kerala.gov.in

Reading Material

Guide for Preparation of Plans of Operation and Maintenance of Irrigation Systems in India. Indian National Committee on Irrigation and Drainage, New Delhi (Mar 1994)
NWRWSKD (2009) Operation and maintenance guidelines for canals. Narmada water resource, water supply, and Kalpsar Department, (NWRWSKD), Government of Gujarat (Apr 2009)

Chapter 8
Canal Operation

Abstract This chapter deals with the regulation of canal hydraulic structures. The procedure for selecting a canal operation method is illustrated using a schematic diagram. The operation methods are classified as proportional control, upstream control, downstream control, constant volume control (BIVAL control), and downstream control on sloping canals (ELFLO control) according to the location of the water level that is maintained by gate regulation. Reasons for water flow control, water level control, and canal control under variable flow conditions are explained. A variety of gates are installed in the water conveyance network. The purpose of these gates in the canal is discussed with an example of weekly irrigation scheduling. To keep all control structures in operating condition, these structures must be exercised and tested at least annually. The exercising and testing procedure for valves and gates is explained. Operating instructions for gates and other facilities should be posted near the control of these facilities and with precautions so that only authorized persons have access. With advances in software and hardware technology, it is now possible to implement computer-aided automatic regulation of canals. The remote control and monitoring of the canal system through programmable computers have greatly decreased the effect of time lag in gate operation with the least operational waste of water. The downstream control of the open channel system has been achieved largely through automatic devices. An example of canal automation in India (Narayanpur Left Bank Canal project in Karnataka) is briefly explained.

8.1 Introduction

The two essential aspects of irrigation system management are (a) efficient organization of the desired water delivery service and (b) operation of hydraulic structures to achieve the desired water delivery service. As discussed in Chap. 9, the water delivery service could be continuous delivery, delivery on demand, and rotational water delivery. A canal is to be operated to satisfy the agreed-upon water delivery to agricultural fields. Inversely, the choice for a given water delivery service depends on the physical and managerial capacities of the existing structures. Details of the three delivery methods are given in Chap. 9.

U. C. Chaube et al., *Canal Irrigation Systems in India*, Water Science and Technology Library 126, https://doi.org/10.1007/978-3-031-42812-8_8

The purpose of water delivery service is to ensure that water is distributed to fields in the required quantities at the proper time, considering soil, crop, and weather conditions. Successful water management minimizes water wastage and satisfies user requirements for irrigation water by implementing efficient canal control logic.

8.2 Operation Methods and Classification

Canal operation may be classified according to the location of the water level that is maintained by gate regulation (NWRWSKD 2009).

- Proportional control, in which the incoming flow is split according to predefined ratios. Thus, there is no constant water level in a canal reach;
- Upstream control, with a constant water level at the upstream side of the regulator as the operational target (Fig. 8.1). The target levels can be fixed or may even be adjustable (Fig. 8.1). Adjustable target levels offer some flexibility in operating canal carrying silt-laden water. This prevents sedimentation during periods of low flow demand, for example, during the early stages of project development.
- Downstream control means constant water level at the downstream side of the regulator as the operational target.
- Constant volume controls (BIVAL control), with a constant water level in the middle of the downstream canal, reach as target (Fig. 8.2).
- The target level in any section can also be between two regulators. When the control point is approximately in the middle of a pool, this method is called the constant volume method.

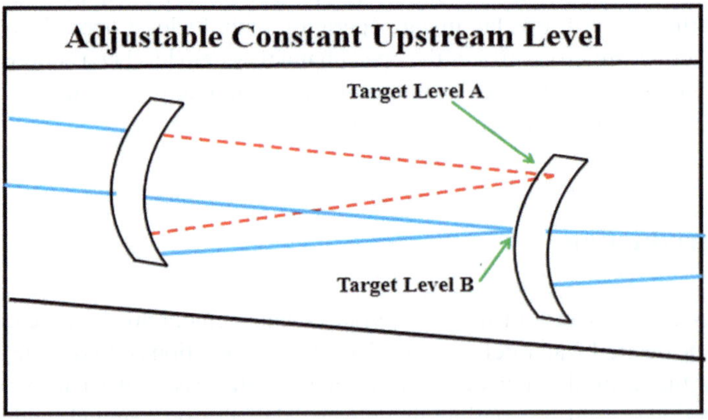

Fig. 8.1 Adjustable constant upstream level

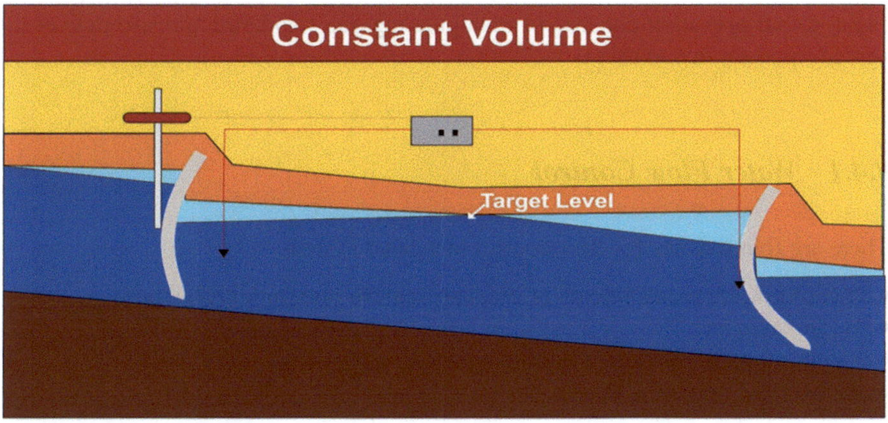

Fig. 8.2 Constant volume

- Wedge Storage: By damming a canal with cross-regulators, the water surface profile varies with discharge from a line parallel to the canal bed at the maximum flow to a horizontal line at zero flow. The volume between the two profiles is called the wedge volume (Fig. 8.2).
- Downstream control on sloping canals (ELFLO control) means a constant water level at the other end of the downstream canal reaches as a target.

8.3 Procedure for Selection of Operating Methods

The procedure consists of

(i) The Decision-Making Procedure: whether it should be supply-based (dictated delivery) or changed water delivery (arranged delivery) or On-demand delivery to the tertiary units. Who makes the decision?
(ii) Method of Water Allocation: How is the water distributed to tertiary units? There are three options for the delivery method (water allocation) to the tertiary unit: split flow, intermittent (on/off) flow, and adjustable flow.
(iii) Method of Water Distribution: How is the water distributed through the main system? The distribution method can be based on 'split flow,' 'intermittent flow', or 'adjustable flow'.

8.4 Operational and Regulation Modes

The objectives of canal operation are related to (i) water flow control; (ii) water level control; and (iii) canal control under variable flow conditions. An operator controls pressure and flows in a piped system and water level and flow in a free-flow canal. Canal operation consists of manipulating gates and structures to deliver the agreed-upon service to users.

8.4.1 Water Flow Control

There are four reasons for flow control in a canal system:

- meeting water requirements of crops to achieve optimum production;
- water savings;
- safety of operation; and
- recovery of operating costs.

Meeting water requirements: Water requirements vary with the type of crop, the stage of crop growth, soil type, seasonal variations in climate, and daily variations in weather, especially rainfall. Both too much and too little water affect crop yield, especially at critical growth stages like tilling and flowering for paddy rice. The response of irrigation supply to demand must be accurate and timely.

Water savings: The need to save water is critical, as water is scarce, and water wastage in the wet season can limit the area under irrigation in both wet and dry seasons. The canal should be operated to accommodate any sudden fall in demand, such as at night or during a sudden rainstorm.

Safety of Operation: To ensure safe operation, the flow released into a canal should not exceed its carrying capacity, decreasing from upstream to downstream. Overtopping of canals can be avoided by means of operating the escape structures to spill excess water into the drainage system. The spilled water can be reused further downstream but at energy cost and with the risk of using water of lesser quality.

Recovery of Operating Costs: The fourth reason for accurate flow control is to provide a basis for sound cost recovery. Farmers can be charged on a volumetric basis for irrigation services, and the application of a volumetric quota is feasible.

8.4.2 Water Level Control

Water level control is a basic concept that has been incorporated into traditional irrigation schemes. In a canal, the water level is controlled through gated cross-regulators (Fig. 8.3).

The reasons for water level control are

- to keep command of the service area through gravity,
- canal safety and protection against deterioration, and
- flow control at off-takes.

Fig. 8.3 Cross regulator. *Source* FAO (2008)

Command Area Under Gravity Flow: The higher the level in a canal, the larger the area which can be irrigated by gravity without pumping is an important consideration for the users.

Canals are run only at full supply to ensure a constant water level at each off-take. However, canals must be operated under variable flow conditions to meet modern irrigation requirements. Since there is a direct relationship between flow and water level in a canal under uniform flow conditions, steps have to be taken to raise the canal's level by artificial means.

Canal safety and Protection Against Deterioration: The next reason for controlling water level is canal safety and protection. The water level should not exceed an upper limit to avoid overtopping and wasting water through escape structures. If the water level drops too low and too fast, canals may deteriorate because of backpressure on canal lining and the instability of canal slopes in difficult soils. Therefore, it is important to keep the water level between a maximum, which is imposed for safety reasons, and a minimum, which is needed to supply water to the land by gravity.

Flow Control at Off-Takes: The next reason for controlling water level variations in a canal is to facilitate flow control and measurement of water at off-takes.

8.4.3 Canal Control Under Variable Flow Conditions

The basic design of an irrigation system operated continually with the variable flow is to divide canals into successive pools and locate the regulators and the off-takes so that the differential water level variations remain within certain limits (Fig. 8.4). The flow delivered through an off-take depends on the water level upstream. Therefore, limiting variations in the water level at canal off-takes which facilitates the control of flow releases, is a crucial task.

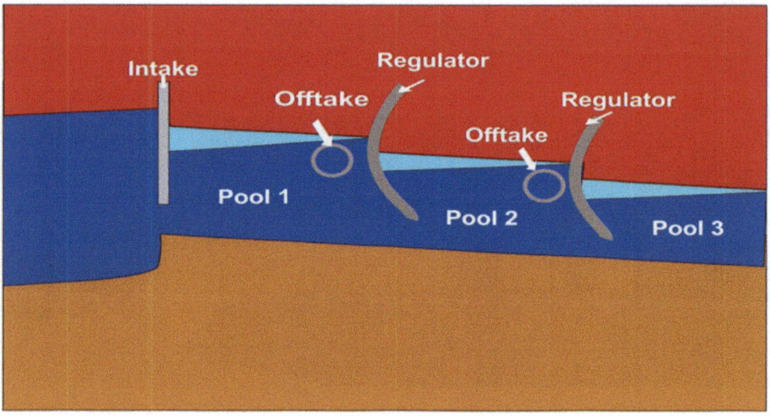

Fig. 8.4 Pools and off takes

8.5 Regulation of Gates for Weekly Irrigation Scheduling

The discharge at each minor off-take is regulated through an adjustable flow-control device. The discharge at each distributary's off-take is regulated with a vertical slide gate. A gate at the distributary's head allows for the adjustment of flow size entering that distributary to the sum of the flow rates delivered through the minor's off-takes. Similarly, a gate at the head of the branch or sub-branch canal allows for the adjustment of flow size entering the branch or sub-branch canal to the sum of flow rates delivered through the distributary's off-takes.

8.5.1 Example: Procedure for Weekly Irrigation Scheduling

The operators add up daily or weekly water needs for each minor, whether requested by the farmers or determined by the project management. The information on total demand is then transmitted further upstream to the distributary's Head Regulator.

The daily demands for each minor are transmitted SEVEN days in advance to the Section Officers responsible for the distributary operation. These Section Officers, in turn, report to the next higher level considering the distribution efficiency. The process is repeated to the head regulator at the Head of the Branch Canal, accounting for the conveyance efficiency. Estimates are then made of the transmission time needed for the change in flow rate to travel from the head regulator to various off-takes (Fig. 8.5). These computations must also account for the change in wedge storage volumes. The flow rate is then changed at the branch canal's head regulator.

If the cross regulators are manually operated, each one should be adjusted at frequent intervals as the flow change moves downstream. Many small adjustments

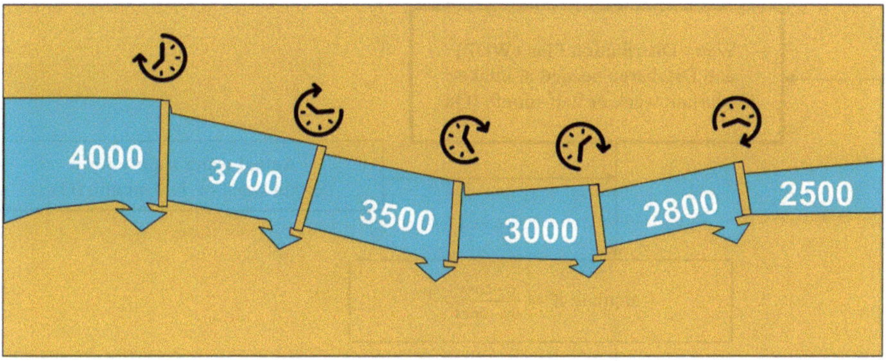

Fig. 8.5 Change in flow rate from head regulator to various off takes

may be necessary because the change in flow rate arrives gradually at various regulators. It is difficult to predict the necessary adjustments at control gates because of the large number of hydraulic variables involved. It is almost impossible to set the flow released at the head regulator of the branch canal to exactly the amount needed to meet the cumulative demand and compensate for seepage and evaporation losses on the way. To ensure that the lowest off-take is adequately supplied, the flow released at the head regulator of the branch canal must include an additional amount as a safety margin. When there is an unscheduled fall in demand due, for example, to a rainstorm, the off-takes are closed to avoid crop damage from over-watering. Water stored in the wedge volumes is then lost through the tail escape to the drains.

8.5.2 Operation Procedure for Minor Offtake and Turnout (Outlet) Gate

Minor offtake and turnout gates are operated according to the planned water distribution plan (WDP), actual water supply, and operation rules, as shown in Fig. 8.6. Adjustment in gate opening is required to account for various losses when water supply at the head of a minor or at the outlet may be different from the discharge needed as per planned WDP.

Adjustments in gate openings would also be required during heavy rainfall and severe water shortage/drought. The actual water supply to off-takes and intakes is calculated by summing up the relevant allocated turnout discharge, including the adjusted ones, and adding their corresponding canal conveyance losses.

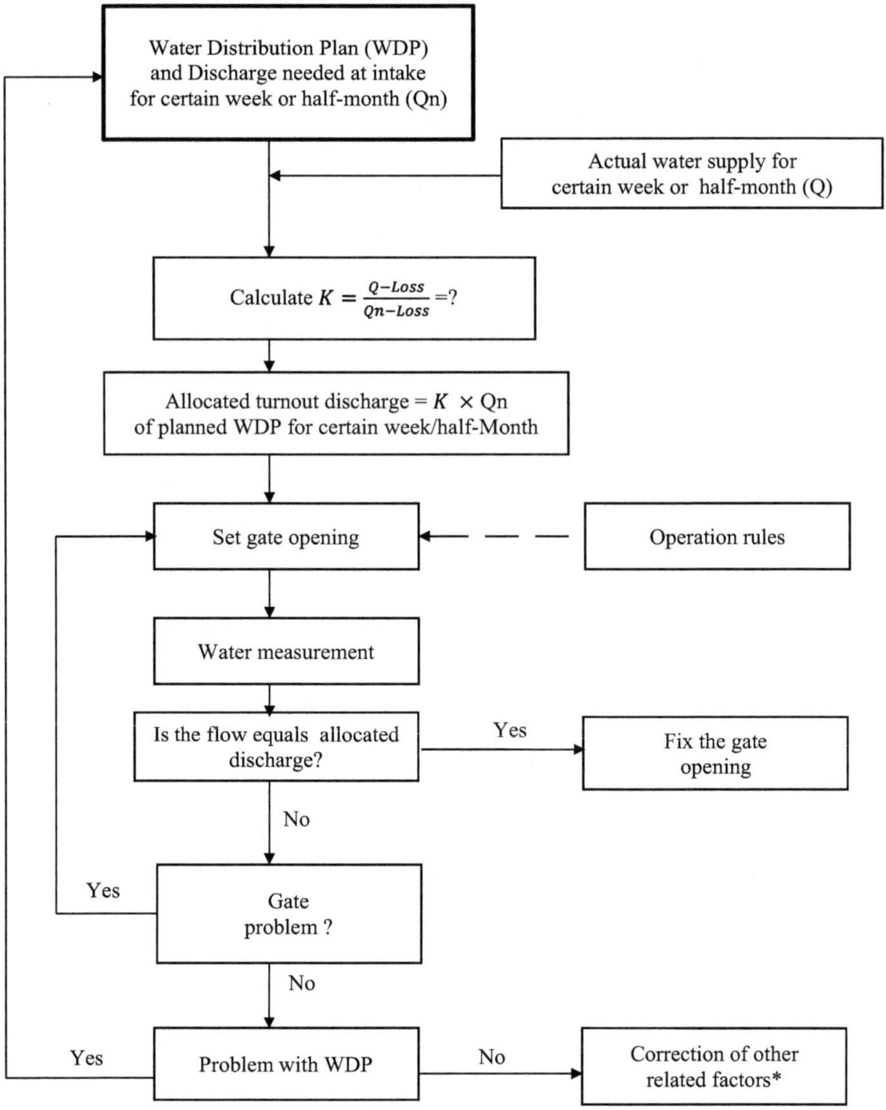

*Such as taking too much water and entering foreign water upstream etc.

Fig. 8.6 Operation procedure for minor off-take and turnout (outlet) gates

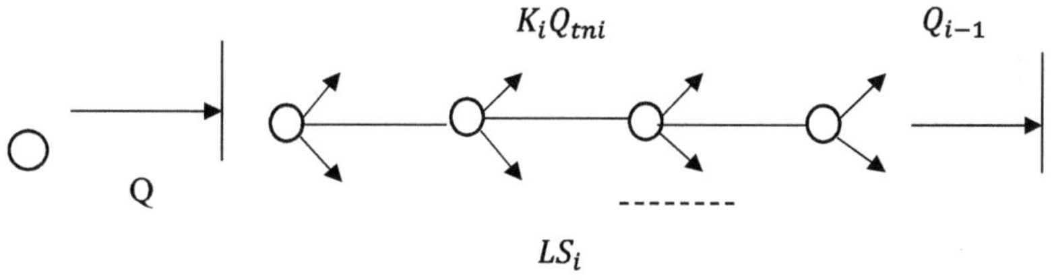

It can also be expressed by the following formula:

K_i, K_{i-1}	Correlation factor at certain canal reach or canal system.
Q_i, Q_{i-1}	Actual water supply to certain offtake or intake i, i − 1 in the planned WDP (lps).
Q_{ni}, Q_{ni-1}	Normal (n) water supply to certain offtake or intake i, i − 1 in the planned WDP (lps).
Q_{tpi}, Q_{tpi-1}	Allocated (p) water supply to certain set of turnouts (lps).
Q_{tni}, Q_{tni-1}	Normal water supply to certain set of turnouts in the planned WDP (lps).
L_{si}, L_{si-1}	Canal conveyance loss at certain reach or canal system (lps).

$$Q_{i-1} = \sum_{1}^{i-1} \left(K_{i-1} Q_{tn_{i-1}} - L_{si-1} \right)$$

$$K_i = \frac{Q_{tp_i}}{Q_{tn_i}} = \frac{Q_i - L_{si} - Q_{i-1}}{Q_{ni} - L_{si} - Q_{ni-1}}$$

$$Q_i = Q_{i-1} + K_i Q_{tn_i} + L_{si}$$

During ordinary conditions, the actual water supply to intake is equal to or higher/lower than the total irrigation demand, and the discharge to each tertiary turnout is allocated based on the planned WDP and the discharge correction factor, K, as described previously.

However, during severe water shortage/drought, rotation within the secondary/primary canal or even within the primary canal need to be implemented with consideration of canal capacity.

8.6 Operational Testing and Exercising

It is necessary to keep all control structures in operating condition. The regulating gates shall be exercised through a complete opening and closing cycle at least annually under a balanced-head condition or with the outlet works dewatered.

The safety of the structure and good operation and maintenance practices require that each gate (of the head regulator, cross regulator, escape, and any other outlet) be tested to confirm that it will operate as designed. Each gate (emergency and regulating gates) will be exercised.

Circumstances at each structure will govern the extent and frequency of testing. Any site-specific variances from these exercise requirements will be reflected and documented.

Exercising and testing machinery should be done using *normal* and *auxiliary* power sources to ensure operation. All exercising and testing results should be recorded and dated in the *Operating Log* of the structure to serve as a permanent record for reference purposes. Reference may be made to the following document available on the internet NWRWSKD (2009) for details on the procedure for testing and exercising:

Operation and Maintenance Guidelines for Canals. Narmada Water Resource Water Supply and Kalpsar Department (NWRWSKD), Government of Gujarat, April 2009.

8.7 Posting of Operating Instructions

As a general rule, operating instructions for gates and generator sets should be posted near the control of these facilities. This ensures that operating personnel use the facilities according to prescribed procedures and provide instructions to others less familiar with who may have to operate facilities during emergencies. Operating instructions should be posted within protective coverings, such as clear plastics, near associated equipment so that equipment can be operated as intended. Each operating device should be permanently and clearly marked for easy identification.

However, the posting of operating instructions may be potentially dangerous if unauthorized personnel gain access to controls and then have operating instructions. To minimize this possibility, precautions should be taken to prevent access to controls by such persons. In any event, instructions should be clear, in readable condition, and in a logical sequence. The printed matter often deteriorates rapidly in an adverse environment.

Standby power facility: The need for and the adequacy of standby power should be evaluated periodically. Periodic trials should be made to ensure reliability. Fuel supply should be checked for volume and freshness. An exterior exhaust system is necessary for all standby systems.

Operation during an emergency: Detailed guidelines are issued for the operation of canals during an emergency situation. Each canal shall have its detailed Emergency Preparedness Plan (EPP).

8.8 Canal Automation

8.8.1 Manual and Automated Operation of Gates

With advances in electronics and communications, it has now been possible to implement computer-aided automatic regulation of canals. The two important requirements of canal regulation are the control of discharge and the head. The problem of regulation in open channels is complicated due to the transients hydraulic effect, wave travel time (lag time), and the randomness in demands. Regulators are often provided to localize these effects.

Manual operation of canals is usually restricted to local upstream control. The supply-oriented (operator-oriented upstream control) canal operation methods are characterized by inflexibility and inequitable distribution of water. Large amounts of water may go to waste and water users may suffer consequences of (i) lag time, (ii) slow response of the control structures to flow changes; and (iii) human errors. Even experienced canal operators are restricted in their ability to provide efficient control of the canal system because of the problem of accounting for a large member of variables that are required to be nalysed for proper control action. Operators are usually familiar with a relatively limited operational range.

8.8.2 Meaning of Automation

Automatic regulation (control) of canals is defined as a procedure or method used to regulate a water system by mechanical, hydraulic, or electronic equipment that takes the place of human observations, effort, and decisions. It is the process of rendering the canal system to be self-measuring, self-adjusting, and self-controlling with direct human intervention.

The automation viz. remote control and monitoring of the canal system through programmable computers, can greatly decrease the effect of time lag, compensate for inaccuracy, and generally provide better service to water users with the least operational waste of water. Downstream control of the open channel system has been achieved largely through automatic devices.

8.8.3 Downstream Control

The upstream control of the canal is inflexible and is not suitable for a demand delivery system. A control system termed “downstream control” was evolved in France. The main features of this system are:

1. Regulators are automatic devices that adjust, depending on the level at the end of the downstream reach.
2. The gate is so designed that the water level immediately on the downstream of the gate is constant for all discharges.
3. At the maximum demand within the downstream reach, the gate opening is maximum and the water surface is parallel to the sloping bed of the canal. If there is no withdrawal, the water surface in the reach reaches a horizontal state, and the gate adjusts to a minimum position.
4. The canal bank must be designed with a level top corresponding to the horizontal water surface (with no withdrawal).
5. As a result of the permissible range of water surface in the reach corresponding to $Q = 0$ and $Q = Q_{max}$, a triangular wedge space is always available as channel storage. As a result, the canal responds more quickly to changes in demand.
6. The above-mentioned factors permit unconstrained withdrawals of water by farmers. Thus, the desired freedom and flexibility to the water user is assured.
7. Since gates are automatic in operation, manpower requirements are reduced, and the use of unfair means to get water is eliminated.

However, the limitations of this control system are:

1. The canal slope should be less than 50 cm/km. The steeply sloping canals cannot be converted to downstream control, as it becomes uneconomical.
2. The water level fluctuates the most at the downstream end, wherefrom the withdrawal is usually effected. This induces instability.
3. There is an unsteady flow in the canal, and the alteration of discharge results in wave formation.
4. In the case of a canal breach, the sudden outflow in a reach causes the gate on the upstream side to open fully rather than close. This is a major operational problem.
5. The automatic functioning of hydro-mechanical gates may be hampered due to the floating matter.
6. Initial cost inputs are sustainable.
7. Since the canal has to cater to a random demand, the canal size is substantially greater than the upstream control canals.
8. There is a possibility of gate oscillations in the case of frequent changes in demand.

Some of the ways to overcome the above limitations are to employ a composite system where the first ½ or 2/3rd length of the canal is made with an upstream control and the rest with a downstream control.

8.8.4 *Control Volume Concept of Operation in California Aqueduct System*

The California aqueduct system was designed primarily to

1. Meet the contractual agreements with water contractors.
2. Provide for a minimum on-peak operation of pumping plants to reduce power costs.
3. React immediately to adverse operating conditions or emergencies.
4. Minimize adverse hydraulic transients during flow changes.

Due to this special feature and the limitations of the downstream control concept, the conventional control was modified, and the controlled volume concept of operation was evolved for this system.

To overcome the instability, a system term as EL-FLO PLUS RESET was tried in Corning canal in U.S.A. It was essentially a downstream control system with gate automation achieved through electronic feedback from the downstream end level of the reach. The feedback was achieved through electronic sensors and a conveyance system of cables ending in an electronic circuitry, which actuates and establishes the gate movement. In order to have stabilized operation, the reset circuitry was introduced. The target depths depend upon the estimated programming of the water delivery volumes. The introduction of electronic feedback in control removes many of the defects associated with the hydro-mechanical feedback arrangement in the level top canals. California Aqueduct, a remote-controlled computerized system, is a complex conveyance system with a maximum discharge of 370 cumecs and more than 700 km of total length. The canal is operated from a project control centre in Sacramento with five other regional centres all geared to computerized automation.

8.8.5 *Dynamic Regulation*

Simultaneously in France, a regulation system termed 'Dynamic Regulation' was evolved by the canal de Provence water authority for its canals. This system though basically similar in concept to EL-FLO PLUS RESET is more sophisticated. Dynamic regulation is basically different from conventional downstream control systems, with local automation, as regards hydraulic operation. It enables overall control of the volumes of water conveyed through various reaches of the canal according to the estimated evolution of demands to be fulfilled and considering the flow time of water (Zaghlool 1993).

The electronic sensors are placed at regular intervals along the entire length of the canal, and the data from these sensors are transmitted through a remote transmission network to a centrally located remote control centre, which houses a network control panel that is computer controlled. Besides water levels, relevant data, such as flow rates, and the demand rates in the reach, are fed into the computer. The computer processes this data continuously for every reach, compares it with the water demand forecasts made with a mathematical simulation model and calculates the gate positions at each canal reach, and transmits the instructions for the adjustment of gate position to the desired value. The main control centre is aided by local sub-centers in its operation. For remote transmission, either communication lines or telephone or telegraph network or micro wave transmission is used. This system is thus used-oriented, and responds quickly to changes rather than requiring an advance scheduling (as in the case California Aqueduct System). Every seven seconds measurements are scanned and at every 15 min, water use predictions (evaluated through statistics) are updated and a small computer program decides the gate movements.

Canal de Provence in France is a remote controlled complex conveyance system with a discharge of 40 cumecs and 250 km of galleries and canals and 300 km of underground pipes. The hydraulic operations of the conveyance system are controlled by the General Remote Control Centre at Le Thonet using Dynamic regulation with an industrial mini-computer.

8.9 Narayanpur Left Bank Canal Automation Project

The Narayanpur dam is on Krishna River in Bijapur District of Karnataka. The reservoir supplies water to Narayanpur Left Bank Canal (NLBC) which is the biggest and the main artery of the canal network about 77 km long. The NLBC Automation System, is a SCADA (Supervisory Control and Data Acquisition) based automation system.

8.9.1 Automation of Existing HR and CR Gates

There are existing 41 radial gates available at HR and CR locations within the NLBC network system, which are needed for canal automation and regulation of water demand in the canal network. These radial gates are updated as Supervisory Control and Data Acquisition (SCADA) based electrical and mechanical retrofit in the project. Also, the CR control rooms were constructed near to these gates by installing panel instruments, for instance, (a) encoders for gate positioning, (b) up and down limit proxy, (c) local control panel for local mode gate operation, (d) u/s and d/s level sensor to calculate discharge, (e) gate control panel with touch screen display, and (f) network equipment for data transmission and reception.

Solar powered integrated gate is specially designed with stainless steel material to avoid erosion and corrosion in the field for combined flow control and metering application within the canal network system of NLBC. The integrated gate has inbuilt various components, such as (a) accurate gate control system, (b) accurate gate measurement system, (c) flow measurement device, (d) u/s and d/s water level measurement system, (e) wireless communication system, (f) self-sufficient solar based power supply system, and (g) CCTV cameras.

The installed integrated gate can be operated and monitored on field as a standalone gate or operated remotely from master control centre when connected to SCADA communication network. It is an intelligent device and uses M2M communication to connect to any TCP IP/SCADA/Internet of Things network. The integrated gate controls the water discharge by gate operation based on the desired set-points, or demand based. The installed integrated gate also has an outer vandalism proof cage, installed for establishing the hybrid wireless data communication network within the system.

8.9.2 Information Kiosk with Farmer Dashboard

Information Kiosk with farmer dashboard has become a powerful tool for e-governance and all information in one place intended for the farmer. The following information is collected and stored for the usefulness of farmers.

(a) Details of farmers with canal command jurisdiction offices for solving grievances.
(b) Information about irrigation schedule for better planning of crop water releases Farmers can raise their water demand, depending upon cropping pattern and area for irrigation within the command area.
(c) Revenue collection being made easy and accessible through crop—water billing.
(d) Weather observed and forecast details on one a single click.
(e) State wide commodity rates are accessible right from the kiosk.

8.9.3 SCADA System Software for Controlling and Monitoring

SCADA system software is implemented for the monitoring and controlling of integrated gates and HR/CR gates of canal network system through wireless data communication network. The water demand data is fed into SCADA software to monitor and control automatically the canal regulation points and to make their proper analytical reports which enable the decision maker to understand situation in the whole command area of NLBC network system.

8.9.4 Master Control Station

Master control station is located at Narayanpur, which caters to all command area related information and control and regulates the information. Master control station is equipped with the following equipment:

- SCADA Application Server
- INMIS Application Server
- GIS Application Server
- Engineering Station
- Wireless Data Communication Master Station Equipment

  10 Mbps Lease line for Web Connectivity

- UPS and Generator set for Power backup

8.9.5 Remote Monitoring Station

The remote monitoring station is meant for monitoring the canal command area, not for any control. The total of ten monitoring stations have been established for monitoring flow and other canal-related activity at Sub-division and Division level within the NLBC network system.

8.9.6 Overall System Architecture

Overall canal automation system comprises various components, the system architecture, and data flow.

(a) Farmers/WUCS list their crop and area which they are willing to cultivate for the particular crop season.
(b) All the water demand goes to the respective office and complete canal command area water report is generated and analyzed through INMIS and GIS.
(c) Irrigation schedule is prepared as per the demand collected.
(d) Irrigation schedule is published in the farmer dashboard.
(e) Analyzed and approved irrigation schedule is fed into canal control and SCADA Software.
(f) All the necessary commands and set points are generated by the software and transmitted to the HR/CR and Integrated Gates.
(g) All the feedback taken into SCADA is also given to the INMIS and Crop—Water Use analyzed into INMIS and GIS.
(h) Water bills are generated through INMIS.
(i) Crop Pattern and Crop Yield report is generated through Remote Sensing data for particular crop season.

Questions

1. What do you understand by efficient organization of a desired water delivery service?
2. Explain the reasons for (i) water flow control; (ii) water level control; and (iii) canal control under variable flow conditions.
3. Discuss various types of gates used in canal regulation.
4. Explain the procedure for weekly irrigation scheduling through regulation of gates.
5. Explain the procedure for operational testing and exercising the gates.
6. What is the purpose of posting operating instructions?
7. What precautions should be taken so that posting instructions are not misused by unauthorized persons?
8. List various warning signs and notices.
9. Draw figures to depict canal control under variable flow condition.
10. Draw figure(s) to depict water level control.
11. Draw figures and explain the meaning of local manual, local automatic, centralized manual, and centralized automatic gate operation.
12. What do you understand by downstream control and what are its limitations.
13. Discuss the salient features of Narayanpur Left Bank Canal Automation Project.

References

FAO (2008) Canal operation techniques-module II "Operation Techniques in Canal Systems by Daniel Renault, Food, and Land Division of Food and Agriculture Organization (FAO)

NWRWSKD (2009) Operation and maintenance guidelines for canals. Narmada Water Resource Water Supply and Kalpsar Department (NWRWSKD), Government of Gujarat (Apr 2009)

Website of Karnataka Government(India): KBJNL website portal: http://www.kbjnl.karnataka.gov.in/

Zaghlool AS (1993) Simulation study of automatic regulation of canal. Ph.D. Thesis, of Zaglool Ashraf Saad, WRDTC, University of Roorkee, Sept 1993

Chapter 9
Water Distribution Planning

Abstract Three different methods of water delivery, namely (i) continuous delivery, (ii) delivery on demand, and (iii) rotational delivery, are discussed. Rotational delivery (known as Warabandi in northern India) allows the sharing of deficit water supply equitably and efficiently. The advantages of rotational delivery are explained. Parameters of rotational water delivery (water allowance, duty, stream size, rotation period etc.) are explained through an example. During the operation stage, water allowance computation is based on the actual conveyance capacity of a water course, sanctioned or as an existing cropping pattern, and irrigation efficiencies attained/ realistically attainable. Examples of computation of water allowance, duty, etc., are given. Irrigation time allocation to farmers depends on the total time of rotation, farm area, and water allowance. This delivery time has to be adjusted for time lost in filling the canal, irrigation time available while emptying a canal, and time equivalent of conveyance loss (seepage, leakage, spill) in the canal. Examples of time schedules for rotational water delivery considering filling, emptying, and conveyance losses are given. Examples of prevalent rotational water delivery practices in the Pochampad Project (Telangana), and Upper Ganga Canal (Uttar Pradesh) are discussed. Improvements in traditional rotational delivery (Warabandi) are suggested A case study on irrigation time scheduling in Cilutung minor canal project (Indonesia) is included to explain water scheduling among blocks and to consider losses. A critical assessment of field practice shows that the success of rotational water delivery depends on (i) consolidation of scattered or small holdings, (ii) watercourse reaching each farmer field, (iii) constant flow through outlets at predetermined times, and (iv) a climate of stable water scarcity condition. The waste of irrigation water at night could be very high in large canals if night irrigation is not practiced. Waste of irrigation water at night could be reduced by avoiding/reducing irrigation at night. Methods to prevent or reduce irrigation and to improve irrigation are briefly indicated. The discussion is based on the experience of irrigation managers and the opinions of farmers. Field research is needed to quantify the waste of irrigation water during the night and to find solutions to related social and managerial problems. The peak irrigation demand can be reduced by dividing the command area into groups and staggering the crop calendar over the groups. The use of the group concept in system design results in lower capacities of the canal network at head. During the operation stage, it helps equitable sharing of deficit water supplies. Irrigation scheduling based on crop water

U. C. Chaube et al., *Canal Irrigation Systems in India*, Water Science and Technology Library 126, https://doi.org/10.1007/978-3-031-42812-8_9

Fig. 9.1 Outlets, command area of outlet and watercourse, and field ditch in the project command area

demand is influenced by factors such as depth of root zone, extraction pattern, and the relation between moisture content, stress, and crop yield. The details about Cropwat Software for Irrigation Schedulling is also provided

9.1 Introduction

Equity in the distribution and adequate, assured, and timely water supply are the essential characteristics of good project management. Irrigation Department operates the main conveyance network from headwork (reservoir, barrage) to outlets (headgates). Below outlets, the water distribution is managed by farmers. Outlets are usually located on distributary and minor canals. Conventionally, the outlets are numbered from head to tail of the canal along with the reference of the right (R) or left (L) position of the outlet command area. The schematic depiction of the water conveyance network below the outlet is shown in Fig. 9.1. Thus, the outlets in the figure are 1L, 2R, 3L, 4R, 5L, 6R, 7L, and 8R. An outlet command area (command area of outlet 3L) and the layout of the watercourse and field drain in an outlet command area (command area of outlet 8R) are also depicted in the figure.

The layout of the main conveyance network and associated structures of a tank irrigation project is schematically depicted in Fig. 9.2 (Maheri Tank Irrigation Project in Madhya Pradesh, India). Depending on the outlet's location in the command area, the outlet command may receive too much/adequate/little or no water at all (Fig. 9.3). In general, outlets in the head reach receive too much water, and outlets in the tail reach receive too little water.

9.2 Water Delivery Methods

The water delivery methods can be classified as:

(1) Continuous delivery,
(2) Delivery on-demand, and
(3) Rotational delivery.

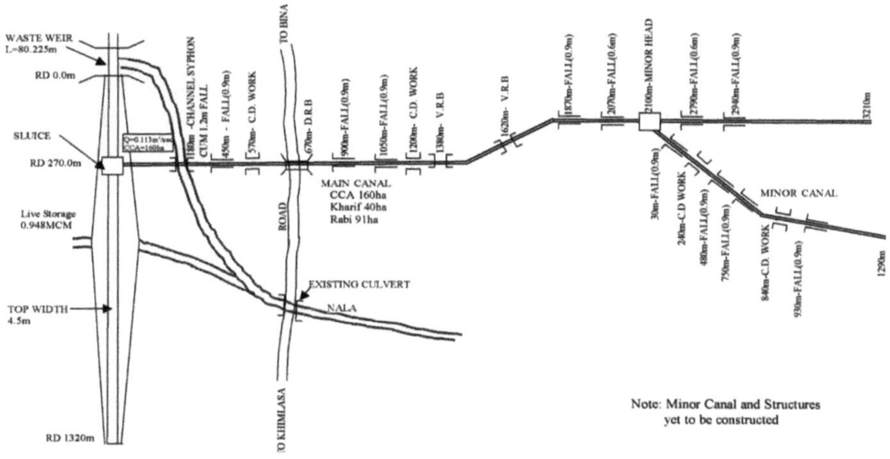

Fig. 9.2 Main conveyance network of Maheri tank irrigation project. *Note* C.D.: cross drainage; V.R.B.: Village road bridge; D.R.B.: District road bridge; CCA: Cultivable command area

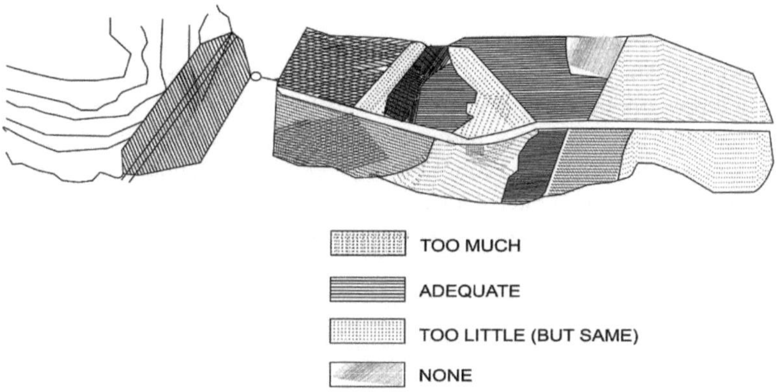

Fig. 9.3 Too much/adequate/little or no water at all in outlet commands

9.2.1 Continuous Delivery

Too large, concentrated cropped areas, such as those held by trusts or large corporations, water delivery can be done continuously instead of by rotation. Also, when the farmers are few and/or the supply is more than the demand (as in the rainy season), the system can be operated on a continuous basis.

Continuous delivery can be used efficiently if proper drainage at the farm level can be ensured. Water supplied on a continuous basis can then be redistributed to different fields according to the soil and crop conditions. The canal's operation under continuous flow is simpler if the canal is small in length and the water level is maintained at the design level above the outlet. This method is not suitable when the holding is small, and irrigation water is to be supplied to a large number of farmers within an outlet command.

Under this system, though each farmer can continuously receive his supplies, he may use his share of water intermittently according to the crop's requirement. Therefore, the channel should have a number of escapes to dispose of surplus water in the channels when the demand for water is less than anticipated.

9.2.2 Delivery on Demand

Continuous water delivery and rotational water delivery are supply-based water delivery methods, whereas the method of water delivery on demand is based on crop water demand. This method allows a farmer to receive supply of water according to crop water demand but is limited to the capacity of the conveyance system. When the water rate is high, the demand method is preferred, as it enables to use the water economically and more closely according to crop requirement. During peak demand, when the system cannot supply water simultaneously to the farmers according to their requirements, the demand method can still be adopted but in a modified form by introducing rotation. For the rest of the irrigation season, however, delivery on demand can be adopted. The system is not generally suitable for most parts of India, as it requires a large contingent of well-trained staff for operation. During the period of preparation and transplanting of paddy, the demand method has an edge over other methods when a large flow for short periods is required, and demand of each cannot be foreseen, and the area under paddy is only a small percentage of the total area. Further details of water delivery on demand are given in Sect. 9.9.

9.2.3 Concept of Rotation Delivery

A rotational delivery system could be defined as a system of equitable water distribution by turns according to a predetermined schedule specifying the day, time, and duration of supply to each farmer in proportion to his area in the outlet command. The rotation in each outlet command is integrated with the rotation in the canal system above outlet. Rostering of minors and pipe outlets is the responsibility of the government, but the rotation below the pipe outlets is managed by the farmers according to the specified schedule. The schedule may be changed in the emergency but will be resumed no sooner after the emergency is over. The rotation delivery system would result in the following improvements:

i. Discipline and economy in the use of water.
ii. Reduction in the wastage of water and additional coverage of area under irrigation.
iii. Farmers' awareness of legitimate share of water.

iv. Farmers' keenness to receive the water timely and their understanding of the significance of proper land development and water application.
v. Avoidance of water disputes.
vi. The upward shift in agricultural production.

Warabandi/Osrabandi/Shejpali: Rotational water delivery is known as Warabandi and Osrabandi in the northern part of India. It is a simple traditional way of ensuring discipline in using irrigation water from the surface irrigation project. Warabandi has been used traditionally in the states of Punjab, Haryana, Western Uttar Pradesh, and in-district Ganga Nagar in Rajasthan for decades. In these areas, the allocation of a fixed period each week for water application is calculated based on the area of the farm irrespective of crop grown and the distance from the outlet. In Maharashtra, Karnataka, and Gujarat, water distribution is also rotational (known as Shejpali), but the allocation of water is based on cropped areas sanctioned for irrigation. However, there are many irrigation systems in India where Warabandi has not been followed effectively, resulting in inefficient irrigation water use.

Example—Problems Before Implementation of Rotational Water Delivery in Chambal Project

The irrigation system in the Chambal project in Rajasthan suffered from the following problems before the implementation of rotational water delivery.

1. It was found that in a typical 14 km long minor canal that was broadly surveyed, water never reached beyond about 8 km. Irrigation was available at FSL only in the head reach, and water fell lower and lower than the FSL as one went down the minor canal.
2. As the water was not available at FSL, farmers resorted to raising the water level by placing obstructions in the minor canal, thus disrupting the entire system.
3. The distribution of water was arranged through pipe outlets. As FSL was not ensured, the full quantity of water could not be delivered to the chaks (outlet command areas). The farmers resorted to installing additional pipes, large pipes, or holes without pipes to ensure large water flow.
4. Farmers close to minors installed their unauthorized outlets or cuts in the embankment for a direct supply of water from the minor canal without utilizing the watercourses laid down by the Irrigation Department.
5. When not needed by the farmers in the head reach, the water flow was observed to flow into the borrow pits and the minor canal instead of the pipe outlet being closed.
6. All the fields were left open for the free flow of water without any regulation. The head reaches received much more water than needed for crops, resulting in waterlogging and salinity in their fields, while fields in the lower reaches were starved for water. It was found that a large number of fields at the tail end received no water at all in a year of water supply.
7. Without assured water availability, the farmers hesitated to risk sowing high-yielding crop varieties or investing in inputs, resulting in a huge loss in production compared with the potential.

9.3 Water Delivery Parameters

There cannot be a uniform design specification of rotational water delivery for all irrigation projects, cropping patterns, soil type, agro-climatic conditions, and farmers' field irrigation practices that vary from one irrigation project to another. The following parameters must be considered in the design and implementation of the rotational water delivery schedule at the field level.

9.3.1 Water Allowance (WA) and Duty

Water allowance (WA): is the continuous flow requirement during a period to irrigate a unit area (based on cropping pattern and crop water requirement). It is expressed in liters per second per hectare. Design WA is the water allowance for which the system is designed. For each outlet, minor, distributary, branch, and main canal, we can ascertain from the project report the maximum water allowance in litres/second/hectare of irrigation area for which the system is designed.

During the operation stage, the WA is modified based on actual conveyance capacity, sanctioned or as existing cropping pattern, and irrigation efficiencies attained/realistically attainable. The WA in the operation stage should not exceed its design value.

Duty: is defined as the area that can be irrigated with unit discharge continuously supplied during a period. Therefore, by definition:

$$Duty = \frac{1}{Water\,Allowance(WA)}$$

Design duty and its value in the operation stage depend on the same factors as does WA.

9.3.2 Stream Size

The stream size is the rate at which water is delivered at the outlet. The stream size should be such that it can be easily handled by farmers, considering the size of their holdings, crops grown, and soil characteristics.

In the case of sandy soil in the flat area, higher discharge (up to 70 L/s) can be handled, but in heavy soil with a low infiltration rate and rolling topography, a smaller discharge should be preferred.

However, the appropriate stream size is likely to be in the range of 20–40 L/s. The outlet command's field channels should have adequate capacity to carry peak period flow for a design interval of irrigation inclusive of the conveyance losses.

9.3.3 Rotation Period

The rotation period is the time interval between irrigations. The rotation period depends upon the depth of water applied and daily consumptive use by the crops grown. Different crops require water at different intervals. Crops, such as wheat, require a fortnightly water application, and tobacco even requires a tri-weekly water application.

The rotation period should also be based on the crop most sensitive to moisture stress (as more than one crop is raised in the outlet command) and the period of highest consumptive use (as the consumptive requirement varies with the stage of growth). For convenience in operation and regulation, it is desirable if the interval of irrigation is one week (days are easier to remember than dates) or its multiple.

The following formula relates the depth of water in one irrigation and rotation period:

$$W = \frac{QIT}{A}(1 - L)$$

where

W the water depth of one irrigation in meters; it usually varies from 0.05 to 0.10 m;
Q the outlet discharge in cumecs;
A the area irrigated in m^2;
I the interval (frequency) for irrigation or the rotation period in days; it is usually a week or its multiple;
T the time of irrigation per day in seconds; and
L the conveyance loss in farm distribution in decimal.

9.3.4 Example: Water Allowance Computation

In a canal irrigation project, sanctioned crops for irrigation are wheat, potato, tobacco, oil seeds, and vegetables. The average duration of the crop season is 120 days, farm distribution efficiency is 80%, and field application efficiency is 85%. This example illustrates the computation of:

(i) Seasonal and fortnightly water requirements and water allowance;
(ii) Seasonal and fortnightly duty;
(iii) Water allowance at canal head and outlet level; and
(iv) Area irrigated considering water allowance, stream size, and days for which canal is running

Seasonal and fortnightly water requirements are computed, as shown in Table 9.1.

Seasonal and fortnightly water allowance (WA) is calculated as below:

Seasonal WA is the flow requirement per hectare on a seasonal basis

Table 9.1 Computation of seasonal and fortnightly water requirements

Crop	Area (%)	Seasonal water requirement (mm)	Peak fortnight water requirement (mm)	Seasonal volume of water required per hectare (m^3/ha)	Peak fortnight volume of water required per hectare (m^3/ha)
Wheat	50	450	116	2250	580
Potato	20	525	115	1050	230
Oil-seeds	10	375	71	375	71
Vegetables	10	600	104	600	104
Tobacco	10	525	102	525	102
Total				4800	1087

$$\text{Seasonal WA} = \frac{\left(4800\,\text{m}^3\right) \times (1000\,\text{L/s})}{(120\,\text{day}) \times (24\,\text{h}) \times (3600\,\text{s}) \times 0.8 \times 0.85}$$
$$= 0.680\,\text{L/s/h}$$

Similarly, fortnightly WA = flow requirement per hectare on a fortnightly basis:

$$\text{Fortnighly WA} = \frac{1087 \times 1000}{15 \times 24 \times 3600 \times 0.8 \times 0.85};$$
$$= 1.23\,\text{L/s/ha}$$

taking 1 cusec = 28.30 L/s

Fortnightly duty (area irrigated by 1 cusec continuous flow for 15 days) = 28.30/1.23 = 23.008 ha/cusec or 821.40 ha/cumec.

Similarly, seasonal duty (area irrigated by 1 cusec continuous flow for 15 days) = 28.30/0.680 = 41.55 ha/cusec or 1484 ha/cumec.

Changes can be made to cropping patterns (allocation of crops for irrigation) in such a way that the flow requirement (L/s/ha) does not exceed the designed value of water allowance. The actual conveyance efficiency must be considered in the computation.

The computed water allowance (L/s/ha) will be highest at the head of the main canal and lowest at the outlet because of water losses in the conveyance network, e.g.,

Water allowance at canal head = 0.75 L/s/ha

Conveyance efficiency from canal head to outlet = 80 %

So, water allowance at the outlet = 0.75 × 0.80

= 0.60 L/s/ha

This means that if, at the outlet, water is released continuously at 0.60 L/s/ha of irrigated land, the farmer will be receiving 0.60 L/s at the outlet.

If stream size is fixed at 30 L/s, the area that can be irrigated = $\frac{30}{0.6}$ = 50 ha with continuous supply for 14 days.

If the canal is running for 7 days and then closed for 7 days, the area irrigated in 1 rotation schedule for 14 days will be 25 ha.

9.4 Formulation of Time Schedule

9.4.1 Unit Time and Basic Time

The allocation of the volume of water to each farmer is proportional to his farm area in the outlet command. The water availability per unit area of the outlet command per rotation is called "unit time".

$$\text{Unit time} = \frac{\text{Time of irrigation water availability per rotation}}{\text{command area}}$$

For weekly rotation, the time of irrigation water availability is $24 \times 7 = 168$ h.

For two-week rotation, it will be $24 \times 7 \times 2 = 336$ h.

A weekly rotation with an irrigation schedule only during the day (12 h) will mean 84 h of water availability.

Basic time of irrigation allocated (hour) to a farmer = unit time (h/ha) $\times$ farm area (ha).

9.4.2 Conveyance Loss and Loss/Gain in Irrigation Time

The losses of water during conveyance and travel time (transport time) are directly proportional to the length of the channel to be traversed by water before reaching a farmer field. Adjustment needs to be made in the basic time allocated to a farmer (computed as above) for conveyance loss, loss of transport time or filling time, and gain in irrigation time during the emptying of channel.

Coveyance loss: Water lost due to seepage, leakage and evaporation during conveyance. In north India it is known as *Chusai* loss. The conveyance losses are to be allocated equitably to all farmers in accordance with the distance of their field from the outlet.

Loss of transport time or filling time: This refers to the time required for attaining the full supply level in the channel. In north India, it is known as *Bharai* loss. This time loss is debited to the common pool.

Gain in irrigation time during emptying of the channel: The volume of water available in the channel when the gate at the head of the channel is closed, provides additional irrigation. In north India, it is known as *Jharai*.

Consider field A and field B in Fig. 9.4. Field A is being irrigated first, and field B is to be irrigated after field A. Conveyance loss during supply to field B depends

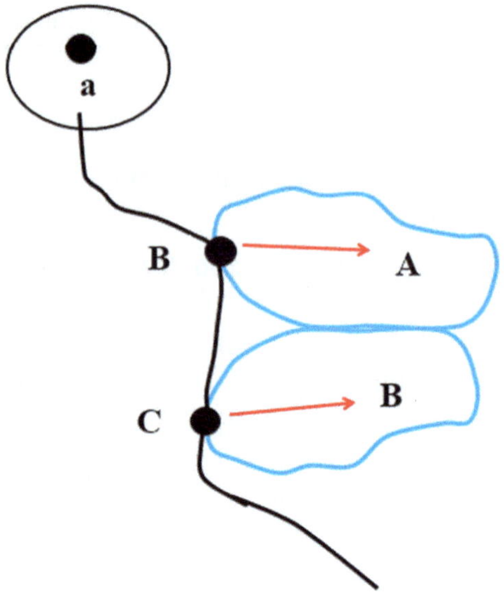

Fig. 9.4 Farm A and farm B

upon the channel length denoted as a-C. The travel time loss depends upon length B-C. The length B-C will depend upon the sequence of turn. The travel time loss will depend upon the flow velocity (or slope) and length. The adjustment for conveyance loss and travel time loss is in time units.

The last shareholder may have enough water volume (in the channel itself) for his field even if the supply is shut off at outlet 'a'. A schedule indicating the day and time from the beginning to the end of irrigation for each farmer is then worked out. An essential input in implementing Warabandi is to seek, secure, and maintain beneficiaries' participation to the fullest extent possible.

9.4.3 Example: Computation of Outlet Capacity and Canal Capacity

This example illustrates a procedure to compute the capacities of outlets and canal capacity considering seepage loss.

A minor receives water for 3 weeks on a 'on' and 1 week 'off' basis The proposed rabi irrigation intensity is 75% in the command area of 118 ha (ha). The conveyance loss is 10% of diversion in a minor canal for 700 m length. Irrigation efficiency below the outlet is 65%.

There are four outlets on this minor, as per the details below. All outlets receive water simultaneously.

Outlet number	Distance from head of minor (m)	Command area of outlet (hectare)
1/R	200	26
2/L	400	37
3/R	550	31
4/L (tail outlet)	700	24

Crop	Cropping pattern (%)	Monthly water requirement (mm)	Water requirement (m^3/ha)
Wheat	25	202	505
Potato	7	200	140
Vegetables	7	167	116.9
Tobacco	40	166	664
Cotton	3	212	63.6
Lucern	2	177	35.4
Total	84		1524.9

Volumetric water requirement of wheat crop in 1 ha command area is = 0.25 × 202 × 10,000/1000 = 505 m^3/ha. Similarly it is computed for other crops.

Total water requirement with 84% irrigation intensity is 1524.9 m^3/ha.

Therefore, water requirement with 75% irrigation intensity = 1524.9 × 0.75/0.84 = 1361.52 m^3/ha.

Monthly water allowance at field level = (1361.52 (m^3/ha) × 1000 (litres))/(31 days × 24 h × 3600 s) = 0.5083 lps/ha.

Monthly water allowance at outlet level = 0.5083/0.65 = 0.78 lps/ha.

Monthly water allowance at the outlet based on actual running days = 0.782 × 31/21 = 1.1544 lps/ha.

Computation of Outlet Capacities

Capacity of outlet 1/R = 26 × 1.1544 = 30 lps.
Capacity of outlet 2/L = 37 × 1.1544 = 42.7 lps say 43 lps.
Capacity of outlet 3/R = 31 × 1.1544 = 35.8 lps say 35 lps.
Capacity of outlet 4/L = 24 × 1.1544 = 27.7lps say 28 lps.

Computation of Seepage Loss in the Minor Canal and Its Capacity at Head

Outlet number	Distance from head of minor (m)	Outlet discharge (capacity) (lps)	Seepage loss in proportion of 10% for 700 m length (lps)	Discharge required at head of canal (lps)
1/R	200	30	30 × 0.1 × 200/ 700 = 0.86	30.86

(continued)

(continued)

Outlet number	Distance from head of minor (m)	Outlet discharge (capacity) (lps)	Seepage loss in proportion of 10% for 700 m length (lps)	Discharge required at head of canal (lps)
2/L	400	43	$43 \times 0.1 \times 400/700 = 2.46$	45.46
3/R	550	36	$36 \times 0.1 \times 550/700 = 2.83$	38.83
4/L (tail outlet)	700	28	$28 \times 0.1 \times 700/700 = 2.8$	30.8
Total		137	8.95	145.95 say 146 lps

Thus, the total seepage loss is 8.95 lps, and the capacity of the minor canal is 146 lps.

Adjustment for Filling Time

Since all the outlets have to receive water simultaneously, the filling time (FT) is required to bring the canal water level to the full supply level (FSL). So the canal has to be opened in advance by FT hours before outlets are opened.

Adjustment for Emptying Time

Adjustment for emptying time is possible only if there are regulating gates in the minor canal at locations just downstream of each outlet.

For example, the gate downstream (d/s) of outlet 3/R is closed t hour ahead so that water available in the canal reach from 550 to 700 m is emptied in this time, and this water is let into outlet 4/L for the duration of t hours.

Tutorial Exercise: *Workout water allowances at the outlet level, outlet capacities and capacity of the canal at its head.*

S.No	*Crop*	*% area*	*Peak fortnightly requirement* (mm)
1	*Wheat*	*30*	*85*
2	*Vegetable*	*5*	*65*
3	*Tobacco*	*45*	*95*
4	*Cotton*	*10*	*108*

Outlet No	*Outlet area* (ha)	*Distance from head of canal* (m)
1R	*35*	*200*
2L	*30*	*400*
3R	*37*	*650*

Desired irrigation intensity = 80%
Watercourse conveyance efficiency = 75%
Field application efficiency = 90%
Seepage loss in canal = 10% of diversion per 1000 m length of the canal.

9.4.4 Example: Block (Group) Based Water Delivery in a Canal (Indonesia)

This case study of Celutung canal (Indonesia) explains rotational water delivery among blocks (Suleman 1996). The command area is divided into six blocks. Each bock has 6–10 farmer fields. A stepwise procedure for rotational planning among the blocks is discussed below.

Table 9.1 Design details of cilutung irrigation system

Name of project	Cilutung
Rotation period	7 days
Day irrigation	14 h each day
Number of blocks	6
Conveyance losses	30% per 1000 m
Total time of irrigation water availability	$7 \times 14 = 98$ h
Total area	335 ha
Unit time	$98/335 = 0.29$ h./ha
Basic time allocated in hours: (column 8) For example	
Block CL1	$50 \times 0.29 = 14.50$
Block CL2	$29 \times 0.29 = 8.41$ etc
For slope 1:1000	Velocity of flows is 0.40 m/s
Column 9 Transport time in hours	Column 7/0. 40
Column 10: Total transmission losses in %	(col7) $\times$ 30/1000
Column 11: Total transmission losses (In hours)	$\frac{column\,10}{100\,\mathrm{m}} \times column\,8$
Column 12; Total time required compensated in hours	$Column\,8 + column\,11$
Column 13: Adjusted time per plot/block in hours:	$\frac{Total\,column\,8 - Total\,column\,9}{Total\,column\,12} \times column\,12$
Column 14: Final time per Plot (block) in hours:	$Column\,9 + column\,13$

Step in Designing the Time Schedule

Step 1 Obtain the ownership plan and design plan of the pipe outlet command area, including field channel layout and distribution boxes.

Step 2 Form groups within the command area having 6–10 farmers in a group. Discuss grouping with farmers. (columns 1 and 2 of the table).

Step 3 The area of each plot has to be earmarked. If one farmer holds more than one plot within a group, each plot has to be dealt with separately (Column 3).

Step 4 Since water requires time to travel from one irrigation box to the next, this time is shared as loss by all farmers in the command area. The traveling time is a function of slope (column 4).

Step 5 Losses during transport (seepage, spill, losses from boxes etc.) of water to the individual field also have to borne by all farmers (column 5).

Step 6 The slope of each field channel stretch is taken from the map (column 6).

Step 7 For easier calculation, the slope of 1:1000 has been adjusted as if it were a slope of 1:800. The adjusted length is 1.5 times longer if a slope of 1.1000 is considered because the losses are 1.5 times higher (column 7).

Step 8 The transport time is calculated now by taking the appropriate slope into consideration (column 9).

Step 9 Each farmer in the command area is allocated a 'basic time' equal to the time of irrigation available divided by the total number or acres ha in the command area. Since the irrigation is performed for 14 h a day in 7 days, in a week 98 h are available for Irrigation.

Step 10 The transport time can be calculated by using the previous length of field channel (column 9).

Step 11 The transmission losses during the transport of water in the field channel is 10% for a slope 1:500 and 15% for a slope 1:1000.

Step 12 Following the determination of transmission losses in percent, these have to be converted into time.

Step 13 The calculation now ignores for some time that only 98 h per week are available for irrigation and each individual farmer is compensated for the time losses due to transmission losses (column 13).

Step 14 Since every plot is compensated for the time for transmission loss, the transport time of column 9 can be added to column 10 resulting into column 14. This is the final time.

Step 15 Column 15 converts decimal hours of column 14 into hours (step 16). All that remains to be done now is to design the time schedule.

Calculation of time schedule for rotational water delivery among blocks is shown in table below.

Table 9.1 Calculation of time schedules for rotational water delivery

S. No.	Name of block	Area in ha	Length of previous wet D.B in M		Slopes of previous D.B	Adjusted length for calculation in m	Basic time allocated in hours	Transport time in hours	Total transmission losses in %
			Individual	Cumulative					
col1	col2	col3	col4	col5	col6	col7	col8 = col3 × 0.29	col9 = col7/0.4	col10 = (col7) × 30/1000
1	CL.1	50	160	160	0.0005	160	14.50	0.11	4.80
2	CL.2	29	220	380	0.0005	380	8.41	0.26	11.40
3	CL.3	98	340	720	0.0005	720	28.42	0.50	21.60
4	CL.4	64	420	1140	0.0005	1140	18.56	0.79	34.20
5	CL.5	59	180	1320	0.0005	1320	17.11	0.92	39.60
6	CL.6	35	250	1550	0.0005	1570	10.15	1.09	47.10
	Total	355					97.15	3.67	

S. No.	Name of block	Total transmission losses in hours	Total time required compensated in hours	Adjusted time per plot in hours $\frac{\sum 8-\sum 9}{\sum 12} \times 12$	Final time per block Considering 14 h irrigation per day	
					Hours	Day and time
1	2	col11 = col8 × col10/100	col12 = col8 + col11	col13	col14 = col9 + col13	col15
1	CL.1	0.70	15.20	11.57	11.68	5 am to 4:30 pm on Monday
2	CL.2	0.96	9.37	7.13	7.39	4:30 pm to 7 pm on Monday and 5 am to 10 am on Tuesday
2	CL.3	6.14	34.56	26.30	26.80	10 am to 7 pm on Tuesday and 5 am to 7 pm on Wednesday and 5 am to 9 am on Thursday
4	CL.4	6.35	24.91	18.95	19.74	9 am to 7 pm on Thursday and 5 am to 3 pm Friday
5	C L.5	6.77	23.88	18.17	19.09	3 pm to 7 pm Friday and 5 am to7 pm on Saturday and 5 am to 6 am on Sunday
6	CL.6	4.78	14.93	11.36	12.45	6 am to 7 pm on Sunday
	Total	25.70	122.85	93.48	97.15	

9.5 Examples of Rotational Water Delivery in India

Ministry of Irrigation, Government of India, had brought out a Manual of Warabandi in Irrigation Project (GOI 1992). This manual provides a detailed description of rotational water delivery in India along with case studies.

9.5.1 Group Warabandi in Pochampad Project in Telangana

A new approach to weekly Warabadi has been developed in the Pochampad project command area in Telangana (GOAP 1979). The system is based essentially on weekly water distribution on a wet or day basis, similar to the traditional Warabadi system in North India. The following improved features, however, have resulted in a great deal of improvement in working.

1. **Group Warabandi**: The individual farmers have been grouped into convenient groups of 6–8 farmers, with an average of 8–10 groups per chak. Besides the indication of individual wari, the waris of the group also have been fixed. The wari is primarily enforced between the groups, the internal wari for individual farmers within a group being left to be managed by the group. Group leaders are elected to serve as members of the pipe or chak committee. This group mechanism has been found to be effective for the enforcement of warabandi, thus eliminating the irrigation department's role in the enforcement of warabandi among individuals. The flexibility among individuals within the group is permitted.
2. **Signboard indicating Warabadi**: Instead of the parchas (pamphlet) distribution to the individual farmer, the warabandi in the Pochampad project area has been indicated on the signboard installed at every chak or pipe outlet. This ensures that warabandi is permanently published for information to all concerned and is an improvement on individual parchas.
3. **Identification of group**: Each group of the Pochampad project command area has been assigned its own colour. This colour is indicated on the Warabandi board on the one hand and all division boxes on the other. This has helped with easy understanding by farmers about the flow of water to the group at the given time per hour according to colour assigned to the group. For the convenience of understanding the warabandi arrangement by the concerned farmer, the group formation could be considered broadly according to the day of the week. The groups were numbered I–VII from Monday to Sunday.

These improvements made the enforcement of warabandi much more effective and easier. The system is easily replicable and can be introduced without too much difficulty.

9.5.2 Traditional Osrabandi in Upper Ganga Canal System

Osarabandi is a method of distributing the shares of water of outlets among cultivators. Northern India Canal and Drainage Act 1973 confers the right of distribution of water in the outlet command to the beneficiaries, who are supposed to fix and apportion their shares by mutual arrangement. If they disagree, they apply to Executive Engineer (EE) for Osarabadi under the Act, which lays down its enforcement rules. EE can also promulgate Osarabandi at his initiative.

Osarabandi, once enforced, is applicable for 10 years. Revisions can be done when there is a change in the outlet or alteration in command or when cultivated land becomes uncultivated or vice-versa. Zilledar (an irrigation department official) prepares Osarabandi under the supervision of the Deputy Revenue Officer. All formalities are normally completed within 3 months. Details of Osrabandi are given in GOI (1992). The procedure is briefly discussed below.

Procedure:

1. Fix the command under the outlet after the spot visit and approval by Division Canal Officer. Areas unfit/debarred or belonging to those who do not want irrigation water are excluded.
2. Prepare a map showing village boundaries, outlet location, field channel alignment, canal irrigated, and well-irrigated fields.
3. Prepare Form 57 H. This form indicates the number and area at each field, and the owner or tenant's name.

Table 9.1 Example: form 57 H

Village Pergana Dist	Field No. as per settlement	Area (acre) flow lift	Name of cultivator	Included in Wara-bandi	Irrigated by canal	Irrigated by other sources	Cultivated (Un-irrigated)	Un cultivated	Un-fit	Remarks
Muzaffarnagar	1825	0.030	Water course	–	–	–	–		0–30	–
	1620	0.015	Chack road	–	–	–	–		2–15	–
	1827	0.350	Nor Mohamm	2–35	2–35	–	–	–	–	–

4. Prepare Form 58 H. Only those fields which are included in Osarabandi are listed. Field numbers are arranged in serial order by the villagers. Thoks and cultivators and abstract by villagers, and thoks have the same twelve hours for irrigation. The members of one family or tenants of adjacent fields form one thok and select their thokdar. The consent of cultivators included in a thok is taken in writing.

Table 9.1 Example: form 58 H

Name of village	Name of thokdar	Name of cultivator	Details of the field with area		Cultivator wise total area (acre)		Thokwise total area (acre)	
			Flow	Lift	No. of field	Area	No. of fields	Area
Kutesra	Normuha-mmad	Shrfoo S/ o Budhu	$\frac{1228}{2.47acre}$	–	1	2.47	3	10.6
	Jabar Din	–	–	–	–	–	4	12.6
	Naseebu	–	–	–	–	–	10	16.85
	Ismail	–	–	–	–	–	10	10.85
			Total		**31**	**50.9**	**31**	**50.9**

5. Zilladar sends Forms 57 H and 58 H and his detailed report to the Deputy Revenue officer, who calls a meeting of all cultivators at a convenient place. Necessary corrections/modifications are made, if required.

Form 57 H, as finally drawn-up, is signed by cultivators and thokdars. If any tenant disagrees, the case is submitted to Divisional Revenue Officer for order. A copy of approved Form 58 H is pasted at a conspicuous place in each village (see sample farm).

6. The Deputy Revenue officer or Zilladar prepares Form 59 H (see sample farm).

Table 9.1 Example: Form 59 H

Name of Thok	Name of Thokdar	Name of cultivator no his field, time allocated and area	Resident			Name of the village	Detail of Sara Part A				Total Time	
			Vill	Perg	Dist		From		To		hour	Min
							Day	Hour	Day	Hour		
1	Noor Muhammed	See note below@					PM	1.09	AM	0.07	34	58
2	Jabar Din										41	34
3	Naseebu										55	35
4	Ismail										55	53
						Total					**168**	**0**

Note @ Sharfoo S/o Buddu (no of field 1, time 8 h. 9 min, area 2.47 Acre)

7. Final parcha is prepared and issued to thokdars (see sample farm).

Example: Final Purcha.
By Order of Executive Engineer- Division Muzaffarnger Ganga Canal,
Mathura Minor, Left Bank, Outlet No. 19, Location Mile 3, Furlong1, Feet 3,
Name of Thokdar: Nor Muhammad;Thok Area 10.60 Acre.
Rotation: 7 Days … No. of Thok…0.1…

No. of Thok	Name of Thokdar	Area Acre	Time		Part A		Part B	
			Hour	Min	Form	To	From	To
1	Sharfoo S/o Buddu	2.74	8	9	1.09 PM Thursday	0.07 AM Saturday	2.11 PM Wednesday	1.09 AM Thursday
2	Noor Muhammed S/o Buddu	2.35	7	44	–	–	–	–
3	Gafoor S/o Umriddin	5.58	19	5	–	–	–	–
	Total	**10.67**	**34**	**58**				

9.5.3 Improved Warabandi in Upper Ganga Canal System

Kartaatmaja (1994) has carried out a study on the improvement in the design and implementation of Rotational Water Delivery on Lakhnauta Minor. The proposed rotational water distribution schedule (RWS) is an improved version of the warabandi method, popular in Northern India. Under this system, all farmers secure their fair share of water in accordance with their size of land holding, irrespective of their position on field channel, and irrespective of their social and economic status. Irrigation efficiencies improve on account of specified time schedules and therefore more area can be brought under irrigation by making productive use of water which otherwise was wasted in the traditional method. In the proposed RWS, irrigation is done from tail to head and water is given at a constant rate of flow on each outlet with fixed area, fixed day, and fixed time.

The farmers are informed about their turn in advance, so there is less scope for mis-use and wastage of water. Special features of improved RWS are compared with traditional irrigation as below:

	Traditional RWS	Improved RWS
1	Outlet capacity is fixed on the basis of thumb rule. Normally pipe outlet is provided having discharge capacity of 1 cusec	The outlet capacity is based on water allowance based on most likely cropping pattern in the outlet area and irrigation efficiencies
2	Irrigation is done from head to tail of the channel	Irrigation is done from tail to head of the channel
3	Conveyance loss and time adjustment during filling and emptying is not considered	Day, time and duration of irrigation of each field are adjusted considering various losses (filling time, emptying time, conveyance losses) and actual cropping pattern
4	Supply to tail end farmers is not assured due to various reasons	Assured supply to all farmers

(continued)

(continued)

	Traditional RWS	Improved RWS
5	Pilot project study is normally not done to understand specific problems of the area	Introduction is based on pilot project study of representative minors/chaks

The following steps have been followed for the preparation of improved RWS:

(a) Work out the evapotranspiration requirement (ET crop) for each crop daily.
(b) Work out fortnightly the net irrigation water requirement for proposed cropping pattern (NIR).
(c) Calculate field irrigation requirement (FIR) at the outlet.

FIR = NIR/ea. Application efficiency (ea) is determined through field tests.

(d) Work out area irrigated by 1 cusec (28.311 lps) stream size continuously available in a fortnight.
(e) Work out the required discharge in LPS/ha.
(f) Work out basic time required per hectare, assuming continuous flow for 168 h during the week (day and night). It is assumed that the fortnight requirements will be met in one week duration.

$$Basic\ time = \frac{168}{Chak\ area(\text{ha})}$$

(g) Allocate the time to each field in accordance with the field area and the basic time calculated above.

$$Allocated\ time\ for\ a\ field = Field\ Area \times Basic\ time$$

(h) Calculate Jharai time (TJharai) i.e. the time equivalent for the volume of water which a farmer gets due to emptying of the water in the length of the watercourse/field channel into his field when it is closed at the u/s of that length by the other cultivator for his turn. It is calculated through the field experiment, usually Jharai time is 3 min/67 m.

$$TJharai\ for\ a\ field\ (\text{min}) = Jharai\ length\ (\text{m}) \times Jharai(3\ \text{min}\ /67\ \text{m})$$

(i) Calculate transmission losses (Tchusai). It is the percentage basic time lost per 100 m run of water while irrigating a field.

Let L be the length of the field channel from turnout to field outlet (m)
Let S be the percent loss of basic time per 100 m length of the channel

$$Tchusai(\text{min}) = basic\ time(\text{min}) \times L \times S/100$$

(j) Compute the time required for each field (R Time)

$$R\,Time = Basic\,Time + T\,chusai$$

(k) Compute Filling Time T Bharai. Bharai is defined as the time required to fill a 67 m length of water course/field channel.

$$T\,Bharai = (Distance\ of\ turnout\ from\ the\ head\ plus\ the\ distance\ of\ field\ outlet\ from\ turnout) \times Bharai/67$$

Once a watercourse is filled up to the tail, it will not require bharai time unless it bi-furcates in between. But when one sub-chak gets irrigation water after the other chak, it needs filling of its field channel length comprising of the distance of field outlet from the turnout. Therefore, this also needs to be accounted for in Tbharai. So this is also computed for each field.

(l) Compute the adjustment factor (Factor) for the entire week.

$$= (168 - Tbharai + Tjharai)/R\,Time$$

(m) Compute adjusted time (A time) for each field

$$A\,Time = R\,Time + Factor$$

(n) Compute find time (F time) for each field F Time

$$= A\,Time + T\,Bharai - T\,Jharai$$

(o) Compute the corrected time (C Time) by rounding the F Time to its nearest five minute. Usually, day and time for the start of the rotation period is 8 AM on Monday. So add C Time to the starting time of the rotation to allocate the day and duration of irrigation to each field in sequence.
(p) The schedule is advanced by 12 h every year so that the farmers who feel the hardship of night irrigation during a year get their turn for irrigation in day time.

9.6 Preconditions for Success of Rotational Water Delivery

The success of rotational water delivery below the outlet depends on the following preconditions.

1. Accurate identification of land holdings and landholders for the consolidation of scattered or small holdings. With many small fragments, a warabandi schedule is liable to be complicated to draw up and difficult to implement.

2. The watercourse/field channel should take water up to the farmer's field and no field to field irrigation should be allowed.
3. There should be stable water scarcity. When water is not scarce there is no need for strict timed turns and no incentive for farmers to enforce them. Moreover, the scarcity must be stable. After heavy rains, farmers do not want to take water. Once a schedule has been disrupted by rain, re-establishing it requires a special effort. Rotational water delivery in non-rainy season (winter and summer crop seasons) is more successful than in rainy season. Scarcity is vital for warabandi at night in day and night (24 h) warabandi. Farmers do not want to irrigate at night if they can avoid it.
4. Day and night time allocations to farmers should be rotated at regular intervals of years.
5. Watercourse losses should be taken into account in the allotment of time. Depending on the water course's soil type and condition, a farmer at the tail end of a 1 km long unlined watercourse may be getting 25–40% less water than a farmer near the outlet due to watercourse losses. He should therefore be given a weightage in time to compensate for the losses.
6. The allotment of basic time on the basis of area of farmer's field gives more benefit of water use to large farmers who are in a position to derive more benefit. The small farmer should be given a weightage in the allotment of time to derive greater benefit per unit area of the small farm size.
7. The Shejpali system's rationale is the optimum utilization of available water for a planned cropping pattern. Left to the farmers choice, they usually go in for high value and high water-consuming crops like paddy and sugarcane. Therefore, prior notification and rigid enforcement of allotted crops are necessary.
8. In the Shejpali system, the farmer is authorized to irrigate a fixed area of a specified crop. The onus of ensuring that the water is not diverted to another crop then falls on the irrigation manager. Attempts are now being made to enforce a fixed turn of specified duration in the Shejapli system also.

9.7 Field Irrigation at Night

The waste of canal water during night is widespread in South and Southeast Asia. Darkness, fear, and desire for sleep deter irrigation staff, farmers, and labourers from activities at night. At the farm level, irrigation at night entails extra labour and costs. The wastage of water in large canals can be reduced by (a) reducing irrigation at night or (b) improving irrigation at night. Chambers (1988) discussed the issues involved in night irrigation.

9.7.1 Visibility and Darkness

Visibility for purposes of field irrigation starts earlier than the time of sunrise and persists even after sunset. The average hours of darkness in North India (Delhi) could range from 9 h 20 min at midsummer to 13 h at midwinter, and in South India (Chennai) darkness could be from 10 h 30 min to 12 h. Therefore, in the situation where canal water is going to waste at night due to no irrigation, there is ample scope for reducing the waste of water at night. In the northwest part of India, where warabandi is practiced at night, water wastage is not so much as in those parts of India where night irrigation is not included in warabandi.

9.7.2 Methods to Reduce Field Irrigation at Night

Reducing irrigation at night can be done either with or without saving water in tanks for subsequent use.

With Water Saving in Storage

Storage within canals: On some canal systems in India, such as Mahi-Kadana in Gujarat and Jayakwadi in Maharashtra, canals are also used for night storage.

Storage in intermediate tanks along canal: Night storage in intermediate reservoirs is an attractive idea. It can reduce night irrigation, save water, and introduce flexibility into management. However, the scope for night storage reservoirs is limited by problems of topography, cost, and operation. On Periyar-Vaigai in Tamil Nadu the new canal system supplies numerous old small tanks, which are sometimes used to retain night flows.

Without Water Saving

(i) Stopping a river diversion at night can eliminate night irrigation. However, it is possible only in a small length of canals.
(ii) Change in rotation delivery schedule on a yearly basis. Those farmers who have their night turn to receive water this year will have day turn to receive water next year.

9.7.3 Methods to Improve Irrigation at Night

Irrigation at Night Can Be Improved in Five Main Ways

Making flows predictable and manageable: Where conditions for warabandi exist, the advantages are obvious. Each farmer knows when his time will come and, therefore when he needs to be up during the night, and does not need to recruit labour to capture and guard his supply. The acceptability of such a system to farmers is confirmed by warabandi in Ganganagar District in Rajasthan in northwest India.

Improving convenience efficiency: Night irrigation is often physically inconvenient and inefficient. These defects can be mitigated in several ways:

(a) Good lighting; and
(b) Installing and maintaining structures, channels, fields, and water-application methods with low requirements for observation and adjustment.

Choosing easy crops: Night irrigation can be improved by choosing crops which are tolerant of flooding and variable water conditions, besides being easy to irrigate. Paddy, trees, and young sugarcane are probably among the easiest.

Zoning for night flows: Where a farm or an outlet receives water throughout 24 h, the day flows can be used (as they usually are) for the more difficult irrigation and crops, and the night flows for those which are easier. In South India (for example, on the Sriramasagar Project) it is common for an outlet's day flows to be used in a more controlled manner for upland crops in red soils near the outlet, and the night flows to go in a less controlled manner into low-lying paddy on black soils further from the outlet.

9.7.4 Need for Field Research on Night Irrigation

The above discussion is based on the experience of irrigation managers and opinion from farmers. More systematic knowledge through field research is required on the following:

Measurement: Research to develop cost-effective 24-h measuring devices, coupled with methods of communication which would help in quantifying the wastage of water. In the Philippines, a measuring device that took automatic readings every 16 min was installed. On its first night, it showed that at 2045 h the level in a channel went up one foot, and then dropped back again to its former level at about 0340. The device was said to have paid for itself one night.

Waste and saving: Assessments of the extent of wastage of water at night, bearing in mind the difference between 'waste' and unavoidable loss, and the benefits of re-use lower down.

Design options, costs, and benefits: Analysis of design options and their costs and benefits related to reducing and improving irrigation at night.

Diagnosis: A diagnostic field study will indicate social and management related problems of the specific area.

9.8 Group System of Water Distribution to Reduce Peak Demand

Water distribution is an important component of irrigation system design and implementation. Each irrigation project has its own specific water distribution plan to achieve the project target. In Indonesia, the water distribution plan is based on a group system. The peak irrigation demand is reduced due to the staggering crop calendar in golongans. For example, in the case of the Batutegi Project (Indonesia), peak demand is 5003 m^3/month/ha (without Golongan System) it could be reduced to 3686 m^3/month/ha by staggering crop calendar in six Golongans. Thus, the concept of Golongan is useful both in irrigation system design and in irrigation system operation. The use of the Golongan concept in irrigation scheduling results in lower capacities of canal network at head and helps in equitable sharing of deficit water supplies during the operation stage.

Figure 9.5 shows the crop calendar and staggering Golongan to lower the peak water requirement for the Batutegi project in Indonesia. It also helps in lowering the peak labor requirement.

In the rainy season, the land preparation in Golongan 1 starts in the second half of October and with a time lag of half a month, the other Golongan starts one by one. In the dry season, paddy land in golongan 1 is prepared in the first half of March by Golongan 1 and 3.

A: Nursery + Land Preparation (1 month).

B: Planting + Growing (2.5 months).

C: Graining (0.5 month).

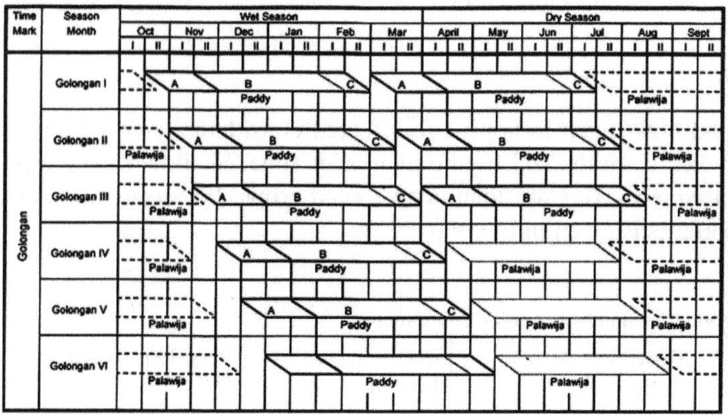

Month	Monthly water requirement (m^3/month/ha)	
	Without system golongan	Apply system golongan
OCT	1529	418
NOV	5003	2761
DEC	3188	3501
JAN	3340	3487
FEB	855	2425
MAR	3551	2198
APR	4147	3686
MAY	3888	3243
JUN	3188	3110
JUL	0	1157
AUG	0	0
SEP	0	0

Fig. 9.5 Crop calendar and irrigation scheduling in groups (Golongan)

9.9 Irrigation Schedulling Based on Demand

Irrigation scheduling means the timing and depth of irrigation application during crop growth as per crop requirement. Thus, it is relevant for water delivery on demand as against rotational water delivery or continuous water delivery which is supply oriented delivery method. Irrigation scheduling on demand is mainly related to

i. Soil moisture content within the effective root zone of a crop and storage capacity within the root zone
ii. Crop evaporation
iii. Relationship between soil moisture content and stress
iv. Relationship between crop yield and soil moisture stress

Irrigation scheduling on demand is influenced by factors such as depth of root zone, extraction pattern, and relation between moisture content, stress, and crop yield. During the operation stage, observed field data can be used to prepare an irrigation schedule. With the availability of software, such as CROPWAT, it is possible to carry out irrigation scheduling on demand in a more scientific manner.

Development of Root Zone

The effective root zone is depth from which the roots of an average plant are capable of extracting soil moisture, which needs to be replaced by irrigation. The rooting depth depends on the nature of crop, but can be influenced by restricting conditions e.g. shallow soil layer or high water table. The effective root zone depths of some of the more common crops are given in Table 9.2.

Measurements at the demonstration farm of IIT Roorkee showed the development of effective root zone for wheat by almost 1 cm/day for the first 70 days, and negligible growth afterwards.

Extraction Pattern

When to irrigate and how much to apply are affected considerably by where and when water is removed from the soil by the plant root zone. Shallow rooted crops will require more frequent irrigation than deep-rooted crops. For normal irrigated soils the average patterns is given in Fig. 9.6.

Table 9.2 Effective root zone depths (on full development)

Shallow rooted 60 cm	Moderately deep 90 cm	Deep-rooted 120 cm	Very deep-rooted 180 cm
Rice	Wheat	Maize	Sugar cane
Potato	Tobacco	Cotton	Citrus
Cauliflower	Groundnut	Soyabean	Apples
Cabbage	Carrots	Sugar beat	Coffee
Onion	Beans	Tomato	Grape vines
	Chillies		Safflower

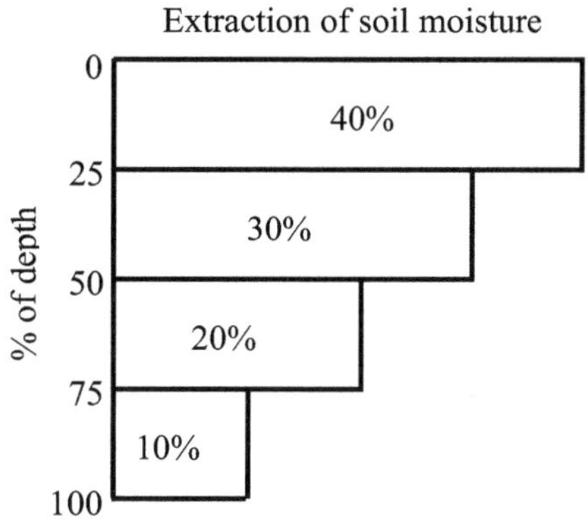

Fig. 9.6 Average extraction of soil moisture by plant roots

Influence of Stage of Growth

The growth of all plants can be divided into three stages with regard to irrigation practice, vegetative, flowering, and fruiting. During the vegetative stage, consumptive use continues to increase. Flowering occurs near and during the peak of consumptive use. Different crops are harvested at different stages, depending on their utilization e.g., Green fodder-and leafy vegetables at the vegetative stage, cauliflower at the flowering stage, tomatoes, green peas, sugarcane, and other fruits at the wet fruit stage, and cereals, pulses, and cotton at the dry fruit stage. The amount of water applied and the frequency of irrigation must be adjusted to the actual consumptive use of crop, the water holding capacity of soil, and depth of rooting. Since consumptive use is at or near maximum during the flowering stage, care must be exercised to ensure adequate moisture in the root zone.

Relationship Between Soil Moisture Content (SMC) and Stress

SMC is related to the soil moisture stress by the soil moisture retention function. A decrease in SMC generally results in an increase in stress. For the same soil moisture deficiency, the stress is higher in clayey soils and less in sandy soils.

Effect of Soil Moisture Stress on Yield

It is generally accepted that up to 50 percent depletion of available capacity, the stress level is such that it does not have an appreciable effect on yield. Also, up to this much soil moisture depletion, the actual evapotranspiration is only marginally less than potential evapotranspiration. In the dry fruiting stage, depletion up to 75% is acceptable. Example: Experimental data on the effect of delay in irrigation on the yield of wheat crops in the demonstration farm of IIT Roorkee is given below.

Soil-clay loam-crop wheat.

Zero stress yield 40 ql/ha (Table 9.3).

Soil Moisture Content

The moisture deficiency Δ at a particular time is given by

Table 9.3 Effect of delay in irrigation on the yield of wheat crop in demonstration farm

Stage	Crown rooting		Flowering	
Delay (days)	5	8	5	8
Yield (ql/ha)	37.5	31.8	38.5	32.7
% loss of yield	6.25	20.5	3.75	19

$$\Delta = \left(w_f - w\right) * Z * \gamma_{ds} \tag{9.1}$$

where

Δ is the required water depth;
w_f is the moisture content at field capacity as a proportion of dry soil weight;
w is the moisture content at the time in question;
Z is the effective root zone depth; and.
γds is the specific dry unit weight of the soil.

If w is the moisture content at the wilting point, the deficiency is expressed as a percentage, P, as below:

$$\begin{aligned} \frac{P}{100} &= \left(w_f - w\right) \\ &= \left(w_f - w_p\right)\left\{1 - \frac{w - w_p}{w_f - w_p}\right\} \end{aligned} \tag{9.2}$$

Equations (9.1) and (9.2) give moisture deficiency to be made up by irrigation. Knowing soil properties, w_f, w_p, and γ_{ds}, the effective root zone depth Z, and permissible stress or permissible depletion, the depth of irrigation watering can be computed. Slight variations in actual technique are possible as follows:

(a) Continuously monitor soil moisture content or soil moisture stress in the root zone. Irrigate when moisture deficiency or stress reaches the maximum permissible limit.
(b) Time interval to next irrigation, T, is estimated by

$$T = \frac{\Delta}{ET_a}$$

where ETa is the actual evapo-transpiration per unit of time. The depth of irrigation Δ is determined by equation (9.1).

(c) Assuming that last irrigation left the soil moisture at field capacity, daily loss can be computed by daily ETa, and irrigation applied when permissible deficiency is estimated to have been reached by such computation.

Table 9.4 Critical stages of wheat days after sowing

Stage for wheat	No. of days from sowing	Order of priority
Crown rooting	20–25	1
Tillering	40–45	4
Jointing	60–65	3
Flowering	75–85	2
Milking	100–105	5
Dough formation	115–120	6

Physiological Growth Stages

Certain stages of plant growth are more sensitive to water requirement than others. As far as possible, there should be no stress as the stages, e.g. for wheat the critical stages and days after sowing are shown in the Table 9.4:

Thus, optimally six waterings are needed. As far as possible, watering should be made to coincide with the more important critical growth stages, e.g., crown rooting and flowering for wheat.

Illustrative Example

Dominant crop wheat.

Effective root zone depth	90 cm
Field capacity	30%
Wilting point	13%
Storage capacity	17%
Specify dry weight	1.6.

The soil profile is uniform over the root zone depth. Irrigation is needed when the average deficiency in the root zone is 50%. Alternatively, moisture content can be monitored in all the three horizons and the average taken. Watering will be needed when the average moisture deficiency is 50% and effective "depth" to be supplied to the root zone will be

$$\frac{0.17}{2} * 90 * 1.6 = 12.24\,\text{cm}$$

If field application efficiency is 80% and loss in water course 25%, water depth required at outlet will be

$$\frac{12.24}{0.8 * 0.75} = 20.4\,\text{cm}$$

Appendix: Cropwat Software for Irrigation Schedulling

"Cropwat for Windows" is software developed by Clarke et al. (1998). It uses Penman–Monteith method for calculating reference crop evapotranspiration. These estimates are used for the computation of crop water requirements and irrigation scheduling.

Cropping patterns can be designed with several crops (up to 30 crops) and with staggered planting dates. Irrigation schedules can be calculated for individual blocks of each crop. Time base can be daily, weekly, or monthly. The program uses monthly climatic data only. The program allows user defined irrigation events and has option to add adjustments to the calculated soil moisture deficit. This provides a flexible fool for managing soil moisture during the growing season. Table 9.5 shows input and output on irrigation scheduling using CROPWAT.

Table 9.5 Input and output on irrigation scheduling using CROPWAT

Input	Output
a. **Crop water requirement** **Data (Monthly basis)** • Climate (temperature, humidity, wind speed, sunshine) • ETo if climate data not available or if ETo is user specified • Rainfall (choice of four effective rainfall models USDA SCS method by default) • Crops, planting dates of crops • Cropping pattern • Crops coefficient data (optional) Software contains typical crop coefficients data	• ETo (penman Monteith method and fixed net/field irr. requirement • Crop area • Crop Kc (Growth stage) • Crop water requirement • Effective rainfall • Field irrigation requirement (mm/day, L/s/ha) • Climatic graph
b. **Irrigation schedulling**	
• All data as above in (a) • Soil type info. Otherwise info on a set of typical soil types is available in CROPWAT • Start of scheduling (planting date by default) • Initial soil moisture • Irrigation scheduling criteria: different options on depth and timing • Specific % of a. Specific % of Readily available or total soil moisture depleted or specified soil moisture depletion occurs b. Fixed interval or variable interval c. Fixed/variable depth or refill to specified % of RAM d. Application efficiency • By default "optimal/criteria is adopted	• Table/graphs on all above • Crop data, soil data, scheduling criteria • Total available moisture, readily available moisture, soil moisture deficit (SMD) • Irrigation interval • Irrigation depth applied • Water loss as surface runoff /deep percolation • Adjustment made to soil moisture deficit (SMD) • % Reduction in yield in different growth stages, Total yield reduction e. Daily soil moisture balance

Scheduling Criteria Option

Several options are available for specifying irrigation schedule as stated below.

If no option is given (default option); then optimal irrigation schedule from the first planting date of the crop is automatically adopted by the program. Irrigation is calculated to take place when all the readily available soil moisture has been used so the crop never becomes stressed. The amount of irrigation is calculated to refill the soil moisture store i.e. irrigate to return the soil to field capacity (100% of the readily available moisture is replaced). Scheduling begins from the earlier planting date of each crop.

Option for Application Timing

(a) Irrigate at fixed intervals (days).
(b) Irrigate when a specified percentage of readily available soil moisture depletion occurs.
(c) Irrigate when a specified percentage of total soil moisture depletion occurs.
(d) Irrigate at variable intervals (user defined) (days).

Option for Application Depths

(a) Fixed depths
(b) Refill to a specified percentage of readily available soil moisture
(c) Variable depths (user specified) (mm)

Once the criteria are defined, the program calculates the irrigation data and amounts.

Irrigation Scheduling Report

It can appear in two ways: either Irrigation Schedule or Daily Soil Moisture Balance. An illustration of the irrigation scheduling report is given by Clarke et al. (1998).

Questions

1. Define the following terms:

Continuous delivery, demand delivery, rotational delivery, water allowance, peak requirement, duty of canal water, conveyance emptying, stream size, rotation period, unit time, basic time, filling time, emptying time, and group Warabandi.

2. Why is it difficult to implement delivery on-demand and continuous basis in an irrigation project in India?
3. What are the advantages of rotational delivery?
4. Design water allowance and actual existing water allowance could be significantly different. Give reasons.

5. Water allowance at canal head is 0.8 L/s/ha. Conveyance efficiency is 75%, stream size is 28 L/s/ha how much area can be irrigated with (i) continuous supply for 14 days (ii) water is supplied on a one week "off" one week on the basis?
6. How the filling time, emptying time, and seepage losses are adjusted in time allotment for irrigation to farmers.
7. Compare the rotational delivery model for the Chambal Project, Pochampad Project, and Upper Ganga Canal Project.
8. Workout water allowances at the outlet level, outlet capacities, and capacity of canal at its head.

S.No	Crop	% area	Peak fortnightly requirement (mm)
1	Wheat	30	85
2	Vegetable	5	65
3	Tobacco	45	95
4	Cotton	10	108

Outlet No	Outlet area (ha)	Distance from the head of canal (m)
1R	35	200
2L	30	400
3R	37	650

Desired irrigation intensity = 80%

Watercourse conveyance efficiency = 75%

Field application efficiency = 90%

Seepage loss in canal = 10% of diversion per 1000 m length of canal.

9. Discuss the factors affecting irrigation scheduling based on crop water demand.
10. What are the possible applications of CROPWAT software?
11. How does group (block) formation help in reducing peak water demand? Illustrate with example irrigation.
12. Can group (block) formation help in reducing peak labour requirement? How?
13. Under what conditions waste of irrigation water at night could be avoided or reduced?
14. Under what conditions night irrigation efficiency could be improved >
15. Identify the areas needing field research on night irrigation.
16. Explain the meaning of block irrigation.
17. Explain how the conveyance loss and filling, emptying time of canal are adjusted in irrigation time scheduling among blocks in Cilutung project (Indonesia).

References

Chambers R (1988) Managing canal irrigation: practical analysis from South Asia. Cambridge University Press, Cambridge [England]

Clarke D, Smith M, El-Askari K (1998) CropWat for windows: user guide. University of Southampton, Southampton, pp 1–43

GOI (1992) "Manual of Warabandi in irrigation project", Ministry of Irrigation, Government of India, New Delhi (Apr 1992)

GOAP (1979) Project report and guidelines for introduction of Warabandi system in Sri Rama Sagar Project, Andhra Pradesh, Government of Andhra Pradesh, C.A.D. Department (January)

Kartaatmaja SS (1994) Improvement in design and implementation of rotational water delivery on Lakhnauta Minor. Unpublished M.Tech Dissertation Report, supervised by Prof U.C. Chaube, Deptt of WRD&M, IIT Roorkee, India

Sulaiman E (1996) water distribution plan and on-farm water management. Report of unpublished M.Tech. Dissertation work by Eman Sulaiman, supervised by Prof U C Chaube, Deptt of WRD&M, IIT Roorkee, Roorkee, India

Reading Material

Kathpalia GN (1090) "Rotational System of Canal Irrigation and Warabandi"' C.B.I.P. Publication 146, New Delhi

Malhotra SP (1982) "The Warabandi and its infrastructure. C.B.I.P. Publication No. 157, India

Sicler D (1981) The new era of irrigation management in India. Unpublished Report

Chapter 10
Measurement of Flow and Sediment in Canals

Abstract Each unit of water let into the canal and conveyed to the field has an associated cost of investment and cost of operation and maintenance of facilities. Efficient water use for irrigation depends mainly on the reliable measurement of water. Flow in the canal needs to be measured at (i) the head of the main canal, branch, distributary, and minor (i.e., offtake points); (ii) at suitable intervals of main canals and branch canals; (iii) at offtake points of watercourses; iv) within the lower third of watercourses; and (v) at the head of the canal reach under the jurisdiction of Canal Inspector, Section Officer, and Sub-divisional Officer when such points are not located at offtake. A brief description of commonly used measuring devices and measurement methods is provided. Current meters and Parshall flumes are used in large-size channels. Parshall, Cutthroat, Replogle flumes, V notch, and Rectangular weir are used to measure flow in small channels. Flow measurement is accurate under free-flow conditions. Free flow condition (modular limit) is defined by the ratio of downstream depth and upstream depth. It is different for each type of flume. Flume discharge formula and correction for submerged flow conditions are given. Relative advantages and constraints are also explained in brief. V-notch and rectangular weirs are commonly used in field channels where falls are available. Flumes, weirs, notches are mostly used in canals carrying small discharges, such as below the canal outlets. The velocity area method is the most common method of flow measurement in large canals. Three methods of velocity measurement are explained with examples (current meter, float method, chemical dilution method). Water level at the time of flow measurement is also made. Staff gage, bubble gage, float type automatic recorders are commonly used for gage measurement. A procedure to develop stage-discharge relation (graphically and mathematically) is explained. An illustrative example of state discharge relation is also provided. Sediment movement in the canal takes on two forms. Suspended sediment is the load of finer particles that are held in suspension by the eddy currents in the flowing stream. Larger solid particles roll along the canal bed and are called the bed load. Generally, the entry of bed load in canal does not occur. A brief description of the methods for measuring suspended sediment and estimating bed load is also provided.

U. C. Chaube et al., *Canal Irrigation Systems in India*, Water Science and Technology Library 126, https://doi.org/10.1007/978-3-031-42812-8_10

10.1 Introduction

Each unit of water let into the canal and conveyed to the field has an associated cost of investment and operation and maintenance of facilities. The performance of irrigation facilities is based on canal capacity, the efficiency of conveyance, canal capacity utilization, water budget, duty, and delta. It should be borne in mind that efficient water use for irrigation and all purposes depend mainly on reliable water measurement.

The water released through irrigation outlets to the watercourse and water conveyed to the tail reach of the watercourse should always be measured. By observing these discharges, the water conveyance efficiency above and below the outlets can be determined accurately, and farmers can be assured of irrigation water supplies according to their allocation. These measurements become more significant when the irrigation water supply is based on a fixed supply period with a fixed stream size. When channels are unlined, the allocation of time for irrigating unit-cropped areas should be based on stream measurement at the upstream half and downstream half of the rotational area. In general, water measurements at the following critical locations are required:

(i) Head of each main canal, branch, distributary, and minor (i.e., offtake points);
(ii) At suitable intervals of main canals and branch canals;
(iii) At the offtake point of watercourses;
(iv) Within the lower third of watercourses; and
(v) At the jurisdiction head of the irrigation conveyance system of Canal Inspector, Section Officer, and Sub-divisional Officer when such points are not located at offtake.

Water measuring devices, such as flumes and weirs, are based on the measurement of the head and reference to calibration charts, tables, and equations. Design, construction, and maintenance of details of flumes, notches, and weirs are available in various documents, such as WALMI (1987), BOS (1976, 1989). Current meters are the most popular measurement devices based on the velocity area principle. Methods suitable for large and small channels are given below:

1. Measurement of flow in a large channel
   - Velocity area method using the current meter for velocity measurement
   - Parshall flumes
2. Measurement of flow in small channels
   - Parshall, cutthroat, Replogle flumes
   - V notch
   - Rectangular weir
   - Velocity area method using pygmy current meter

The canal reach should be straight, and a small length of canal at the gauging site should be lined. To avoid errors in the measurement of flow, current meters should be used when the velocity of flow is more than 20 cm/s. The stage-discharge relationship should cover the entire range of canal discharges. The stage or depth of flow should be measured after the canal flow is allowed to stabilize. When there are large fluctuations in the water levels in the canal, and the flow is more than 15 m^3/s, automatic gauge recorders should be installed to give a continuous water depth record (GOI 2003).

10.2 Flow Measurements

10.2.1 Parshall Flume

Components of Parshall Flume

The Parshall flume is a special type of venturi flume and can be either free-flow type or submerged. This flume was developed by Parshall (1926) at Colorado State University, USA, and is so named. Twenty-two flumes of various sizes ranging from 2.54 cm. throat width to 15.24 m throat width have been developed and calibrated (BOS 1976, 1989). Each flume is an entirely different device. For example, the length of the throat section for a 6″ (15.2 cm) flume is 30.5 cm but the length of the throat portion of 9″ (22.8 cm) flume is not proportionately increased to 45.75 cm (1.5 × 30.5 cm), but it is 30.5 cm only. The dimensions for each of these flumes are available in the literature (Bos 1976). The measurement by Parshall flume consists of upstream piezometric head *ha* and downstream piezometric head *hb* at specified locations. The relationship between upstream head (h) and discharge (Q) is represented by a power equation:

$$Q = Kha^u$$

e.g. for 6″ (15.2 cm) Parshall flume $Q = 0.3812\ \mathrm{ha}^{1.58}$

where K is a dimensional factor and depends on the width at the throat. The value of ‘u’ varies from 1.522 to 1.6. The K and ‘u’ values are available in tabular form in the literature. Q is the discharge (free flow) in m^3/s. The flumes can be used to measure discharge from 0.09 L/s to 93.04 m^3/s.

The above formula is valid for the following modular limits i.e., hb/ha should be less than or equal to values given in Table 10.1:

If this limit is exceeded for the Parshall flume in use, a reduction in the modular discharge (as computed by the above formula) due to submergence will have to be made. The submerged flow discharge (Q_s) is computed as:

$$Q_s = Q - Q_c$$

where

Table 10.1 Modular limits for free flow condition in Parshall Flume

Parshall flume size (throat width)	Modular limit (hb/ha)
2.54 cm, 5.08 cm, 7.62 cm	0.5
15.24 cm, 22.86 cm	0.6
30.48 cm to 2.44 m	0.7
3.05 m to 15.24 m	0.8

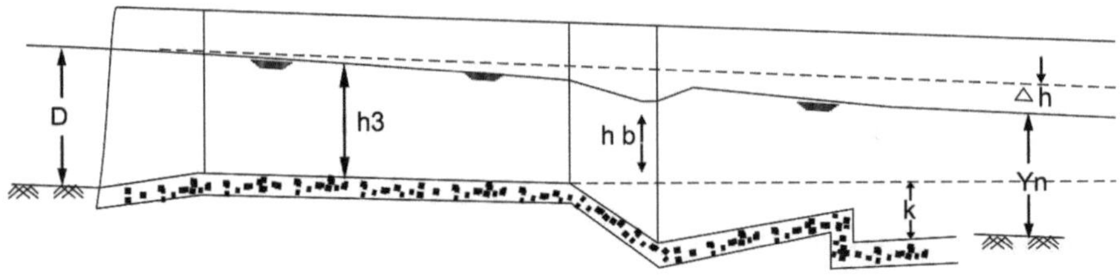

Fig. 10.1 Section of a Parshall flume illustrating the determination of height of setting the free flow

Q the free-flow discharge, and
Q_c depends on ha and hb/ha.

Setting of Parshall Flume for Free Flow Condition

Generally, the Parshall flume is installed, if possible, for the free-flow condition. For this purpose, the flume's sill is kept at a higher elevation than the channel bed level (see Fig. 10.1). The difference in channel bed level and sill level of the flume is known as the height of the setting and is denoted by K in Fig. 10.1.

i. The flume's size, full supply discharge (FSD), and normal depth Yn at full supply discharge should be specified.
ii. From the table, read ha required for full supply discharge.
iii. Compute hb for the maximum submergence allowed for free flow:

$$HB = 0.6 \times \text{ha}$$

iv.

$$SettingK = Yn - hb$$

v. Raise the sill of the flume by K above the channel bed level.
vi. Upstream water surface above the sill is given by D = ha + k.
vii. Check the upstream water surface for the freeboard available. In case the freeboard is not sufficient, raise the freeboard for a considerable distance upstream of the Parshall flume.

Construction of Parshall Flume

Brick, stone, or concrete can be used to construct Parshall flumes. Care is taken to see that the final dimensions of the finished work tally with the standard dimensions. Thus, while laying out the work, the thickness of plastering, etc., is taken into account. Precast concrete flume in separate components is available in the market. They can be bolted together and plastered. In this case, there is every possibility of distortion with time, and the dimensions do not remain as desired. Prefabricated metal flumes are also available. They are usually sturdy and accurate, though somewhat costly. While installing these flumes at the site, care must be taken that the sill (floor of the converging section) is perfectly leveled and the setting's desired height is attained.

Maintenance of Parshall Flumes

After the Parshall flume has been duly installed, periodic maintenance is necessary for its satisfactory operation. This maintenance includes:

- Clearing of the bed in the upstream entrance portion;
- Checking if there is any erosion in the diverging portion, and the portion may be provided with a concrete bed if necessary;
- Removal of moss on the inside surfaces;
- Painting the gauges;
- If the flume is made of metal sheet, removal of corrections and painting the flume with red oxide paint;
- Checking the levels and carrying out the correction, if any;
- Checking the tilting of the flume and correcting the same; and
- Occasionally calibrating the flume if some settlement and tilting are observed and cannot be rectified.

With the above precautions during the installation and maintenance, the Parshall flume yields sufficiently accurate measurements of flow during the system's day-to-day operation.

10.2.2 Cut-Throat Flume

Cutthroat flumes have some advantages compared to Parshall flume.

(a) It is easy to construct. Convergence and divergence angles are the same for all sizes;
(b) the rating of intermediate sizes can be developed from the available rating equations; and
(c) it gives nearly as accurate measurements as Parshall flume for smaller throat widths ($\approx 5\%$ errors)

Figure 10.2 shows a sketch of cutthroat flume.

L is the total length of the flume; W is the throat width; L1 and L2 are the lengths of converging and diverging sections; and A and B are the upstream and downstream gauge locations.

The size of a cutthroat flume is mentioned in terms of its throat width W and length L, i.e., as W × L. Generally, in field channels, 10 cm × 90 cm and 20 cm × 90 cm flumes are used to measure a discharge up to 54 lps and 115 lps, respectively, under free-flow conditions. Dimensions of 10 cm × 90 cm and 20 cm × 90 cm cut throat flumes are given in Table 10.2.

The cut-throat flumes are not suitable for measuring discharges under submerged flow conditions, i.e., for Hb/Ha more than 65.3%.

The discharge equation under free-flow conditions can be expressed as:

$$Q = CHa^n$$

where

Q the discharge in m^3/s;
Ha the upstream gauge reading in m;
C the discharge coefficient, and
n the exponent depending upon the flume length.

The value of discharge coefficient C can be calculated as follows:

$$C = KW^{1.025}$$

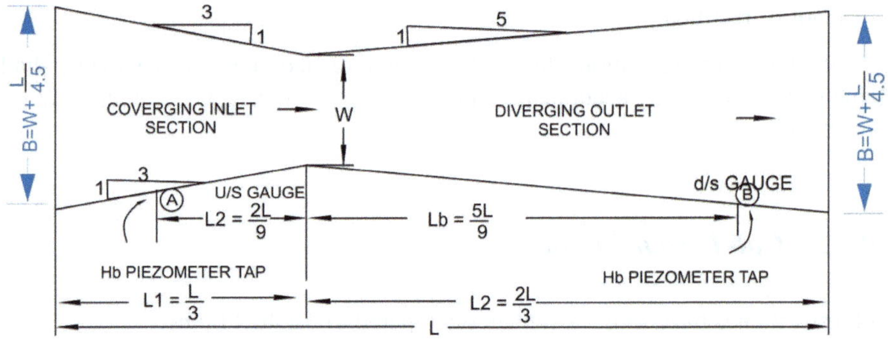

Fig. 10.2 Sketch of cut-throat flume (WALMI 1987)

Table 10.2 Dimensions of 10 cm × 90 cm and 20 cm × 90 cm cut throat flumes

Throat width W (cm)	L (cm)	L1 (cm)	L2 (cm)	La (cm)	Lb (cm)	B (cm)
10	90	30	60	20	50	30
20	90	30	60	20	50	40

K is the coefficient, depending upon the length of the flume, and W is the throat width in m.

For a 90 cm long flume, the value of exponent n and coefficient K are 1.943 and 3.89, respectively. The value of n and K for other lengths of flume is available in the literature.

The discharge equation for a 10 cm × 90 cm flume is:

$$Q = 0.3672\,\text{ha}^{1.8443}[M.K.S.units]$$

The discharge equation for 20 cm × 90 cm flume is:

$$Q = 0.7473\,\text{ha}^{1.843}[M.K.S.units]$$

For achieving the free flow condition, the bottom of the flume is kept at a higher elevation than the channel bed level, and the setting (height) can be calculated in the same manner as in the case of the Parshall flume.

All the precautions of installation and maintenance for Parshall flumes are also applicable to cut-throat flumes.

10.2.3 Replogle Flume

The Replogle flume is a broad crested weir (Fig. 10.3). The channel sides are also the sides of the flume. The Replogle flume operates satisfactorily when the downstream water head is less than 85% of the upstream water head.

The advantages of this flume are:

- It is easy to construct and cheap;
- the upstream approach transition is not required; and
- the free flow limit is higher (up to 85%) than for the Parshall flume and cut-throat flume.

It can be constructed either in masonry or in concrete.

In the figure, different abbreviations used are as follows:

d	the construction depth of the channel;
D1	the upstream water depth,
F1	the actual freeboard,
Y1	the sill reference flow depth or gauge reading,
L	the sill length,
Y1	the loss of head caused by the flume,
S	the sill height,
dm	the normal depth of flow,
3 × S	the three times the sill height (ramp length),
B1	the channel bottom width,

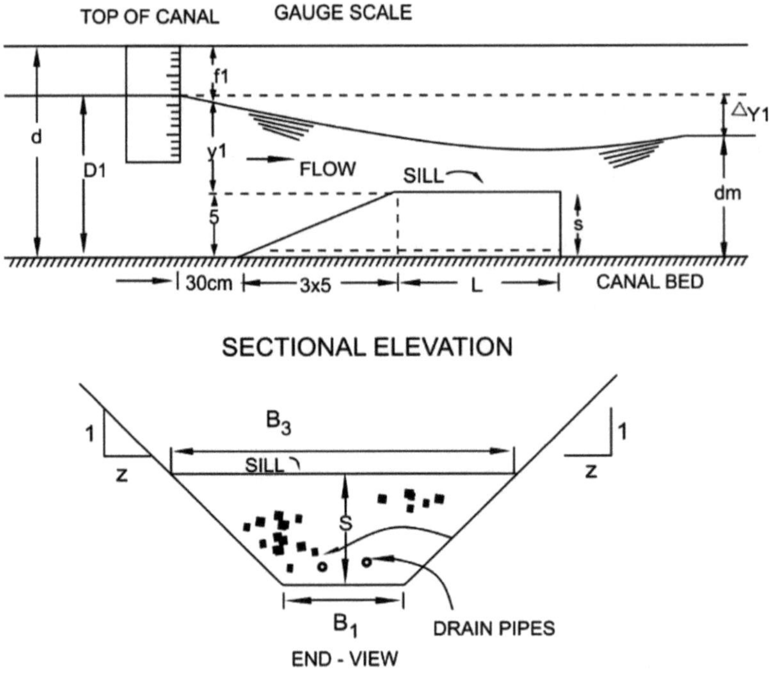

Fig. 10.3 Details of Replogle flume (BOS 1976, 1989)

B3 channel width at sill level or sill width, and
z the channel side slope.

The gauge is at a distance of 30 cm upstream of the ramp, and its zero at the sill level.

Table 10.3 shows discharges for the corresponding gauge readings (Y1) for three sets of sill heights (S) and sill lengths (L) for a channel with a bottom width of 30 cm and a side slope of 1:1.

10.2.4 Thin Plate V-Notch

Generally, 90° V-notch is used in field channels where falls are available (a drop of about 45–50 cm). The advantages of V-notch are its low cost and ease of construction. Secondly, for a small variation of discharges, the head variation is comparatively more, and discharge, therefore, can be accurately measured. Figure 10.4 shows the fixing of 90° V-notch in a small irrigation channel (field channel). The formula for thin plate V-notch is

Table 10.3 Discharge table for Replogle flume

Discharge (lps)	Channel shape B1 = 30.0 cm, Z = 1.1		
	B3 = 76.2 cm L = 61.0 cm S = 22.9 cm	B3 = 91.4 cm L = 45.7 cm S = 30.5 cm	B3 = 106.6 cm L = 30.5 cm S = 38.1 cm
	Gauge reading Y1 (cm)		
10	3.89	3.48	3.15
11	4.14	3.70	3.35
12	4.38	3.92	3.55
13	4.61	4.13	3.74
14	4.84	4.33	3.92
15	5.05	4.53	4.10
16	5.27	4.72	4.27
17	5.47	4.90	4.44
18	5.68	5.09	4.61
19	5.88	5.26	4.77
20	6.07	5.44	4.93
21	6.26	5.61	5.08
22	6.44	5.78	5.24
23	6.62	5.94	5.39
24	6.80	6.10	5.54
25	6.98	6.26	5.68
26	7.15	6.42	5.83
27	7.32	6.57	5.97
28	7.48	6.72	6.11
29	7.65	6.87	6.25
30	7.81	7.02	6.38
31	7.97	7.17	6.52
32	8.12	7.31	6.65
33	8.28	7.45	6.78
34	8.43	7.59	6.91
35	8.58	7.73	7.04
36	8.73	7.87	7.16
37	8.87	8.04	7.29
38	9.02	8.14	7.41
39	9.16	8.27	7.45
40	9.31	8.40	7.66

(continued)

Table 10.3 (continued)

Discharge (lps)	Channel shape B1 = 30.0 cm, Z = 1.1		
	B3 = 76.2 cm L = 61.0 cm S = 22.9 cm	B3 = 91.4 cm L = 45.7 cm S = 30.5 cm	B3 = 106.6 cm L = 30.5 cm S = 38.1 cm
	Gauge reading Y1 (cm)		
41	9.45	8.53	7.78
42	9.59	8.66	7.90
43	9.72	8.79	8.01
44	9.86	8.91	8.13
45	9.99	9.04	8.25
46	10.13	9.16	8.36
47	10.26	9.29	8.47
48	10.39	9.41	8.59
49	10.52	9.53	8.70
50	10.65	9.65	8.81
51	10.78	9.76	8.92
52	10.90	9.88	9.03
53	11.03	10.00	9.14
54	11.15	10.11	9.24
55	11.28	10.23	9.35
56	11.40	10.34	9.46
57	11.52	10.45	9.56
58	11.64	10.57	9.67
59	11.76	10.68	9.77
60	11.88	10.79	9.87

Source BOS (1976, 1989)

$$Q = Cex\left(\frac{8}{15}\right)\left(\tan\frac{\alpha}{2}\right)\left(\sqrt{2g}\right)\left(he^{\frac{5}{2}}\right)$$

Here, α is the angle of the notch. For a 90° notch, the formula becomes

$$Q = Cex\left(\sqrt{2g}\right)\left(he^{\frac{5}{2}}\right)\left(\tan\frac{\alpha}{2} = \tan 45^o = 1\right)$$

where Ce is the coefficient of discharge, and he is the effective head = h + kh.

For a 90° notch, kh = 0.85 mm or 0.085 cm.

For a 90° notch, the coefficient of discharge Ce is a function of two variables h/p and p/B as below:

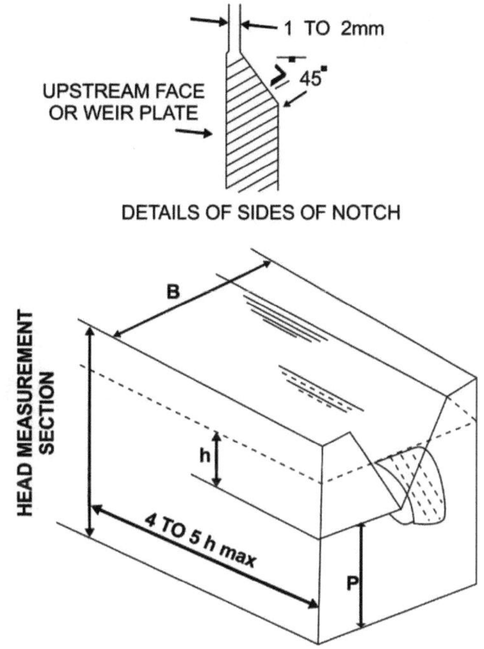

Fig. 10.4 Sharp-crested 90° V-notch

$$Ce = f\,(h/p, p/B)$$

where P is the height of the vertex of the notch above the channel bed, B is the width of the approach channel, and h is the head above the vertex of the notch measured at a distance 4–5 times hmax u/s of the notch.

For a 90° V-notch, Ce varies from 0.58 to 0.61.

10.2.5 Rectangular Weir

The rectangular weir (Fig. 10.5) is a horizontal weir with vertical sides and is used in field channels for the condition where discharge is more than 50–60 lps. Like the V-notch, the rectangular weir is also installed at the fall site. This weir requires comparatively high heads, considerable maintenance of the weir and the stilling pool, and protection of the downstream channel. The accuracy of measurements is good, provided free flow conditions exist.

Discharge Formula

The formula for the rectangular weir is:

$$\mathrm{Q} = (\mathrm{Ce})\,(2/3)\left(\sqrt{2}\mathrm{g}\right)(\mathrm{be})\left(\mathrm{he}^{3/2}\right)$$

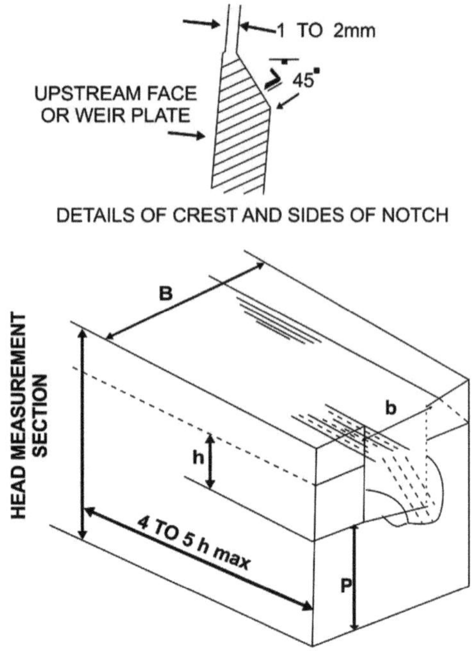

Fig. 10.5 Sharp-crested rectangular weir

where

Ce the coefficient of discharge,
be the effective width, and
he the effective head.

The coefficient Ce is a function of b/B and h/p. The effective width and head are given by the following equations:

be b + kb and
he h + kh

in which kb and kh are the experimentally determined quantities. The value of kb depends on the b/B ratio and is given below:

b/B	kb (mm)
0.4	2.7
0.6	3.6
0.8	4.2

The value of kh can be taken as a constant value equal to 1 mm. The value of coefficient of discharge Ce can be computed as shown below:

b/B	Ce
0.4	0.591 + 0.058 h/p
0.6	0.593 + 0.018 h/p
0.8	0.596 + 0.045 h/p

With the above conditions, the flow over the weir is partially contracted. The weirs are not suitable, if a suitable natural drop site is not available on the field channel.

10.2.6 Flow Measurement in Large Canals

Flumes, weirs, and notches are mostly used in canals carrying small discharges, such as below the canal outlets. The main disadvantages of using the discharge measuring structures (flumes or weirs) in a large canal are:

(a) These structures prove to be very costly in medium to large canals;
(b) The sections where the amount of sediment load or debris movement is higher, these structures cannot be used. Since the bed conditions change in such circumstances, therefore the discharge relationship, with respect to the structure, no longer remains the same and thus poses calibration problems;
(c) These structures create backwater effects in the upstream portion of the stream which may not be desirable in many cases; and
(d) These structures operate under a certain range of heads which should be available in the section. If this condition is not satisfied at some sites, the structure may not be employed.

The velocity area method is the most common method for flow measurement in large canals. The gauging stations are established on the canal network as explained in Sect. 10.1.

10.3 Water Level (Stage or Gauge Height) Measurement

Bureau of Indian Standards Code IS 15119 (BIS 2002) explains the procedure for establishing and operating a gauging station.

10.3.1 Gauging Site

An ideal gauge site satisfies the following criteria:

(a) The general course of the canal is straight for about 100 m upstream and downstream from the gauge site;

(b) The stream bed is not subject to scour or deposition of silt and is free from weeds;
(c) The site is readily accessible for ease of installation and operation of the gauging station; and
(d) The site should be free from weed growth, vortices, dead water, and excessive silting.

10.3.2 Measurement of Stage

Stage (gauge height or water level) is measured with reference to a recognized datum, such as mean sea level, or an arbitrary datum plane chosen for convenience.

Staff Gauge: The most generally used equipment at non-recording stations consists of a graduated staff or float tape gauge. Staff gauges are either placed vertically or in an inclined fashion. Vertical staff gauges may consist of porcelain enameled iron sections, each 150 mm wide; one meter long, and graduated every 10 mm (Fig. 10.6a). Float-Tape Gauge: A float-tape gauge consists of a float, graduated steel tape, counterweight, and pulley (Fig. 10.6b). The float pulley is usually 0.150 m in diameter, grooved on the circumference to accommodate the tape, and is mounted on a support (Fig. 10.7).

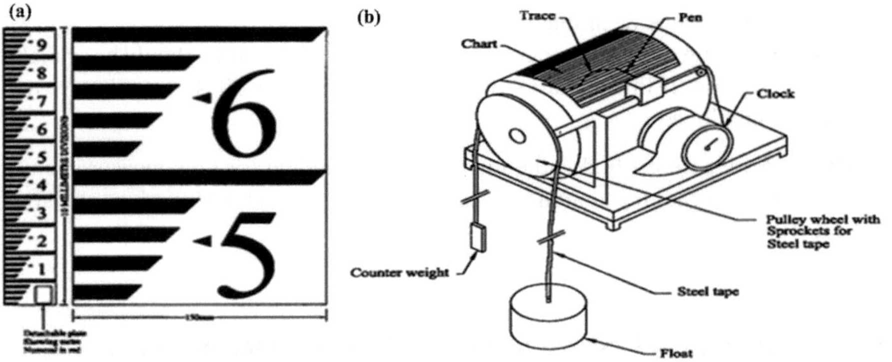

Fig. 10.6 **a** Close up of typical staff gauge. **b** Typical float, pulley and counterweight water level sensor

SAQ 2

(a) How the water depths, which are more than the height of one staff gauge, are measured in a stream?
(b) Why is a counterweight used in float-tape gauge?

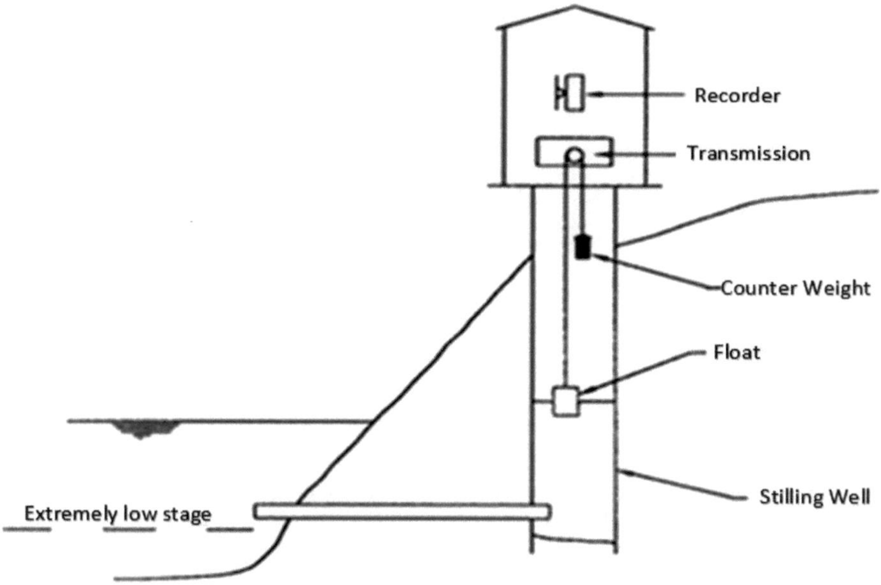

Fig. 10.7 Float type automatic water level recorder

Answer SAQ 2

(a) When it is expected that the depths of water in a stream would be more than the height of one staff gauge, then additional gauges are established along the cross–section of the stream so that there is sufficient overlap of around 0.5 m between consecutive gauges. When the water level goes above one gauge, then the next gauge is read, which is put on a higher level towards the banks.

(b) When the float rises due to rising water level, the tape slackens, and does not transform its vertical movement into the corresponding rotational motion of the pulley. To avoid such slackening and to develop a winding effect, a counterweight on the other side of the tape is used. This counterweight balances the weight of the float.

Recording Stream Gauging Station: The depth of water surface is sensed for automatic recording by a stilling well (Fig. 10.8), or by a gas-purge system that transmits the pressure head of water (due to the prevailing depth of water) in a stream to a manometer. The gas-purge system, which does not require a stilling well, is known as a bubble gauge. In India, mostly float sensors are used since they are comparatively cheap.

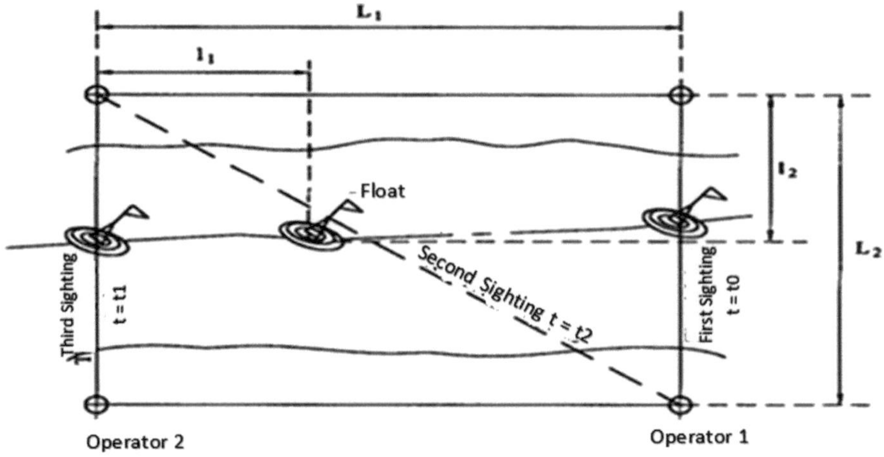

Fig. 10.8 Sketch to illustrate float positioning using the double stopwatch method 1995 (*Source* Streamflow measurement by Herschy)

SAQ 3

Study the literature and enumerate the merits and demerits of the bubble gauge, vis-à-vis recording type float gauge.

Answer SAQ 3

The bubble gauge has the following specific advantages over a float gauge:

(i) There is no need for any stilling well, which is costly to build,
(ii) A larger range of change in stage, as high as 30 m, may be measured,
(iii) The recorder assembly can be far away from the sensing point, and thus a small room may suffice, which may be built beyond the banks at a safer location, and
(iv) Since the air constantly bleeds from the nozzle, there is less likelihood of the inlet getting blocked or chocked.

The float gauge, on the other hand, has some advantages as follows:

(i) The mechanism of the float gauge is very simple and may be handled by a semi-skilled person, whereas the bubble gauge requires to be handled by a skilled person, and
(ii) There are simple mechanical parts in a float gauge compared to a bubble gauge. Therefore, the maintenance of the float gauge is rather simple and inexpensive.

10.4 Methods of Discharge Measurement

Discharge is the flow of water in items of volume passing through a certain section of a channel in a unit period. It is expressed in cubic meters per second (cumec) or cubic feet per second (cusec). The following methods are given in the literature.

(a) **Direct Determination of Discharge**

i. Area velocity methods: The basic procedure involves measuring the flow velocity and the flow area, and the multiplication of these two gives discharge.
ii. Dilution techniques,
iii. Electromagnetic method, and
iv. Ultrasonic method.

The current meter method and dilution technique for velocity measurement are discussed below. For other methods, reference may be made to manuals and textbooks.

(b) **Indirect Determination of Discharge**

Under this category fall those methods that use the relationship between discharge and flow depth (i.e., the rating curve) at specified locations. Field measurement, in these methods, is restricted to the measurement of depths only. Broadly speaking, indirect methods can be classified as:

i. Those using already existing flow measuring structures on the stream, and
ii. Slope area method.

10.5 Measurement of Velocity

10.5.1 Using Floats

For a float measurement, two cross-sections are selected on a straight reach of the given channel (Fig. 10.8).

The velocity of float is equal to the distance between the two cross-sections divided by the time taken by the float to travel from the upstream cross-section to the downstream one. The mean velocity in the vertical is equal to the float velocity multiplied by a coefficient whose value is dependent on the shape of the vertical—velocity profile of the stream and on a portion (strip) of the cross-section. A coefficient of 0.85–0.90 is commonly used to convert the velocity of a surface float to the mean velocity in the vertical.

In India, floats are still being used under normal circumstances and on a regular basis, and measurement of discharge on many streams is still being done using floats. An accuracy of about 90% may be achieved in the estimation of discharge based on measurement by floats.

SAQ4

The velocity of float in a canal was observed to be 3.0 m/s. Compute the average flow velocity corresponding to this observation. Adopt a reasonable value of the required coefficient and explain why this coefficient is needed.

Answer SAQ4

A simple float moving on the stream surface is called a surface float. These surface floats are affected by surface winds and thus do not yield accurate observations. Thus, to estimate the mean velocity, the observed velocity is multiplied by a reduction coefficient. A reduction factor of 0.85–0.90 is normally used to get the mean velocity from the surface velocity.

Hence, for the given observed velocity of 3.0 m/s the average velocity should be equal to say $0.90 \times 3.0 = 2.7$ m/s.

10.5.2 Using Current Meter

The most commonly used instrument in hydrometry to measure the velocity of flow at a point in the flow cross-section is the current meter. It essentially consists of a rotating element which rotates due to the reaction of the stream current. It employs the principle of proportionality between the velocity of water and the resulting angular velocity, which is acquired by the rotor of the meter. By placing a current meter at a point in a stream and counting the number of revolutions of the rotor during a measured time interval, the velocity of water at that point is determined. Current meters are generally of two types: horizontal axis rotors and vertical axis rotors. Vertical-axis meters consist of conical cups mounted around the vertical axis. The cups rotate in a horizontal plane, and a cam attached to the vertical axial spindle records the generated signals proportional to the revolutions of the cup assembly. The normal range of velocities measured by current meters is from 0.15 to 4.0 m/s. The relationship between flow velocity, v, and the number of revolutions per second N is given by

$$\mathrm{v} = \mathrm{a} + \mathrm{bN}$$

where v is the constant of proportionality, and a is the starting velocity required to overcome friction.

Horizontal-axis meters consist of a propeller mounted at one end of a horizontal shaft (Fig. 10.9). These are available in a wide variety of sizes with propeller diameters in the range 6–12 cm and can register velocities in the range of 0.15–4.0 m/s. Off-Neyrtec and Watt-type meters are typical instruments of this kind.

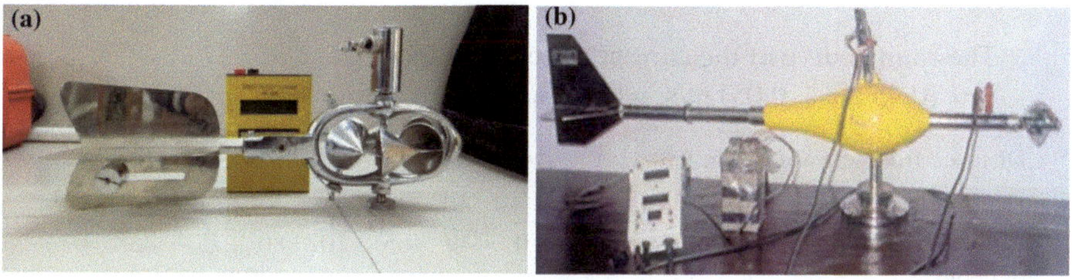

Fig. 10.9 **a** Cup types current meter. **b** Propeller type current meter

Fig. 10.10 Sketch illustrating the velocity area method

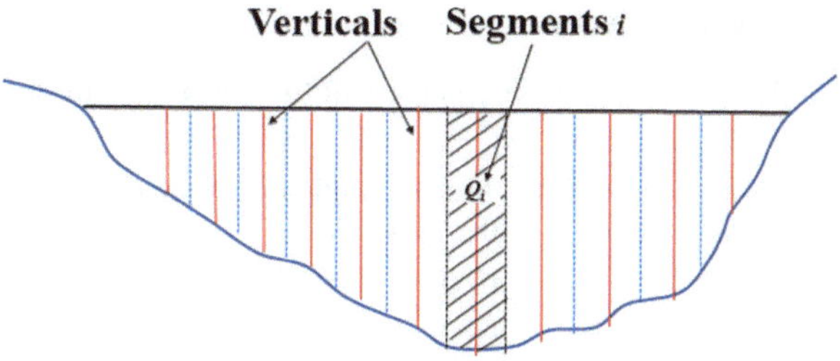

$$Q = \int_0^n v.dA \approx \sum_{i=1}^{n} V_i A_i = \sum_{i=1}^{n} q_i$$

10.6 Computation of Discharge

After the cross-section has been selected, the width of the stream is divided into the adequate number of particular sections (strips or segments, Fig. 10.10) to have lesser variation between two adjacent verticals (Subramanya 2017).

Figure 10.10 shows the cross-section of a channel in which N-1 verticals are drawn.

The computation of total discharge (Q) is done by summing up the discharges in the individual verticals. The discharge in a vertical is equal to the product of mean velocity (V_i) in the vertical and area (a_i) represented by that vertical. It may be expressed as:

$$Q = \sum_{i=1}^{n} (V_i x a_i)$$

SAQ5

The data related to a stream gauging operation are given below. Calculate the discharge flowing in the stream.

The rating curve of the current meter is defined by:
$v = (0.45\ N_s + 0.05)$ m/s, where N_s, is in revolutions per second.

Distance from left water edge (m)	0.0	2.0	4.5	5.5	8.0	11.0	13.0	15.0
Depth (m)	0.0	2.1	3.0	3.5	3.2	2.7	2.0	0.0
Number revolution if a current meter at 0.6 of depth	0	49	65	102	80	75	35	0
Duration of observation(s)	0	125	115	160	120	110	120	0

Answer SAQ 5

The calculations are shown in the following table:

Distance from left water edge (m)	Average width of the section (m)	Depth (m)	Velocity[a] (m/s)	Segmental discharge (m^3/s)
0.0	0.00	0.0	0.00	0.00
2.0	2.25	2.1	0.23	1.09
4.5	1.75	3.0	0.30	1.58
5.5	1.75	3.5	0.34	2.08
8.0	2.75	3.2	0.35	3.08
11.0	2.50	2.7	0.36	2.43
13.0	2.00	2.0	0.18	0.72
15.0	0.00	0.0	0.00	0.00

[a]From the given rating-curve equation

From the analysis of data, as tabulated above, we can see that the total discharge, as required, is equal to 10.98 cumecs.

10.7 Dilution Technique of Flow Measurement

The dilution method of flow measurement, also known as the chemical method, depends upon the principle of continuity applied to a tracer which is allowed to mix completely with flow. Let C_o be the small initial concentration (background concentration) of the tracer in the stream flow. At Sect. 10.1 a small quantity (volume V_1) of high concentration (C_1) is added. Let Sect. 10.2 be sufficiently far away, downstream of Sect. 10.1, so that the tracer mixes thoroughly with the fluid due to turbulent mixing while passing down through the reach. The concentration at Sect. 10.2 will have a base value C_o in the stream flow at the start of the experiment (t_1); the concentration must increase to a peak value (less than C_1), and then gradually reach a base value of C_2 at time t_2. The stream flow is assumed to be steady. By the continuity of the tracer material, we may write:

$$M_1 = mass\ of\ tracer\ added\ at\ section\ 1 = V_1 \times C_1$$

$$\int_{t_1}^{t_2} Q(C - C_0)dt + \frac{V_1}{t_2 - t_1} \int_{t_1}^{t_2} (C - C_0)dt$$

Neglecting the second term on the right-hand side as insignificantly small, we get

$$Q = \frac{V_1 C_1}{\int_{t_1}^{t_2} (C - C_0)dt}$$

Thus, the discharge, Q, in the stream can be estimated for a known value of M1, knowing the variation of C with time at Sect. 10.2 and Co. This method is known as a sudden injection, gulp, or integration method.

Another way of using the dilution principle is to inject the tracer of known concentration C_1 at a constant rate Q_t at Sect. 10.1. At Sect. 10.2, the concentration gradually rises from the background value of C_o at time t to a constant value C_2. At this steady state, the continuity equation for the tracer is written as:

$$Q_t C_1 + Q C_o = (Q + Q_t) C_1$$

$$Q = \frac{Q_t (C_1 - C_2)}{(C_2 - A_0)}$$

This technique in which Q is estimated by knowing $C_1 + C_2$, C_o, and C_t is known as the constant rate injection method or plateau gauging.

SAQ 6

A bag of 100 kg of common salt was suddenly emptied at a certain section of the stream. The background concentration of the salt in the stream was 0.1 in 1000. After a lapse of 5 min, the salt concentration started increasing at some downstream sections. This concentration increased linearly to a value of 3 in 1000 in the next 5 min and decreased linearly to the background value in the next 5 min. Calculate the flow rate in the stream?

Answer SAQ 6

The mass of the tracer added is equal to 100 kg.

The mass of the tracer reaching the downstream section is the integration of the mass observed over the time the concentration was greater than the background concentration.

That is to say (considering the area under the concentration curve):

$$100 = (0.5 \times (0.003 - 0.0001) \\ \times 5 + 0.5 \times (0.003 - 0.0001) \times 5 \times Q = 0.0145Q$$

Hence, Q implies the flow in the stream.

Therefore,

$$Q = \frac{100 \times 100\,\text{kg}}{0.0145} = 6896.5\,\text{m}^3/\text{min}$$

or

$$Q = \frac{6896.5}{60} = 115\,\text{cumec}$$

10.8 Stage-Discharge Relationship

The relationship between stage and discharge is a single-valued relationship that is expressed as

$$Q = a(H - H_0)^b$$

where Q is the stream discharge, H is the gauge height (stage), H_0 is a constant that represents the gauge reading corresponding to zero discharge, and a and b are rating curve constants. Taking logarithms of both sides of the equation:

$$\log Q = \log a + b \log(H - H_o)$$

or

$$Y = mX + C$$

in which the dependent variable Y is log Q, the independent variable X is log $(H–H_o)$ and c is log a. The values of the coefficients for the best-fit straight line using data of N observations of X and Y are:

$$m = \frac{N\sum XY - \sum X \sum Y}{N\sum X^2 - \left(\sum X\right)^2}$$

and

$$c = \sum Y - \frac{m\sum X}{N}$$

and the coefficient of correlation is given by:

$$r = \frac{N\sum(XY) - \left(\sum X\right)\left(\sum Y\right)}{\sqrt{N.\left(\sum X^2\right) - \left(\sum X\right)^2}\sqrt{N\left(\sum Y^2\right) - \left(\sum Y\right)^2}}$$

If r is between 0.6 and 1.0, it is generally taken as a good correlation. Reference may be made to methods given in textbooks on hydrology for the estimation of H_o.

SAQ7

(a) Establish the stage-discharge relationship using the values of the observed stages and the corresponding discharges from the following table. Assume the stage at zero-discharge as 100.0 m.

Stage (m)	101.5	102.7	103.8	105.6	106.2	107.4	109.9
Discharge (cumec)	615.5	707.8	826.2	956.5	1045.3	1165.3	1335.2

(b) Using the relationship established above, estimate the discharge if the stage is 104.5 m.

Answer SAQ 7

(a) Let the gauge-discharge Eq. (11) be

$$Y = mX + c$$

where Y = log Q, X = log ($H–H_o$) and c = log a; and H_o = 100.0 m.

For the given values of H and Q, we can construct a table as given below. From the table, we have:

$$\Sigma X = 4.5927,\ \Sigma Y = 20.7480,\ \Sigma XY = 13.8115$$

$$\Sigma X^2 = 3.4877,\ \Sigma Y^2 = 61.5783,\ (\Sigma X)^2 = 21.0929$$

$$(\Sigma Y)^2 = 430.4795 \text{ and } N = 7$$

Gauge H (m)	(H–H$_o$) (m)	X = log (H–H$_o$)	X^2	Q(m^3/s)	Y = log Q	Y^2	X × Y
101.5	1.5	0.1761	0.0310	615.5	2.7892	7.7796	0.4912
102.7	2.7	0.4314	0.1861	707.8	2.8499	8.1219	1.2294
103.8	3.8	0.5798	0.3362	826.2	2.9171	8.5095	1.6913
105.6	5.6	0.7482	0.5598	956.5	2.9807	8.8846	2.2302
106.2	6.2	0.7924	0.6279	1045.3	3.0192	9.1156	2.3924
107.4	7.4	0.8692	0.7555	1165.3	3.0664	9.4028	2.6653
109.9	9.9	0.9956	0.9912	1335.2	3.1255	9.7688	3.1117

The values of coefficients for the best-fit straight line using data of N observations of X and Y are:

$$m = \frac{N\sum XY \sum X \sum Y}{N\sum X^2 - \left(\sum X\right)^2}$$

or $m = \frac{7\times13.8115-4.5927\times20.7480}{7\times3.4877-21.0929} = \frac{1.3912}{3.321} = 0.4189$

$$c = \frac{\sum Y - m\sum X}{N} \quad C = \frac{20.748 - 0.4189 \times 4.5925}{7} = 2.6892$$

$$c = \frac{N\sum(XY) - (\sum X)(\sum Y)}{\sqrt{N(\sum X^2) - (\sum X)^2 N(\sum Y^2) - (\sum Y)^2}}$$

or $r = \frac{7\times13.8115-4.5927\times20.7480}{\sqrt{7\times3.4877-21.0929},\sqrt{7\times61.5783-430.4795}} = 0.986$

Since the value of r is very near to 1.0, the correlation is very good. Hence the stage-discharge rating curve is given by the relationship:

$$Q = 488.877(H - H_0)^{0.4189}$$

(b) The relationship established in SAQ 10 (i) above is given as follows:

$$Q = 488.877(H - H_0)^{0.4189}$$

Gauge reading for zero discharge (H_o) is 100.0 sm.
Therefore, for the gauge reading equal to 104.5 m the discharge would be:
$Q = 488.877\ (104.5–100.00)^{0.4189} = 917.97$ cumecs.
Thus, the discharge at a stage of 104.5 m would be 917.97 cumecs.

10.9 Measurement of Suspended Sediment and Bed Load

Sediment movement in the canal takes on two forms. Suspended sediment is the load of finer particles which are held in suspension by the eddy currents in the flowing stream, which only settle down when the flow velocity decreases, such as when the canal bed becomes flatter, or the canal discharges into a pond or lake. Larger solid particles roll along the canal bed and are called the bed load. Generally, the entry of bed load in the canal does not occur. A brief description of the methods for measuring suspended sediment and estimating bed load is given in Appendix.

Appendix: Sediment Transport Measurement

Sediment movement in the canal takes on two forms. Suspended sediment is the load of finer particles which are held in suspension by the eddy currents in the flowing stream, which only settle down when the stream velocity decreases, such as when the streambed becomes flatter, or the stream discharges into a pond or lake. Larger solid particles roll along the streambed and are called the bed load. Generally, the entry of bed load in the canal does not occur.

Measurement of Suspended Sediment

Sediment Discharge

Estimates of the total sediment discharge in streams may be made by combining estimates of the sediment concentration with the velocity of flow. The *Velocity/area method* explains how velocity varies at different places in the stream, and how an average velocity may be calculated from a series of measurements. The sediment concentration also varies, being usually greater near the bottom, and so the amount of total sediment discharge is the product of these two variables, as shown in Fig. 10.11.

Sampling of Suspended Sediment

Sediment sampling in a given stream is the surest method of obtaining accurate information or the sediment load that is being carried by a stream at a particular location. This method is preferred for determining the sediment load at a given point. Results of periodic sampling are made available as a rate in tons per day of discharge when sampling is obtained, and results of daily sampling are again made available as the average daily sediment load in tons per day. From this, it is possible to obtain the average annual sediment yield at a given site in tons per year. The sediment sampling results in parts per million can be converted to a rate in tons per day by the following formula.

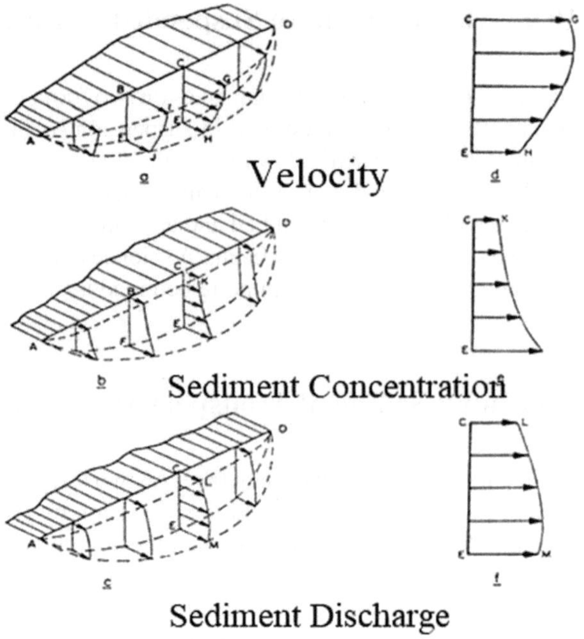

Fig. 10.11 Velocity, sediment concentration, sediment discharge in streams

$$Q_s = 0.0864 \times C \times Q_w \ tons/day$$

where

Q_s Sedient load in metric tons per day
C silt concentrations in parts per million (dry weight of sediment/dry weight of seiment and water in sample $\times 10^6$)
Q_w Average daily discharge in cumec

Grab Samples

The simplest way of taking a sample of suspended sediment is to collect the sediment-laden flow in a bottle/bucket. Care must be taken to ensure that flow enters into the sampling bottle/bucket at the same velocity as that of flow.

Depth-Integrating Samplers

Depth integrating samplers are designed to continuously extract a sample as they are lowered from the water surface to the streambed and returned at a constant rate of travel. Ascending and descending speeds need not be the same, but the rate of travel must be constant in each direction. Thus, it gives a sample that reflects the entire water column but does not indicate the distribution of contents within the column.

Point-Integrating Samplers

Point integrating samplers are designed to be lowered to a specific depth within the water column, at which time a valve is triggered to open for a pre-determined period to allow the sample to accumulate. By this means, a sample is taken that represents only that specific point in the water column which also allows for analysis of particle distribution throughout the water column.

Opening and closing of the valves of the sampler are controlled from the surface electrically or by cables. Samples should be taken at a number of depths at each of several vertical sections, as described in the section *Velocity/area methods* in Sect. 10.3, for the gauging of streams by the current meter method, so these two operations are often carried out at the same time.

With the control valve fixed in the open position, these samplers are also used to obtain depth-integrated samples. One-way depth integrated samples may be obtained by opening the valve with the sampler at the water surface and lowering it to the streambed at a constant speed. This permits sampling to greater depths.

To give a truly representative sample, suspended-sediment samplers should be designed to sample isokinetic ally which means that water is entering the nozzle at the same speed as the water would be traveling if the sampler wasn't there.

Continuous Samplers

The pattern of rising and falling flow rates and the variations of sediment concentration at different flow rates can be best achieved by a system of continuous measurement of sediment concentration using the optical turbid meter. An optical turbid meter passes a beam of light through sediment-laden water from a source on one side of a channel to a sensor on the other side. The sensor may either measure the extent to which light is absorbed by the sediment particles, and this is called the attenuation principle, or measure the extent to which the light is scattered by the suspended particles, called the scatter principle. Like pumping samplers, they can be programmed to react to the combinations of the depth of flow and time, but since the data are recorded in a Read-Only-Memory (ROM) device, any amount of data can be stored.

Estimating Bed Load

Direct Measurements

The simplest way to estimate bed load is to dig a hole in the streambed and remove and weigh the material that drops into it. Now a days, portable bed load traps are also available. One such example is the USFS Portable Bedload Trap developed for representative sampling initiation of gravel motion and coarse bedload transport rates in wadable gravel and cobble bed mountain streams.

Table 10.4 Maddock's classification for estimation of the bed load

Suspended sediment concentration parts per million	River bed material	Suspended elements texture	Bed load discharge expressed as % of suspended sediment discharge
Less than 1000	Sand	Similar to the river bed	25–150
Less than 1000	Gravel, rocks, hard clay	Low sand content	5–12
1000–7500	Sand	Similar to the river bed	10–35
1000–7500	Gravel, rocks, hard clay	25% sand or less	5–12
more than 7500	Sand	Similar to the river bed	5–15
More than 7500	Gravel, rocks, hard clay	25% sand or less	2–8

Samplers

The bedload samplers are designed to sample sand, silt, gravel, or rock debris carried on or immediately above the stream/canal bed. Estimates of bed load may be obtained from samples caught in a device that is lowered to the bed for a measured time and then brought up for weighing the catch. Bedload samplers differ from point and depth integrating samplers in that they do not capture a water sample, but rather use a sample bag made of a nylon mesh, which acts as a filter, allowing the water to pass through it while trapping the particulate matter.

Empirical Estimations

The difficulty of obtaining reliable measurements of bed load has led to some attempts to calculate it from more easily measured parameters, but these are not widely used. A very simple method based on knowing the suspended sediment concentration and the texture of both suspended material and bed material is given in Table 10.4. A sophisticated approach was developed by Einstein (1950) and has been later modified and improved. There are many other theoretical formulas, and there has been much debate on their accuracy and reliability.

Questions

In addition to self-assessment questions given in different sections, answer the following questions:

1. Explain the importance of flow measurement at (i) offtake of minor canal, (ii) head of canal reach under jurisdiction of a junior Engineer, and (iii) offtake of a branch canal.

2. Which flow measuring device is suitable for the offtake of a branch canal and head of main canal?
3. Compare the advantages and disadvantages of (a) Parshall flume and Replogle flume and (b) Parshall flume and Cutthroat flume.
4. Explain the usefulness of developing a stage-discharge relationship (rating curve).
5. What precautions should be taken in developing the stage-discharge relationship at allocation along a canal and using it?

References

BOS MG (1976, 1989) Discharge measurement structures. International Institute of Land Reclamation and Improvement, Netherland

Einstein HA (1950) The bed-load function for sediment transportation in open channel flows. USDA Tech Bull 1026. Washington DC

GOI (2003) Volume 4 "Hydrometry" of the "Manual on hydrological field measurements and data processing". Hydrology Project, Government of India & Government of The Netherlands (Jan 2003)

Hershy RW (1995) Streamflow measurement, 2nd edn. E & FN Spon, London

Internet source: www.rickly.com

Indian Standard IS15119 (Part 1) 2002 Measurement of liquid flow in open channels: part 1 establishment and operation of a gauging station

Parshall RL (1926) The improved Venturi flume. Trans ASCE 89:841–851

Subramanya K (2017) Engineering hydrology, 4th edn. Text Book Published by McGraw Hill Education (India) Private Limited, New Delhi

WALMI (1987) On-Farm Development Works (1987) Publication 12, Water and Land Management Institute (WALMI), Aurangabad, Maharashtra

Reading Material

FAO Irrigation and Drainage Paper 26/2

Chapter 11
Performance Evaluation

Abstract Performance evaluation of an irrigation system is vital for seeking improvement (betterment of performance) in operation and maintenance. The objectives of this chapter are to: gain knowledge of performance parameters related to the water conveyance system, on-farm system, and cropping system that impact agricultural production; understand the importance of these parameters; increase our understanding of a particular irrigation system as it operates based on performance parameters; and learn the application of essential parameters and data analysis through a case study.

A distinction is made between objectives and criteria. Objectives are the effects or impacts irrigation is intended to have. Criteria are standards or measures of performance. Different people (agronomists, irrigation engineers, sociologists, etc.) judge performance differently because they have different objectives. Criteria of good system performance according to the type of objects are indicated. Criteria generally used for judging performance are water delivery-related efficiency, economic or financial efficiency, productivity equity, and stability.

A large number of performance parameters have been suggested in the literature. The definition and significance of some of these parameters are explained. Examples are included in the hydraulic performance of the water distribution system. The selection of parameters for the performance evaluation of a particular irrigation system depends on the availability and reliability of field data, method of water delivery, and field irrigation practice.

Checklists are given to visually observe and make a preliminary assessment of the performance of the canal and related structures. A Khairana tank irrigation project case study is presented to illustrate the procedure for evaluating parameters. Molden and Gates (1990) have suggested the performance standards for adequacy, equity, timeliness, and sustainability parameters. These standards have been used to classify the performance of the Khairana irrigation project.

A field-based study on the performance evaluation of the Right Main Distributary of the Upper Ganga Canal system (Uttar Pradesh) by Prasad (2004) is also provided.

U. C. Chaube et al., *Canal Irrigation Systems in India*, Water Science and Technology Library 126, https://doi.org/10.1007/978-3-031-42812-8_11

11.1 Introduction

The success of an irrigation water delivery system can be measured by how well it meets the irrigation objectives, i.e., delivering an adequate and dependable water supply in an equitable, efficient, and timely manner to the user served by the system. In many irrigation systems, inequitable and unreliable water distributions are major water management problems.

A water delivery system comprises structural and management components. The structural components include the facilities for the conveyance, regulation, and measurement of water. The management component is responsible for efficiently operating and maintaining hydraulic structures to implement water delivery schedules that specify the timing and delivery. System performance suffers if either or both components are deficient. Therefore, evaluating the contributions of the above four factors to the system's overall performance is necessary. Here discussion is limited to technical performance at the hydraulic level of the water distribution system. Hence, the success of an irrigation system can be measured by how it supplies the required quantity of water at the right time in an equitable manner to users served by the system.

11.2 Objectives and Criteria of Performance Evaluation

11.2.1 Distinction Between Objectives and Criteria

A distinction can be made between objectives and criteria. Objectives are the 'ends' of irrigation and the effects or impacts it is intended to have. Criteria are standards or measures of performance. Common and useful performance criteria are measurable and include productivity, especially of water, equity, especially distribution of water, and stability over time, all at low cost. Criteria come earlier on causal chains, and objectives later. Criteria are more specific, easier to measure, and more concerned with physical performance. Objectives are more general, harder to measure, and more concerned with human impacts.

At a general level, all might agree that irrigation and irrigation management should contribute to development. Most people might agree, in some sense, that the objective of well-being, especially for those who are less well off, is that livelihoods for people with low incomes should be objective; that those who are deprived should have more of what they want and need. An overarching objective that all might accept might then be phrased as 'optimizing human well-being'. The word 'optimizing' can accommodate trade-offs between objectives, and 'well-being' can embrace many elements, including food security, income, nutritional status, health, education, amenity, social harmony, and self-respect.

Different people judge performance in different ways. If asked to give their first criterion of good performance of a canal irrigation system, different people are

Table 11.1 Criteria of good system performance according to type of person

Type of person	Possible first criterion of good system performance
Landless labourer	Increased labour demand, days of working, and wages
Farmer	Delivery to his or her farm of an adequate, convenient, predictable, and timely water supply for preferred farming practices
Irrigation engineer	Efficient delivery of water from head works to the outlet
Agricultural engineer	Efficient delivery and field application of irrigation water from the outlet to the root zone of the crop
Agronomist	Creation and maintenance of the optimum moisture regime and plant growth and, in particular (maximizing) production of that part of the plant, which is the harvestable product
Agricultural economist	High and stable farm production and incomes
General economist	A high internal rate of return
Political economist	Equitable distribution of benefits, especially to disadvantaged groups
Sociologist	Participation of irrigators in management

likely to provide different initial answers, reflecting their personal and professional interests, such as shown in Table 11.1.

11.2.2 Efficiency Criteria

Efficiency is also sometimes treated as a single comprehensive criterion. For water, efficiency is the ratio of water at a point of issue to water at a point of delivery. Water efficiencies can be defined as follows (Bos 2000):

$$Field\ Application\ Efficency = \frac{Crop\ Irrigation\ Water\ Requirement}{Water\ Deliverd\ to\ Field}$$

$$Distribution\ Efficency = \frac{Total\ Water\ Delivery\ To\ Fields}{Total\ Inflow\ Into\ The\ Delivery\ System}$$

$$Coveyance\ Efficency = \frac{Total\ Water\ Supplied\ By\ the\ Conveyance\ System}{Total\ Inflow\ Into\ the\ Conveyance\ System}$$

$$Overall\ Project\ Efficency = \frac{Crop\ Irrigation\ Water\ Requirements}{Total\ Inflow\ Into\ The\ Canal\ System}$$

$$Water\ Delivery\ Performance \rightleftarrows = \frac{Actual\ Volume}{Target\ Volume}$$

11.2.3 Economic/Financial Criteria

The internal rate of return or whatever other criterion in cost–benefit analysis is used may also be treated as the single measure of economic performance. The performance of an existing canal irrigation system may also be assessed in other single numerical forms, as implied by the statement that 'the main objective of management is to ensure a proper return on investment.' The financial viability during the O&M stage can be evaluated by the following criteria:

$$Total\ Financial\ Viability = \frac{Actual\ O\&M\ Allocation}{Total\ O\&M\ Requirements}$$

$$Fee\ Collection\ Performance = \frac{Irrigation\ Fees\ Collected}{Irrigation\ Fees\ Due}$$

11.2.4 Focus on Single Criterion

Four reasons explain the human tendency to focus on a single criterion: production or productivity, efficiency, or internal rate of return. The first is the limited capacity of the irrigation managers to handle multiple objectives and criteria. Irrigation managers tend to simplify a reality that they recognize, at another level, to be more complex. Second, for practical purposes, a single criterion helps decision-making: to know whether to invest and what it is, or to understand how well a project or its management is doing. Third, many professionals need, and also feel a need, for measurement. Production, productivity, efficiency, and social costs and benefits are measurable and permit comparisons between projects, parts of projects, and between different points in time. Finally, each criterion corresponds to that aspect of performance with which a different discipline or profession is concerned: agronomy with production; engineering with efficiency; and economics with costs and benefits.

Though widely approved and used, these criteria have limitations, as these are for aspects of operational performance, not the indicators of achieving ultimate objectives. An irrigation scheme serves a range of wider social objectives, such as import substitution, export promotion, development of the backward region, social security of weaker sections of society, etc.

Some consequences and effects are presented in Fig. 11.1.

For an ex-post evaluation and learning lessons, it can be useful to compile and use a checklist against which performance can be assessed, not necessarily with measurement, but it can also lead to excessive demands for data. The search has, instead, to be for cost-effective compromises. The question is whether a shorter list of criteria can be proposed to capture much that is useful and indicative without seeking to discover and understand everything (Fig. 11.2).

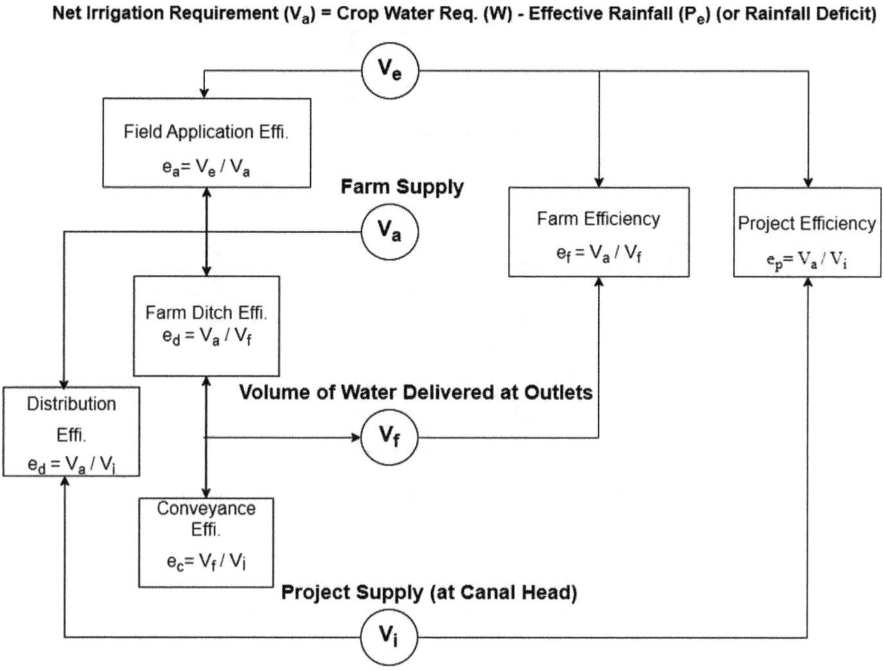

Fig. 11.1 Water efficiencies

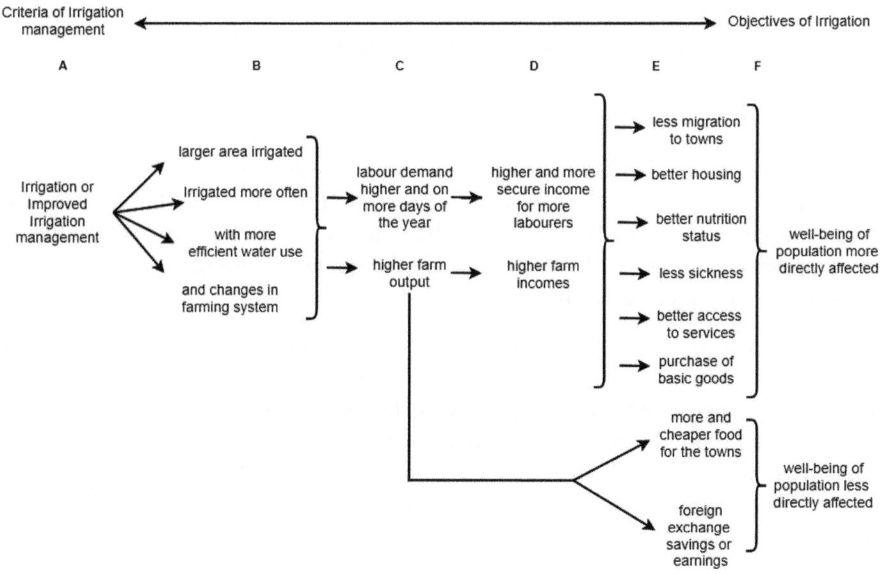

Fig. 11.2 Criteria, objectives, and beneficial causal chains

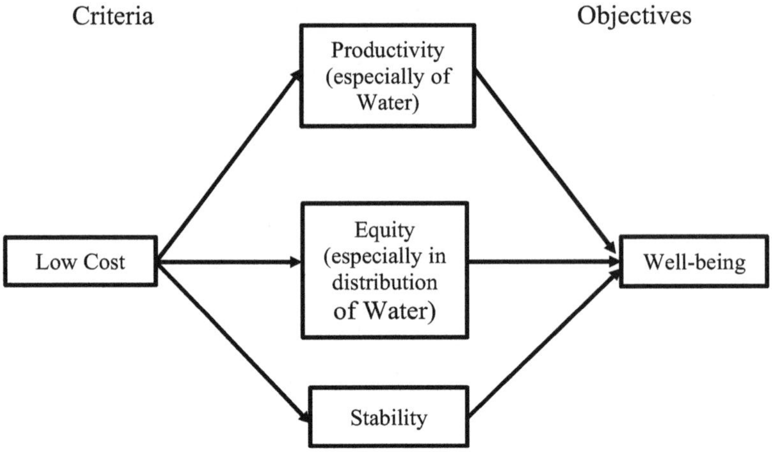

Fig. 11.3 Productivity, equity, stability criteria, and well-being as the objective

Productivity (of agricultural output), equity (fairness in the distribution of resources and benefits), and stability (of output over time) (Fig. 11.3) are performance criteria, not in themselves ultimate objectives. These three contribute to the fourth category—well-being, which is an overarching objective.

11.2.5 Utility to Irrigators and Carrying Capacity

Other criteria can also be used. Two important ones are:

Utility to irrigators: This criterion for assessing a water supply captures a key set of relationships between what farmers want or find useful, and what an irrigation system supplies to them. The utility can be defined in various ways, depending on what farmers want, but will usually include the quantity, timing, and predictability of the water they receive or obtain.

Carrying capacity: This refers to the performance of an irrigation system in making possible adequate, secure, and decent livelihoods. One way of describing it is livelihood-intensity, the ratio of acceptable livelihoods generated or supported to water or other scarce resources. The importance of this aspect depends on the conditions. In a prosperous, labour-scarce environment, it could be irrelevant in an environment of labour surplus, widespread poverty, and distress migration; it is likely to be very important.

11.3 Water Delivery Related Performance Parameters

11.3.1 Adequacy

A fundamental concern of a water delivery system is adequately delivering the water required to irrigate crops. The adequacy of delivery depends on water supply, specified delivery schedules, the capacity of hydraulic structures to deliver water according to the schedules, and the operation and maintenance of hydraulic structures.

Knowledge of the relative water supply (RWS) could be used to specify characteristics, irrigation organization, and system management. The degree to which the system capability is actually realized is called the intensity of management. By varying intensity, it is possible to match supply with demand.

In a physical sense, RWS is the ratio of water supplied to an irrigation unit and the demand for water in that unit over a period of time.

$$RWS = \frac{Water\ supplied\ to\ a\ Command\ Area}{Demand\ of\ Water\ in\ that\ area} = \frac{IR + RN}{ET + S\&P}$$

where IR = the mean irrigation application (mm/day);
RN = the mean rainfall (mm/day);
ET = the mean evapotranspiration (mm/day); and
S&P = the seepage and deep percolation (mm/day).

$$Water\ Delivery\ Performance = \frac{Amount\ of\ water\ actually\ delivered}{Amount\ of\ water\ intended\ to\ be\ delivered}$$

If the amount of water actually delivered is based on frequent flow measurements, then there is a greater likelihood that managers can match actually to intended/scheduled flows. If the water delivery performance ratio is nearly unity, then the management inputs must be effective.

11.3.2 Equity (Delivery of Fair Amount)

The equity related to water delivery systems can be defined as delivering a fair share of water to users throughout the system. A share of water represents a right to use a specified amount. There are many factors that determine the meaning of a "fair share," because a fair share is often interpreted subjectively. However, it is important to define measures relating to equity so that systems can be designed or rehabilitated to deliver water impartially to users served by the system.

Interquartile ratio (IQR) is defined as the ratio of volume delivered respectively to the most favored upper 25% of the command area and the least favored lowest 25% of the command area:

$$IQR = \frac{Volume\ (ham)\ delivered\ to\ the\ most\ favoured\ upper\ (25\%)\ of\ CCA}{Volume\ (ham)\ delivered\ to\ the\ least\ favoured\ lowest\ (25\%)\ of\ CCA}$$

$$Equity\ ratio = \frac{1}{IQR}$$

There is some degree of correlation between the incidence of inequity and irregularity in supply. The areas at the tail reach are most likely to receive their water erratically.

11.3.3 Timeliness (Uniform Delivery Over Time)

The simplest method to access timeliness is to calculate the relative productivity potential. Monthly data appear to give a good indication of whether the discharge is more or less timely supplied. The concept of potential productivity, as a numerical parameter, enables us to link adequacy and timeliness into one number or indicator of performance:

$$Relative\ Productivity\ Potential = \frac{Crop\ Production\ (\mathrm{Kg})}{Water\ Supply\ (\mathrm{m})^3}$$

The irrigation system is meant to provide water to crops to raise the productivity of land and water. The excess or deficit of water supply impacts crop yield, though the effect varies with the crop stage. The effect of water shortage may be reflected within a season, whereas an excess application may not have an immediate impact unless it creates oxygen deficiency conditions. We can calculate a ratio between the output that can be achieved under this water supply and the output that would have been achieved if the water supply had been ideally matched to the needs of crops. This ratio is called the potential productivity of this particular water delivery pattern.

The potential productivity index is worked out, based on weekly or ten-day averages of water inputs and outputs, with an allowance for field seepage loss based on whatever local observations may be obtainable:

$$Production\ Performance = \frac{Total\ Production}{Target\ Production}$$

$$Water\ Productivity\ Performance = \frac{Actual\ Water\ Productivity}{Target\ Water\ Productivity}$$

11.3.4 Reliability

Reliability means the degree to which the irrigation system and its water deliveries conform to the users' prior expectations. Reliability relates to a fairly constant rate of supply.

11.3.5 Dependability

A system, which performs consistently, may be considered dependable. The dependability of water delivery is important to farmers because it allows for proper planning. A system that dependably delivers an inadequate amount of water may be more desirable than one that delivers an adequate yet unpredictable supply on average.

$$Dependability\ of\ Supply = \frac{Actual\ Duration\ of\ Water\ Delivery}{Planned\ Duration\ of\ Water\ Delivery}$$

11.3.6 Sustainability

Sustainability means long-lasting or long existence or long-term development of performance. The aspects of physical sustainability that can be affected by irrigation managers relate primarily to over- or under-supply of irrigation water leading to waterlogging or salinity.

$$Sustainability = \frac{Current\ Irrigable\ Area}{Initial\ Total\ Irrigable\ Area}$$

The initial area refers to the total irrigable area in the system's design or the latest rehabilitation. The current irrigable area is reduced because of salinity, water shortage, and low agricultural profitability due to urban and industrial development.

The sustainability of irrigation is also determined by the ratio as follows:

$$Relative\ Groundwater\ Depth = \frac{Actual\ Groundwater\ Depth}{Critical\ Groundwater\ Depth}$$

The critical groundwater depth mostly depends on the effective rooting depth of the crop. Waterlogging occurs when the water table rises, and soil in the plant's root zone gets saturated, and air circulation is stopped. The crop yield is usually affected when the depth of the water table below the land surface is equal to or less than the value given below for different crops, i.e., below effective rooting depth. The average effective rooting depths of paddy, wheat, and soyabean crops are 0.6 m, 0.6 m, and 0.5 m, respectively.

11.3.7 Performance Standards

Molden and Gates (1990) suggested the following performance standards (Table 11.2).

11.4 Examples

An experiment was conducted on an agricultural plot growing wheat crops. Data is given below. Calculate irrigation efficiencies.

1	Name of outlet	Outlet no. 4 R
2	Location of outlet	The right side of Gadarjudda minor
3	Village	Gadarjudda
4	Name of crop	Wheat
5	Method of irrigation	Check basin by border irrigation
6	Size of field	$11.80\ \text{m} \times 10.40\ \text{m} = 122.72\ \text{m}^2$
7	Soil type	Sandy loam
8	Crop age	55 days
9	Volume of water delivered (discharge measured using cut throat flume)	$6.27\ \text{m}^3$
10	Delivery time of water	15 min

Calculation of efficiencies conducted for a wheat plot:

Volume of water delivered (D) $= 6.27\ \text{m}^3$

Area of the plot $= 11.80 \times 10.40 = 122.72\ \text{m}^2$

Table 11.2 Performance standards

Measures	Performance class		
	Good	Fair	Poor
Adequacy	0.9–1.0	0.7–0..9	<0.7
Equity	0.85–1.0	0.7–0.85	<0.7
Timeliness	0.75–1.0	0.25–0.75	<0.25
Sustainability	0.95–1.0	0.85–0.95	<0.85

Depth of water applied = $\frac{6.27}{122.72}$ = 0.05105 m = 51.05 mm.

Average depth of infiltration = 19.20 mm.

Volume of infiltration = $\frac{19.20}{1000}$ × 122.72 = 2.35 m^3

Adjustment factor = $\frac{6.27}{2.35}$ = 2.66.

Soil moisture content (SMC) before test = 19.65%

Field capacity = FC = 21%

Soil moisture deficit = SMD = $\frac{FC-SMC}{100}$ × BD × Z.

where

BD = the bulk density of soil = 1.45 m.

Z = the root zone depth = 0.90 m.

∴ SMD = $\frac{21-19.65}{100}$ × 1.45 × 0.90 = 0.0176 m

Volume of soil moisture deficit (R) = 0.0176 × 122.72

= 2.15 m^3

Deficit after irrigation in root zone (B) = 0.

11.4.1 Field Irrigation Efficiencies

1. Storage efficiency (Es) = $\frac{R-B}{R}$ × 100

   Es = $\frac{2.15-0}{2.15}$ × 100 = 100%

2. Percolation efficiency (Ep) = $\frac{R-B}{T}$ × 100

   where, T = D − R = 6.27 – 2.15 = 4.12 m^3

   Ep = $\frac{2.15-0}{4.12}$ × 100 = 52%

3. Delivery efficiency (Ed) = $\frac{D-R}{D}$ × 100

   Ed = $\frac{6.27-2.15}{6.27}$ × 100 = 65.70%

11.4.2 Ponding Test to Find Seepage Loss

The ponding method of measuring irrigation channel water loss involves filling a closed channel reach at both ends for ponding the water and measuring the rate at which the ponded water seeps and leaks out of reach.

(a) Materials:

A channel reach with a length of 30 m, staff gauge, wooden plot/plank, gauge reading, datasheet, stopwatch, tape measures, and hammer.

(b) Procedures:

- Ponding loss measurements are made by installing leak-free dams at the beginning and end of channel reach. The length of the channel reach is 30 m (Fig. 11.4).
- Filling the section with water to above the normal operating level.
- Record how quickly the water level drops and then take water level readings periodically for at least 24 h.
- Measure the average water surface top width at the section's operational level, which is the only other data required to determine the steady-state loss rate from the reach.
- Plot the staff gauge reading versus time and measure the slope of a curve fitted to the data at the operational level.

(c) Calculation of loss rate: From the curve of gauge reading versus time, the loss rate is:

$$Q_L = \frac{dd}{dt} \times TW \times \frac{1}{3600} \times 1000 \times 100 \times \frac{1}{1000}$$

$$or\ Q_L = \frac{dd}{dt} \times TW \times \frac{1}{36}$$

where

Q_L = the loss rate in (lps/km);

dd = the change in water surface level per unit time (cm/hr); and

Fig. 11.4 Layout for ponding measurements

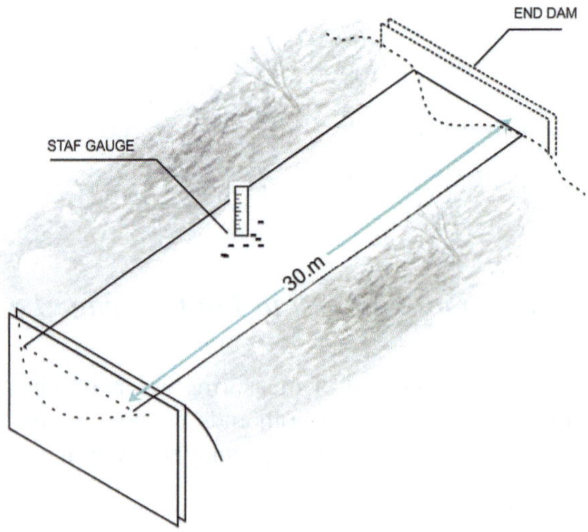

TW = the average top width at the same water surface level (cm).

(d) Example of an experiment

The Gadarjudda minor is one of the distributaries of the Upper Ganga Canal system (UGC). Its command area lies in the head reach of UGC. A thirty-meter length of channel reach of the Gadarjudda minor canal was selected for the ponding experiment.

- The operational water level was determined by inquiry, which was varied by the flow calculation.
- Depth of optional water level = 65 cm.

The observed data are given below:

Table 11.2 Water surface recession data

Day	Time	Gauge reading (cm)	Staff reading (cm)
1	12:05	87.67	70.00
1	12:25	87.21	69.54
1	12:45	86.34	68.67
1	13:05	86.07	68.40
1	13:45	85.51	67.84
1	14:15	85.43	67.76
1	14:30	85.05	67.38
1	15:00	84.90	67.23
1	15:30	84.00	66.23
1	16:00	83.25	65.58
1	17:30	81.50	63.83
2	9:45	74.34	56.67
2	11:45	74.34	56.42

The width of water surface measured at 5 different sections in the reach during the ponding test is given below:

Table 11.2 Channel section water surface measurement

Date	Time	Gauge	Water surface width (m)					Average width (m)
			I	II	III	IV	V	
1	12:05	87.67	1.43	1.50	1.50	1.60	1.57	1.52
1	16:00	83.25	1.42	1.49	1.45	1.60	1.53	1.49
2	11:45	74.34	1.32	1.45	1.40	1.52	1.50	1.43

Figure 11.5 shows a plot of gauge reading versus time. From water surface recession versus time, the seepage loss is given by:

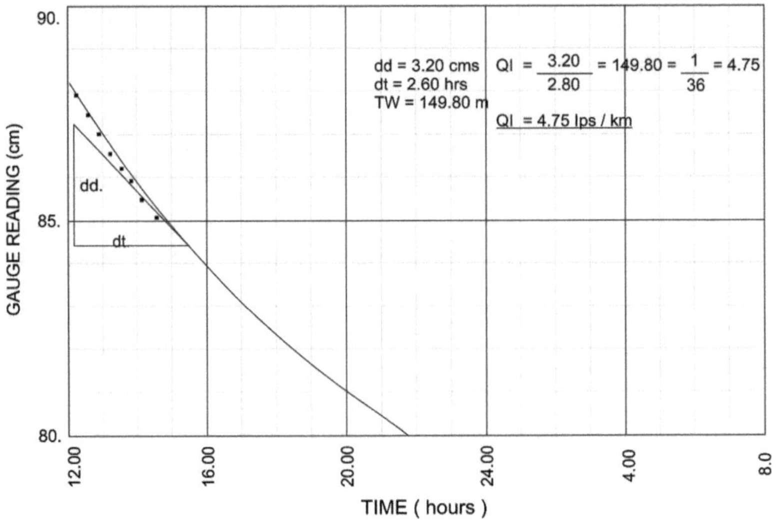

Fig. 11.5 Graph of water level change with time

$$Q_L = \frac{dd}{dt} \times TW \times \frac{1}{36}$$

$$\text{dd} = 3.2\ \text{cm};\ \text{dt} = 2.8\ \text{cm so dd/dt} = 1.1428$$

$$\text{TW} = 149.8\,\text{cm}$$

$$Q_L = 1.1428 \times 149.8 \times \frac{1}{36}$$

Hence,
seepage loss rate (Q_L) is = 4.75 lps/km.

11.4.3 Seepage Loss by Inflow-Outflow Method

The inflow-outflow method of measuring conveyance loss in a channel involves the measurement of the amount of water flowing into the channel at a section. Discharge at the head as well as at the tail of the section is measured by a current meter, and the discharge passing through all the outlets between the two points of measurement is also measured at the same time.

In this case, the discharge of Godarjuda minor is calculated at the head (at Ch. 30 m) and tail (at Ch. 1460 m). The loss can be expressed in the form of a rate of decrease in inflow rate per unit length of the channel. Thus, the loss rate is given by

$$Q_L = \frac{Q_{in} - Q_{out}}{L} \text{ per unit length}$$

where

Q_L = the loss rate per unit length (lps/km),
Q_{in} = the discharge at the head (lps),
Q_{out} = the discharge at tail (lps), and
L = the length of the section (km).
The loss rate can also be expressed in percent per km.

$$Q_L = \frac{Q_L}{Q_{in}} \times 100 = \frac{Q_{in} - Q_{out}}{L \times Q_{in}} \times 100 = (\%)$$

The discharges calculated at the head and tail are tabulated below:
Perimeter and area of channel section:
At head (i) Perimeter: $0.64 + 0.64 \times 1.67 = 2.95$ m.
(ii) Area: $\frac{1.67+2.09}{2} \times 0.615 = 1.1562\,\text{m}$.

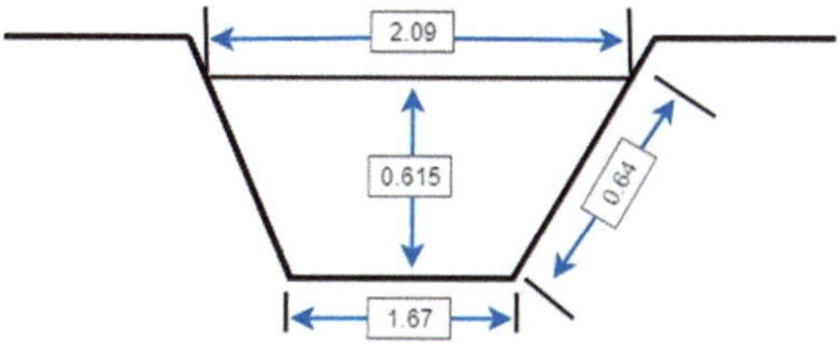

At tail (i) Perimeter: $0.64 + 0.64 \times 1.14 = 2.34$ m
(ii) $\frac{1.72+1.347}{2} \times 0.533 = 0.762\,\text{m}$

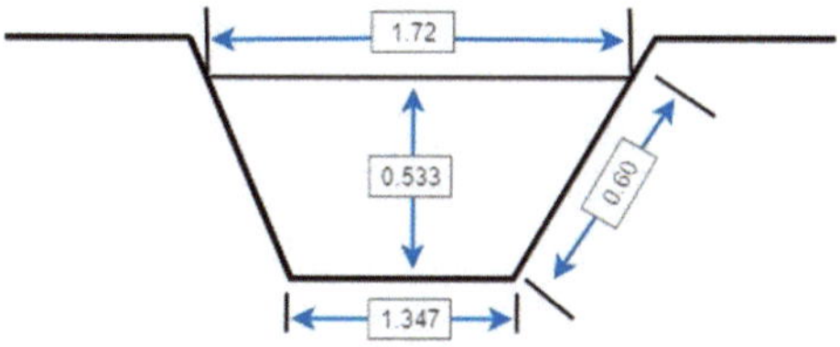

Velocity and discharge (current meter)

1. At the tail

Date	Water	Depth	Current meter reading						Av rev	Vel m/s	Discharge m^3/s
			Left side		Centre		Right side				
			Rev	Time	Rev	Time	Rev	Time			
7.2.89	0.533		50	54	50	55	50	53	–	–	–
			50	53	50	53.6	50	52.7			
			50	53.47	50	53.7	50	53.6			
			50	54.1	50	54	50	53			
	Average		50	53.7	50	54.07	50	53	50	0.6437	0.490

2. At the head

Date	Water	Depth	Current meter reading						Av rev	Vel m/s	Discharge m^3/s
			Left side		Centre		Right side				
			Rev	Time	Rev	Time	Rev	Time			
7.2.89	0.615		40	53	40	54	40	51	–	–	–
			40	55	40	51	40	50			
			40	55	40	51	40	51.5			
	Average		40	54.3	40	52	40	50.8	40	52.37	0.5257

Total discharge at head = 604 lps.
Total discharge at tail = $\frac{490\ \text{lps}}{114\ \text{lps}(balance)}$.
Discharge through the outlet = 35 lps.
Net loss = 79 lps per 1430 m.
Hence, loss per km = $Q_L = \frac{Q_{in}-(Q_{out}+Q_{oL})}{1.43km} = \frac{Q_{net}}{1.43}$

$$Q_L = \frac{604-(490+35)}{1.43} = 55.24\ \text{lps}$$

11.4.4 Check List of Conveyance Canal Above Outlet

Check List for Canal

1. What is the condition of the canal?

   (a) Lined
   (b) unlined

2. Is there a cavity behind the lining?
   (a) Yes
   (b) No
3. If yes, how is the cavity condition?
   (a) Big, needs urgent maintenance.
   (b) Small
4. Are there any cracks on the panel of lining?
   (a) Yes
   (b) No
5. What is the condition of joints between the panels of lining?
   (a) Worse, seepage is occurring through it.
   (b) Working good, and no seepage is occurring?
6. Are the pressure release valves behind the lining working properly?
   (a) Yes
   (b) No
7. Is there vegetation and aquatic growth in the canal?
   (a) Yes
   (b) No
8. Is there siltation occurring in the canal?
   (a) Yes
   (b) No
9. Is there any erosion from the inner slope of the embankment to the canal?
   (a) Yes in the chainages:
   (b) No
10. Is there any unauthorized cutting of the bank and unauthorized withdrawal of water?
    (a) Yes, from chainages:
    (b) No
11. Is there any unauthorized entry of animals into canal to bath or to cross the canal?
    (a) Yes
    (b) No
12. Is the cross slope of the service road well-maintained and free from holes?
    (a) Yes
    (b) No, there is a problem in the Chainages:

Table 11.3 Questionnaire for canal structures

S. No.	Questions	Conditions	In details if any	Remarks
1	Are the gate **well greased?**	Yes		
		No, Not at		
2	Are the gate parts **free from rust?**	Yes		
		No, Not at		
3	Is there any **leakages from gates and shutters** due to worn out seals ?	Yes at		
		No		
4	Is there any **unathourized removal gate/metal parts?**	Yes at		
		No		
5	Are the **joint sealing material** at structure working properly?	Yes		
		No, Not at		
		Str. Ch:		
6	Are the **Trashracks at inlet of Canal CD structure** working properly?	Yes		
		No, Not at		
		Str. Ch:		

13. Is there any unauthorized use of the service road by cars and trucks disturbing inspection works?
    (a) Yes
    (b) No
14. Is there unauthorized removal of lining panels or blocks from canals?
    (a) Yes
    (b) No

Check List for Canal Structures: Questionnaire is shown in Table 11.3.

11.5 Case Study on Performance Evaluation of an Irrigation Project

Chaube (2006) conducted a study on evaluating four tank irrigation projects in the Sagar District of Madhya Pradesh, India. Subsequently, using the data from this report and other documents, Nissanka (2007) carried out a performance evaluation of tank irrigation projects. The discussion in what follows is based on these studies.

The Khairana tank irrigation project in the Sagar district of Madhya Pradesh state in India is selected for a case study on performance evaluation. The main purpose of this case study is to illustrate the procedure for evaluating performance parameters.

Tanks are designed for the storage of monsoon runoff of local streams. Irrigation is proposed in Kharif (wet) and Rabi (dry) seasons, mainly emphasizing Rabi irrigation.

The following parameters have been calculated based on the available data:

- Relative water supply (an indicator of adequacy);
- Relative productivity potential (an indicator of timeliness);
- Production performance (an indicator of timeliness);
- Crop income performance (an indicator of timeliness);
- Relative to groundwater depth (an indicator of sustainability);
- Relative to irrigation area (an indicator of sustainability);
- Delivery of fair amount (an indicator of equity);
- Water delivery scheduling performance; and
- Cropping intensity performance.

11.5.1 Adequacy

$$Relative\ Water\ Supply = \frac{Irrigation\ Application + R.F.}{Etc + P + S} = \frac{Water\ Supply}{Water\ Demand}$$

- Assume 75% of the live storage is available for the Rabi season.

A. Adequacy as per design stage:

$$= \frac{1.466MCM \times 0.75}{1.76MCM} = 0.62$$

B. Adequacy as per the existing stage

$$= \frac{1.466MCM \times 0.75}{1.70MCM} = 0.65$$

Illustrate (compute) each of above indices using data.

11.5.2 Timeliness

$$Relative\ Productivity\ Potential = \frac{Crop\ Production\ in\ Kg}{Water\ Supply\ in\ (\mathrm{m})^3}$$

A. Relative Productivity Potential in Design condition

$$= \frac{481250}{1.466 \times 0.75 \times 10^6} = 0.44\ \mathrm{Kg/m^3}$$

Table 11.4 Average ground water depth for Khairana project command

S. No.	Item	Pre-monsoon	Post monsoon	Rabi season	Water table fluctuations	
		(May) (m)	(Nov) (m)	(Jan) (m)	May–Nov (m)	May–Jan (m)
1	Water table before the irrigation project (BGL)	12.56	6.44	7.57	6.12	4.99
2	Water table after irrigation project (BGL)	9.47	5.23	6.78	4.26	2.7

B. Relative Productivity Potential in Existing condition

$$= \frac{213375}{1.466 \times 0.75 \times 10^6} = 0.19 \text{ Kg/m}^3$$

11.5.3 Ground Water Sustainability

Six years of data on groundwater level observations in the wells located at Kachhi Piparia, Chandpur, Chirari, and Gourjhamar in the command area of Khairana Project Command have been analyzed. Table 11.4 shows the average groundwater depth (pre- and post-monsoon and in January) in the Khairana Project Command.

$$Relative\ Groundwater\ Depth = \frac{Actual\ GW\ Depth}{Critical\ GW\ Depth}$$

$$\text{Critical Ground Water Depth} = \text{Root zone depth of crops} = 0.60 \text{ m}$$

$$= \frac{5.23}{0.60} = 8.72$$

11.5.4 Irrigable Area Sustainability

Current irrigation area data for each project have been shown in Chap. 7

$$Sustainability = \frac{Current\ Irrigable\ Area}{Initial\ Total\ Irrigable\ Area}$$
$$= \frac{80}{275} = 0.29$$

11.5.5 Cropping Intensity Performance

The cropping pattern and cropping intensity were computed.

Crop	Khairana CCA-308 ha	
	Design %	Existing %
Kharif		
Soybean	–	69.0
Pulses	–	0.3
Vegetable	–	–
Paddy	71.43	0.3
Rabi		
Gram	–	49.0
Wheat	89.29	31.0
Pulses	–	17.0
Vegetable	–	–
Cropping intensity %	160.71	166.6

$$Cropping\ Intensity\ Performance = \frac{Actual\ Cropping\ Intensity}{Target\ Cropping\ Intensity}$$
$$= \frac{166.6}{160.71} = 1.04$$

11.5.6 Summary Results of Case Study

Adequacy: The parameter is 0.65. It means poor performance.

Timeliness

(i) It is 43% (poor class), as indicated by the relative productive potential (RPP).
(ii) It is 44%, as indicated by crop production.

Project	Total crop production (metric ton)-Rabi		Performance %
	Design stage	Existing stage	
Khairana	481.25	213.38	44.0

(iii) Indicated by crop income: In US$

Project	Design stage		Existing stage		%	RPP	Performance class
	Income (US$)	US$/m^3 Irrigation	Income (US$)	US$/m^3 Irrigation			
Khairana	175,000	0.15	133,232	0.11	76%	73%	Fair

Sustainability

i. Indicated by the relative groundwater depth. The average groundwater depth is 5.23 m, and the index is 8.72. It means no waterlogging before the start of irrigation in the command area. However, water logging in the command has occurred due to the excessive irrigation practice in black cotton soil, which is poorly drained.
ii. According to the sustainability performance class, the Khairana project belongs to the poor category as indicated by the irrigation area achieved: The design irrigation area in Rabi season is 275 ha. The average irrigation area achieved in Rabi season is 80 ha. The sustainability index is 0.29.

Cropping Intensity Performance: Design and existing and design cropping intensities are 166.6 and 160.71. The ratio is 1.04.

Appendix: Performance Evaluation of Right Main Distributary of Upper Ganga Canal

The Upper Ganga Canal (UGC) was commissioned in 1854 by diverting water from the River Ganga at Haridwar into the UGC system. The initial design discharge of 189 cumecs has been gradually increased to the present-day capacity of 297 cumecs. The total command area of UGC is 2.023 M ha, out of which the CCA on outlets is 1.011 M ha. This is spread over the western part of Uttar Pradesh and some parts of Uttarakhand State in India.

The original planning (drought protection) and design objectives of the scheme are not relevant in the context of present-day extensive agriculture and ever-increasing irrigation demands existing for all the time in a year. A major part of the water supply is going through the extensive sugarcane cultivation in the upper part of the project area.

Table 11.5 Salient features of right main distributary

S. No.	Name of canal	Off-take points (Km)	Length (Km)	CCA (Ha)	PPA (Rabi)	PPA (Kharif)	Authorised discharge in Cumecs	No of direct outlets including tail cluster	Remarks
1	Right main Dy	7.26 km. of Deoband branch	17.70	2111	866	1393	2.69	54	2111 Ha direct irrigation

Right Main Distributary

It takes off at 7.26 km from Deoband Branch. It has two nos of distributaries and three nos of minors, as shown in the schematic diagram (Fig. 1).The command area lies between latitude 29°50′N and 29°55′N and longitude 77°50′E and 77°55″E. It has a CCA of 2111 ha from direct outlets of this distributary and a total of 5890 ha from the whole of Right Main Dy system. Table 11.5 shows the salient features.

A field-based study of performance evaluation of the right main distributary was taken up to examine and evaluate, with recent practices about the deficiency of engineering infrastructure, existing efficiency in operation, the present status of equity in the distribution of water, use of available water resources, improvement in irrigation infrastructure, the overall yield of crops, and impact of socio-economic aspect in command of the canal.

Canal Operation Plan

Within the diversion headwork operating rules framework, water distribution has been guided by the following principles.

(i) The main canal operates continuously during land preparation, crop establishment, and flowering stage of the crop and intermittently during nursery and post-flowering stages.
(ii) Canal runs at FSL during continuous /intermittent supply as far as possible.
(iii) During October–January for vegetables and November–March for Berseem and Potato, partial water supply becomes necessary in all the channels, which is being supplied equitably.
(iv) The off-takes from the main canal have been operated in an “ON & OFF” mode.
(v) The off-takes of minors have been proportionately distributed without gates. They are operated in an “ON & OFF” mode.

(vi) The outlets in distributaries and minors are provided with proportional distribution without gate.

(vii) Farmer's committee has not been formed for any outlets to watch and distribute the water among the farmers equitably.

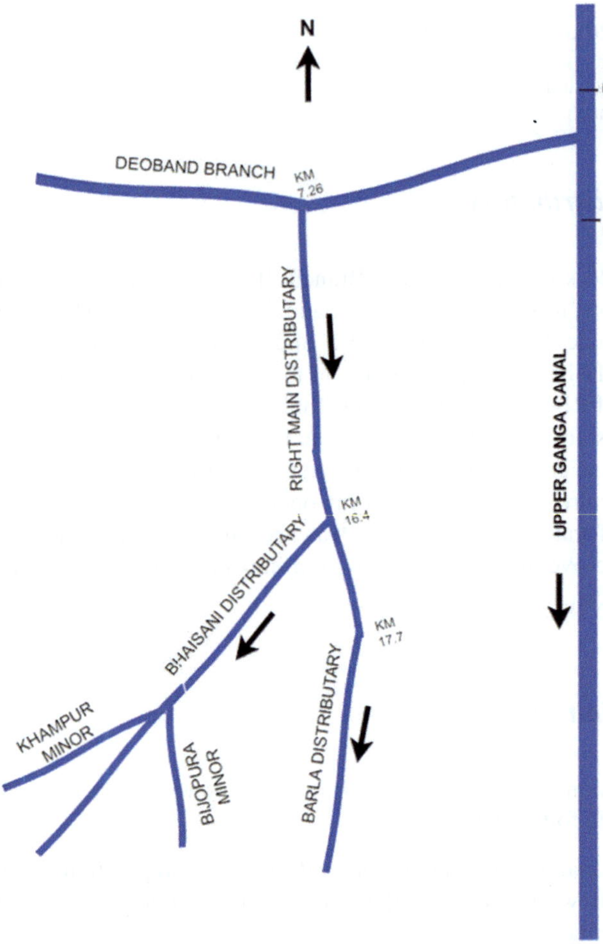

Fig. 11.6 Index plan of right main distributary of Upper Ganga Canal

Parameters of Irrigation Performance

Relative Water Supply as a Measure of Adequacy

The early measures of adequacy revolved around the duty of water. Abernethy (1990) mentioned that duty is especially suited for a water deficit environment. It is approximately related to water requirements only. The water requirement for a given agricultural policy is a function of land area, crop consumptive use, application losses, and leaching requirement.

In the physical sense, Relative Water Supply (RWS) is the ratio of water supplied to an irrigation unit and the demand for water in that unit over a period of time.

Knowledge of relative water supply (RWS) could be used to specify characteristics, irrigation organization, and system management. The degree to which capability is actually realized is called intensity of management. By varying intensity, it is possible to match supply with demand. Relative Water Supply (RWS):

$$RWS = \frac{Water\ supplied\ to\ a\ command\ area}{Demand\ of\ water\ in\ that\ area} = \frac{IR + RN}{ET + S\&P}$$

where

IR = the mean irrigation application (mm/day);

RN = the mean rainfall (mm/day);

ET = the mean evapotranspiration (mm/day); and

S & P = the seepage and deep percolation (mm/day).

Hence, RWS is the inverse of water use efficiency. When RWS is low, it represents water scarcity giving rise to higher tension.

Abernethy (1990) argued that including percolation losses into demand affected the meaning of RWS. Because of losses (may be due to poor water management practices) where more demand, would be more and hence RWS would be low, indicating a better management.

Molden and Gates (1990) developed a procedure to determine the relative contributions of hydraulic structures and management procedures to water delivery adequacy. As per Clemmens and Bos (1990):

$$Water\ delivery\ performance = \frac{Amount\ of\ water\ actually\ delivered}{Amount\ of\ water\ intended\ to\ be\ delivered}$$

If the amount of water actually delivered is based on frequent flow measurements, there is a greater likelihood that managers can match actually to intended/scheduled flows. If water delivery performance ratio nears unity, then the management inputs must be effective.

Besides the water delivery performance ratio, the following also measures adequacy.

(i) $Field\ Application\ Ratio = \frac{(ET_p - Pe)}{Volume\ of\ Water\ delivered\ at\ field}$

where

ETp = the evapotranspiration by irrigated crop;

Pe = the effective part of precipitation; and

ETp − Pe = the volume of irrigation water needed to avoid undesirable stress in the throughout the growing cycle.

(ii) $Overall\ Performance\ Ratio = \frac{ETP-Pe}{Vc+Vi}$

where

Vc = the volume of irrigation water delivered from river/reservoir, and

Vi = the inflow from other sources to conveyance system.

Inter-Quartile Ratio as Measure of Equity

Abernethy (1990) proposed the use of inter-quartile ratio (IQR) defined as the ratio of volume delivered, respectively, to the most favored upper 25% of the command area and the least favored lowest 25% of the command area.

The meaning of IQR is readily grasped by a layman, so it facilitates communication among those involved in the inequity problem. The desirable density of measurements for the definition of IQR is in the range 20–50 measurement points. Within the land area supplied from each measuring point, it is necessary to make some assumptions about the degree of equity that exists downstream. The common value of IQR = 3. There is some degree of correlation between the incidence of inequity and irregularity in supply. The areas (near canal tails) which are at the bottom of an equity curve are the most likely to receive their water erratically. Their season totals of water received may include periods of excess supply (for instance, after a mid-season rainfall when everyone's requirements are temporarily satisfied).

Relative Productivity Potential as Measure of Timeliness

Timeliness means the correspondence of water supply to crop water requirements, and it relates to the distribution of water supply across the season to some utility-based standard. For many irrigation activities, the flow rate must be near the intended flow rate for water use to be effective (Clemmens and Bos 1990).

The simplest method to assess timeliness is to calculate relative productivity potential.

Monthly data appear to give a good indication of whether the discharge is more or less timely supplied.

Sustainability

Aspects of physical sustainability that can be affected by irrigation managers relate primarily to the over or undersupply of irrigation water leading to water logging or salinity.

Sustainability = Current irrigable area/Initial total irrigable area.

The sustainability of irrigation is also determined by the ratio as follows:

Relative Ground Water Depth = Actual groundwater depth/Critical groundwater depth.

The critical groundwater depth mostly depends on the (effective rooting depth) of the crop. If the actual groundwater depth is near the critical depth, the time interval between readings of the ratio should be near one month. One year is suitable for most other purposes.

Concept of Potential Productivity

(i) It is useful to estimate the effect of water deficiency on crop yield and differentiate between effects, on ultimate yield, of water shortage that may occur at different stages in crop history. From this kind of information, we can calculate a ratio between the output that can be achieved under this water supply and the output that would have been achieved if the water supply had been ideally matched to the needs of the crop. This ratio is called the potential productivity of this particular pattern of water delivery (Abernethy 1986). It produces a number in the range 0 to 1, which would be obtainable under a given water supply regime. The execution of this calculation should take into account soil characteristics. Because, in practice, water is supplied irregularly, and what matters is not precise timeliness of delivery but the state of the moisture content in the root zone.

The potential productivity index is worked out based on weekly or ten-day averages of water inputs and outputs, with allowances for field seepage loss based on whatever local observations may be obtainable. Such an index would, in one value, sum up the system's performance in meeting crop demand fairly well if relevant data for adequacy and timeliness could be collected.

(ii) Productivity means production per unit of input. We may have land productivity (tonnes/hectare) or yield. We may have water productivity (kg of crop/m^3 of water) or water use efficiency. Hence, input means hectare of land or m^3 of water in the case of land or water, respectively. Where water management is a focus, water may be the dominant constraint on output.

As a numerical parameter, the concept of potential productivity enables us to link adequacy and timeliness into one number or indicator of performance.

Results of Analysis

The tables given below show the performance parameters.

Table 11.6 Relative Water Supply (RWS) in Different Channels (May to October).

Table 11.7 Relative Water Supply (RWS) in Different Channels (November to April).

Table 11.8 Relative Water Supply (Season Wise).

Table 11.6 Relative water supply (RWS) in different channels (May to October)

S No.	Name of canal	Description	May		June		July		August		September		October		Total May to October	
			Total reach	Tail reach	Total reach	Tail reach	Total reach	Tail reach	Total reach	Tail reach	Total reach	Tail reach	Total reach	Tail reach	Total reach	Tail reach
1	Right main Dy	Water supply (ham)	147.5	–	117	–	117.6	–	104.2	–	76.2	–	61.5	–	624	–
		Water demand (ham)	449.5	–	585.4	–	599.7	–	558.9	–	729.3	–	443.3	–	3366.1	–
		RWS	0.33	–	0.20	–	0.20	–	0.19	–	0.10	–	0.14	–	0.19	–

Table 11.7 Relative water supply (RWS) in different channels (November to April)

Sl. No.	Name of canal	Description	November		December		January		February		March		April		Total Nov to April	
			Total reach	Tail reach	Total reach	Tail reach	Total reach	Tail reach	Total reach	Tail reach	Total reach	Tail reach	Total reach	Tail reach	Total reach	Tail reach
1	Right main Dy	Water supply (ham)	91.8	–	39.5	–	68.4	–	36.4	–	70.3	–	119.7	–	426.1	–
		Water demand (ham)	340.4	–	227.7	–	207.46	–	282.4	–	377	–	320.9	–	1755.86	–
		RWS	0.27	–	0.17	–	0.33	–	0.13	–	0.19	–	0.37	–	0.24	–

Table 11.8 Relative water supply (season wise)

S. No.	Name of canal	Location along canal				Average	Remarks
		Total reach		Tail reach			
		May to Oct	Nov. to April	May to Oct	Nov. to April		
1	Right main Dy	0.19	0.24	–	–	0.22	
2	Bhaisani Dy	0.21	0.21	–	–	0.21	
3	Barla Dy	0.26	0.31	0.10	0.09	0.19	Lowest
4	Khoja Nagla Dy	0.37	0.39	0.32	0.50	0.40	Highest
5	Bijopura Mr	0.27	0.23	0.44	0.41	0.34	
6	Khampur Mr	0.15	0.11	0.09	0.36	0.18	
7	Total of Right Main Dy System	0.26	0.35	–	–	0.31	

Table 11.9 Water Delivery Performance (Relative Productivity Potential) Period: May to October.

Table 11.10 Water Delivery Performance (Relative Productivity Potential) Period: November to April.

Table 11.11 Abstract of Water Delivery Performance in Terms of Relative Productivity Potential.

Table 11.12 Inter-Quartile Ratio.

Table 11.13 Sustainability of Command Area.

Recommendation

The following parameters should also be evaluated. Financial viability and social viability are more important than economic viability. Due to the paucity of data, these parameters could not be evaluated in the present study.

(a) **Economic Viability of Indicator**

The system manager is most likely concerned with the financial resources available at the system level and the source of those funds. He is less concerned with the overall profitability of agriculture and least concerned with the overall profitability of the irrigation project that created the systems.

(b) **Financial Viability of Irrigation Systems**

One set of indicators concerns efforts to raise revenues from water users that help support management, operation, and maintenance costs as follows:

Table 11.9 Water delivery performance (relative productivity potential) period: May to October

S. No.	Name of canal	Description	Unit	Location along the canal							
				Total reach				Tail reach			
				Paddy	Other Kharif	Sugarcane (12 months)	Total	Paddy	Other Kharif	Sugarcane (12 months)	Total
1	Right main Dy	Crop production	10^4 kg	101	208.5	8939	9248.5	–	–	–	–
		Water supply	10^4 m^3	421.8 × 0.19 = 80.14	225.6 × 0.19 = 42.86	(2718.7 + 1030.5) × 0.22 = 824.82	947.82	–	–	–	–
		Relative productivity potential	Kg/m^3	1.26	4.86	10.84	9.76	–	–	–	–

Remark Total area irrigated by right main dy paddy = 202 ha other kharif = 417 ha sugarcane = 1277 ha

Table 11.10 Water delivery performance (relative productivity potential) period: November to April

S. No.	Name of canal	Description	Unit	Location along the canal					
				Total reach			Tail reach		
				Wheat	Barseem	Total	Wheat	Barseem	Total
1	Right main Dy	Crop production	10^4 kg	233.6	1004	1237.6	–	–	–
		Water supply	10^4 m^3	93.07	81.02	174.09	–	–	–
		Relative productivity potential	Kg/m^3	2.51	12.39	7.11	–	–	–

Remark Total area irrigated by Right Main Dy Wheat = 584 ha Other Rabi = 251 ha

Table 11.11 Abstract of water delivery performance in terms of relative productivity potential

S. No.	Name of canal	Location along canal				Remarks i.e., average
		Total reach		Tail reach		
		May to Oct	Nov. to April	May to Oct	Nov. to April	
1	Right main Dy	9.76	7.11	–	–	8.44

(i) Total financial viability = Actual MO & M allocation/Total MO & M requirements; and
(ii) Return on investment = Gross value of output/Investment for the irrigation system.

(c) **Social Viability**

If long-term sustainability of irrigation is the objective and improving and maintaining social well-being is ultimately important, then social viability is relevant from a strategic management perspective.

(i) Administrative performance = (Equivalent water rights paid current years + Equivalent water rights paid previous years)/Total equivalent water rights due to the current year

Equivalent water rights are expressed in m′/ha. Delayed payment of water charges may cause major annual fluctuation in the above indicator of "administrative viability" which is quantified as a moving average over 3–5 years.

Irrigation employment generation = (Annual day/ha labour by scheme)/(Annual number of official working days).

$$Equity\ Ratio = \frac{1}{(Inter\ Quartile\ Ratio)} = \frac{1}{2.15} = 0.465$$

Table 11.12 Inter-quartile ratio

		Design discharge (Cumec)	Location along canal				Inter-quartile ratio		Average
			Volume (ha-m) delivered to the most favoured upper 25% of CCAA		Volume (ha-m) delivered to the most favoured lowest 25% of CCA				
			May to Oct	Nov. to April	May to Oct	Nov. to April	May to October	November to April	
1	2	3	4	5	6	7	8 = 4/6	9 = 5/7	10 = (8 + 9)/2
1	Bhaisani Dy	1.22	145.30	72.30	49.84	25.50	2.92	2.84	2.88
2	Barla Dy	0.34	62.80	36.80	23.20	9.80	2.71	3.76	3.23
3	Khoja Nagla Mr	0.25	35.15	20	17.10	16.60	2.06	1.20	1.63
4	Bijopura Mr	0.09	4.60	2.20	3.50	1.76	1.31	1.25	1.28
5	Khampur Mr	0.09	3.76	1.38	2.13	0.80	1.77	1.73	1.75
Average							2.15	2.15	2.15

Table 11.13 Sustainability of command area

S. No.	Name of canal	Year	Initial total irrigable area (Ha)	Current irrigable area (ha)	Agricultural land lost (Ha)	Sustainability of irrigable area
1	2	3	4	5	6 = 4–5	7 = 5/4
1	Right main Dy	1998–99	2111	1911	200	0.905
		1999–00	2111	1763	348	0.835
		2000–01	2111	1846	265	0.874
		2001–02	2111	1885	226	0.893
		2002–03	2111	1962	149	0.929
		2003–04	2111	2005	106	0.950
Average			2111	1895.3	215.7	0.898

Questions

1. Identify the parameters for performance related to (i) adequacy of water supply, (ii) timeliness of water supply, and (iii) equity in water supply.
2. Identify the data required for the evaluation of sustainability parameters.

3. List the data required for the evaluation of water conveyance efficiency and application efficiency.
4. Explain the use of ground water level data in the evaluation of performance of a canal irrigation project.
5. Explain the use of crop-related data in the evaluation of performance of a canal irrigation project.
6. Explain the use of area-related data in the evaluation of performance of a canal irrigation project.
7. Critically examine the evaluation of performance parameters of Right Main Distributary of Upper Ganga Canal.
8. Critically examine the Relative Water Supply parameter of Right Main Distributary of Upper Ganga Canal
9. Critically examine the Relative Productivity Potential parameter of Right Main Distributary of Upper Ganga Canal
10. What data are required to evaluate (a) social viability, financial viability of an irrigation project?

References

Abernethy CL (1986) Performance measurement in canal irrigation. IIMI/ODI. Irrigation Management Network, Paper 86/2d

Abernethy CL (1990) Indicators of the performance of irrigation water distribution systems. In: Memiographed paper, International Irrigation Management Institute, Colombo, Sri Lanka, pp 1–22

Bos R (2000) Guidelines on performance assessment. Report on workshop international commission of irrigation and drainage

Chaube (2006) Report on evaluation of rural infrastructure (irrigation) projects in Sagar District of Madhya Pradesh; Sponsored Research by National Bank for Agriculture and Rural Development (NABARD), Mumbai, India

Clemmens AJ, Bos MG (1990) Statistical methods for irrigation system water delivery performance evaluation. Inig Drainage Syst 4(4):345–365

Molden DJ, Gates TK (1990) Performances measures for evaluation of irrigation water delivery systems. J Irrig Drainage 116:804–823. https://doi.org/10.1061/(ASCE)0733-9437(1990)116:6(804)

Nissanka N (2007) O & M aspects of small tank irrigation projects—some case studies. Unpublished report of M.Tech Dissertation, supervised by Prof. U C Chaube, IIT Roorkee, Roorkee

Prasad R (2004) Study on performance evaluation of right main distributary in Upper Ganga Canal System. Unpublished report of M.Tech Dissertation, supervised by Prof. U C Chaube, IIT Roorkee

Reading Material

Abernethy CL (1987) The objectives of irrigation water management and methods of measuring their achievements. In: Främji KK (ed) Improvement in inigation management with special reference to developing countries. LC.I.D., New Delhi, State of the Art No 4, pp 55–67

Asian Development Bank (ADB) (2006) Rehabilitation and management of tanks in India. A study of select states, Publication Stock No. 122605. Manila

Chambers R (1984) Irrigation management: ends means and opportunities. In: Pant N (ed) Productivity and equity in irrigation system. Ashish Publishing House, New Delhi

Chapter 12
Use of Ground Water in Canal Command Area

Abstract Depending upon the purpose, a number of ground water use patterns are possible in the area served by a canal network. The concept and positive and negative factors involved in the use of ground water in canal areas are explained. The scope of ground water use in canal commands includes natural and artificial recharge (monsoon canals), use of shallow or deep tube wells by farmers and by public agency for supplementary irrigation during periods of low canal supplies or canal closures, and managing water supply quality. Augmentation tubewells along the Western Yamuna Canal (Haryana) are an excellent example of planned conjunctive use. A detailed discussion is made of the four possible alternatives for operating ground water facilities, depending upon field conditions. Guidelines for minimum and maximum limits on groundwater extraction, total groundwater extraction, and the quantum of groundwater use for irrigation are discussed. There are certain issues that need to be resolved for efficient operation of ground water facilities. These relate to (i) water prices from canal and tube well sources; (ii) formation of water users association for equitable distribution of canal water and ground water; (iii) control over the development and use of groundwater through private tube-wells; (iv) coordination at field level among activities of various government agencies; and (v) conflict interfaces. A case study on ground water recharge and its withdrawal in Khairana tank irrigation project is disused. The study explains (i) ground water recharge, (ii) canal water budget, (iii) design of wells, and (iv) design of pumping facility.

12.1 Introduction

Use of groundwater in canal command areas has long been practiced in one form or the other in several countries. The Irrigation Commission (GOI 1972) and the National Commission on Agriculture (1976) have elaborated on conjunctive use in detail. The Central Water Commission (CWC 1995) has prepared a document "Guidelines for Planning Conjunctive Use of Surface Water and Groundwater in Irrigation Projects." This chapter provides useful information for investigating and assessing groundwater and its use in canal command areas. Irrigation planners and

U. C. Chaube et al., *Canal Irrigation Systems in India*, Water Science and Technology Library 126, https://doi.org/10.1007/978-3-031-42812-8_12

managers need to follow these guidelines to introduce groundwater use in irrigation commands during the O&M stage of the existing projects.

Conventionally a canal irrigation scheme is designed based on the dependable availability of surface water only. A number of ground water use patterns are possible in the area served by the canal network, depending upon the purpose. These are:

i. Dugwells, dug-cum-bore wells, and shallow private tubewells by farmers.
ii. Direct irrigation water supply from deep public tubewells by a public agency.
iii. Augmentation of tubewells for supplementing canal supplies by agency owning canal network.
iv. Monsoon canals for the dual purpose of providing Kharif irrigation and recharge to groundwater bodies for extensive use in the Rabi period.
v. Vertical drainage to lower the ground water level below the root zone of crops.
vi. Use of saline groundwater by mixing it with canal waters.
vii. Induced recharge from rivers during floods.

12.2 Conjunctive Use: Positive and Negative Factors

Conjunctive Use Concept

Conjunctive use management implies coordinated and harmonious development and use of surface water and ground water for meeting water requirements in time and space. Conjunctive use is a result of the interaction between surface water and groundwater in time and space. The conjunctive use concept takes on the form of (i) recharge of the groundwater from rainfall; (ii) recharge due to seepage from canals and field irrigation; (iii) use of shallow or deep tube wells for supplementary irrigation during periods of low canal supplies or canal closures. It can also take on the form of irrigating pockets exclusively with groundwater in a canal command; and (iv) use of augmentation tube wells discharging directly into the canal, and thereby supplementing supplies. A parcel of land under canal irrigation induces additional groundwater recharge. If this recharge is extracted in another parcel of land, it is also conjunctive use.

The concept gainfully extends to the dilution of fresh surface water with saline or brackish groundwater in suitable proportion to maintain requisite water standards for the purpose for which the water is intended to be used.

The following are the positive and negative factors of conjunctive use (Chaube 2000).

Positive Factors of Conjunctive Use

(i) Operation of both surface and groundwater reservoirs provides for larger water storage, and hence greater water conservation can be achieved.
(ii) Groundwater storage can be pumped to use during a series of dry periods to supplement the scarce surface water supply. Hence, the surface storage reservoir and surface distribution system can be made comparatively smaller.

Thus, there is a reduction in capital investments and operational expenditures on surface water storage and conveyance.

(iii) Seepage from canals recharges the groundwater reservoir, and part of the seepage losses can be recovered by pumping. It reduces the need for canal lining in smaller distributory channels and, thereby, the cost of lining. If canal water is of low quality, it can be "recycled" by pumping purer water from aquifers adjacent to canals.

(iv) Pumping out groundwater for utilization reduces and controls the rise of the water table. Hence, the drainage systems can be made comparatively smaller; thus, conjunctive use reduces drainage capacity and cost.

(v) Surface water projects require a long gestation period for completion, which delays the utilisation or reduces the potential of the project. The project's full potential can be created in a short time when groundwater extraction is combined with surface water supply.

(vi) Surface wastes, such as conveyance loss and evaporation loss, and subsurface outflow are reduced by conjunctive use. Hence, conjunctive use provides higher water utilization.

(vii) Conjunctive use provides high reliability of supply and hence greater confidence to use. With the increase in reliability, high-value crops (sugarcane and vegetables) are taken up by farmers.

Negative Factors of Conjunctive Use

(i) Varying water supplies from two different sources (surface and groundwater) requires extensive analysis to determine equitable water rates; hence, cost allocation becomes difficult.

(ii) Reducing groundwater levels by over-pumping may promote land subsidence, particularly in confined aquifers, which can damage canals and other surface works.

(iii) A reliable and adequate energy supply is necessary to operate pumps to extract groundwater.

12.3 Examples of Planned Conjunctive Use

12.3.1 Augmentation Tube Wells Along Western Yamuna Canal

The construction and operation of augmentation tube wells in the Western Yamuna Canal (Haryana) irrigation area is an example of government-sponsored scheme of conjunctive use (Chaube 2000). 256 augmentation tubewells along the Western Yamuna Canal (WYC) in Haryana were installed in the early nineteen fifties. The main object was to save the fertile agricultural land from waterlogging by pumping seepage water into the Western Yamuna Canal (WYC). Later, 128 additional augmentation tubewells were installed along this canal during 1967–70. Subsequently, the application was extended to other areas of the state served by the Bhakra Main Branch (BMB) and WYC systems to serve twin purposes—one to save the land along canals from getting waterlogged and the other to supplement the canal supply and narrow down the period of rotation of supply in the distribution system so that different areas are benefitted. Till April 1983, 1593 augmentation tubewells had been energized, supplementing the canal supplies by 956 MCM each year.

The impact of conjunctive use has been appreciable. The canal water supply increased from 85 to 115 cumecs during the critical crop period. This reflects that whenever the surface water is scanty, augmentation tubewells can play a key role in building up an alternative supply source.

12.3.2 Conjunctive Use with Saline Ground Water in Haryana State

Conjunctive use of surface and groundwater has a larger scope for its application in some areas of Haryana state in India (Chaube 2000). In these areas, the quality of groundwater is brackish or saline (i.e. electrical conductivity is between 2000 and 15,000 microsiemen/cm) and the area is covered with good network of carrier channels. In order to evolve suitable management techniques to combat various problems intimately connected with the continuous water table rise, a project on studies for the use of saline water in the command area of irrigation projects was taken up in Haryana State.

Groundwater balance studies show that over the years, the storage had been steadily building up to about 26,000 km^2 area. Against the total annual recharge of about 65,000 ha-m, the draft is only 5000 ha-m. The continuous rise in water levels at the variable rate of 0.15–0.90 m each year has brought 6400 km^2 area under waterlogged conditions and in another 7000 km^2 area, the water table depth is between 3 and 9 m.

It has also been estimated that out of the total surplus annual recharge potential of 60.0 MCM, only 40% of it can be exploited for conjunctive use as, to this extent, only the water could be led into and mixed with the nearby canal supplies.

The conjunctive use of saline groundwater in Haryana creates certain problems as the canal water is also extensively used for Rural Water Supply purposes. It, thus, requires that the mixing ratio should be adjusted in such a way that the prescribed public health standards for drinking water are not exceeded. The sensitive crops are likely to be damaged with the use of mixed water for pre-sowing irrigation; as such, it is required that the running schedule of augmentation tubewells is fixed in such a way as to avoid damage to sensitive crops and to reduce the salt accumulation in soils during summer months when canal water is required frequently. It has been suggested that augmentation tubewells should run between 15th October and 31st March and from 1st July to 15th September only. In conjunctive use, the mixing ratio is required to be decided more carefully as the application of mixed water also depends on the soil conditions. The studies conducted on samples of saline groundwater during individual zone tests and mixing them with nearby canal water in different proportions show the dilution ratio from 1:10 to 1:55.

12.4 Estimation of Ground Water Resources for Conjunctive Use

The first requirement in planning the conjunctive use of surface and groundwater is the knowledge of their availability and distribution. It is also essential to work out a water budget for all sources of water for each area/sub-basin/basin/region. In addition, knowledge of land resources, rainfall patterns, soil types, cropping patterns, etc., is also essential to properly evaluate the potential of conjunctive use. Surface water is easily accessible and can be quantified more easily than ground water. The seasonal availability and aerial distribution of surface water resources are generally known.

In the case of ground water, because of its being concealed under the earth and its occurrence, availability, and distribution being controlled by complex hydrogeological and hydrological factors, like sub-surface lithology, nature and properties of aquifers, sources of recharge, pattern, and quantity of rainfall, nature of the infiltrating medium, etc., knowledge is comparatively scanty particularly for hard rock areas. The quantification of groundwater potential is also based mostly on empirical norms.

The actual utilization of the total available ground water resources would depend on the extent to which the lowering of the water table is permissible to eliminate the ground water runoff in the form of effluent discharge (base flow) and loss from the ground water body in the form of evapotranspiration. Since it is difficult to make any precise calculations, certain percentage factors are usually adopted for different regions to compute the usable ground water resources. Groundwater balance studies for 'with and without conjunctive use' conditions are briefly explained below (CWC 1995).

From the estimate of the usable ground water resources, the present withdrawal use has to be subtracted to compute the net ground water resources available for further utilisation. The present withdrawal use may be calculated either based on existing figures of gross cropped irrigated area multiplied by the average total depth of irrigation per crop or by preparing an inventory of existing works and calculating the total withdrawal based on the number of works of different categories and the average draft per unit of work under each category.

All the available ground water resources, however, cannot be utilised because of various factors, such as (a) evapotranspiration losses from the forest, and waterlogged and marshy areas; (b) inevitable effluent discharge to the rivers; and (c) the maintenance of minimum base flow required in the rivers for committed usage on the existing canal systems and for ecological reasons.

Inflow and Outflow:

The inflows into groundwater may consist of:

1. Rainfall infiltration could be as much as 15–20% of rainfall in alluvial areas and only up to 2 or 3% for certain massive hard rock areas. The guidelines and experience of the Central Ground Water Board (CGWB)/State Ground Water Directorates based on field observations specific to the region may be used to make a decision.
2. Deep percolation from irrigation in the field and seepage from canals and tanks.
   a. For an unlined canal in a major project, seepage loss could be around 50% of the head deliveries. For a medium project, the loss could be nearly 30%, and for minor projects, including state tubewells irrigating areas of 100 ha, this loss could be 20%. The corresponding figures for a fully lined system were lined up to around 20 ha blocks and could be about half of the figures mentioned for the unlined canals.
   b. Field channels usually are unlined, and seepage could be as much as 10–30% of the deliveries at the outlet, depending on the site conditions.
   c. Deep percolation loss for all dry crops could be about 10–15% of the water supplied to the field. For ponded crops, particularly for paddy, deep percolation is unavoidable. The standard rate for percolation through paddy is 3 mm per day to 16 mm per day.

Outflows from groundwater would normally consist of:

(a) Base flows or return flows into the surface stream network;
(b) Direct evapotranspiration via the capillary rise. It could be around 10% of the total outflow;
(c) Withdrawal of groundwater by pumping for irrigation and other purposes; and
(d) In a semi-confined aquifer, there could be a net outflow from the bottom of the aquifer.

Withdrawal can be estimated on the basis of the operation of water withdrawal structures.

12.5 Limits on Ground Water Withdrawal

The Central Water Commission (CWC 1995) has specified "minimum necessary" and the "maximum permissible" withdrawals (Table 12.1).

Table 12.1 Minimum and maximum withdrawals

Present groundwater Depth of groundwater	Status trend	Minimum necessary additional withdrawal as a percentage of the additional recharge caused by the project (%)	Maximum permissible withdrawal as a percentage of the additional recharge caused by the project (%)
Less than 2 m	Rising	70	100
-do-	Generally steady	50	80
-do-	Falling	30	60
2–6 m	Rising	60	90
-do-	Steady	40	70
-do-	Falling	20	60
More than 6 m	Rising	50	80
-do-	Steady	30	60
-do-	Falling	0	40

Note For this table, a general long-term rise or fall of more than 0.2 m/s year in the case of alluvial condition and of more than 0.5 m/year in the case of hard rock areas would qualify for reclassifying the trend as "rising" or "falling"

12.6 Four Alternatives for Ground Water Use

The following alternatives or their combinations are possible (CWC 1995):

Alternative 1. Irrigate separate parcels of land with surface water and groundwater.

Alternative 2. Surface water is used in one season and groundwater is used in other season(s).

Alternative 3. Combination of alternatives 1 and 2 in which:

(a) specified fields are always irrigated with surface water;
(b) specified fields are always irrigated with ground-water; and
(c) specified fields are supplied with surface water in one season and groundwater in another.

Alternative 4. This is a variation of Alternative (3). In this, for fields in (c) in which both groundwater and surface water are used, the intra-annual regime of uses varies from year to year to take advantage of the stabler regime of groundwater. This could involve groundwater partly for carryover purposes. Also, it may require more extensive use of surface water in years of surplus surface flows.

12.7 Issues Related to Conjunctive Use

The conjunctive use of ground water in the canal command area implies unified control, regulation, and efficient use of surface water and groundwater for achieving specific objectives. Conjunctive use is a process influenced by several factors, technical, legal, socio-economic, and organizational. It could become a complex and long drawn out process of adjustment if the issues involved are not properly dealt with during the operation stage. Some of the issues are listed below (Khan 1993).

(i) A large disparity in water prices from canal and tube well sources would discourage the use of more expensive groundwater. Further, water charges for the conjunctive use of surface water and groundwater depend on the form the conjunctive use may take.
(ii) For proper implementation of conjunctive use in canal command, farmers association is required for the equitable distribution of water among farmers and its efficient application in the field.
(iii) In the absence of groundwater acts, the government has no control over the development and use of groundwater through private tubewells.
(iv) As many as four government departments may be involved in managing irrigation schemes and agriculture production [Irrigation Department, Ground Water Organization, Agriculture Department, State Electricity Board, and Command Area Development Authority (CADA)]. There is a lack of coordination at the field level among the activities of these government departments
(v) A farmer belongs to several social groups. There can be many conflict interfaces, such as between WUA—Individual Farmer; WUA—Project Authority; WUA—Religions/Cast Group; WUA—Electricity Department; and Project Authority-private tube well owners. Therefore, in framing legal Acts, it is first necessary to recognize multiple interfaces and the types of conflicts.

A Case Study on Ground Water Recharge and its Withdrawal

Khairana tank irrigation project in the Sagar district of Madhya Pradesh state in India is selected as a case study. The study was carried out by Chaube (2006) as a field research project sponsored by National Bank for Agriculture and Rural Development (NABARD). A detailed analysis of the cropping system and farm income under various situations was carried out. Brief details of the study are given in the Annexure Appendix: Case Study of Khairana Tank Irrigation Project. A to this chapter. The study covers (i) ground water recharge in the canal command, (ii) canal water budget in command area, (iii) design of shallow and deep wells, and (iv) pumping facility for ground water wells (estimation of number of wells, calculation of pumping head, number of pumping units, pumping rate etc.). A detailed discussion of the study is given in Appendix Appendix: Case Study of Khairana Tank Irrigation Project.

Appendix: Case Study of Khairana Tank Irrigation Project

General

Conjunctive use of ground water in canal commands implies coordinated and harmonious use of surface water and ground water for meeting the water requirements in time and space. Its scope includes (i) recharge of ground water from natural surface water; (ii) artificial and induced recharge from surface water to ground water; (iii) use of shallow or deep wells in canal command areas for supplementary irrigation during periods of low canal supplies or canal closures. It can also take on the form of irrigating pockets exclusively with ground water in a canal command; (iv) use of augmentation tube wells discharging directly into the canal, thereby supplementing the canal supplies, and vertical drainage to lower the ground water level and thus control waterlogging.

Ground water is the main source of water supply for agriculture and drinking purposes in rural areas of India. With the increase in population and the need for more water for irrigation, water demand is on a constantly increasing trend. Several of the existing canal irrigation systems suffer either from inadequate supplies or waterlogging due to excessive irrigation. In fact, even within a command area, there are pockets of water-logged areas and pockets of water deficit areas. Over the years of a continuous canal operation, the water table keeps on rising in the command area, and therefore it becomes necessary to pump put the water.

Chaube (2006) carried out a study on the evaluation of four tank irrigation projects in Sagar District of Madhya Pradesh, India. Subsequently, using the data of this report and other documents, Nissanka (2007) carried out a study on the use of ground water in canal command of Khairana tank irrigation project in Sagar district of Madhya Pradesh state of India. The study covers (i) groundwater recharge in the canal command, (ii) canal water budget in the command area, (iii) design of shallow and deep wells, and (iv) pumping facility for ground water wells. The discussion in what follows is based on these studies.

Ground Water Related Characteristics of Khairana

The characteristics relevant to ground water study in the command area of the four projects (geology, geomorphology, soil type, suitability for artificial recharge etc.) are shown in Table 12.2. The command area of the Khairana project is in Deccan Trap. The characteristic geomorphology is Deccan Plateau and denudational hills.

Table 12.2 Ground water related characteristics of Khairana

Item & Unit	Khairana
Geology	Deccan trap
Geomorphology	Deccan plateau + Denudational hills clayloam, gravel sandyloam
Soil infiltration rate (cm/h)	1.9
Saturated hydraulic conductivity cm/h	0.55
Pre-monsoon GW fluctuation between 1985 and 2000	2–3 m rise
Pre to post monsoon rise in G.W due to rainfall (average)	4–6 m
Suitability for artificial recharge	Suitable

Ground Water Recharge

Available ground water data for observation wells in the command area of the tank irrigation project and in the vicinity has been collected from WRD. These data have been analysed to study the impact of tank projects on ground water recharge, as discussed below.

The ground water data of four observation wells in the vicinity of Khairana project command for six years was analyzed. Canal irrigation in the command area is causing a rise in the water table. The average groundwater level in the area has increased by 3.07 m, 1.21 m, and 0.79 m during pre-monsoon (May), post-monsoon (November), and Rabi season (January), respectively, after the project came into existence (Table 12.3). The average ground water fluctuation between pre- and post-monsoon periods is decreased by 1.86 m and pre-monsoon to Rabi season is decreased by 2.7 m.

Table 12.3 Groundwater data for Khairana project command

S. no.	Item	Pre monsoon	Post monsoon	Rabi season	Water table fluctuations	
		(May) (m)	(Nov) (m)	(Jan) (m)	May–Nov (m)	May–Jan (m)
1	Water table before irrigation project (BGL)	12.56	6.44	7.57	6.12	4.99
2	Water table after irrigation project (BGL)	9.47	5.23	6.78	4.26	2.7

The underground aquifers are supplemented by sources other than rainfall, such as seepage from canals and field canals, ponds, tanks, influent drainage from rivers, deep percolation from irrigated fields etc. In most canal irrigated areas, a substantial component of the water applied to fields percolates below the root zone and contributes to the groundwater potential. A comprehensive and complete assessment of groundwater recharge from canal seepage is possible only through the compilation and analysis of reliable data from systematic scientific studies. Based on annual average rainfall, lengths, and wetted perimeter irrigation canals, area of water bodies, and application efficiency of the irrigation water, the total possible annul recharge in command of Khairana is estimated, as shown in Table 12.4.

As per the preliminary survey conducted by the National Institute of Hydrology, the water table available at a depth of 7–10 m is suitable for shallow tube wells or dug wells.

Estimation of Canal Seepage and Canal Water Budget

Various methods are currently available for the estimation of canal seepage. Estimating canal seepage from lined and unlined canals provides a key input to the economic analysis of canal lining.

Direct Determination of Canal Seepage

Inflow–Outflow Method: The actual inflow entering into one end of the selected reach of the canal and outflow going out of the other end of the reach is measured with the help of existing weir or flumes or by area velocity method with the help of current meter. The difference between inflow and outflow is the seepage. The limitation is that the number of measurements is to be taken to find the average value, and the error could be $\pm 10\%$.

Ponding Method: It is a more accurate and popular method and is generally used. In this method, a reach of the canal is isolated by making suitable water-tight barriers at both ends. Water is filled in the reach to form a pond. The fall in water level is noted at regular intervals or lost water is replaced to maintain a constant level. It is continued till a steady state is reached. It is corrected for evaporation and precipitation, if any. Thus, the rate of seepage per unit wetted area is evaluated.

Its limitations are.

1. Canal operation has to be discontinued for the experiment.
2. Deposition of sediment and growth of algae and fungi during the experiment period reduce the seepage rate.
3. Changes in the groundwater table due to the stoppage of the canal for the experiment will change the seepage rate.

Table 12.4 Total annual recharge calculation—Khariana project

A. Monsoon recharge							
	i. Area sq.km						3.587
	ii. Water table fluctuation (WTF) in m						4.26
	iii. Specific yield (Sy) (%)						9
	iv. Rainfall (annual average) in mm						1394
	v. Gross Kharif Draft (DW), Mcm/year						0
	Monsoon Recharge, Mcm/year						
	(A × WTF × Sy) + Dw − (Ris + Rigw + RS)						1.42
	A × WTF × Sy		= Suitable area Sq.km × WL rise × S. Yield				1.37
	Dw		= Gross Kharif Draft (Dw) Mcm/year				0.10
	Rs		= Monsoon seepage from canal and tanks				0.05
	Rigw		= Monsoon seepage from ground water irrigation				0
	Ris		= Monsoon seepage from surface water irrigation				0
B. Recharge from surface sources							
1	**Recharge from canals**						
	Applied seepage factors			Unlined	2.5 Cumec/M m^2		
				Lined	0.3 Cumec/M m^2		
	Wetted perimeter Mm^2		Average running days		Seepage Mcm/year		Total
	Unlined	Lined	Non-Monsoon	Monsoon	Non-monsoon	Monsoon	0.12
	0.0055	–	100	–	0.12	–	
2	**Recharge from surface water irrigation**						
	Irrigation water applied		= Area irrigated × Average depth of water applied				
	Seepage factor is 35% of the water applied						
	Area irrigated ha		Average water depth applied m		Seepage Mcm/year		Total
	Non monsoon	Monsoon	Non monsoon	Monsoon	Non monsoon	Monsoon	0.25
	232	–	0.311	–	0.25	–	
3	**Recharge from surface water body**						
		Average water spread area km^2		Seepage factor m	Seepage Mcm/year		Total
					Non monsoon	Monsoon	0.23
		0.38		0.6	–	0.23	
C. Total annual recharge Mcm							

(continued)

Table 12.4 (continued)

	Type of recharge		From rain fall	From irrigation sources	Total
	a. Monsoon recharge		1.42	0	1.42
	b. Recharge from surface water sources				
		i. Recharge from canal	0	0.12	0.12
		ii. Recharge from surface water irrigation	0	0.25	0.25
		iii. Recharge from surface water body	0	0.23	0.23
		Total			2.02

Empirical Methods

These are only approximate relations and give only a rough estimate. The number of such relations is available, but only the method used in Maharashtra is discussed.

Studies on the main canals in Maharashtra have given the following results. Considering the similarities in geography and other factors, this method is used to calculate seepage loss in irrigation canals in the study area.

Seepage loss in the unlined canal is 8 cfs/Msft of the wetted canal perimeter.

Seepage loss in the lined canal is 2 cfs/Msft of the wetted canal perimeter.

Canal Water Budget for Khairana Project

The canal water budget for the Khairana project is estimated as shown in Table 12.5.

Computations are explained below

Table 12.5 Canal water budget in Khairana project

Sl.	Description	Without canal lining (Mcm)	With canal lining (Mcm)
01	Live storage	1.466	1.466
02	Loss (evaporation and percolation)	0.293	0.293
03	Available to canal	1.173	1.173
04	Operational losses	0.234	0.176
05	Seepage losses	0.137	0.029
06	Sub total	0.802	0.968
07	Water course losses	0.080	0.097
08	Sub total	0.722	0.871
09	Application losses	0.144	0.174
10	Availability for crop	0.578	0.697
11	Overall efficiency	49.3	59.4
12	Water supply factor (%)	54	65.06

Line 02—Water losses due to evaporation and percolation from the water body 20% of line 01.

Line 04—20% of line 3 for unlined canals and 15% for lined canal

Line 05—Seepage Losses
Losses (Mm^3) – Flow loss rate (ft^3/s) (days operation) (seconds/day)*10^{-6}
Loss rate
Unlined Canal—8 Cfs/Msf
Lined Canal—2 Cfs/Msf
Unlined wetted perimeter—6482 m^2
Lined wetted perimeter—5499 m^2
Number of days canal operation in Rabi season = 100 days
Seepage loss from the unlined canal
$= 8 \times 6482 \times 100 \times 24 \times 60 \times 60 \times 10^{-12}/3.28$
$= 0.137\ Mm^3$
Seepage losses from the lined canal
$= 2 \times 5499 \times 100 \times 24 \times 60 \times 60 \times 10^{-12}/3.28$
$= 0.029\ Mm^3$

Line 07—Water course losses
10% of line 06.

Line 09—Application Losses
20% of line 08.

Line 10—Available for Crop
Line8–line9.

Line 11—Overall efficiency
Line 10/Line 3 × 100.

Line 12—Existing canal irrigation cropping intensity 54% (Table 4.1).
Irrigation cropping intensity after concrete lining
54/49.3*59.4 = 65.06%.

Water Withdrawal Using Shallow Wells

Calculation of Head of Pumping

1. Static Head = Depth of water table average 9.0 m.
2. Delivery Head = 1.25 m.
3. Drawdown calculation from the formula

$$Q_p = \frac{2\pi T S}{2.3 \log\left(\frac{R_e}{R_w}\right)}$$

where Q_p = the discharge of pumping in m^3/sec;
T = the coefficient of transmissivity in m^2/s = 0.0069 m^2/s;
S = the drawdown in well in m;
R_e = the radius of the depression cone, taken 150 m; and
R_w = the radius of pumping well = taken 3 m.

$$S = \frac{2.3 * Q_p * \log\left(\frac{R_e}{R_w}\right)}{2\pi * T}$$

Taking Q = 15 lits/s (small farmer) = 0.015 m^3/s

$$S = \frac{2.3 * 0.015 * \log\left(\frac{150}{3}\right)}{2 * 3.14 * 0.0069}$$
$$= 1.35 \text{ m}$$

4. Velocity head: limiting the velocity at the delivery side to 2.0 m/s

$$= \frac{v^2}{2g}$$
$$= 2^2/2 \times 9.81$$
$$= 0.204 \text{ m}$$

5. Friction head:

$$Head\, loss\, due\, to\, friction\, in\, the\, pipe = \frac{4flv^2}{2gd}$$

where l = total length of pipeline in meter;
V = the velocity of flow in the pipe in m/s, taken as 2 m/s;
D = the diameter of the pipe line in meter, taken as 0.125 m; and
F = the coefficient of friction, generally assumed as 0.006.

$$h_f = \frac{4 * 0.006 * 11.6 * 2^2}{2 * 9.81 * 0.125}$$

h_f = 0.45 m

L losses at entry and bends taken as 25% of friction losses in the pipe line h_b = 0.11 m

$$\therefore Total head of lifting = 9.0 + 1.25 + 1.35 + 0.204 + 0.45 + 0.11 \text{ m}$$

$$= 12.364\,\text{m} \quad \text{Say } 12.40\,\text{m}$$

Calculation of Pumping Unit

$$Horse\ power\ of\ a\ motor = \frac{W * Q * H}{75 * \eta}$$

where

W = the unit weight of water in kg/m^3;
η = the efficiency of the set, which is generally taken as 60%;
Q = the discharge to be delivered in cumec; and
H = the total head in meters against which the motor has to operate.

$$HP = \frac{1000 * 0.015 * 12.4}{75 * 0.60}$$

Considering an extra cover of 20% for the seasonal variation of load on motor $HPofmotor = 4.13 \times 1.2 = 4.956 = Say5HP.$

Calculation of Number of Shallow Wells Required

Assuming 1000 hours working in a year,

Annual yield per well in a year = 0.015 × 60 × 60 × 1000/1000,000 Mcm = 0.054 Mcm

Current water duty at farm outlet level = 0.0043 Mcm/ha

Area in ha proposed to irrigate under GW = 76 ha

Groundwater requirement = 0.33 Mcm

Number of wells required = 6.11

(with extra a well to facilitate sharing excess with neighboring farmers) = Say 8 Nos.

Water Withdrawal Using Tubewells

Tube well supplies differ from canal supplies in several aspects. Tube wells usually provide a steady water supply for irrigation. It is easy to provide irrigation at optimum times with reference to the stage of growth of crops, thus maximizing production. Canal supplies fluctuation, and canals often remain closed for long periods. In the case of canals, the annual working expenses remain the same in spite of the variations in the volume of water supplied. On the other hand, in tube wells, the cost of power consumed in pumping and, to some extent, the maintenance cost is related to the volume of water pumped. In tube wells, it is also possible to conveniently measure the discharge rate with a simple V- notch and relate the volumetric discharge with the power consumed. This facility and the high cost of tube wells and pumps make it desirable that the tube well water is charged based on water supplied to individual farmers.

Calculation of Pumping Rate

Present water supply factor = 75.33%

Proposed water supply factor with conjunctive use = 100%

CCA = 308 ha.

Irrigation duty at farm out let = 0.43 m.

Extra quantity of water required = $0.2467 \times 308 \times 0.43 \times 10^4/10^6 = 0.3267$ Mm^3

Conveyance losses (20%) = 0.0653 Mm^3

Total pumping quantity = 0.3920 Mm^3

Assuming Number of pumping hours as 1000 h.

Required pumping rate = $0.3920 \times 10^6/1000 \times 60 \times 60 = 0.1089$ Cumec.

Well Design

Coarse sand strata (assumed) = 40–70 m.

Thickness of confined aquifer = 70–40 m = 30 m.

Discharge required = 6534 Lit/min.

i. **Casing and well screen diameter**

Casing diameter = 45 cm (Table 14.1 Varshney-1997).

Optimum diameter of well screen = 30 cm (Fig. 14.12 Varshney-1997).

ii. **Length of strainer and slot size**

- 70–80% of aquifer depth may be screened for confined well
- Assume 75% length of screen = $30 \times 75/100 = 22.5$ m
- Slot size = 2.0 mm.

iii. **Discharge of well**

The maximum depression head permitted is between 9 and 40 m

$$q = \frac{2\pi kb(H - h_w)}{2.3\log\left(\frac{R}{rw}\right)}$$

where $K = 0.09$ cm/s $= 9 \times 10^{-4}$ m/s

$b = 22.5$ m.

$H - h_w = 31$ m.

$R = 400$ m (assumed).

$R_w = 0.15$ m.

$Q = 2 \times \pi \times 9 \times 10^{-4} \times 22.5 \times 31/(2.3 \times \log_{10}(400/0.15)) = 0.5005$ cumec = 30,003 Lit/min.

The well is capable of discharging much higher discharge than required.

iv. **Screen entrance velocity**

Assume 20% of slot area, the screen entrance velocity when discharging 0.11 cumec is given by

$\pi \times 30 \times 20/100 \times 22.5 \times 100 \times V = 110 \times 1000$

$V = 2.59$ m/s.

Since screen entrance velocity is within the admissible limit (i.e. <3 m/s), the assembly is O.K.

H.P. of Motor

$$Horse\ power\ of\ a\ motor\ is\ given\ by = H.P. = \frac{W * Q * H}{75 * \eta}$$

Velocity through casing pipe $= 0.11 \times 4/(\pi \times 0.30^2) = 1.56$ m/s.

Velocity head $= 1.56^2/(2 \times 9.81) = 0.12$ m

$$Loss\ of\ head\ due\ to\ friction\ in\ pipe = \frac{4flv^2}{2gd}$$

where f = the coefficient of friction assumed as 0.006

l = the length of pipe including horizontal length = 75 m.

d = the diameter of pipe in m = 0.30 m.

$h_f = 4 \times 0.006 \times 75 \times 1.56^2/(2 \times 9.81 \times 0.3) = 0.74$ m.

Assuming entry losses in the strainer and bend as 25% of friction loss = 0.19 m.

Total head loss $= 9 + 31 + 012 + 0.74 + 0.19 = 41.0$ m.

H.P. required $= 1000 \times 0.11 \times 41/(75 \times 0.7) = 85.9$ Say 90 Hp.

Financial Performance

Results of the financial performance pertaining to possible future investment and their implications are discussed as follows.

The financial performance of the Khairana project in connection with concrete lining and conjunctive use of surface and groundwater was also analyzed by Chaube (2000) as part of the sponsored research project. Results are given in the table below.

Source	PWC	PWB	NPV	B/C	IRR
Surface irrigation	124.66	79.65	−45.01	0.64	10.34
Concrete lining	19.28	15.981	3.299	0.83	12.18
Shallow well	35.735	44.078	8.36	1.23	19.05
Tube well	33.725	44.078	10.355	1.31	21.60
Combination of concrete lining and tube well	47.509	49.689	2.179	1.05	15.8

PWC present worth of cost; *PWB* present worth of benefit; *NPV* net present worth; *BC* benefit–cost ratio; *IRR* internal rate of return

Financial parameters, such as B/C ratio and IRR for concrete lining, at 15% discount rate are 0.83 and 12.18, respectively. Therefore, concrete lining in the entire canal length is infeasible. Limited concrete lining where seepage losses are high along with some management improvements such as rotational water issues can further improve financial parameters.

The conjunctive use of surface and ground water in shallow wells, Public Tube wells, and a combination of concrete lining and tube wells has become economically feasible. The B/C ratio and IRR at a discount rate 15% corresponding to the above three cases are 1.23, 1.31, 1.05, and 19.05%, 21.6%, 15.8%, respectively.

Inequity, unreliability, and untimeliness are the major issues in surface irrigation projects. The conjunctive use of surface water and groundwater can improve efficiencies in lower reaches where surface water is used to irrigate land close to the canal and groundwater to irrigate land which is further away.

The reduced uncertainty of poor or irregular supply from surface water allows farmers to risk investment in water-intensive and higher-value crops, HYV seeds, and associated inputs like fertilizer and pesticides.

Questions

1. Explain the following terms:
   Surface water—groundwater interaction, natural recharge, artificial recharge, water charge, augmentation tube well, and unplanned conjunctive use.
2. Explain various advantages of an irrigation project planned on the basis of conjunctive use of surface water and groundwater.
3. Explain alternative strategies of conjunctive use of surface water and groundwater.
4. Write a note on artificial recharge and its use for irrigation.
5. Explain why the charges for the supply of canal water and groundwater for irrigation are different?
6. Why the irrigation charges are different for different crops?
7. What is the basis for the rationalization of irrigation charges?

8. From your experience what could be the interface problems between (i) farmers and electricity departments; and (ii) farmers and canal irrigation department.
9. Identify various government agencies involved in the conjunctive use of surface water and groundwater.
10. Explain the role of water user association in conjunctive use management.
11. Explain the need for a comprehensive Groundwater Act.
12. Referring to appendix Sect. "Water Withdrawal Using Shallow Wells".@@

    Calculate the total head of lifting and HP of the pump if static head = 8 m, delivery head = 1.4 m; other data being same.

    If working hours are 1100 in a year, how many shallow wells are required.
13. Referring to Appendix "Water Withdrawal Using Tubewells" calculate the required pumping rate if the water supply factor is 78% and the number of pumping hours in a year are 1100; other data are the same.

References

Chaube UC (2000) Conjunctive use management of surface water and ground water course package (unpublished). Centre for Continuing Education IIT Roorkee

Chaube (2006) Report on evaluation of rural infrastructure (irrigation) projects in Sagar District of Madhya Pradesh; sponsored research by National Bank for Agriculture and Rural Development (NABARD), Mumbai, India

CWC(1995) Guidelines for planning conjunctive use of surface and ground water in irrigation project. Central Water commission, New Delhi

GOI (1972) Report of the irrigation Commission, Ministry of Irrigation & Power, Government of India, New Delhi

GOI (1976) Report of the National Commission on Agriculture, Ministry of Agriculture, Government of India, New Delhi

Khan S (1993) Implementation of conjunctive use management plan—a case study. M.E. dissertation report, WRDTC, University of Roorkee, Roorkee

Reading Material

Chawla AS (1991) "Evaluation and improvement of irrigation technology in tube well commands" a report, WRDTC, University of Roorkee, Roorkee

Goel MMK (2003) Spatially distributed simulation of an irrigation system. Ph.D. research thesis under the supervision of Prof. U. C. Chaube, Indian Institute of Technology, Roorkee

IAHS (1986) Proceeding volume of budapest symposium on conjunctive water use. IAHS Publication, No. 156

Chapter 13
Maintenance of Irrigation Systems

Abstract During the transition and at the beginning of the operation phase, irrigation managers and farmers may need to undergo a maintenance learning process. The objectives of this chapter are to: understand various types of maintenance and repair priorities for improving hydraulic performance and understand the scope of various maintenance plans. A maintenance plan is prepared to carry out various maintenance activities in accordance with priorities under given constraints of time, financial budget, and available machinery/manpower. Maintenance plans are classified as (a) Essential Structural Maintenance Plan (ESM); (b) Normal Maintenance Plan; (c) Catch-Up Maintenance Plan; and (d) Preventive Maintenance Plan. The scope of the Maintenance for each of these plans is discussed in detail. Functions of engineering staff and work-charged staff in relation to maintenance practices are discussed. The literature available on the websites of the Ministry of Water Resources (MOWR), Government of India (www.mowr.gov.in), and Department of Water Resources Government of Odisha (www.dowrodisha.gov.in) is reviewed and referenced for preparation of this chapter.

13.1 Introduction

Irrigation structures are created through massive investment. The structures need to be kept in good health, for which a well-established maintenance programme needs to be followed. Good health means that design parameters of head works, canals, and related structures are maintained for desired hydraulic operation.

There has been a gradual deterioration of the maintenance standards in irrigation projects in several developing countries, partly due to inadequate finances for maintenance expenditure and partly due to the ignorance/negligence of maintenance-related functions, such as hydraulic survey and discharge measurements by irrigation managers.

Maintenance Learning Process

The O&M agency is seldom well-staffed initially, and the staff will need to undergo a maintenance learning process (Skogerboe 1990). Farmers also need to be trained

U. C. Chaube et al., *Canal Irrigation Systems in India*, Water Science and Technology Library 126, https://doi.org/10.1007/978-3-031-42812-8_13

in methods of irrigation and maintenance of structures (particularly below outlets). Therefore, sometimes 3 to 5 years are required in the transition from rainfed area to fully irrigated area (irrigation transition) and from traditional cultivation practices to improved agricultural practices (agriculture transition).

Maintenance learning process (Skogerboe 1990) is meant to develop appreciation, create awareness, and equip the irrigation managers and farmers with the background of the maintenance process. The maintenance learning process includes (a) identifying problems (diagnostic analysis), (b) learning to develop and testing solutions, and (c) making use of O&M experience in other projects for further improvements in the implementation of solutions.

13.2 Operation and Maintenance Activities During the Operation Phase of a Project

The O&M activities are depicted in a flow chart, as shown in Fig. 13.1. Some activities are briefly described below.

Flow control structures, discharge measurement: Identify flow control structures used for discharge measurement (falls, drops, culverts, regulators, gates at the dam, and barrage). Develop discharge ratings in the field for flow control structures. Estimate channel losses by the inflow-outflow method or ponding test for the reaches suspected to have unusual seepage losses.

Conduct 'Walk-thru' maintenance survey: This should be preceded by (i) hydraulic survey, (ii) measuring channel losses, and (iii) obtaining maintenance information from farmers. 'Walk-thru' rather than 'drive-thru' is very important. This helps understand the physical phenomena causing the problems.

Prepare field notes: Detailed field notes should be made for (1) reaches requiring sediment removal, (2) reaches experiencing scour and the causes thereof, (3) areas of vegetative growth, (4) grading of roadway, (5) repair or replacement of structures, (6) cavities at bridge abutments and gate structures, (7) cavities along concrete-lined reaches along with an estimate of cavity volume, (8) failed lining panels, and (9) holes in lining panels.

13.3 Steps in Formulation of Maintenance Plan

The maintenance of irrigation facilities (water control structures) is carried out according to the maintenance plan for the following purposes:

(i) To improve the hydraulic operation of irrigation structures.
(ii) To remove scour and sedimentation problems.

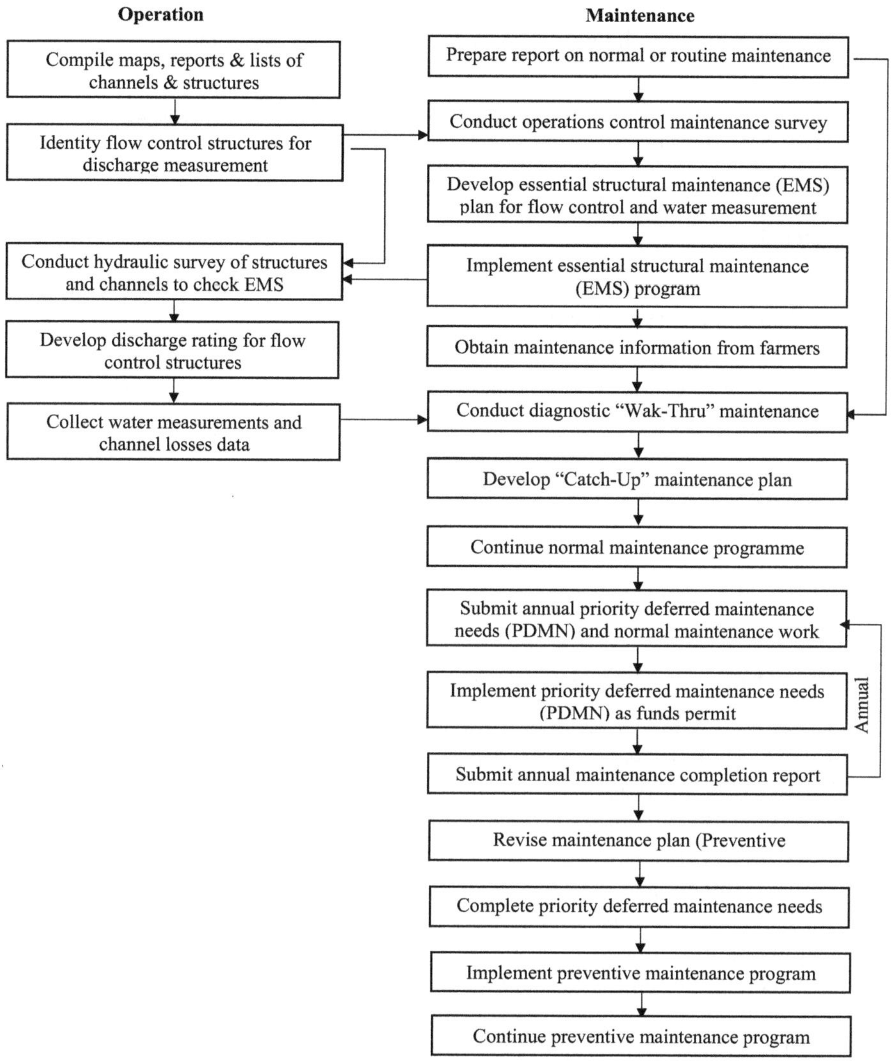

Fig. 13.1 O&M activities during the operation phase of the project

(iii) To develop discharge ratings for flow control structures.
(iv) To minimize irrigation channel losses.

The following steps are followed for the formulation of a maintenance program:

The steps involved in the formulation of a maintenance program are depicted in Fig. 13.2.

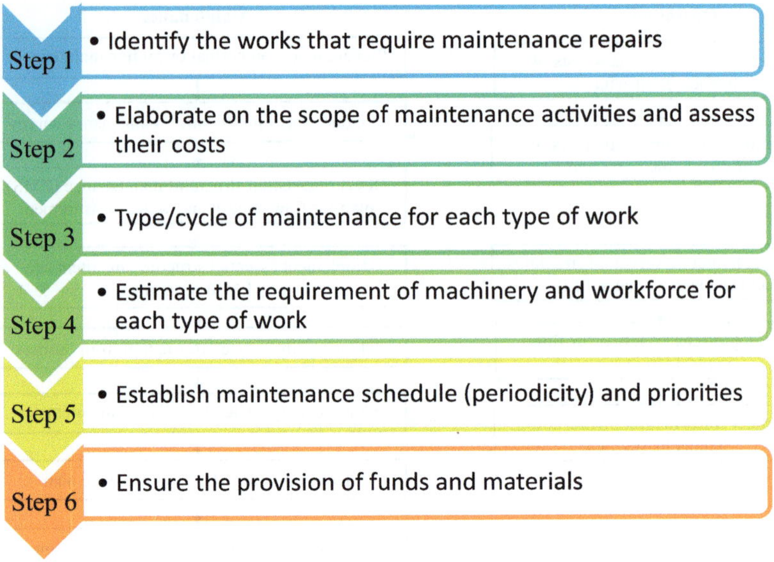

Fig. 13.2 Six steps in formulation of maintenance program

Provision of Budget Allocation for Operation and Maintenance:

Annual financial budgets are made for O&M, but experience shows that a very significant portion (80–90%) of this budget is usually spent on operational activities (mainly staff salaries) (Chaube 2000).

The reason for this lopsided allocation of budget is that maintenance has no immediate direct impact on the objective of an irrigation system, that is, in providing the timely, adequate, and appropriate delivery of irrigation water. With the lack of glamour and attention, it is not surprising that maintenance is most often neglected in managing an irrigation system. There is another more compelling reason, which is that the economic returns on maintenance are only attractive when it is deferred to justify rehabilitation. The supportive role of maintenance is significant, but only as far as it affects the irrigation system's operational effectiveness.

13.4 General Priorities for Maintenance Repairs

The following structures and facilities may have maintenance needs/problems:

1. Headwork's, i.e., dam or barrage
2. Cross regulators
3. Escapes
4. Irrigation canal network above the outlet
5. Tertiary irrigation system (water distribution below outlets)

6. Drainage network
7. Communication
8. Any other important constituent specific to the irrigation scheme.

Under the constraints of financial resources and time available, priority for maintenance repairs needs to be fixed. General priorities for repair work are indicated in Table 13.1. The reasons for repair work are also explained in Table 13.1.

Importance of Documentation of Maintenance Work

To document the costs of various maintenance activities, the irrigation department needs to have good records of field inspection notes and actual costs, including the

Table 13.1 General priorities for maintenance repairs and reasons thereof

Sl. no.	Priority type	Reasons
1	Dam, barrage, weir and intake	Failure of this structure would have serious consequences for the operation of the system. The headwork has to have top priority for maintenance
2	Leakage, unauthorized offtakes and overtopping	Losses of water from the canal system due to leakage through banks, unauthorized off-takes, and overtopping of the canal embankments are a serious matter because the loss results in inequitable water distribution to tertiary outlets
3	Gates and control structures	Without gates in head work, canal, control and regulation of water is difficult. The irrigation systems can be operated efficiently only if control structures are in good condition
4	Masonry repair	Repairing cracks in masonry is necessary to stop water from behind the masonry and causing cavities, leading to the ultimate collapse of the masonry
5	Embankment Protection	Protection of canal embankments takes several forms that are protection from: • erosion by canal water • gullying caused by low spots and crab holes • removal by farmers for cultivation close to or even on top of embankments • erosion by human and animal traffic across the canal
6	Measuring structures	Inefficient and incorrect water management will result from having measuring structures in poor condition Measurements will be incorrect, water allocation planning will be wrong, and water distribution will be inequitable
7	Silt removal	Silt removal upstream of measuring structures should have a very high priority (equal to priority 1 or 2). Excessive silt build-up in the bed of the canal causes a rise in the canal's water level or drain, leaving an inadequate operating freeboard
8	Vegetation removal	Vegetation removal, such as cutting grass on embankments, is very low priority. Removing grass from cracks in masonry and removing strong roots in shrubs and trees from the vicinity of structures is more important

variability for each type of activity. The advantages of documentation of maintenance work are as follows (WAPCOS 1989):

- This provides the necessary records for obtaining appropriate funding levels for the maintenance of irrigation projects.
- It is very useful for future planning to improve the maintenance of other irrigation projects.
- It helps with the inspection of the work by the Superintending Engineer and Chief Engineer to ensure quality control in the maintenance procedures and by the Executive Engineer to quickly review the program being made.
- The maintenance documentation also facilitates the evaluation by experts concerned with improving irrigation system maintenance practices.

13.5 Essential Structural Maintenance Plan

Report on the Essential Structural Maintenance Plan (ESM) includes the following (Incid 1994).

13.5.1 Physical Description of Irrigation System

A general description of the irrigation project should be provided, including a small location map and a larger map of the irrigation system showing the main canal(s), branch canals, distributaries, minors, and structures, including outlet structures, each properly labelled. Each minor, distributary, branch canal, and main canal's lengths can be provided in tables, along with their design discharge capacity and culturable command area (CCA).

13.5.2 Proposed Flow Measurement Programme

This section of the ESM plan must show on the large map and list the structures selected for flow control and water measurement in a table. The structure listing should state the structure type, important dimensions pertinent to flow measurement, and the design discharge capacity. The role of various structures in controlling discharges in the system and equitably delivering water to the outlets serving the tertiary systems needs to be clearly presented.

Next, the schedule for developing discharge ratings for flow control structures should be presented. This phase should state how the system will be operated during discharge measurements, how many measurements are required for each type of flow control structure, and how many days will be needed to complete the discharge ratings in each channel reach. The schedule for completing this work is, first, dependent on the qualified personnel available for doing the work and, secondly, the time lag involved in reaching steady-state flow conditions in the channel reaches where discharge measurements are to be made, which can vary from a few hours to a few days each time the discharge rate in the channel network is changed.

13.5.3 Proposed Programme for Evaluating Channel Losses

Assuming the completion of essential structural maintenance and the development of discharge ratings for flow control structures, an operational programme for evaluating seepage loss rates in various reaches of the main canal, branch canals, distributaries, and minors using the inflow-outflow method should be prepared. First, serious consideration needs to be given to operating the system at different water levels, either by adjusting gate settings, varying discharge rates, or combining the two. Suppose there are some reaches of the irrigation channel network where there is some doubt as to whether the inflow-outflow method will provide satisfactory data. In that case, these reaches should be delineated, and consideration should be given to whether or not the ponding method will be used in these reaches. A time schedule for conducting the channel loss measurements should be given; this requires some thought as to how much of the system could be evaluated on a single day, considering the availability of personnel. Also, for large-scale projects, it may be preferable to evaluate an entire distributary for one discharge on a single day, which would require three days for three discharges, not necessarily consecutive days, to collect the necessary field measurements so that the variation in seepage loss rate with water surface elevation will be known.

13.5.4 Essential Structural Maintenance

In this section, the results of the field survey are summarized. The major categories of maintenance requirements can be listed under (1) in-place maintenance requirements; (2) workshop maintenance requirements; and (3) replacement of gates, gears, gate rods, and gate frames. Then, under these three major categories, the specific maintenance activities can be listed with the required quantities.

13.5.5 Cost of Essential Structural Maintenance

This is a relatively short section that utilizes the list of required maintenance activities and the quantities from the previous section of the ESM Plan. Then, unit costs are applied to these activities to arrive at the estimated total cost for each activity and the total cost for essential structural maintenance. If there are any particular problems with some of the unit costs being used, they should be presented in the written plan.

13.5.6 ESM Implementation Plan

A proposed time schedule for completing the essential structural maintenance should be presented. The discussion should state any particular problems that might disrupt the proposed schedule. The statement should also be made as to the possibilities of completing the work in a shorter time period and under the conditions to be stated. The proposed schedule will dictate the cost schedule. Consideration should also be given to how much of these costs could be met with present project resources, the remaining being a request for additional funds.

13.5.7 Field Notes and Sketches

Preferably, a copy of the field written notes and sketches (edited if necessary) should be appended to the ESM Plan.

13.6 Normal Maintenance Plan

The report includes the following.

13.6.1 General Description of Irrigation System

A general description of the irrigation system should be presented, including a small map showing storage facilities, main and canal branch canals, significant drains, and command areas. The amount of available storage and discharge capacity at the canal headworks should be cited. A table could be used to present the discharge capacity at the head and length for each canal, branch canal, distributary, and major.

13.6.2 Available Maintenance Manpower

Different levels of staff and their numbers should be listed for job-related functions in normal maintenance. Then, the level of effort in maintenance activities by these staff should be listed. Also, the number of temporary laborers employed throughout the year should be stated, and the total number of days of work and how much of this effort is on maintenance activities should be specified.

13.6.3 Available Maintenance Equipment

A listing of all equipment presently used for the maintenance of the irrigation system should be listed. The number of hours or days each piece of equipment is required to be used during a year should be given, and how much of this use is for maintenance activities. Finally, the condition of each piece of equipment should be stated.

13.6.4 Present Maintenance Activities

This is an extremely important section that will also provide important guidance in developing the "Catch Up" Maintenance Plan and the Preventive Maintenance Plan. For each major activity listed below, it is important to indicate the quantity of maintenance work done, and it's the total cost, as well as list unit cost for each maintenance activity.

Watershed and Reservoir: Any activities by other departments, such as Forestry or Agriculture, regarding the planting of trees or other vegetation in the catchment area should be described. Maintenance problems in the reservoir area should be described, such as debris or vegetative and aquatic growth. Any maintenance activities associated with these problems should be listed, including quantities, costs, and unit costs. Also, any problems related to the enrichment of the reservoir should be stated, along with any measures being taken to avoid or discourage encroachment.

Masonry Dam: The quantities, unit cost, and annual cost should be listed for any of the activities listed below that are included in normal maintenance:

- Maintenance for non-overflow and overflow dams, their bucket/apron, wing walls, and head sluices.
- Inspection gallery—any maintenance needed for lighting, pumping, and ventilation.
- Maintenance of spillway gates.

Earthen Dams: Some of the major maintenance activities for earthen dams are listed below. However, there may be other activities as well. List all maintenance activities, the quantities of work, unit costs, and total costs.

- Jungle clearance and earthwork for the earthen dam to bring its top level and slopes as near to the designed parameters as possible. Filling of raincuts, rodent holes, etc., should be done.

- Maintenance, if any, is needed for the upstream pitching and downstream filter toe.
- Maintenance, if any, is needed for the head outlet sluices and gates, painting and greasing of gates, etc.
- Maintenance for the spillway, including bucket and apron, wing walls, etc.

Earthen Channels: There are numerous maintenance activities associated with earthen channels. For each activity, list the quantities of work, unit costs, and total costs. Some of the major maintenance activities are listed below:

- Jungle clearance, de-weeding, and desilting from the canal bed.
- Jungle clearance, de-weeding, and earthwork on canal slopes.
- Jungle clearance and earthwork on canal berms, dowels, and service roads.

Lined Channels: Some of the major activities are listed below:

- Desilting of the bed.
- Filling of cracks, pockets in the concrete, or repairs of the masonry lining in the bed and slopes.
- Jungle clearance on canal berms, dowels, and service roads.

For each activity, list the quantity of work, unit cost, and total cost.

Irrigation structures: The maintenance needs of all structures ought to be listed. Such structures are bridges, aqueducts, syphons, falls, head and cross regulators, and measuring flumes. The quantities of work, unit costs, and total costs should be cited in the report.

Drain Network: Mostly, this activity would involve the removal of vegetation from the network of open drains, usually by the labourer. List the hours or days spent on this activity, unit costs, and total costs for various drains sizes.

Communication System: Inspection roads, telegraphs, and telephones or wireless systems come under communication. Any repairs or other maintenance needed for these facilities should be listed in the report.

Ancillary Works: Buildings constructed in the project area, both residential and non-residential, as well as tree plantation, are part of the ancillary works. The annual maintenance costs for these works are worked out.

13.6.5 Major Maintenance Difficulties

This section's primary purpose is to draw upon project experience in delineating the difficulties associated with accomplishing the annual maintenance activities. This will be of considerable value in preparing various maintenance plans.

Physical: The last portion and present maintenance activities, when combined with project experience, should provide ample material to give an excellent description of the physical facilities and their degree of maintenance, including some indications of deferred maintenance. Notice that the emphasis is on MAJOR difficulties, not on small details.

Manpower: The emphasis should be on the staff's technical capability for doing maintenance work and whether sufficient staff is available. Also, if there are difficulties with the staff's willingness to undertake the necessary maintenance work, this should be stated, and any suggestions for correcting the situation should be made.

Equipment: Any difficulties associated with accomplishing the normal maintenance with existing equipment should be reported. A distinction should be made between the maintenance of existing equipment, expensive repairs, and replacement.

Budget: Almost every irrigation project suffers from a lack of funds for maintaining the irrigation system. This is why the degree of deferred maintenance is increasing. But the important question is, how can existing funds provided to the project be used cost-effectively to accomplish much more maintenance. An indication of how much deferred maintenance might be alleviated by various means with existing project funds should be indicated.

Administrative: Major difficulties with present staff and staffing conditions should be reported. Also, any constraints associated with higher levels of administration should be cited. Finally, any difficulties in doing the necessary maintenance work as a result of political interference should be added to this report on the normal maintenance program.

13.7 Catch-Up Maintenance Plan

The report covers the following.

13.7.1 Physical Description of Irrigation Project

The material contained in the first section of the Essential Structural Maintenance (ESM) plan should be included here. Also, more background about channel size might be provided in a list included in a table or series of tables. Also, the number of outlets along different minors and distributaries should be listed, along various reaches of other channels if they exist.

13.7.2 Essential Structural Maintenance

The material for this section of the “Catch-up” maintenance plan would be taken directly from the ESM Plan. If there have been any modifications to the original plan resulting from the experience gained while implementing the essential structural maintenance for flow control structures, then this should be presented.

13.7.3 Status and Costs of ESM Plan

First of all, the present status of the ESM Plan should be presented. Has it been approved? If so, what fraction of the work has been completed? Then, the cost of ESM can be presented; if the program is underway or completed, then revised cost estimates can be presented.

13.7.4 Inventory of Required Maintenance

This is a summary of field notes written during the diagnostic “Walk-Thru” maintenance survey. All of the required maintenance can be summarized in preparing a funding request for maintenance.

13.7.5 Maintenance Costs

Each irrigation project has unit costs for various maintenance activities commonly used in preparing a budget request. In preparing the estimated costs for this section of the “Catch-Up” maintenance plan, there may be some difficulty if, for example, the project field staff are not experienced in certain maintenance activities. If this is the case, then the Executive Engineer doing this maintenance activity should be consulted. The unit costs are used with the tables listing maintenance/requirements to arrive at the estimated costs for catching up on deferred maintenance.

13.7.6 Priority Maintenance Needs and Costs

In most cases, the degree of deferred maintenance will exceed the irrigation project’s capacity to complete it in a single year. Thus, there is a need to prioritize the proposed maintenance activities so that the highest priority maintenance (PMN) works are completed during the first year following the approval of the “Catch-up” maintenance

Plan, followed by the second highest PMN for the second year, and so on, until all of the deferred maintenance can be alleviated.

13.7.7 Maintenance Equipment Requirements

In general, most irrigation projects will have to undertake these deferred maintenance activities without receiving additional equipment. If it is obvious that the proposed maintenance activities cannot be economically done without some additional items of equipment, then a request for the required equipment should be presented and a strong justification provided.

All of the presently available equipment should be listed. Then, the capacity of this equipment for completing PMN should be stated. For example, how much concrete can be placed in a year, how much sediment can be removed, how much turfing (sodding) can be accomplished, and what percentage of the drains can be cleaned annually?

13.7.8 Maintenance Manpower Requirements

First of all, the present irrigation project staff's capacity to undertake the deferred maintenance activities should be presented, including any deficiencies in staff capability or unwillingness to undertake the necessary maintenance activities. Suppose certain staff vacancies exist that would seriously hamper the implementation of the "Catch-up" maintenance plan. In that case, this deficiency should be clearly stated, and good justification provided for filling such vacancies. Finally, the amount of temporary labor usually employed by the irrigation project and the annual capacity of this temporary labour to undertake the proposed maintenance activities should be presented. If additional temporary labour will be required, then the number of man-days and costs should be presented for the first year, the second year, and so on until the deferred maintenance is proposed to be completed.

13.7.9 Maintenance Plan

1. **Maintenance Issues**: Maintenance Plan should address, as a minimum, the following four issues:
   - achieving "regime" in the irrigation channels;
   - eliminating sediment sources;
   - minimizing seepage losses; and
   - reducing deterioration of channel lining.

For most irrigation channels, a key issue is the sediment movement through the irrigation delivery subsystem onto the fields. This is a very complex hydraulic phenomenon that is extremely difficult to evaluate. Indeed, large quantities of sediment deposition annually in a channel reach indicate the channel is not in "regime". Careful analysis is required in evaluating why an irrigation channel is not in the regime. Sediment removal is treating the symptoms of the problem rather than the causes.

The second issue, eliminating sediment sources, should come from field notes while conducting the diagnostic "Walk-Thru" maintenance survey. However, it is important that the "Catch-Up" maintenance plan clearly describes this issue so that when the deferred maintenance is completed, the quantities of sediment entering along the irrigation channels afterward will only be a small fraction of what is presently entering the channels.

The third issue, minimizing seepage losses, will become increasingly important with each passing year. Much can be accomplished with improved maintenance practices. Also, some seepage losses can be reduced by improved operation practices. However, if the seepage losses are very high, then irrigation system improvements may be required to reduce these losses significantly.

The fourth issue, reducing deterioration of channel lining, should also clearly address how the completion of the "Catch-Up" maintenance plan will significantly decrease the degree of cavity formation behind lined channels and irrigation structures. Again, the intent is that, eventually, the level of required maintenance for lined channels and structures should be significantly less than the present situation.

2. **Methods of Implementation**: Three primary methods can be used in completing the proposed maintenance activities:

   - departmental maintenance by the Irrigation Department;
   - maintenance through contractual agencies: and
   - contracting the main system, distributary maintenance to the water users.

The method of implementation should be stated. Also, if a combination of methods will be employed, then the amount of maintenance to be accomplished by each method should be presented.

3. **Implementation Schedule**: The number of years proposed for completing the "Catch-Up" Maintenance Plan should be stated. Then, summarized versions of the proposed work plan for each year should be presented. The summary for the first-year work plan should be based on the preparation of a detailed work plan, which is a separate funding request document that could be forwarded following the review and approval of the "Catch-Up" maintenance plan. The summarized work plans for the succeeding years would not be based on so much detail but would be based on subdividing the total maintenance requirements and costs into annual workloads that can be accommodated by the irrigation project staff by any of the implementation methods or a combination thereof.

13.7.10 Field Notes and Sketches

Based on the original field notes and sketches, this work would be recopied and appended to the "Catch-Up" maintenance plan. This material can be recopied by hand to save time, or the field notes can be typed, and the sketches drafted. Although appending all of the field notes and sketches to the "Catch-Up" maintenance plan adds a lot of bulk, it is extremely important to document them for use during the implementation of the "Catch-Up" maintenance plan.

13.8 Preventive Maintenance Plan

The report covers the following.

13.8.1 Physical Causes of Maintenance Problems

The irrigation project field staff should know the causes of maintenance problems that they visualize.

13.8.2 Anticipated Extent of Maintenance Problems

This section of the preventive maintenance plan should describe the anticipated or expected maintenance requirements for the project after completion of the priority maintenance needs. The type of maintenance activities required and the annual magnitude of these activities should be presented. Each activity's magnitude should be a quantitative listing for the reaches of the main and branch canals and each distributary and minor canals.

13.8.3 Maintenance Equipment Requirements

The equipment available for correcting priority maintenance needs is expected to be sufficient for implementing the preventive maintenance plan. However, some equipment may need special repair or replacement. Also, some pieces of equipment can be identified, which would make the preventive maintenance programme more cost-effective; if so, then the justification for purchasing such equipment should be presented.

13.8.4 Maintenance Manpower Requirements

The expectations are that the preventive programme will require less manpower. Certainly, all of the permanent irrigation department field staff will continue to function with the maintenance activities as before, but the amount of time they spend will be slightly less, and the amount of time required for temporary employees will be less, and for part-time labourers much less. The manpower requirements and annual costs for the preventive maintenance program should be presented and then compared with the manpower requirements for completing the priority maintenance needs and previously required for normal maintenance.

13.8.5 Maintenance Requirements for Water Users

When preparing this preventive maintenance plan, the irrigation project field staff will have spent some time collecting field data in the tertiary systems below some outlet structures. If not, a diagnostic "Walk-Thru" maintenance survey be conducted on a sample of tertiary system channels to make a preliminary evaluation of the extent of maintenance problems and requirements for alleviating these problems by the water users (farmers). The preventive maintenance plan should discuss how the irrigation project field staff might provide technical assistance to the water users so they can do a better job, if required, in maintaining their tertiary channels. Since the tertiary channels are most likely unlined, field data from measuring irrigation channels losses will help assess seepage and leakage losses, which would indicate the need for improved maintenance practices.

13.8.6 Estimated Annual Maintenance Costs

An estimate of the annual maintenance costs under a preventive maintenance programme should be presented. The format required in preparing a funding request should be used.

13.8.7 Preventive Maintenance Plan

a. Maintenance Issues: The maintenance issues addressed in preparing the "Catch-Up" maintenance plan should also be addressed in the preventive maintenance plan. These are:

   - Achieving "regime" in irrigation channels;
   - Eliminating sediment sources;

- Minimizing seepage losses; and
- Reducing deterioration of channel lining.

This section should state the degree to which these issues have been accomplished to complete the priority maintenance needs. Then, the role of preventive maintenance programme in addressing these issues should be presented.

b. Preventive Maintenance Activities: A summary of sub-sections A, B, and E should be presented. This would state the causes of maintenance problems in the irrigation project, the anticipated extent of preventive maintenance activities, and the proposed programme for project staff to provide technical assistance to water users to improve their maintenance practices in the tertiary channels.
c. Preventive Maintenance Requirements: A summary of sub-sections C, D, and F should be presented. This is a summary of the equipment manpower and annual budget requirement for the preventive maintenance programme.

13.9 Activities of Engineering Staff

The field maintenance activities are done by the engineering staff in the irrigation division and subdivisions, with 3 to 5 sections under each subdivision. Each section, headed by a Junior Engineer, has a crew of masonry workers, mates, gauge readers, and beldars. Their functions in relation to irrigation system maintenance practices are described below.

13.9.1 Maintenance Survey and Discharge Rating of Flow Control Structures

The Assistant Engineer in charge of operation and maintenance lists flow control structures to be calibrated for discharge measurement. He provides this list of structures to each Junior Engineer for the area under his charge. The Junior Engineer conducts the maintenance survey accompanied by the mistry/supervisor when that portion of the reach is serviced by that mistry/supervisor. The Junior Engineer with the gauge reader under his supervision will do the discharge rating in his responsibility area.

13.9.2 Essential Structure Maintenance Plan

(a) Junior Engineers each give their plan of work to be done along with a cost estimate with necessary explanatory notes and sketches to the Assistant Engineer;

(b) The Assistant Engineer compiles and prepares a complete plan and submits it to the Executive Engineer,
(c) The Executive Engineer technically approves the Essential Structural Maintenance Plan (ESPM) or submits it to the Superintending Engineer if it is beyond his powers to approve.

13.9.3 Collect Water Measurement and Channel Loss Data

The Assistant Engineer will make out schedules for conducting the inflow-outflow tests to evaluate channel losses for various reaches in the irrigation channel network. This may have to be supplemented with ponding tests for some reaches. While conducting these tests, the Assistant Engineer will coordinate the main canals, distributaries, and minor canals with field data collection.

13.9.4 Obtain Maintenance Information from Farmers

(a) The Junior Engineers, for their respective sections, will be responsible for convening the maintenance committee meeting with farmers on a fixed day, fixed place, and fixed time. The Assistant Engineer will be the convener of the committee.
(b) Each Junior Engineer will enter: (i) the problems/suggestions of the farmers in a page number-based register on the left half of the page; and (i) record the action taken on each in the right half of the page by the time the next meeting takes place at the same station.
(c) The Assistant Engineer will see each item for the action taken and record his instructions, if any, for further action.
(d) The Executive Engineer should also attend at least one such meeting a month and watch for the farmers' problems/suggestions.

13.9.5 Prepare Normal Maintenance Programme Report

(a) Each Junior Engineer will compile a listing of routine maintenance activities and the cost activity for the command area under his jurisdiction of the past three budget years.
(b) The Assistant Engineer will compile and edit the materials received from the Junior Engineer under his supervision.
(c) The Executive Engineer will be responsible for preparing the Normal Maintenance Programme Report. The completed report should be forwarded to the Superintending Engineer for review and comments.

(d) The Superintending Engineer should review the normal Maintenance Programme Report and make comments to the Executive Engineer if further improvements are necessary. When the Superintending Engineer approves, the report should be forwarded to the appropriate Chief Engineer for information.

13.9.6 "Catch-Up" Maintenance Plan

(a) Based on the suggestions of hydraulic surveys by maintenance committees and the diagnostic 'walk through' survey, the Junior Engineer should prepare a listing of priority maintenance needs for the portion of the system under his charge. He will also prepare the cost estimate for each priority maintenance need.
(b) The Assistant Engineer will compile the reports from each of the Junior Engineers under- his supervision and prepare a single compilation of all reports.
(c) The groups of Assistant Engineers will prepare a 'catch up,' or system deficiency maintenance plan based on the above surveys and prepare the total project's cost estimates. They will also write the other sections of the 'catch-up' maintenance plan under frequent consultation with the Executive Engineer.
(d) The Executive Engineer will advise the Assistant Engineers on the requirements for preparing a 'Catch-up' Maintenance Plan and will review each section of the plan as it is prepared. The Executive Engineer is responsible for the final editing of the plan.
(e) The 'Catch-up' Maintenance Plan should be forwarded to the Superintending Engineer for review and comments (if the Executive Engineer requires changes). Upon approval, the plan should be forwarded to the appropriate Chief Engineer for information and to arrange a budget.

13.9.7 Annual Maintenance Work Plan

(a) Each Junior Engineer prepares plans of normal and 'catch-up' maintenance activities to be undertaken during the next budget year. He submits these plans with cost estimates to the Assistant Engineer,
(b) The Assistant Engineer will then prepare separate plans for normal maintenance and priority maintenance needs for the following budget year.
(c) The group of Assistant Engineers will prepare a draft of the annual maintenance work plan in consultation with the Executive Engineer.
(d) The Executive Engineer will do the final editing of the proposed cost estimates of the normal maintenance and Priority Maintenance Needs (PMN) works to be undertaken during the following budget year.
(e) The annual maintenance work plan, along with the cost estimate is then reviewed by the Superintending Engineer and approved.

13.9.8 Annual Maintenance Completion Report

(a) Each Junior Engineer will submit a monthly progress report for his section to the Assistant Engineer for the completed maintenance works, giving details of the men and material used with expenditure details.
(b) The Assistant Engineer would compile all the sections' details and submit the same to the Executive Engineer.
(c) The Executive Engineer would compile the reports from all of the Assistant Engineers into a final annual maintenance completion report, which would be forwarded to the Superintending Engineer.
(d) The Superintending Engineer will review the annual maintenance completion report, finalize it, and forward the report to the appropriate Chief Engineer for information.

Appendix 13.1: Case Study on Maintenance Plan of Gohira Irrigation Scheme

This case study is based on review of the operation and maintenance manual for Gohira irrigation Scheme prepared by the Department of Water Resources (DoWR) of Government of Odissa (Odissa 2016). The manual explains the procedures and processes to define the O&M Plans for operation during the Rabi and Kharif seasons and maintenance during the pre-season periods.

The manual focuses on O&M of the main supply system and also describes the procedures and activities of the Pani Panchayats (PP) for the management of the minor distribution system (i.e., downstream of the head of the minor canals) and how these integrate together.

The Gohira Irrigation Scheme

(i) Gohira Medium Irrigation sub-project (Fig. 13.3) is located in the Brahmani River basin and comes under the northwestern plateau zone (agro-climatic zone) of Orissa, approximately 30 km from the district centre of Deogarh.
(ii) Gohira Dam is located at latitude 21° 28′–10″ N and longitude 84° 34′ 10″ E, at Ambaghat Village, in Deogarh district. The dam intercepts a fan shaped catchment area of 236 km^2 in Deogarh hill range across Rriver Gohira, a right bank tributary of River Brahmani.
(iii) The project has a cultivable command area of 8165 ha, and ensures water for close to 90% rice paddy cropping in Kharif and, in recent years, just over 30% of the command in Rabi. It has been in operation since 1981. The scheme is bisected by the Gohira River, the source of irrigation, and is divided into two major blocks.
(iv) Gohira reservoir has gross and live storage capacities of 73.0 and 65.35 million m^3, respectively. The distribution system consists of left and right main canals,

22.25 km and 26.35 km long, respectively. The total length of minor and sub-minor canals is 140.40 km.

Maintenance Calendar

As shown in the O&M calendar (Fig. 13.4) the bulk of system maintenance is carried out during two periods. In Gohira, the Rabi season ends on 30th April, and the Kharif season begins on 16th June, so the interim period of 46 days, from 1st May to 15th June, should be utilized by the scheme management for the preparation and implementation of a pre-Kharif maintenance plan. Similarly, in the O&M calendar for Gohira, the Kharif season ends on 31st October, and the Rabi season begins on 16th December, so a second period of 45 days, from 1st November to 15th of December, should be utilized for a pre-Rabi maintenance plan.

Walkthrough

The following points describe how the participatory walk-through survey should be conducted:

- It should be a walk-through, not a drive-through survey, ensuring that every key asset is inspected in detail and maintenance needs are identified in a systematic and participatory manner.
- It should cover the entire system, not only selected sites, where assets are already known to need maintenance or the proposed works are expected to match the available budget. The combined cost of all the identified maintenance needs will likely exceed the available budget; however, a full survey is required for effective planning, budgeting, and prioritisation of works, and full documentation will

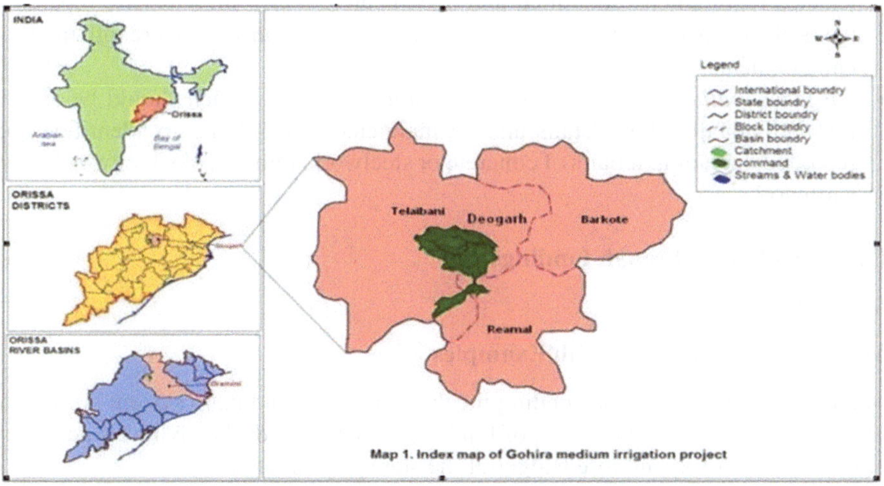

Fig. 13.3 Scheme location map of Gohira medium irrigation project (Odisha)

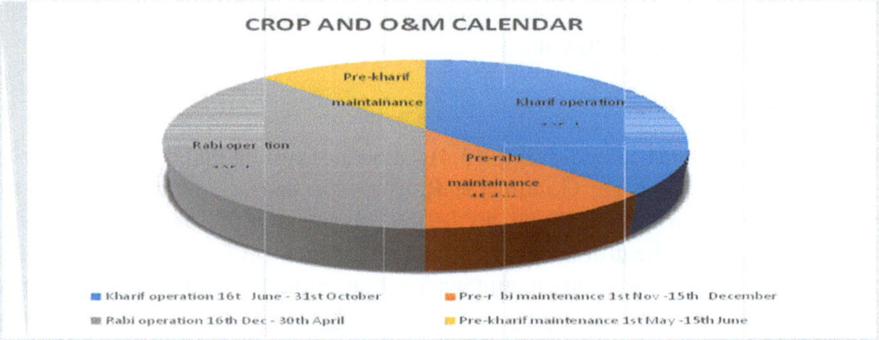

Fig. 13.4 Annual O&M reference calendar—graphical presentation

enhance the accountability of the DoWR staff and PPs, and benefit their financial management in the future.

- Each survey should start from the primary structure at the start of a hydraulic subsystem, such as a head regulator, and work downstream. By opening or closing the gates, the system may be inspected full or empty, depending on whether the focus is to assess the hydraulic performance and availability.
- *Head Regulators*: Records should be made of the location, name of parent canal, command area and PP served etc., then all maintenance requirements should be identified and recorded, which may be repairs of concrete or steelwork and/or silt removal.
- *Canal/Drain reaches*: Each reach should be identified individually, details of location, head regulator, command area, PP etc., should be recorded, and all maintenance requirements identified and documented, which may include repairs to the broken concrete lining, removal of silt, weed clearance, and strengthening of embankments.
- *Off-takes, outlets, Chaks, culverts and other structures*: Each should be identified individually, then details and all maintenance requirements identified and recorded, including repairs of concrete or steelwork, removal of silt or weeds, and earthworks.

Example of Walkthrough Findings

See Table 13.2.

Maintenance Categories with Examples

Maintenance categories and funding for the headworks and main system are shown in Table 13.3. In India, the source of funding for regular O&M activities is generally Non-Plan O&M Expenditure budget of the state government.

Table 13.2 Example of a typical walk-through survey record format

(1) Date & time of P W T done	10th April 2014, Friday, (from 09.00 to 12.00 PM)
(2) No of participants & Key participants	As per guidelines
(3) Name of the sub system P WT done	Menjaribahal sub-minor (from head to tail) of PP No-01
(4) Findings given below for some locations as illustration)	

Sl. no.	Location RD in M, from-to	Structures/or canal reach	Present condition	Findings of P W T	Maintenance work and priority
1	0.00 m	H/R	Damaged condition	H/R is damaged many breaches are found. Gate is there but fully rusted and damaged need replace. The upstream head wall need to re-construction	Re-construction of H.R up stream with new shutter gate
2	50	CD-1	Partially damaged	Silt deposition in the bed and erosion of banks	Silts should be removed from the channel bed and banks are to be raised. **Immediate priority**
3	128	CD-2	Partially damaged	Silt deposition in the bed and erosion of banks	Silts should be removed from the channel bed and banks are to be raised. **Immediate priority**
4	0.00–400 m	Guard wall	In good condition	Channel in this reach looks good, so no maintenance is required	Only jungle clearance on both bank is proposed
5	136 m	1R (NEW)	New	Municipality drain connects to W/C in left side of drain, so it causes pollution & chocking of pipe	Municipality drain should be separated. JE and PP jointly can initiate action with Municipal authority

(continued)

Table 13.2 (continued)

Sl. no.	Location RD in M, from-to	Structures/or canal reach	Present condition	Findings of P W T	Maintenance work and priority
13	500	7R	Partially damaged	Banks on both the sides of W/C are eroded in this section of reach	Raising and strengthening of banks with excavated earth is **Immediate priority**
14	530	Fall	Partially damaged	The sides and top of the fall is partially broken	This fall should be repaired **Immediate priority**
29	140	18L	Partially damaged	Silt deposition in the bed and erosion of banks	Silts should be removed from the channel bed and banks are to be raised. **Immediate priority**
43	2147	Fall	Partially damaged	Silt deposition in the bed and erosion of banks	Silts should be removed from the channel bed and banks are to be raised. **Immediate priority**
44	2180	Tail	Partially damaged	Silt deposition in the bed and erosion of banks	Silts should be removed from the channel bed and banks are to be raised. **Immediate priority**

Questions

1. Briefly explain the following terms:
   Maintenance learning process, objectives, targets, monitoring, implementation, hydraulic performance, deferred maintenance, and catch-up maintenance.
2. What is the purpose of maintenance learning process?
3. Explain the operation activities listed in Fig. 13.2.
4. Explain the maintenance activities listed in Fig. 13.2.
5. Explain the aspects which are emphasized in the maintenance learning process.
6. Explain the steps involved in planning for maintenance.
7. Explain the strategy for improving an existing irrigation system.
8. Why diversion weir and intake have highest priority for maintenance.

Table 13.3 Maintenance categories with examples

Maintenance categories			
Routine/scheduled maintenance (minor or normal)	Periodic maintenance (major or deferred)	Emergency maintenance	Special improvement, upgrading or extension works
Being minor and light works, routine maintenance in the main system is carried out by the operation staff of DoWR (Khalasis), and in the minor system by the PPs (Water Masters) with casual labour hired as required	Maintenance implemented less frequently (e.g. every 2–5 years) to repair deteriorated infrastructure that needs more than routine/ scheduled work, identified during an annual inspection but deferred for cost-effectiveness, and planned for the Rabi, summer closure, when the area is driest. Works are usually carried out by contractors, but sometimes by hiring and managing labour	Emergency repairs are sudden emerging problems and damage not identified at inspections or foreseen in the maintenance plan and which must be undertaken immediately to avoid partial or complete system shutdown (or to restore function as a matter of urgency) and subsequent loss of agricultural production	Maintenance is implemented to address recurring problems with the system, prevent issues from arising and/ or improve the system. These will likely be carried out infrequently, every 10–20 years, as the irrigation environment changes
For example: • Canals: Minor sediment removal at critical sections • Drains: Weed and sediment removal at critical sections • Embankments: Rainfall erosion repairs, grass cutting and removal of vegetation • Structures: Removal of sediment, repairs of cracks and leaks, minor re-plastering and re-pointing • Gates: Repair of gates, lubrication, painting and seals replacement • Water Level Gauges: Cleaning, repair and repainting/replacement • Roads: Repair of pot holes and damage to verges • Buildings: Repair of leaks in roof, painting	*For example*: • Canals: Sediment removal and re-profiling • Drains: Weed and sediment removal and re-profiling • Embankments: Grass cutting and re- profiling • Structures: Major repairs including new walls, bridge decks, etc. and re-plastering/ re-pointing • Gates: replacement of worn parts • Water level gauges: Replacement and re-calibration • Roads: Repair of pot holes and damage to verges, re-grading of gravel roads and resurfacing • Buildings: Major masonry and/or roof repairs, refurbishment and painting	*For example*: • Cyclone damage • Canal or flood embankment breaches • Sudden failure of a structure	*For example*: • Canals: Re-sectioning or lining to address recurring instability, erosion or lack of capacity • Drains: Re-sectioning or lining to address recurring problems or lack of capacity • Embankment: Paving due to higher traffic loads • Structures: Additional hydraulic control/ measurement capabilities • Roads: New or improved bridges, etc. • Buildings: Extensions to house additional facilities

1. Explain in brief the following terms:
   Field note, discharge rating, preventive maintenance, and 'walk-thru' maintenance survey.
2. Why do we need to plan maintenance activities?
3. Explain the difference between the Essential Structural Maintenance Plan and the Normal Maintenance Plan.
4. Explain the difference between Catch-up Maintenance Plan and Preventive Maintenance Plan.
5. Identify and discuss major difficulties in a normal maintenance program for an irrigation project.
6. Compare implementation schedules for various types of maintenance plans.
7. Explain the functions of a Junior Engineer in relation to the maintenance of main conveyance network.
8. Explain the functions of Executive Engineer in relation to the preparation of annual budget estimate for normal maintenance.
9. Explain the functions of Assistant Engineer in relation to the preparation of annual maintenance plan.
10. Discuss the importance of preparing annual report of maintenance work carried out.

References

Chaube UC (2000) Maintenance of irrigation works. Course package (Unpublished), Centre for Continuing Education IIT Roorkee

INCID (1994) Guide for preparation of plans of operation and maintenance of irrigation systems in India. Indian National Committee on Irrigation and Drainage, New Delhi

Odissa (2016) Gohira operation & maintenance manual, volume 1 Main Report), Odissa Z Department of Water Resources Project Management Unit

Skogerboe GV (1990) Development of the irrigation M&O learning process in irrigation and drainage systems, vol 4.Kluwer Academic Publishers, Netherlands, pp 151–169

WAPCOS (1989) Handbook for improving irrigation system maintenance practices. Report of Water and Power Consultancy Services (India) Limited, New Delhi

Chapter 14
Maintenance of Dams, Barrages, and Related Equipment

Abstract The objectives of the chapter are to be aware of physical phenomena causing maintenance problems and understand various inspection and maintenance procedures, and cover the maintenance aspects of dams, barrages, and gates. Structural and operation differences between a dam, a weir, and a barrage are explained. Examples on the checklist of items to be inspected are provided for the earth, masonry, and concrete dams, mechanical equipment and other aspects. Conditions leading to various types of maintenance (critical, regular, condition based) are identified. Maintenance and repair work should be carried out following sound engineering practices specified by expert agencies (Central Water Commission and Bureau of Indian Standards). Every dam has its own specific O&M Manual that includes detailed instructions and schedules for performing periodic and critical maintenance works of the dam, the appurtenant works, and reservoir areas. Items of a typical O&M budget for a dam are listed. Chaube (2006) carried out a study on the evaluation of four tank irrigation projects in the Sagar District of Madhya Pradesh, India. Based on field surveys of the earthen dams (tank bunds), existing deficiencies in maintenance status are discussed. A case study of Bhoothathankettu Barrage in Kerala state of India is discussed to understand the specific maintenance requirements of a barrage. The literature available on the websites of the Ministry of Water Resources (mowr), Government of India (www.mowr.gov.in), State Government of Kerala (www.kerala.gov.in), and other official agencies (www.mwrra.org) is reviewed for the preparation of this chapter.

14.1 Dams and Barrages

14.1.1 Dam

The Central Water Commission of the Government of India has provided guidelines for preparing a manual on the operation and maintenance of dams (CWC 2018). A dam has an overflow and non-overflow sections. The overflow section is commonly referred to as spillway which is made up of concrete. Spillway gates are used to increase the reservoir's storage capacity and regulate the flood outflow. The material

U. C. Chaube et al., *Canal Irrigation Systems in India*, Water Science and Technology Library 126, https://doi.org/10.1007/978-3-031-42812-8_14

Fig. 14.1 Overflow and non-overflow sections of dams

of the non-overflow section could be (i) earth, (ii) earth and rockfill, (iii) masonry, or (iv) concrete, depending on the material availability and site requirement. Examples of these are pictorially depicted in Fig. 14.1.

14.1.2 Weir, Barrage, and Head Works

Weir is a small height control structure (masonry/concrete) across a stream with or without small shutters for raising the water level upstream. A barrage has large vertical gates (instead of shutters) over a low-height weir to raise the water level. Table 14.1 lists the relative merits of weir and barrage.

Figure 14.2 shows a plan view of a barrage showing various components. The main components of a barrage are:

Afflux embankments (bunds)
Upstream and downstream guide banks (bunds)
Low height weir
Vertical gates
Hoisting arrangement
Piers, Bridge
Divide wall

Table 14.1 Relative merits of weir and barrage

Weir	Barrage
Low cost	High cost
Higher level crest	Low crest
About 2 m shutter	High gates
Ponding by raised crest & by the shutter	Ponding by gate
Not possible to combine road or rail bridge with weir	Possible to combine road or rail bridge with a barrage wherever necessary
Silting on the upstream	Silting may be controlled by the judicial operation of gates
Afflux created during a flood is high	The afflux during high floods is low due to the relatively low crest of the weir and as the gates can be lifted even above the high flood level

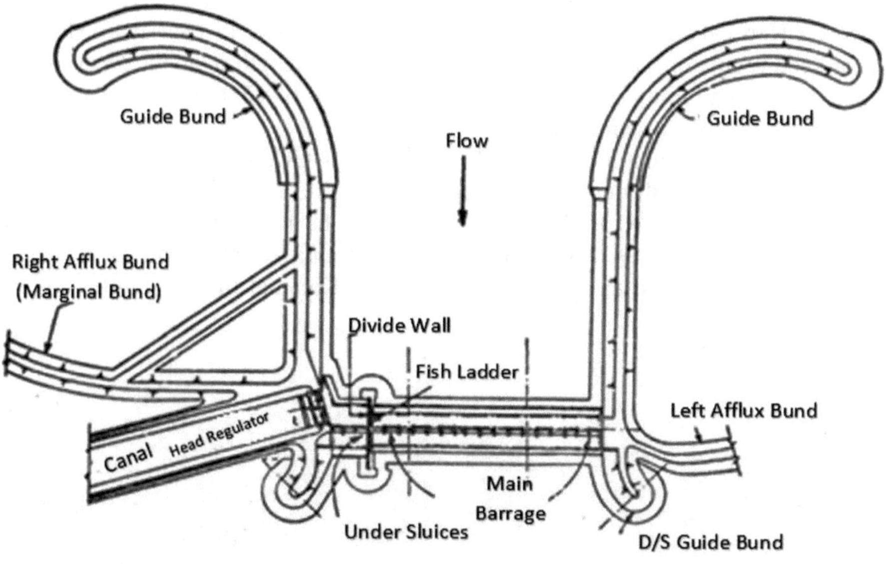

Fig. 14.2 Plan of a barrage showing various components

Under sluice portion
Fish ladder
Head regulator
Flexible protection works upstream and downstream of weir.

14.1.3 Example: Bhimgoda Barrage and Headworks of Upper Ganga Canal (UGC)

Permanent headworks consisting of a weir, under sluices, canal head regulator, and river training works were constructed on the Ganga River at Haridwar from 1971 to 1992. The World Bank has prepared a report on the modernization of the Upper Ganga Canal in its head reaches (Govt. of U.P. 1991). With the development of paddy and intensive use of irrigation as well as the requirement of hydel power stations on the canal, the difficulties in the operation of the Bhimgoda weir were seriously felt. Therefore, a new headworks consisting of the Bhimgoda barrage has been constructed just downstream of the old weir at Hardwar (Fig. 14.3).

The Upper Ganga Canal (UGC) is fed through a link channel that has a capacity of 14,500 cusecs. The old supply channel can carry about 40,000 cusec during floods. Bagh–Rao drainage adds about 15,000 cusecs to the old supply channel. The excess water and the bulk of the silt load coming into it is made to escape at Hardwar dam opposite Har-ki-Pauri (Fig. 14.4). The reach between Har-ki-Pauri and Mayapur canal head regulator has a width of about 400 m and length is about 2 km. In this reach, Lalta Road drainage having an estimated flood of 12,000 cusec, also meets. This reach serves as a sediment trap.

The Mayapur regulator is the head regulator of UGC. This regulator is used to escape excess water through the Mayapur escape and regulate supply into UGC. Therefore, parallel UGC (PUGC) offtake (under the modernization project) was possible only from UGC some distance downstream of Mayapur. UGC passes through an urban built-up area from Mayapur to Kankhal (about 3 miles). A silt

Fig. 14.3 Bhimgoda barrage and headworks of Upper Ganga Canal (UGC)

Fig. 14.4 Tank bed cultivation and embankment material taken from Mahuakheda reservoir bed

ejector is located in this reach, with its escape channel having an outfall in the Ganga. PUGC off-take with a cross regulator has been fixed at miles 3–4 of UGC, keeping in view various problems relating to land acquisition, construction cost, and administration. WAPCOS (1990) has prepared a manual on O&M of the Upper Ganga Canal.

14.2 Dam Inspection Before and During Rainy Season

The Central Water Commission has brought out a proforma for recording various observations of all dam components to ensure proper maintenance and take timely action to correct the defects if any are found (CWC 2016).

14.2.1 Example-Check List Items for Earth Dams

1. Check the emergency stockpiles.
2. Patrolling round the clock should be implemented once the reservoir is near FRL and the anticipated inflow is sizeable.
3. Check for any holes in burrowing animals, cracks, and depressions.
4. Examine the instrumentation data for corrective measures where necessary.
5. Look for boils that may occur downstream in the entire reach.

6. See if any material is being washed out with the seepage water.
7. Look for cracks, both longitudinal and transverse, near the outlet location.

14.2.2 Example-Check List Items for Masonry/Concrete Dams

1. Check for excessive settlement, deflection, seepage uplift pressures, and deterioration of concrete/mortar.
2. Inspect the drainage system in the foundation.
3. Ensure foolproof access to vital parts and adequate lighting and auxiliary facilities.
4. Where there is provision for pumping drainage water, ensure an alternative source of power for uninterrupted pumping to avoid the build-up of pressure.

14.2.3 Example-Check List Items for Mechanical Equipment

1. See that the paint on the gates is in good condition.
2. Clean and lubricate the gate wheels and ensure that the wheels are rotating on the bearings.
3. Check for any foreign materials, boulders, and logs in the gate grooves and inspect all embedded parts to see that they are in proper shape.
4. Examine all wire ropes and chains for rusting, broken stands, and reduction in cross-section, and see that they are in good working order with proper lubrication.
5. Ensure all bearing points, hinges, spur gears, and pinions are appropriately lubricated.
6. Check gear teeth for any cracks and undue wear and replace/repair, if necessary.
7. Check for gear oil in the worm gear reducer and limit switches. Drain oil from boxes, rinse with cleaning oil, and fill them up to the marked level with suitable oil.
8. Ensure that the brake shoe will not drag but will bear tightly when applied.
9. Check the electrical wiring system and ensure no loose contacts/defects.
10. Check all keys and bolts of the hoist mechanism and tighten them if loose.
11. See that the operating crane is in good working condition. See that hoist loads are in the vertical direction only and ensure that the crane's allowable capacity is not exceeded.
12. Ensure that fully trained operators familiar with operating instructions are on site for the operation of gates.
13. See if any unusual phenomena like excessive vibrations and noises are noticed, especially when the gate is operated.
14. Keep adequate stock of the spare parts that are generally required on-site.

15. Ensure power supply from two independent sources. See that diesel-generating sets of adequate capacity to serve as standby are kept in good working order.

14.2.4 Example-Check List Items for Other Aspects

1. See that the data book in complete shape is kept ready for reference, which comprises a complete set of design calculations, assumptions made, construction drawings, etc., at the site regarding the dam and appurtenant structures.
2. Ensure that operating manuals/instructions issued by the manufacturers of the gates and instruments are available at the site.
3. Ensure that the reservoir area is free from potential slips and slides, and floating logs and wood should be closely watched and removed quickly to prevent a triggering action.
4. Ensure that there is a gauge observation site well upstream of the reservoir so that the travel time of water from the gauge site to the dam site is at least 10 to 12 h to serve as a forewarning. Also, ensure a foolproof communication system so that the reservoir can be depleted to a low level to absorb the incoming flow and confirm the spillway is saved from heavily anticipated discharges.
5. Ensure that competent personnel are positioned at the site to take command of the situation and act boldly by making correct decisions and implementing them during an emergency.

14.3 Safety Measures

14.3.1 Restriction on Public Entry and Security

Public entry into chutes, stilling basins, and hoist platforms for gates at head regulators, cross regulators, outlets, siphons, and escapes should be restricted. Operating personnel should ensure that public entry is restricted in the above area. Section Officers shall ensure that the instructions given are promptly adhered to. Cases of tempering with gates will be promptly brought to the notice of local police, and appropriate complaints will be registered.

Adequate physical security should be provided to preclude unauthorized tampering with and operating equipment. Such security may include chaining and locking outside the gate and constructing and locking adequate restrictive barriers, covers, or enclosures.

14.3.2 Safety of Operating Personnel

Particular importance should be attached to tagging controls for equipment serviced to prevent operations that could endanger other employees. Equally important is the requirement that two employees are in attendance while servicing gates to avoid openings or closing that could injure them. Section officers should be familiar with the identity, location, and telephone numbers of nearby hospitals, doctors, ambulances, and rescue units to provide medical assistance. A directory containing all these details should be readily available.

14.4 Dam Maintenance Plan

14.4.1 Catchment Protection to Check Soil Erosion

The cost of maintenance of a watershed does not form part of the normal maintenance budget of the dam. Watershed maintenance works are usually carried out by the Forest and Soil Conservation Departments under their normal watershed management plans. Soil conservation measures within the catchment are the best solution to check or minimize the silt accumulation in a reservoir. Large debris needs to be removed from the reservoir. Regular watch and control are required to check the encroachment in the foreshore areas. Aquatic weeds need control at the source (i.e., in the reservoir itself) by their uprooting and destruction manually.

14.4.2 Critical Maintenance of Dam

The following conditions are critical and call for immediate attention & reservoir lowering.

- A dam about to be overtopped or being overtopped during high floods.
- A dam about to be breached by erosion, slope failure, etc.
- A dam showing signs of piping or internal erosion indicated by increasingly cloudy seepage or other symptoms.
- Inoperable gates of the spillway.
- Evidence of excessive seepage appearing anywhere at the dam site.

14.4.3 Conditions-Based Maintenance of Dam

These may include but are not limited to:

- Fill animal burrows.
- Repair livestock trails and fences to keep livestock off the dam.
- Restore and reseed eroded areas and gullies on embankment dams.
- Repair of defective gates, valves, and other hydro-mechanical equipment.
- Repair any concrete or metal components that have deteriorated.
- Cleaning of the choked drainage holes in the dam body/foundations in concrete/masonry dams.

14.4.4 Record Keeping for Dam Maintenance

Many tasks must be taken up regularly (CWC 2018). A data book for the dam must be maintained for proper upkeep and maintenance of a dam. This book contains details of the location, accessibility, and salient features of the dam and the problems met during construction, and how they were solved. This also includes details of foundations, geological aspects, hydrology, flow searching criteria, detailed drawings, and instrumentation. Knowledge of the design/details goes a long way to assist in properly maintaining a dam.

14.5 Example of Poor Maintenance of an Earth Dam

Chaube (2006) carried out a study on the evaluation of four tank irrigation projects in Sagar District of Madhya Pradesh, India. Subsequently, using the data of this report and other documents, Nissanka (2007) carried out a performance evaluation of the tank irrigation projects. The discussion in what follows is based on these studies.

14.5.1 Reservoir Bed Excavation for Construction Material

According to the norms/technical specifications, no material should be excavated from the reservoir bed up to 10 H distance from the embankment, where H is the height of the embankment from the bed level of the stream. Such excavation endangers the safety of the dam. Figure 14.4 shows that this norm has been violated in the construction of an earthen dam located in the central part of India. The soil has been extensively excavated from the reservoir bed adjacent to the dam and used as an embankment material.

14.5.2 Use of Land Along the Periphery and Bed of the Reservoir for Cultivation

It is a common practice in South Asian countries to construct small tanks for monsoon runoff storage and use it for supplementary irrigation. With the gradual release of stored water for irrigation in the command area, more and more land on the periphery becomes available for cultivation. This land has sufficient moisture and fertile soil. It is debatable whether reservoir bed cultivation should be allowed in Rabi and the summer season. Cultivation on the periphery and in-tank beds should be permitted only if these lands can be protected from soil erosion.

14.5.3 Heavy Seepage from Dam Foundation

Heavy seepage from the dam body and its foundation, particularly in the Nala closure, causes waterlogging in the land downstream. Figure 14.5 shows the fields immediately downstream of an earthen dam became waterlogged due to heavy seepage. Some fields have been rendered unsuitable for cultivation, and in some other fields, rabi cultivation has been delayed abnormally.

Fig. 14.5 Waterlogged fields downstream of Khairana Dam

14.5.4 Jungle Clearance

The jungle attracts burrowing animals, rodents, and stray cattle for grazing and burrowing holes in the body of the dam, causing leakages, and piping, thereby endangering the dam's safety. The embankment and its toes should be kept free of such jungle and facilitate easy inspection of leakages and damages in the bunds.

14.5.5 Top of Bund and Downstream Slope

The top of the bund should be meticulously kept free of potholes. Water should not be allowed to collect in the potholes, thereby entering the body of the dam. Any large-scale widening and deepening of rain cuts or gullies under torrential rains may threaten the dam's safety. The flowing silt from the downstream slopes will choke the toe filters, rendering them ineffective. The rain cuts should be backfilled and compacted systematically in a timely fashion. The top of the bund and roadway should be graveled once or twice a year for the smooth running of inspection vehicles. Frequent inspection should be carried out on the downstream slopes for an indication of slides, rain cuts, cracks, seepage, and leakages, and immediate corrective maintenance and repairs should be undertaken.

14.5.6 Upstream Slope and Pitching

Upstream slopes should be carefully inspected for evidence of slides, cracks, erosion, and displacement of pitching. Pitching should be kept free of vegetation and other materials, such as silt and slipped earth.

14.5.7 Toe Filter

If clear water is coming out of the filter, it means that soil particles are not being dislodged or disturbed. If muddy water is coming out of the filter, it indicates the displacement of soil particles through the body of the dam and acts as a signal toward a possible breach of the dam. The downstream filter should be kept free of vegetation and silt to function effectively and facilitate proper inspection. Figure 14.2 shows vegetation and silt in the toe drain of an earthen dam. Frequent inspections of the filter toe are required during the filling time and when the dam is full.

14.6 Guidelines for Maintenance of Bhoothathankettu Barrage

The maintenance procedure for the barrage is similar to that for a dam. The Bureau of Indian Standards Code of Practice number IS 7349: 2012 BIS (2012) explains guidelines for the maintenance of a barrage.

The barrage constructed across the River Periyar at Bhoothathankettu is the headwork of Periyar Valley Irrigation Project (PVIP). This barrage is located in the Ernakulam district of Kerala state. The government of Kerala has prepared an Operation and Maintenance (O&M) Manual for Bhoothathankettu Barrage (Kerala 2020). The barrage maintenance procedures are given below.

14.6.1 Immediate Maintenance

The following conditions are critical and require immediate action by the project authorities.

- Barrage showing signs of piping or internal erosion indicated by increasingly cloudy seepage or other symptoms.
- An increase in the reservoir level near the top of the barrage. Water overtopping the barrage.
- A barrage being blocked or with some inoperable gates.
- An emergency action plan will be activated when any of the above conditions are noted.

14.6.2 Condition-Based Maintenance

The following maintenance works should be completed as soon as possible after the defective condition is noted. These include but are not limited to:

- Repair of defective gates and any other hydro-mechanical equipment etc.
- Repair of any damages on barrage glacis, piers, training/divide walls, upstream and downstream protection works, etc.
- Removal of vegetation from the barrage and appurtenant structures.

14.6.3 Routine Maintenance

These include but are not limited to the following:

- Maintenance of electrical & mechanical equipment and systems e.g., servicing, lubrication, etc.
- Keeping the gate slots clear of silt/debris.
- Remove any vegetation growing on the barrage and appurtenant structures.
- Ensure the approach road to the barrage is in good condition.
- Test the communication equipment.
- Minor repairs of concrete/masonry can be carried out during routine maintenance. For any major works, the guidance of experienced engineers/panel of experts will need to be taken.

14.6.4 Routine Maintenance of Barrage, Sluice Gates and Hoist

Satisfactory operation of Bhoothathankettu Barrage envisages proper operation of the Gates & Hoisting Equipment of its barrage and irrigation outlets for canals.

- If routine inspection of the hydro-mechanical equipment shows the need for maintenance, the work should be completed as soon as possible.
- The O&M Manuals of the Gates Manufacturers would govern the overall maintenance of Gates & Hoists whenever there is any contradiction with the instructions in the manual.
- The manual describes the procedure of daily inspection, monthly and quarterly maintenance, annual maintenance for the gates, hoists, electrical works, and access roads.

14.6.5 Materials Requirement During Monsoon Period

The details of materials required for the monsoon period are:

HSD—400 L; Grease—50 kg; Oil—90 grade—50 L: Cotton Waste—10 kg; Brush for applying grease—10 nos; Patroling torches—2; Rain coats—4; and Gumboot—4 pairs.

At Barrage, a 24 × 7 h patrol schedule is carried out during the monsoon period. At the same time, additional management requirements during the monsoon period are enhanced.

In addition, norms have been prescribed by the Government of Kerala for their projects during the monsoon period for patrolling duties. Details of the materials required for patrolling duties include:

Torches: 1 for each EE, AEE, AE, and 2 in spare.

Cells: 1 fill and 2 spare sets.

Mobile phone/telephone for communication; Rain coat: 3.

14.6.6 *Instrumentation and Monitoring*

The Automated Data Acquisition System in the barrage is required. A gauge is established to measure the water level upstream of the barrage. Based on the water level on upstream of the barrage, flow to the main canal is regulated. The water levels are recorded on an hourly basis. Also, a rain gauge is established by the Hydrology Wing of the Irrigation Department at the barrage site to measure rainfall events. Rainfall data are recorded automatically and can be accessed online by IDRB. The upstream water level and rainfall data are visually measured by the staff of the Irrigation Department for proper analysis of the performance of the barrage.

Table 14.2 shows a checklist of hydro-meteorological instruments.

Table 14.2 Checklist of various instruments installed

Sl. no.	Name of instruments	Nos.	Location	Since when installed (Month/Year)	Whether in working condition (Yes/No)	Date last calibration Date for next calibration	Observations Maintained (Yes/No)	Agency responsible for data collection and processing	Analysis of data done at field level (Yes/No)	Data sent to DSO regularly? (Yes/No)	Remarks
(1)	(2)		(3)	(4)	(5)	(6)	(7)	(8)	(9)	(10)	(11)
Hydro-meteorological instruments											
1	Rain gauge on Dam										
2	Rain gauge in the catchment										
3	Pan Eeaporimeter										
4	Wind velocity recorder										
5	Wind direction recorder										
6	Wave height recorder										
7	Wet and dry bulb										
8	Barometer										
9	Thermometers for air temp.										
10	Thermometers for reservoir water temp.										
11	Automatic weather station										

(continued)

Table 14.2 (continued)

Sl. no.	Name of instruments	Nos.	Location	Since when installed (Month/ Year)	Whether in working condition (Yes/No)	Date last calibration Date for next calibration	Observations Maintained (Yes/No)	Agency responsible for data collection and processing	Analysis of data done at field level (Yes/No)	Data sent to DSO regularly? (Yes/No)	Remarks
(1)	(2)		(3)	(4)	(5)	(6)	(7)	(8)	(9)	(10)	(11)
12	Reservoir level gate (i) Staff gate (ii) Automatic										
13	Other hydro-meteorological Instruments, if any										

Questions

1. Explain structural differences in weir, barrage and dam.
2. Explain in brief the following terms: aquatic weed, leakage holes, dowel, berm, soding, soil–cement mixture, filter fabric, hydraulic pressure, and discharge.
3. Study manuals/textbooks on watershed management and briefly describe some important soil conservation measures in the catchment.
4. 'Knowledge of the design details goes a long way to assist in proper upkeep and maintenance of the dams.' Illustrate it with examples.
5. What are the priority maintenance works for an earth dam?
6. What lessons can be learnt from the maintenance status of four earthen tank bunds given in the example?
7. Obtain a manual on the maintenance of barrages (from internet) and study the maintenance procedure related to your engineering discipline (civil, electrical, or mechanical). State the question.

References

BIS (2012) IS 7349: 2012, Barrages and weirs—operation and maintenance—guidelines. Bureau of Indian Standards, Manak Bhavan, 9 Bahadur Shah Zafar Marg, New Delhi

Chaube (2006) Report on evaluation of rural infrastructure (irrigation) projects in Sagar District of Madhya Pradesh, Sponsored Research by National Bank for Agriculture and Rural Development (NABARD), Mumbai, India

CWC (2016) Guidelines for developing emergency action plans for dams (Doc No. CDSO_GUD_DS_01_ v2.0). Central Water Commission, Government of India

CWC (2018) Guidelines for preparing operation and maintenance manual for dams (Doc.No.CDSO_GUD_DS_03_v1.0) Central Water Commission, Government of India

Govt. of U.P. (1991) Upper ganga modernization project. I&P (World Bank) Govt. of Uttar Pradesh, Department of Irrigation, Roorkee

Kerala (2020) Operation and maintenance (O&M) manual for Bhoothathankettu Barrage, Government of Kerala. www.irrigation.kerala.gov.in

Nissanka N (2007) O & M aspects of small tank irrigation projects—some case studies. Unpublished report of M.Tech Dissertation, supervised by Prof. U C Chaube, IIT Roorkee, Roorkee

WAPCOS (1990) Manual on O&M of Upper Ganga canal. Water and Power Consultancy Service, New Delhi

Chapter 15
Maintenance of Canals and Related Structures

Abstract The objectives of this chapter are to (a) understand the physical and social phenomena causing maintenance problems of canals and related structures and mechanical and electrical equipment; and (b) learn the techniques for repair and maintenance, and inspection and exercising of equipment. This chapter explains the various solutions to the physical and social phenomena causing maintenance problems of canals and canal structures. High-intensity rains, erosion, vegetative growth, and seepage, and leakage losses are some of the physical phenomena, while cutting banks, putting obstruction in the bed, unauthorized/oversized outlets, bank encroachment, and removal of lining panels are part of social phenomena. The channel section is maintained by controlling the erosion of its banks, removing aquatic weed growth, and removing sediment from the canal bed. Some simple and practical methods are suggested to control soil erosion in canals (roadway and berm, cut banks, animal crossing, human bathing, and washing). Exploration of leakage holes and measures to control seepage and leakage from embankments are discussed. Similarly, illustrative diagrams are used to explain (a) collection of surface drainage from a roadway and conveyance into the canal, (b) reconstruction of the eroded bank, (c) subsurface drainage in the cut area reach of the canal (open drains for a high roadway embankment in a cut area, interceptor drain and a perforated pipeline drain), and (d) procedure for filling cavities behind line d channels). Guidelines for inspection, maintenance, and repair of concrete and masonry works (canal structures and drains) are given. To determine whether maintenance is required for an inverted siphon (or a culvert), a useful technique is to periodically check the coefficient of discharge for the structure. An example is given to determine coefficients of discharge. Remedies to relieve hydraulic pressure at the outlet transition for a drop structure are illustrated. The tertiary subsystem is generally the most neglected portion of irrigation projects in South Asia. Technical guidance needs to be provided to farmers particularly for the maintenance of water course structures and drains. Minor canals and water course channels can be best maintained (weed removal, repair of banks) by farmers themselves. Important Do's and don'ts have been stated for the guidance of WUAs. Inspection of concrete/masonry construction is explained. Poor quality materials and short composition of concrete and mortar will be evident by scaling, spalling, pop-outs, crumbling, and other deterioration forms. Repair procedures for concrete surface deterioration, cracks, and joints are discussed.

U. C. Chaube et al., *Canal Irrigation Systems in India*, Water Science and Technology Library 126, https://doi.org/10.1007/978-3-031-42812-8_15

15.1 Canal Distribution System

A typical irrigation canal distribution system is schematically depicted in Fig. 15.1. It consists of a main canal, branch canal, distributary canal, minor canal, and canal structures (head regulator, cross regulator, outlets etc.). Outlets are generally located on minor canals.

Typical Cross-Section of a Minor Canal

Typical cross sections of a minor with head discharge less than 0.3 cumec and head discharge between 0.3 and 0.6 cumec are shown in Fig. 15.2:

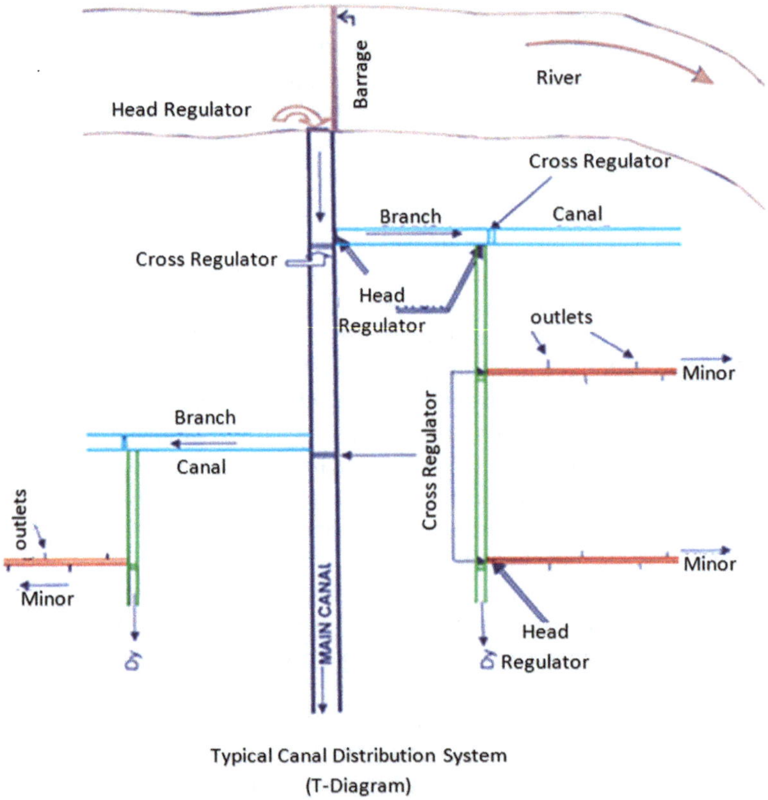

Fig. 15.1 Canal distribution system (*Source* UPWSRP 2008)

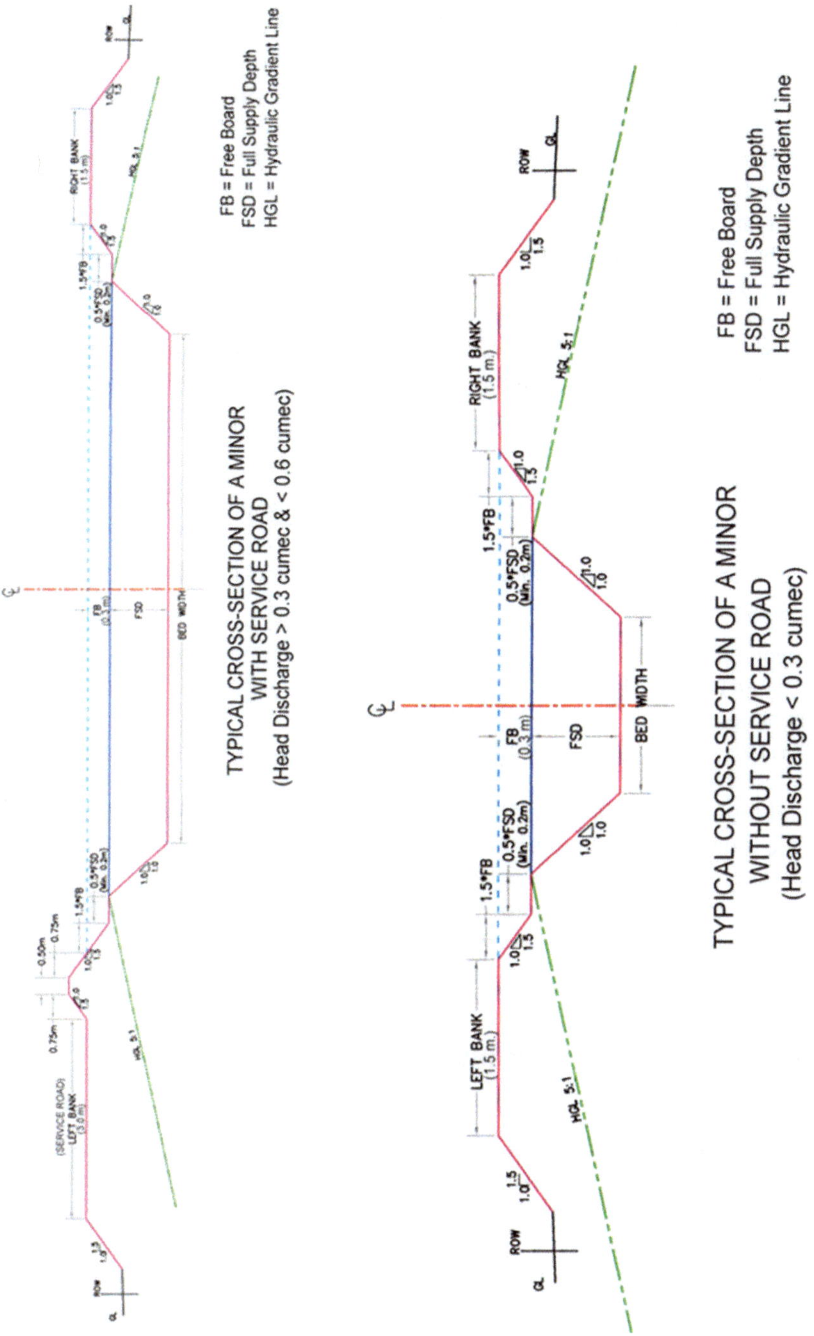

Fig. 15.2 Typical cross sections of minor canal (*Source* UPWSRP 2008)

15.2 Checklist to Assess Extent of Repair and Maintenance

The Department of Irrigation Government of Uttar Pradesh has prepared a manual for the construction and maintenance of Minors (UPWSRP 2008). This manual provides useful guidelines for repair and maintenance, including checklists. A checklist is a useful tool for planning repair and maintenance works. Checklists to assess the extent of repair and maintenance in a minor canal are stated in Appendices. Appendix 15. A. shows the checklist to assess the condition of a minor canal. There are 18 components to be checked. Some of the important components are:

Jungle/weeds, silting, bank section, bank-top cuts, service road, ramps at VRBs/ outlets, longitudinal/cross slopes at bank top, Raptas at outer slopes of banks, seepage drains, the lining of canals, bed bars/profiles, reference pillar, water level gauges.

For each component of the canal, the checklist consists of

(i) location details;
(ii) condition of minor: whether it needs repair and whether it needs rehabilitation; and
(iii) description of maintenance need and whether it is routine or priority maintenance.

Appendix 15. B. shows the checklist to assess the condition of concrete and masonry structures. There are 17 components to be checked. some of the important components are:

U/S, D/S wing wall and Abautment of VRB/fall/weir, the crest of fall/weir, bridge/ regulator abutment, deck slab and joints, parapets, gates of regulator/weir, stilling basin of fall/weir, U/S, and D/s Pitching of VRB/fall/weir, pipe of outlet, earth cover over the foundation of abutment/pier, launching apron/filter in bed at VRB/fall/wier.

For each component, the checklist consists of similar items to the canal.

Maintenance and repair procedures for the canal and related structures are explained below using line sketches (WAPCOS 1989).

15.3 Control of Aquatic and Vegetative Growth

Unlined Canal: Pulling weeds from the bed sediments by wading laborers has proved more effective than cutting by hand blade because the regrowth rate is reduced. There is a significant advantage in periodically closing a channel, if possible, to reduce aquatic growth. Also, at this time, vegetative growth can be uprooted and then gathered and burned. This allows for the inspection of the channel bed and reduces the total water losses from channels over a cropping season.

Lined Canal: Excavate and remove the silt deposit in a lined canal. Any remaining vegetative growth would be found in broken joints and holes in lining panels; therefore, asphalt sealing of holes and repair should eliminate this possibility.

15.4 Erosion and Sedimentation Control

Sediment removal should begin at the lower end of a channel and proceed upstream; if begun upstream and not completed, the excavated materials serve as a sediment trap.

15.4.1 Roadway and Berm Erosion

Lowering roadway and berm embankments to only the required heights would reduce the degree of erosion and result in less maintenance cost. For example, the embankments do not have to be higher than 30 cm above the full supply level (FSL) for earthen channels and 15 cm above the top of the lining for lined channels. The roadways and berms need to be periodically graded (0.5%), so that runoff from rainfall flows away from the irrigation channel; this may be particularly cumbersome on the berms because of narrow widths and sediment that has been removed from the channel piled on the top of the berm, which should never be allowed. Sod should be placed on slopes above FSL and berms. Although the ideal solution for lined channels would be to place the roadway at the same elevation as the top of the lining panels, there are two advantages to placing it about 15 cm higher. First, vehicles will be prevented from riding on the edge of the panels and inflicting damage. Second, there is less likelihood of ponding and subsequent cavity formation behind the lining.

15.4.2 Erosion from Cut-Banks

The most significant maintenance problems usually occur when irrigation channels are constructed in cut areas. A typical cut-bank cross-section is shown in Fig. 15.3. Usually, the earthen toe drain fills with sediment each monsoon season. Removing the sediment from the toe drain is crucial. Otherwise, the surface runoff during and immediately following a rainfall event will transport considerable sediment into the irrigation channel. Also, the berm should be sloped away from the irrigation channel towards the toe drain. Then, the placement and maintenance of vegetative cover on the berm, and even more importantly on the cut-bank, will minimize erosion.

15.4.3 Erosion by Animal Crossing

Earthen channels get damaged by animal crossing. In order to prevent erosion from animal crossings, a brick ramp on a slope of 3:1 is constructed for earthen channels.

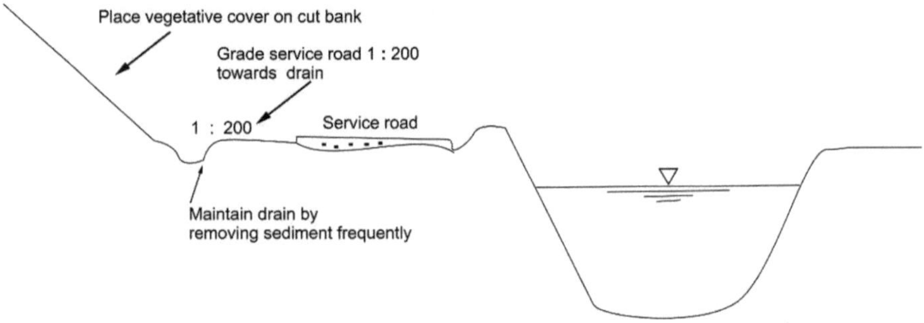

Fig. 15.3 Typical cross-section of a cut-bank

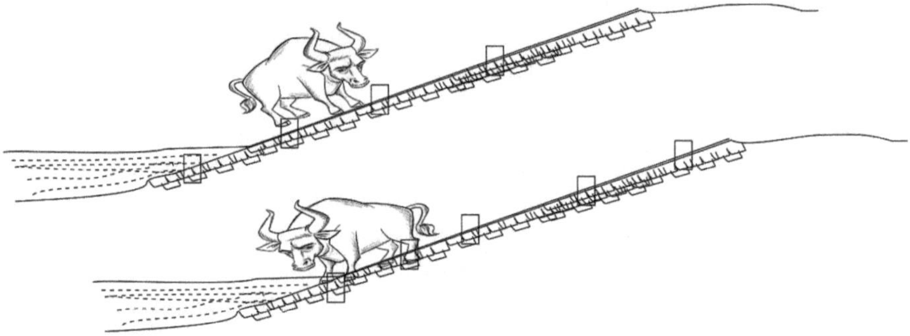

Fig. 15.4 Brick ramp for animal access for bathing and crossing

Periodically, about every meter, a protruding brick is placed that can be used by the animals as a foot-hold, as shown in Fig. 15.4.

For lined channels, erosion resulting from animal crossings could be eliminated for the most part by installing a concrete bridge, but care should be taken to provide adequate measures to prevent surface erosion near abutments. In some cases, farmers may need to be encouraged to use a nearby bridge for animal crossing.

15.4.4 Human Bathing and Washing

Bathing ghats (steps) are usually made of bricks, but concrete could also be used. Without these bathing ghats, erosion will occur in the pathways frequented by people.

15.4.5 Sodding

As part of a long-term effort to eliminate sediment sources flowing into irrigation channels, the provision of vegetative cover, often by sodding, is a primary activity. Once this job is completed for an irrigation project, an essential preventive maintenance activity is one of maintaining the vegetative cover. Appropriate plant species should be selected that are drought resistant. Plants must be grown in relatively fertile soil, periodic fertilizer applications may be required, particularly during the first year of growth, and plants may have to be irrigated to establish good root development. Then, cattle should not be allowed to graze in newly planted areas for at least one year.

15.4.6 Sediment Removal from Bed of Earthen Channels

a. Before starting the work of general silt clearance, silted L- section along with cross-sections shall be measured and designed longitudinal section, and cross-sections should be marked over it to calculate the quantity of silt to be cleared.
b. Canal should always be cleared from downstream to upstream, starting either from the tail or a fall on the canal.
c. Silt should not be cleared below falls, but if outlets in such places are over-drawing due to a rise in the water surface, they should be raised.
d. Any over-excavation in width/depth with respect to designed values of a canal section during de-silting leads to a fall in water level in the canal. This reduces discharge through outlets in such reaches.
e. Silt clearance should, as far as possible, be finished by the time Rabi sowing starts.
f. While de-silting, the canal bed will have a longitudinal gradient, as given in the L-section, and will be kept level transversely. The cross-section shall conform to those given in the drawings (designed) and will be neatly finished.
g. Excavation of silt will be done as per drawing and level/section defined by bed bars. Care shall be taken that bed bars are not damaged during de-silting works.
h. After completion of de-silting, final cross sections of the canal shall be taken at the same locations as that of the initial cross sections. The quantity of excavated silt will be calculated by superimposing final cross sections over initial sections.
i. If the minor takes its authorized discharge, there is no need to clear the silt from such a minor.
j. Silt clearance in a minor shall be done in a manner that FSL is maintained with designed discharge.

15.4.7 *Silt Disposal*

Disposal of silt from bed depends on its quality as given below:

a. Coarse Silt—Coarse silt (sand) may be found in the head reaches of canal. This should be disposed off in filling old borrow pits first and then on long outer slopes of banks; care being taken that the top of the heap is never allowed to come above the bank level.
b. Medium Silt—If taken out before or during the monsoon. It may be utilized in closing leaks, securing outwards of slopes, and widening bank only as a supplement to the berm earth if it proves insufficient.
c. Fine Silt—It is found in the tail reaches of canals. It is valuable for all repairs and should be used like berm earth.
d. In the case of minors taking off from a distributary, generally fine silt is found in head reaches, and no silt is found in mid and tail reaches.
e. Coarse and medium silt should not be spread on canal service road. If medium silt has to be used for any reason, the surface should be covered by a 20 cm thick layer of good loamy soil or fine silt.
f. Bed silt may be thrown on the outer slopes and in hollows. Silt should not be heaped above the bank level.

Before undertaking the sediment removal operation, proper longitudinal and cross-sectional levels should be observed and plotted. The designed cross-sections should be superimposed on the drawing. The depth of sediment deposition along the bed and banks should be ascertained. A proper layout for excavation should be given, and then the sediment removal operation can begin. The earth/silt so excavated should be deposited far away from the canal or in depressions to avoid its re-entry into the canal. The desilting operation should be completed before the start of the cropping season.

Operation at FSL to Control Sediment Deposit

Irrigation channels need to be operated at the Full Supply Level (FSL) in order to minimize the deposition of sediment. Also, during canal closure, the water level in the channels must be lowered slowly to minimize the risk of bank sloughing for earthen channels or failure of channel lining due to significant uplift pressures.

15.5 Control of Seepage and Leakage from Embankments

Visual evidence of the necessity for reconstructing the embankment is numerous leakage holes discharging along the outside bank, usually near the toe of the embankment. It normally begins on the inside bank just above the usual operating water level. Exploration for leakage holes and embankment reconstruction are shown in Fig. 15.5.

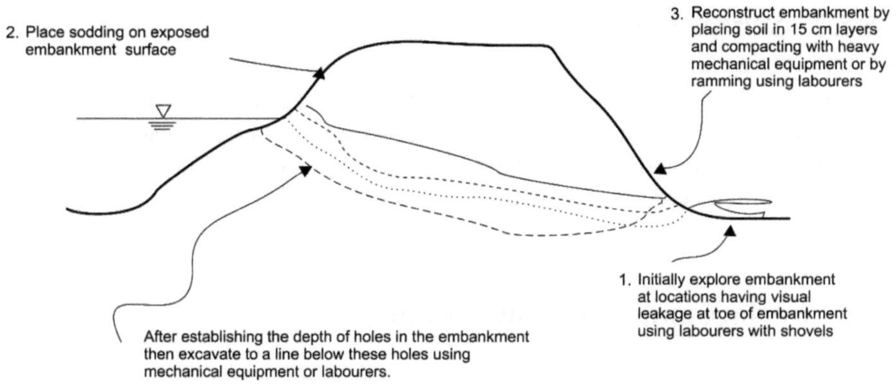

Fig. 15.5 Exploration for Leakage Holes and Embankment Reconstruction

Measuring seepage loss rates with various water levels will disclose the magnitude of seepage and leakage losses and how rapidly the seepage loss rate changes near the usual operating water levels. Rapidly increasing values of seepage loss rate indicates that the embankment should be explored, using laborers, to determine if small holes and caverns exist. This situation is more likely to occur for earthen channels than lined channels, but certainly does occur for either lined or unlined channels.

15.5.1 Compaction

For lined channels, horizontal cracks are evidence of inadequate compaction, usually of the berm embankment. The procedure for reconstructing an inadequately compacted embankment is shown in Fig. 15.6.

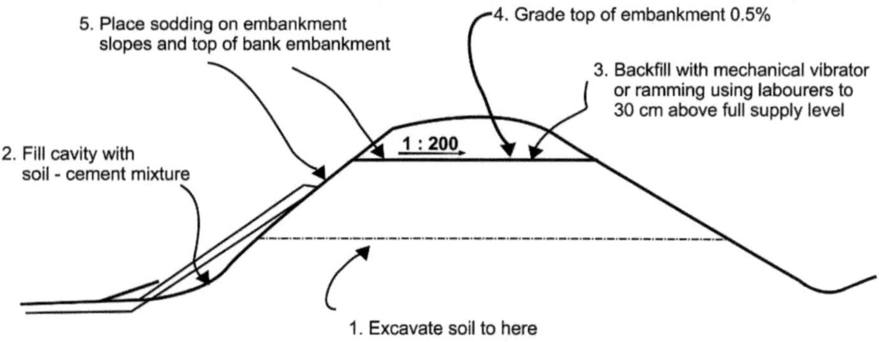

Fig. 15.6 Reconstruction of an Inadequately Compacted Embankment

15.5.2 Service Roads

Potholes, surface irregularities, and ruts should be noted and corrected. The gallows, which consist of two vertical metal poles embedded in a service road and a metal cross-bar on top, should be inserted to be sure that the size of vehicles having access to the service road is properly controlled.

15.6 Surface and Subsurface Drainage

15.6.1 Surface Drainage on Roadway and Berm

To eliminate the sources of sediment entering into irrigation channels, the roadway and berm need to be graded so that surface runoff flows away from the irrigation channel. On-road bends where super-elevation has been purposely constructed to facilitate vehicle traffic, there are two possible solutions for left-turning bends (in the downstream direction) which result in surface runoff flowing into the irrigation channel:

1. Place the curb of earth, brick, concrete, or asphalt on the inside of the bend near the irrigation channel and convey the water from the curb to the irrigation channel in a concrete-lined chute (Fig. 15.7); or
2. Remove the super-elevation and grade the roadway so surface runoff will be conveyed away from the irrigation channel, then place a sign warning motorists to reduce their vehicle's speed.

Sometimes, the roadway's cross-slope is so steep that the surface runoff has a relatively high velocity, which increases even roadway erosion. This problem can be easily corrected by using compacted backfill in the eroded area if required, then grading the cross-slope to be much flatter (1/2 of 1% or less).

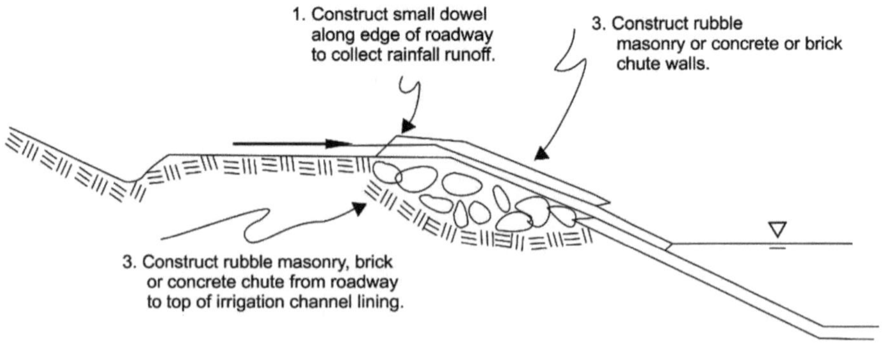

Fig. 15.7 Collection of surface drainage from a roadway and conveyance into the canal

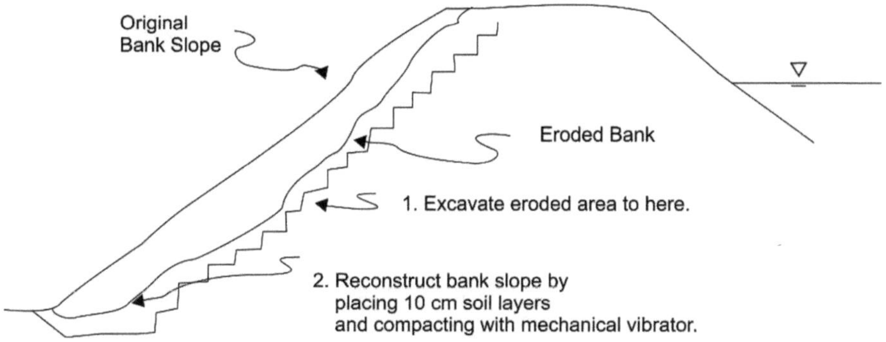

Fig. 15.8 Reconstruction of an Eroded Bank

Sometimes the bank slope is too steep, or too high, or both. This results in bank erosion. The eroded bank needs to be reconstructed, as shown in Fig. 15.8.

To prevent bank erosion, the surface runoff from the embankment surface should be collected and conveyed down the outer bank surface in a concrete-lined, brick, or rubble masonry chute to collect the surface runoff from the roadway or berm. A dowel (curb) will have to be constructed along the outside edge of the top of the embankment. At the outside bank's toe, special measures will be required to avoid erosion, requiring a stilling pond lined with concrete, brick, or rubble masonry.

15.6.2 Subsurface Drainage in the Cut Area Reach of the Canal

Cut areas, in particular, require that the design of the irrigation channel includes adequate subsurface drainage. This is especially true for lined channels but is often the case for earthen channels. This can usually be accomplished by excavating open drains alongside the roadway and berm, provided they are sufficiently deep. Preferably, the invert elevation of the open drain should correspond with the irrigation channel's invert elevation or be even lower.

Suppose the roadway embankment already has asphalt pavement, but the roadway is too high. In that case, it may become too expensive to lower the roadway to provide better surface and subsurface drainage. Alternatives are shown in Fig. 15.9, where open drains are used to provide both surface and subsurface drainage, as shown in Fig. 15.10, which illustrates the use of a surface runoff interceptor drain for surface drainage and a perforated pipeline drain for subsurface drainage. Figure 15.10 also demonstrates using a perforated pipeline drain to provide subsurface drainage for a cut-bank.

For lined channels, weep holes can be used to relieve hydraulic pressure behind lining panels. Also, porous rubble masonry, or any other porous material, can be used. These are lower-cost measures that treat the symptoms rather than the cause of

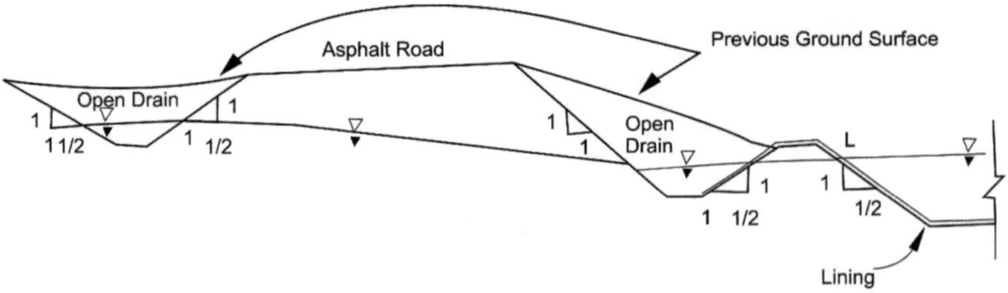

Fig. 15.9 Typical cross-sections for open drains for a high roadway embankment in a cut area

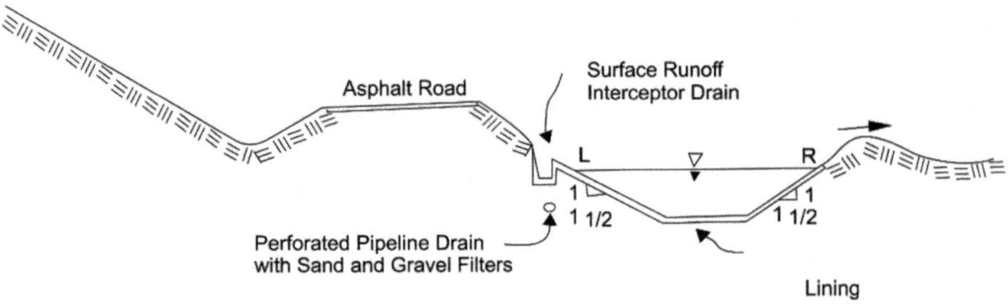

Fig. 15.10 Use of a surface runoff interceptor drain and a perforated pipeline drain

maintenance problems. However, this is very satisfactory if these measures continue to function well with time and do not become clogged or they do not allow soil behind the lining panels to be washed into the irrigation channel.

15.6.3 Filling Cavities Behind Lined Channels

Filling cavities is an expensive maintenance cost. Usually, by the time a cavity is filled, it has become a huge cavernous opening that may extend 1–3 m in width on each side of the joint opening, thereby leaving an extensive area of lining unsupported. Also, there may be a sizeable hole in the lining. If so, one procedure is to remove the broken pieces of lining material so that the size of the cavity can be seen. If the cavern size is quite large, then the soil should be removed from the top of the cavity downward until a large opening has been made. The cavities should be filled, as shown in Fig. 15.11. Sometimes the cavity cannot be detected at the top of the lining wall panels but only by a small opening in the joint. The procedure for filling cavities, as shown in Fig. 15.11, should be followed. Preferably, the joint opening should be sealed with asphalt or some other sealant. If the cavity formation has resulted from the lack of subsurface drainage, and subsurface drainage is not going to be provided, then a weep hole should be placed when the concrete is repaired, as shown in Fig. 15.9.

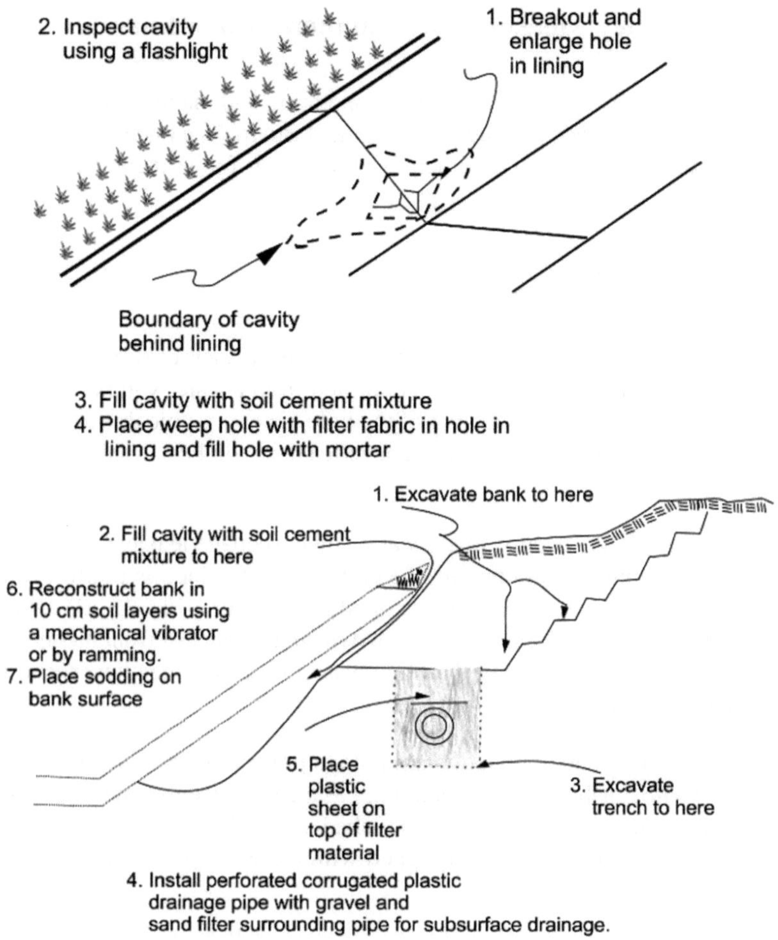

Fig. 15.11 Procedures for filling cavities

If there is extensive damage in a channel reach resulting from the lack of subsurface drainage, then the facilities shown in Figs. 15.7 and 15.8 should be considered. Finally, the roadway bank or berm bank adjacent to the lining walls having cavities must be graded and shaped. Then, sod (vegetation) should be placed on the bank to prevent erosion.

15.7 Emergency Measures for Earthen Embankments

JICA (2017) has prepared a manual for the operation and maintenance of hydraulic structures for the Bangladesh Water Development Board. The manual (pdf file available on internet source) discusses emergency measures for earthen embankments.

15.7.1 Seepage/Leakage

(a) Direct method: If possible, close the hole with a sack closure of old clothes (Fig. 15.12). The hole spot can be identified from a water vortex (eddy).
(b) Indirect method: If direct closing of the hole is not possible due to the depth of the hole spot, then prevention of seepage/leakage will be carried out behind the embankment.

(i) If seepage/leakage occurs just locally, then seepage/leakage water will be collected, placing a drum behind the embankment, which is reinforced by the dumping of compacted clay soil surrounding the drum (Fig. 15.13). This helps in lowering the water level difference between the water in front and behind the embankment. So, the seepage/leakage water force becomes lower, and piping can be overcome.

(ii) If the aforementioned method is not able to prevent piping, then the drum will be filled with crushed stone, functioning as a counterweight and filter. Clay soil is placed compacted on the above-crushed stone, then sandbags are placed on it (Fig. 15.14). Besides, it is equipped with a drain pipe to drain the seepage/leakage water.

(iii) If some leakage/seepage appears along the toe of the embankment, then a timber/woven bamboo mattress is placed surrounding the leakage/seepage. The mattress is reinforced by driven piles and compacted clay soil (Fig. 15.15).

(iv) If the outflow of leakage/seepage water from the steep back side of an embankment will cause sliding of the embankment slope, then that leakage/seepage

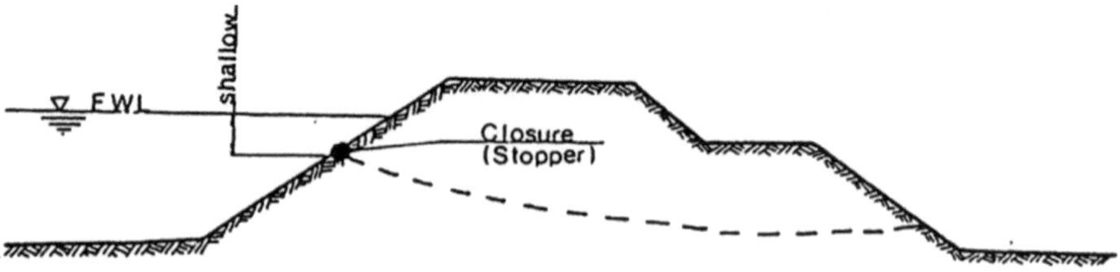

Fig. 15.12 Close the hole with a sack closure of old clothes

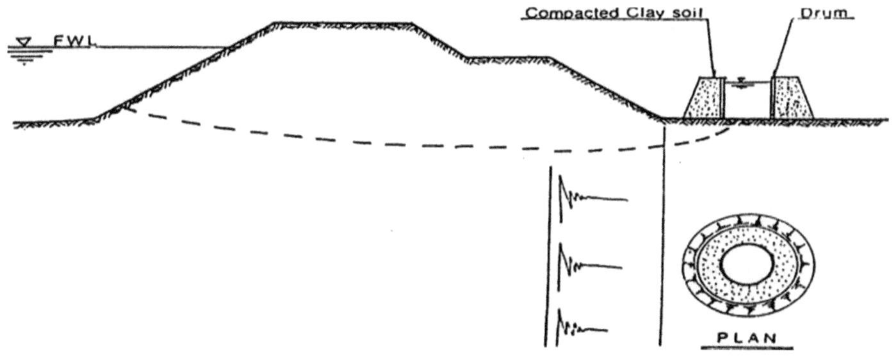

Fig. 15.13 Drum behind the embankment, reinforced by compacted clay soil

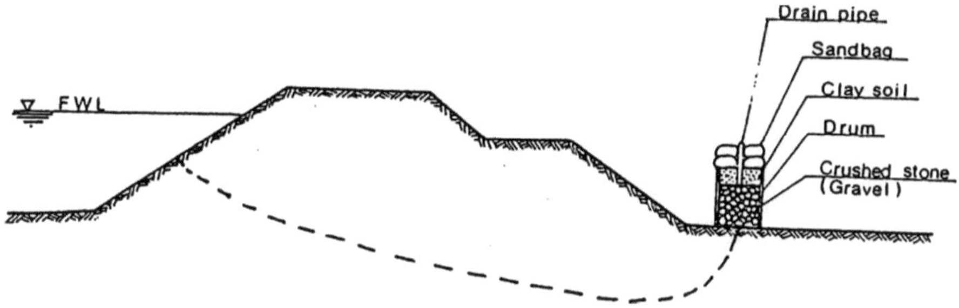

Fig. 15.14 Drum filled with crushed stone

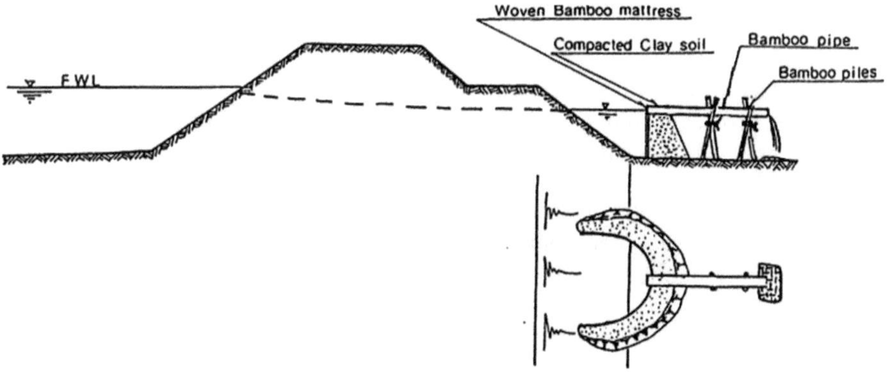

Fig. 15.15 Timber/woven bamboo matters surrounding the leakage/seepage

spot shall be dumped and compacted with clay soil (Fig. 15.16). Besides, it is supported by the woven bamboo mattress fixed at a bamboo pile.

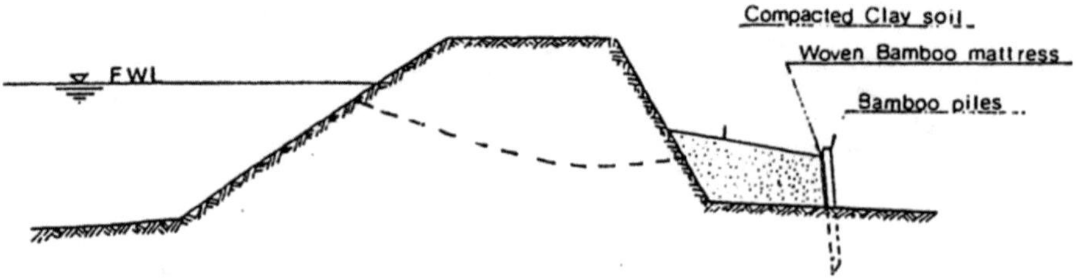

Fig. 15.16 Leakage/seepage spot dumped and compacted with clay soil

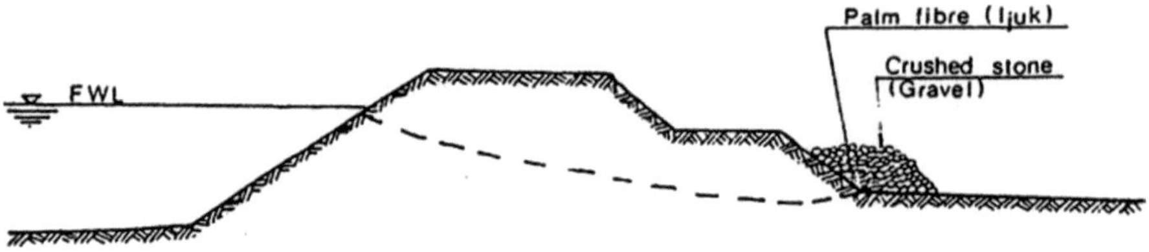

Fig. 15.17 Leakage/seepage closed by palm fibre layer functioning as a filter

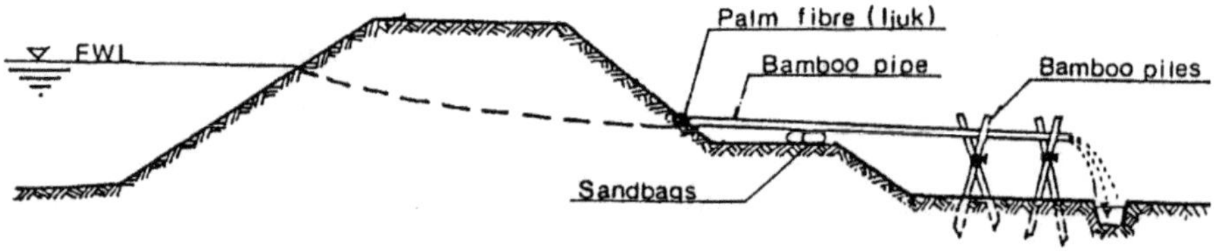

Fig. 15.18 Bamboo pipe to release leakage/seepage water from the toe of the embankment

(v) The place of the leakage/seepage closed by palm fibre layer functioning as a filter to protect the movement of the foundation material (Fig. 15.17). The palm fibre layer is protected by the dumping of crushed stones.

(vi) At the place of leakage/seepage, bamboo pipe will be provided to release leakage/seepage water from the toe of the embankment (Fig. 15.18). At the top of the bamboo pipe, palm fiber will be provided, functioning as a filter.

15.7.2 Scouring

(a) To prevent scouring caused by turbulent flow, a tree stem, including its branches, will be provided at the front side slope of the embankment (Fig. 15.19). At the front end, a counterweight shall be provided to prevent washing out by the waterflow, while the tail end of the tree will be fixed on a pile driven in the embankment body.

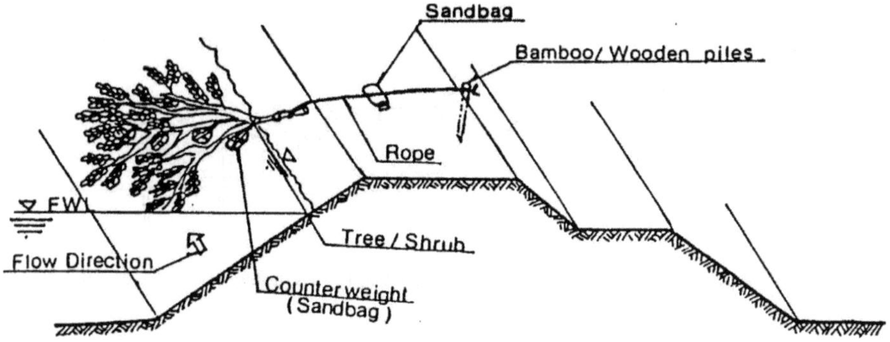

Fig. 15.19 Tree stem and branches at the front side slope of the embankment

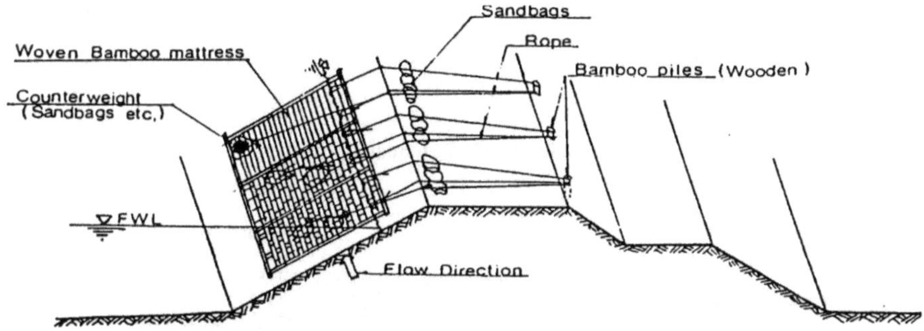

Fig. 15.20 Woven bamboo mattress provided with a counterweight at the tail end

(b) Another method is the use of a woven bamboo mattress provided with a counterweight at the tail end and fix the front end at a pile driven in the embankment body (Fig. 15.20).

15.7.3 Sliding

(a) If sliding occurs at the backside of the embankment, then after dumping compacted soil on it, that place will be provided with a counterweight which consists of sandbags with driven piles (Fig. 15.21).

(b) If sliding occurs at the front side of the embankment, to avoid a breach, the damaged position will be provided with a tree stem, including its branches (Fig. 15.22). The tail end of the tree fixing on a bamboo pile driven in the embankment body, while the front end of the tree is equipped with a stone counterweight. The back side of the embankment should be strengthened with the widening of the embankment crest by dumping compacted soil.

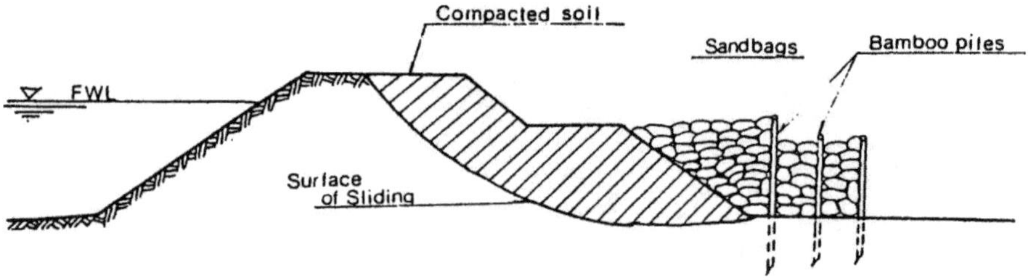

Fig. 15.21 Dumping of compacted soil and a counterweight of sandbags with driven piles

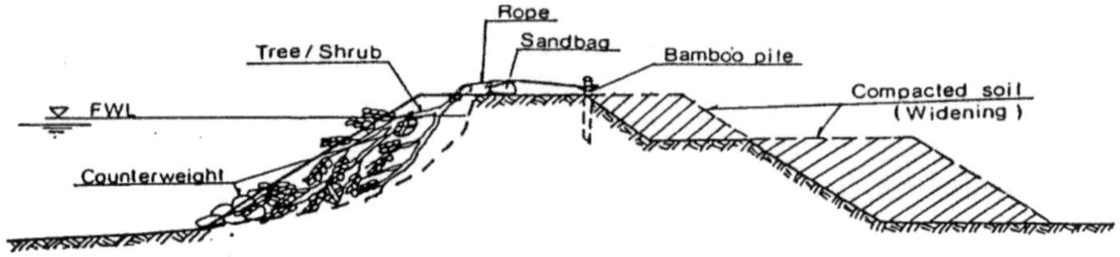

Fig. 15.22 Damaged position provided with a tree stem and branches

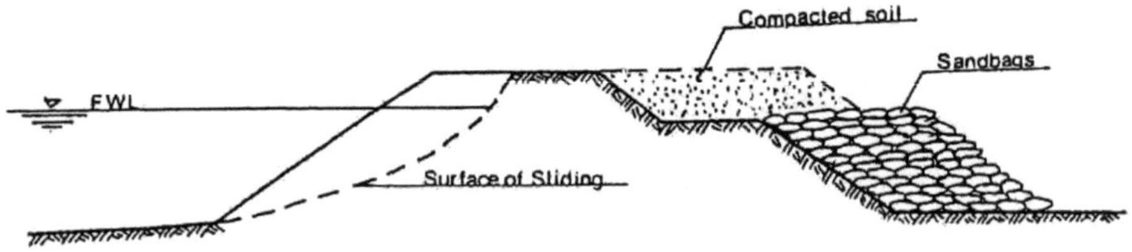

Fig. 15.23 Emergency embankment consisting of sandbags, and compacted clay soil

(c) Sliding at the front side of the embankment, may also be prevented by the construction of an emergency embankment which consists of sandbags, and compacted clay soil, as the same as the original height at the backside of the damaged embankment (Fig. 15.23).

15.7.4 Settlement of Embankment

Place an additional embankment (heightening) wherever settlement is noticed (Fig. 15.24).

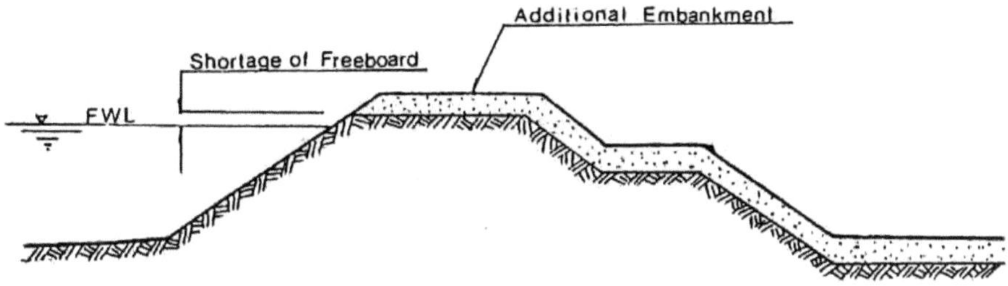

Fig. 15.24 Additional embankment

15.7.5 Breach of Embankment

An emergency embankment is constructed which consists of a staple of sandbags reinforced by driven piles (Fig. 15.25).

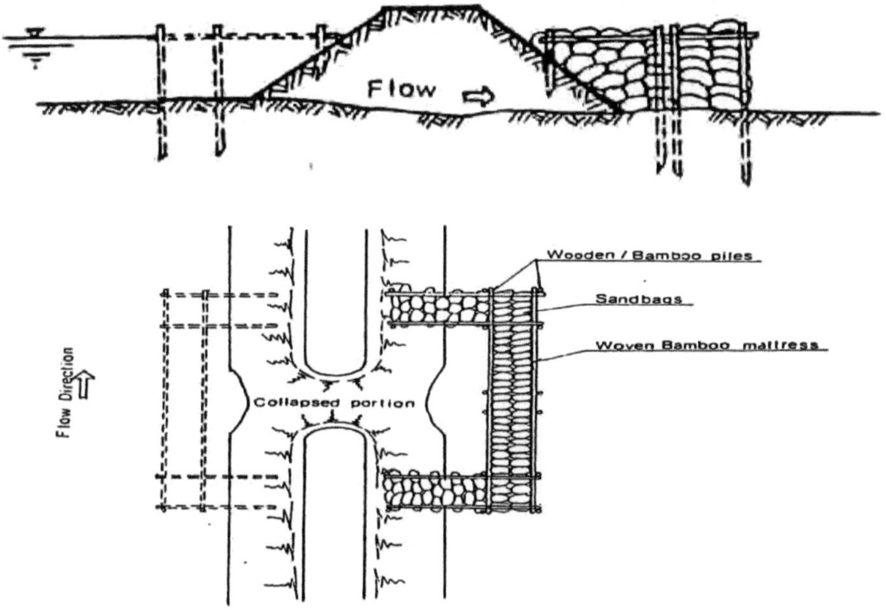

Fig. 15.25 Sandbags reinforced by driven piles

15.8 Maintenance of Canal Structures

Usually, concrete structures will be in excellent structural condition unless they are very old. For both earthen and lined channels, there are problems with flow control structures, primarily with the maintenance of gates.

15.8.1 Under-Canal Structures (Drain Culverts)

The most common maintenance requirement for drain culverts underneath irrigation channels is removing sediment and shrubbery at the inlet and outlet. The fact that shrubs exist, and often the growth is quite dense, is a good indicator that maintenance has been neglected.

In some cases, banks near the headwalls have been erased. This situation should be remedied by refilling and compacting soil in the eroded area, careful grading, and sodding, with particular attention given for some time afterward to assure that good vegetative growth is occurring in the sodded area and that the erosion problem has been corrected.

The drain culverts are valuable structures for serving as outlets for either open drains or pipe interceptor drains constructed along an irrigation channel to provide subsurface drainage. In some cases, it may be desirable to build additional drain culverts to provide outlets or subsurface drainage facilities.

15.8.2 Bridges

Bridges are usually a problem for lined canals. Commonly, surface runoff from rainfall causes erosion at the bridge abutment and the formation of cavities behind the lining panels. The required maintenance is to place rubble masonry, brick, or concrete between the top of the lining and bridge abutments. This remedy should be incorporated into the design criteria for such structures in the future. If the bridge settles, then the brick, concrete, or rubble masonry will crack, and the cracks can be filled with asphalt or mortar.

15.8.3 Drop Structures (Syphons)

The inlet to a drop structure needs to be inspected for cavities along the outside walls and immediately upstream in the transition inlet structure. If cavities exist or erosion is occurring, then the appropriate maintenance remedies should be applied.

Suppose an open channel chute is used between the inlet and outlet for the drop structure. In that case, the walls can be inspected for cavity formation and soil erosion, with appropriate maintenance remedies applied. If a pipeline is used between the inlet and the outlet, then the major concern is water leaking at one or two joints, usually because of soil settlement in the embankment. The visual evidence that significant leaking of the joints is occurring would be a cavity in the soil above the pipeline. If so, the overlying soils need to be excavated, the pipeline inspected and repaired as appropriate, the backfill properly compacted, the top of the embankment properly shaped to minimize the potential for erosion, and then sodded, with the sodding being given special attention during the following year.

Often, a hole will have been created by the concrete's deterioration at the joint opening at the end of the outlet transition structure. This is the result of hydraulic pressure behind the structure. The remedies are shown in Fig. 15.26.

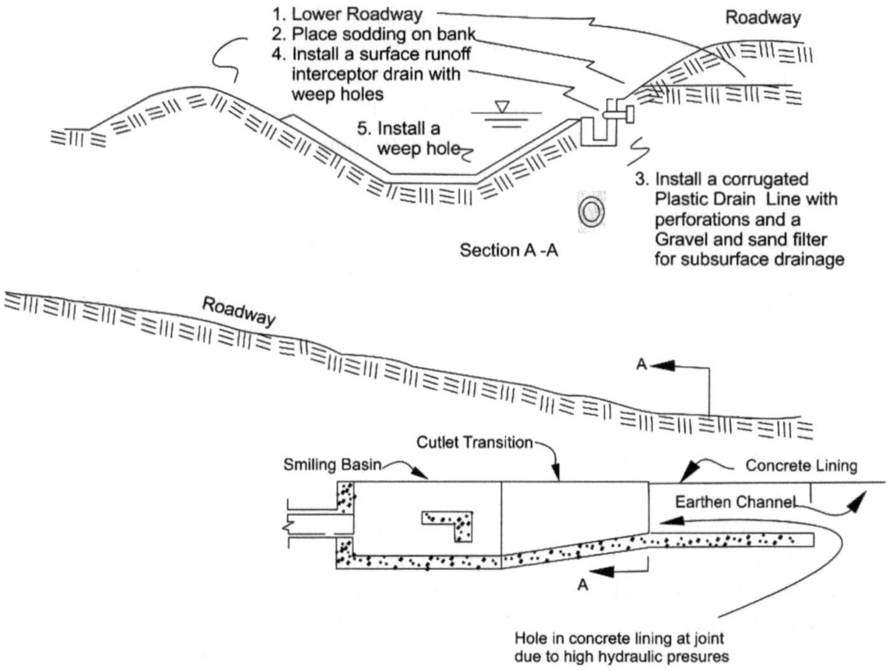

Fig. 15.26 Remedies to relieve hydraulic pressure at the outlet transition for a drop structure

15.8.4 Fall Structures

If scour continues to occur below a falling structure, then a concrete, rubble masonry, or riprap-lined stilling basin should be constructed to dissipate much of the excessive hydraulic energy.

15.8.5 Culverts and Inverted Siphons

Usually, there are only minor maintenance problems with culverts, with the usual maintenance activity being the removal of debris at the culvert inlet. This would also be the case for inverted siphons. The primary maintenance problem for inverted siphons is the accumulation of sediment, gravel, rocks, twigs, branches, or lumber pieces. Maintenance is expensive because after canal closure, the water in the inverted siphon has to be pumped out, and then a laborer enters the inverted siphon to remove the accumulated materials.

To determine whether maintenance is required for an inverted siphon (or a culvert), a useful technique is to check the coefficient of discharge for the structure periodically. This can be done with a single discharge measurement using a current meter, along with the difference between upstream and downstream water surface elevations. If the material is accumulating, either in a culvert or inverted siphon, where submerged outlet control exists, the discharge coefficient will decrease. When the structure's flow capacity has been reduced to the point where downstream operations are affected, the accumulated material should be removed.

15.8.6 Gate Structures

The usual visual indicators of gates and gate frames requiring maintenance are gate leakage, bent gate rods, gears needing lubrication, and poorly installed gate frames. Special attention should be given to correcting these deficiencies and maintaining these gate structures.

If a gate frame was improperly installed, then the frame should be removed and installed properly. In some cases, the gate frame will have to be replaced. If the gate rod is bent, then it should be removed, taken to a workshop, and straightened or replaced. If there is gate leakage, one should attempt, first of all, to correct the problem in the field. If this becomes too difficult, one should remove the gate and gate frame, haul them to an appropriate workshop, and correct the deficiencies.

Gates and gate frames should be cleaned and painted frequently, preferably every year. Commonly, metal gates will corrode, and the metal becomes weak inside without being visible until the gate buckles. This is usually the result of infrequent cleaning and painting.

If scour below the gate structure is substantial, serious consideration should be given to constructing a concrete, rubble masonry, or riprap-lined stilling basin to dissipate the excessive hydraulic energy that will result in lower velocities entering the downstream earthen channel section.

15.8.7 Outlet Structures

If a few outlets have been tampered with, the situation can be technically corrected. Those outlets that are too low should be reset at the appropriate elevation. For those outlets wherein the constriction has been made larger, the outlet opening must be reconstructed. In some cases, for an Adjustable Proportional Module (APM), a cast-iron APM block embedded in concrete can be used, which is extremely useful in avoiding tampering. After correcting any deficiencies in the outlet structures, they should be checked during every season to be sure there has been no further tampering.

If numerous outlets have been tampered with, say 25% or more, the situation will not be corrected by technical measures alone. This becomes a challenging social problem that must be dealt with by all parties concerned, including both Irrigation Department staff and WUAs. Serious consideration should be given to document the discrepancies between actual structural measurements, elevations, and discharge rates compared with authorized or sanctioned measurements, elevations, and discharge rates. Then, this information should be extensively communicated with WUAs in the irrigation system so that, if possible, they can resolve the problem of tampering among themselves, with perhaps the Irrigation Department staff playing an intermediary technical assistance role.

15.9 Maintenance of Minors and Water Courses by Water User Associations

15.9.1 Maintenance Procedures by Farmers

Forming the WUAs (Water User Associations) and handing over the conveyance system (minors) to them after rehabilitation and modernization is an important part of participatory irrigation management (PIM). Maintenance works on the minors and below the outlets are best carried out by Water Users' Associations. Department of Irrigation, Government of Uttar Pradesh (UPWSRP 2008) has brought out a manual to be used by WUA. The manual provides for their duties, responsibilities, functions, authorities, etc.

The system below the outlet consists of a watercourse network, turnout, drops, etc. Its construction and maintenance are the WUA's collective responsibility. To run the authorized full supply discharge, it becomes essential to remove extra silt deposited (generally in head reaches) in the bed of the watercourse and remove berm formation on the sides. Similarly, in the filling reaches, the bed and banks get eroded due to high velocity, and the embankment is weakened. These embankments need strengthening to avoid any breaches or seepage through the banks. Apart from this, weeds and bushes also grow in the watercourse and hamper water flow.

Sometimes, structures are damaged by the water flow. The displacement of stones and the disintegration of joints or cracks need timely repair. Careful watch must be maintained for the undermining of structures. Wooden planks fixed in the turnouts to regulate the direction of the flow of water generally leak, and water is wasted. These leakages can be controlled by placing the earth behind the wood planks.

Farmers generally elect one leader to manage the system's maintenance and repairs to be done collectively by the farmers. This leader is someone who has the respect and confidence of other farmers. He distributes the maintenance quantum of work to each farmer in proportion to their landholdings within the outlet command and directs and supervises the maintenance activities.

Some farmers who cannot participate in maintenance activities due to unavoidable circumstances provide laborers on their behalf. In a few cases, if some farmers neither turn up themselves nor provide laborers, the team leader has their share of the work done by others, who are paid in money, and the money is recovered from the absentee farmers. This activity is generally done once in a crop period before the water is first released in the watercourse. Thereafter, the shareholders maintain the watercourse between the nakka points related to their holdings.

15.9.2 Emphasis on Technical Assistance

The tertiary portion of the irrigation system is quite complex because it contains three irrigation subsystems: (1) water delivery; (2) the farm; and (3) water removal.

The tertiary subsystem is generally the most neglected portion of irrigation projects in South Asia. Increased emphasis on improving the tertiary channel network's performance will enhance the Irrigation Department's communication and credibility with farmers and strengthen WUA's organizational effectiveness.

The initial emphasis will be assisting each WUA in providing more equitable water distribution to individual farmers. This will be accomplished by evaluating channel losses throughout the tertiary network, with participation by leadership in the Water Users Group, and communicating the results in open meetings with all Water Users Group members (farmers). First of all, these results will disclose whether improved maintenance needs to be performed by the WUA. Also, this data will indicate whether any improvements in the tertiary system are needed to provide an adequate water supply or equitably distribute water among farmers. This will be a highly participatory process of farmer involvement, with the department providing technical assistance.

15.9.3 Important Do's and Don'ts for WUAs

Do's
1. *The silt clearance should be finished before the sowing of Rabi crop*
2. *The final cross-sections should be measured jointly by the contractor and PIM team leader after the completion of de-silting works*
3. *The excavated coarse silt should be disposed off in filling old borrow pits and on outer slopes of banks of minor*
4. *The excavated medium silt should be utilized in closing traffic leeks on banks,* and widening *banks as a supplement to berm earth*
5. *The excavated fine silt should be used for all repairs of the earthwork on minor*
6. *The disposal plan should be prepared on Shazara sheet of the area giving Khasra numbers of the fields identified for disposal before start of de-silting*
7. *The surplus excavated silt should be disposed off as per the disposal plan*
8. *The consent of the farmers should be taken before the disposal of silt in their fields*
9. *The earth for repairs should be obtained from specified sources such as internal clearances of minor, beds of nearby drains/guls and borrow pits in bed of minor as well as outside pits*
10. *The top of bank should be dressed in an outward cross slope of 1.25%*
11. *The banks should be 30 cm. above the country level*
12. *Where more than 50% section of bank is damaged, reconstructing whole section afresh should be considered*
13. *All canal gauges should be cleaned periodically*
14. *The displaced boundary pillars should be re-fixed as per land plan of the minor*
15. *A register of masonry/concrete works on minor should be maintained and any improvements in the structure should be recorded regularly*

(continued)

(continued)

16. *A line marking the FSL of minor should be painted on the upstream face of every pucca work on it*
17. *All exposed surfaces of steel gates of head regulator, orifice module and duck bill weir should be kept painted with specified anti-corrosion enamel paint*
18. *The regulation schedule for the minor should be available*
19. *All outlets should be inspected thoroughly for any choking or tampering during closure period of minor*
20. *The outlets should be checked for their position, size and correctness of fixing as per roster*
21. *A register of outlets should be maintained giving details of their size, location, discharge and water head corresponding to FSL of minor*
22. *Water head at outlet should be measured monthly*
23. *All fallen or dead trees should be immediately reported to Engineer-in-Charge/President WUA*
24. *The pruning of trees should be done by sawing in February*
25. *The roots of seedlings should be dug out with a good ball of earth adhering to it when transplanting*
26. *All newly planted trees and the trees less than* 2 m *in height should be properly protected by tree guards with 1* m. *dia and 1* m *height*
27. *Suitable species of trees should be planted at boundary line of the land of the minor*
28. *The branches of a newly planted tree, taking off near its roots should be lopped off*
29. *The aquatic weeds should be completely and regularly removed manually from the internal section of minor*
30. *The routine damages to gates such as rusting of gate leaf, cracks in rubber seal, bending of lifting rod, damage to head post* etc., *should be rectified promptly*
31. *The routine damages to pucca structures such as broken plaster, masonry & concrete, decay of mortar in joints of masonry/pointing and concrete cover to steel, broken parapet and wings should be repaired regularly*
32. *The discharge in the minor should be reduced from the head of minor in case of breaches in large minors*
33. *Joint walkthrough of minor with office bearers of WUA and PIM team should be done before and during rainy season*

Some Don't s

Do not allow the following

1. *Don't allow any over- excavation than designed section of minor during de-silting*
2. *Don't allow that coarse or medium silt is spread on canal service road without a cover of 20* cm *thick layer of good soil*
3. *Don't allow any blockage of existing outlets on minor due to disposal of excavated silt*
4. *Don't allow digging of borrow pits within 5* m *from the outer toe of banks/ramps of bridges*
5. *Don't allow digging of borrow pits within old borrow pits*
6. *Don't allow borrow pits more than 15* cm. *deep in cultivated fields*
7. *Don't allow that daula is cut for repairing the banks*
8. *Don't allow trees to grow on canal berms & daula*
9. *Don't allow that a tube-well is bored within 25* m *of banks of minor*
10. *Don't allow painting of coal tar on steel work*
11. *Don't allow sudden increase or decrease of discharge in minor*
12. *Don't allow any temporary outlet to function beyond sanctioned period*
13. *Don't allow double trees in any tree guard*
14. *Don't allow any floating debris to accumulate in U/s of pucca structures*
15. *Don't allow any obstruction in a drain at U/s of the drainage crossing on the minor*

15.10 Inspection of Concrete/Masonry Construction

The construction considered herein is concrete or masonry, including the combination of concrete and masonry work (UPWSRP 2008).

15.10.1 Surface Deterioration

Poor quality materials and short composition of concrete and mortar will be evident by scaling, spalling, pop-outs, crumbling, and other deterioration forms. Likewise, poor quality construction, such as using too much water in mixes or finishing and/or permitting exposure to freeze conditions before adequate curing and hardening of concrete or mortar, will lead to early and accelerated concrete deterioration.

One should inspect concrete and masonry surfaces annually for deterioration, including scaling, spalling, pop-outs, crumbling, and cracking. This type of deterioration may be relatively slow, and recordkeeping in the form of photographs, sketches, or notes indicating the location, size, and depth of deterioration will be helpful in evaluating the severity and rate of progression. When concrete deterioration reaches embedded reinforcing steel, immediate repair is usually required.

15.10.2 Cracks

Cracking is usual and normal in concrete works. Most cracks are superficial and not very deep or wide, and some cracks in concrete where damp or high humidity atmospheric conditions prevail, as in conduits, may self-heal or seal by chemical action as with calcium carbonate formation. In mild climates, this type of cracking is usually not a problem or detrimental to the structure's integrity. Cracks that are deep or through a structure may be of significant concern as they can be signs of settlement, structural movement, and distress and may permit seepage or water leakage. Sometimes when seepage or leakage occurs, staining of concrete may be noted. This may indicate reinforcing steel corrosion or chemical reaction of water with foundation materials or concrete and may or may not be severe. Seepage and leakage through cracks most commonly occur at portions of structures below the water level and in the case of barrels of a conduit.

15.10.3 Joints

Expansion, contraction, and construction joints may be of concern when they become "open" and permit seepage or water leakage. Opening of joints can be a signal of structural movement or disbanding of concrete at construction joints. In the case of contraction joints, seepage may indicate water stops failure. Some opening or closing of joints occurs typically with temperature changes.

15.10.4 Seepage

Seepage measurements should be performed in the semiannual inspection, and their quantity should be recorded in a logbook. The clarity of seepage (that is, whether clear or carrying sediment) and any staining or discoloration of concrete should also be noted. Seepage may occur not only at cracks and joints but also at the abutments and downstream from the construction.

Appendix 15.A: Checklist for Condition of Minor

Name of canal system: Location with Chainage (Km); Name of Minor: ..Date of inspection…

S. No.	Component	Reach in Km	Location			Condition				Maintenance Needs		Remarks	
			Left bank	Bed	Right bank	Needs Repair		Needs Rehabilitation		Description	Priority		
						Yes	No	Yes	No		Urgent	Routine	
1	2	3	4	5	6	7	8	9	10	11	12	13	14
1	Jungle/Weeds												
2	Silting												
3	Bank section												
4	Bank-top cuts												
5	Service Road												
6	Ramps at VRBs/ outlets												
7	Longitudinal/ Cross slopes at Bank top												
8	Raptas at outer slopes of banks												
9	Filter at outer toe of banks												
10	Seepage drains												
11	Lining of canal												
12	Bed bars/profiles												
13	Reference pillar												
14	KM./ H-meter stones												

(continued)

(continued)

S. No.	Component	Reach in Km	Location			Condition				Maintenance Needs		Remarks	
			Left bank	Bed	Right bank	Needs Repair		Needs Rehabilitation		Description	Priority		
						Yes	No	Yes	No		Urgent	Routine	
15	Water level gauges												
16	Cross bandha												
17	Floating debris/ Fallen tree												
18	Others												

Appendix 15.B Checklist for Canal Structures

Name of structure: .... Name of canal system: ......
Type of structure: concrete/masonry/pitching Location with chainage (Km)

S. No.	Component	Bank		Condition				Maintenance needs			Remark
				Needs repair		Needs rehabilitation		Description	Priority		
		Left	Right	Yes	No	Yes	No		Urgent	Routine	
1	2	3	4	5	6	7	8	9	10	11	12
1	U/S wing wall of VRB/fall/weir										
2	D/S wing wall of VRB/fall/weir										
3	Crest of fall/weir										
4	Abutment of VRB/ fall/weir										
5	Bridge/Regulator Abutment										
6	Bridge/Regulator deck slab & joints										
7	Bridge/Regulator Parapets										
8	Gates of Regulator/ weir										
9	Stop-log grooves										
10	Stilling basin of fall/ weir										
11	U/S Pitching of VRB/fall/ weir										
12	D/S Pitching of VRB/fall/weir										

(continued)

(continued)

S. No.	Component	Bank		Condition				Maintenance needs			Remark
				Needs repair		Needs rehabilitation		Description	Priority		
		Left	Right	Yes	No	Yes	No		Urgent	Routine	
13	Pipe of outlet										
14	Earth cover over foundation of abutment/pier										
15	Launching apron/ filter in bed at VRB/ Fall/ wier										
16	Functioning of drainage holes in deck/crest of fall										
17	Others										

Questions

1. Explain, in brief, the following terms: aquatic weed, leakage holes, dowel, berm, soding, soil–cement mixture, filter fabric, hydraulic pressure, and discharge.
2. Refer to the checklists for assessment of condition and repair and maintenance needs of canal and related structures. Which components are relatively more importatnt as far as canal operation is concerned.
3. Visit a minor canal and identify repair, maintenance needs referring to the check lists
4. What are the sources of sediment in the canal?
5. Discuss the reconstruction of earthen embankment when (a) there is a leakage hole, (b) cavity behind lining, and (c) outer slope is heavily eroded.
6. Discuss various methods for the collection and conveyance of surface runoff from the service road on the embankment.
7. Discuss procedures for filling cavities behind the lining.
8. Discuss maintenance remedies to relieve hydraulic pressure (a) behind the lining and (b) at the outlet transition for a drop structure.
9. Explain the role of farmers in the maintenance of the tertiary system.
10. Explain the use of data on groundwater levels in deciding the maintenance needs of the drain network.
11. Identify the emergency measures for failure of (i) outside slope of canal embankment, (ii) inside slope of canal embankment

References

UPWSRP (2008) Manual for construction and maintenance of minors, Uttar Pradesh water sector restructuring project (UPWSRP), Department of Irrigation, Government of Uttar Pradesh, Lucknow

WAPCOS (1989) Handbook for improving irrigation system maintenance practices, Technical Report No. 19-A Water and Power Consultancy Services (India) Ltd. and Louis Berger Int. Inc

JICA (2017) Operation and maintenance manual, prepared by Japan international cooperation agency for bangladesh water development board. Internet source: https://openjicareport.jica.go.jp/pdf/12292587_02.pdf

BIS Codes

IS: 4839 (Pt. I), Maintenance of Canals

IS: 4839 (Pt. III), Canal Structures, Drains, Outlets, Jungle Clearance, Plantation and Regulation

IS: 4839 (Pt. I), Maintenance of Canals, Appendix-A.

Reading Material

INCID (1994) Guide for preparation of plans of operation and maintenance of irrigation systems in India. Indian National Committee on Irrigation and Drainage, New Delhi

Chaube 2. U.C (2000) Maintenance of irrigation works. Course Package, Centre for Continuing Education IIT Roorkee

Chapter 16
Field Drainage

Abstract This chapter provides a detailed description of field drainage implementation and maintenance of drainage. It also provides removal (drainage) of excess irrigation water from agricultural lands; hence it is part of on-farm development work. Stormwater drainage works for the canal (cross drainage, drain along the canal in cut and fill sections) are part of canal structures and have been covered in Chap. 15. The main purpose of field drainage is to control waterlogging and salinity. The criteria for determining the criticality of waterlogging and salinity are discussed. To illustrate manmade causes of waterlogging, a case study on waterlogging (sample villages of Gidderbaha Tehsil in Muktasar district of Punjab) is presented. The planning and implementation process of physical components of field drainage is explained. This helps in understanding the maintenance need of surface drains which is as important as the maintenance need of irrigation canals. Groundwater monitoring becomes extremely important where groundwater levels are shallow, not only for indicating the impending waterlogging and salinity problems but also to indicate the effectiveness of the open drain network for subsurface drainage. The procedure for maintenance of drains is similar to that for irrigation canals, and the material presented in Chap. 15 may be referred to in this regard. The main focus is on maintaining the hydraulic performance of drains. In addition to the removal of weed and silt from the drains, lining of drains may be required. Types of lining are discussed. Farmer's participation in the maintenance of drains is highlighted through a case study of a drainage cooperative society in Kheda district of Gujarat. Tube wells have been successfully used for vertical drainage of waterlogged areas. Well, hydraulics is explained using a schematic of tubewell drainage. Case studies on deep and shallow tube wells, dug wells for vertical drainage, and water quality management are provided. The concept of bio drainage is given. The advantages and limitations of bio drainage are explained. Despite the advantages outweighing the disadvantages, biodrainage is not yet a popular drainage method in South Asian countries due to some limitations.

U. C. Chaube et al., *Canal Irrigation Systems in India*, Water Science and Technology Library 126, https://doi.org/10.1007/978-3-031-42812-8_16

16.1 Introduction

The definition of land drainage, as given in the constitution of the International Commission on Irrigation and Drainage/ICID (1979), reads:

> Land drainage is the removal of excess surface and subsurface water from the land to enhance crop growth, including the removal of soluble salts from the soil.

Land drainage, as a tool to manage groundwater levels, plays an important role in maintaining and improving crop yield:

- It prevents a decrease in the productivity of arable land due to rising water tables and the accumulation of salts in the root zone.
- A large portion of land currently not being cultivated has waterlogging and salinity problems. Drainage is the only way to reclaim such land.

In the ICID definition of drainage, 'the removal of excess water' indicates that (land) drainage is part of On-Farm Development works as against natural drainage. The On-Farm Development Works include:

- Channels for the conveyance of water i.e., field channels;
- Structures on field channels;
- Drains;
- Structures on drains;
- Graded bunds and grassed outlets; and
- Land forming i.e., land levelling and shaping.

This chapter deals with the drainage of water from the agricultural land. Drainage works for the canal (cross drainage, drain along the canal in cut and fill sections) are part of canal structures and are covered in Chap. 15.

16.2 Waterlogging and Salinisation Criteria

Waterlogging refers to the situation when the moisture content increases beyond the maximum water holding capacity of the soil (field capacity) in the root zone, thereby driving out air whose place is taken by water. There may be an extreme condition when all the pore spaces are filled fully, causing the root zone to be completely devoid of air at the full saturation stage of soil. Waterlogging and salinization both are related phenomena. Waterlogging in the irrigated land is mainly due to the rise of groundwater levels by faulty irrigation management.

The criteria to assess the criticality of waterlogging (Table 16.1) and the severity of salt-affected soil (Table 16.2) are given below (GOI 2006).

Table 16.1 Criteria for the criticality of waterlogging (GOI 2006)

Category	Depth of water table(cm)
Most critical	< 1
Critical	1–2
Less critical	2–3
Non-critical	>3

Table 16.2 Criteria for the severity of salt-affected soil (GOI 2006)

Level 1	Level II	Non-vertisales		
		PH	EC(ds/m)	ESP
Saline	Medium	< 8.5	4–8	< 15
	High	< 8.5	8–10	< 15
	Very high	< 8.5	> 10	< 15
Sodic/Alkaline	Medium	8.5–9.0	< 4	15 to < 25
	High	9.0–10.0	< 4	25 to < 40
	Very high	> 10	< 4	> 40
Saline Sodic	Medium	8.5–9.0	4–8	15 < 25
	High	9.0–10.0	8–10	25 < 40
	Very high	> 10.0	> 10	> 40

EC Electrical conductivity; *ESP* Exchangeable sodium percentage i.e., the ratio of exchangeable sodium to cation exchange capacity expressed as a percentage

16.3 A Case Study: Water Logging in Gidderbaha (Punjab)

Contrary to the problem of depletion of the water table in other parts of Punjab, Gidderbaha tehsil in Muktasar district in the Punjab state of India has been facing an acute water logging problem. In the nineteen fifties, the subsoil water level in this area was about 33 m below the ground level. After the construction of twin canals i.e., Rajasthan Feeder and Sirhind Feeder, and Abohar Branch and BikanerCanal, the subsoil water level started rising from 0.2 m to 1 m annually. The problem has been further aggravated, as the flat topography of the area impedes the surface drainage system. This case study is based on field investigations by Chaube (2008).

16.3.1 Waterlogging in Gidderbaha Tehsil

Due to the rise in the water table, certain parts on the west side of the Rajasthan and Sirhind Feeder canals have become critically waterlogged. In low-lying areas, the land stands submerged and is unfit for cultivation. As a result, not only the

Table 16.3 Pre- and post-monsoon water-logged area in Gidderbaha

Season	Tehsil Area (ha)	Category			
		Water logged area	Sensitive area	Total	Village affected
Pre-monsoon	66,467	228.18(0.34)	158.97(0.24)	387.15(0.58)	11
Post-monsoon	66,467	293.22(0.44)	157.04(0.24)	450.26(0.68)	12

Table 16.4 Pre-monsoon salt affected area in Gidderbaha Tehsil

Season	Tehsil area (ha)	Salt-affected soil ha (%)	Saline soils with patchy soils ha (%)	Total ha (%)	Village affected
Pre-monsoon	66,467	65.26(0.1)	50.52(0.07)	115.78(0.17)	16

agricultural production of this area has declined over the last few years, soil quality has deteriorated quite substantially.

A study using satellite data of pre and post-monsoon season shows that pre-monsoon waterlogged areas and areas sensitive to waterlogging were 387.15 ha in Gidderbaha Tehsil in comparison to 362.72 ha in the Muktsar Tehsil. However, in post-monsoon, the waterlogged and waterlogging sensitive area was 450.26 ha in Gidderbaha compared to 695 ha in Muktsar Tehsil. In Gidderbaha Tehsil, water logging is predominant along the western side of the twin canals (Sirhind and Rajasthan feeder) where the Gidderbaha thermal power plant is situated (Tables 16.3 and 16.4).

16.3.2 Water Logging Condition in Sample Villages

A field survey was carried out in some villages of Gidderbaha Tehsil in the Muktasar District of Punjab. Figures 16.1, 16.2 and 16.3 show waterlogging and salinity conditions in the study area. Table 16.5 shows a rise in groundwater level over 28 years at Fakkarsar Theri, Gidderbaha, and Malaut, which are located in the study area. Fakkarsar Theri and Malaut are on the west side of Sirhind Feeder Canal, with Fakkarsar Theri being the nearest ($\approx$1 km) to the project area. Gidderbaha is on the east side of the Sirhand and Rajasthan Feeder Canal.

As per the criteria for the criticality of waterlogging (GOI 2006), the study area was critically waterlogged in 1990, but the water table in recent years has remained between 2 and 3 m bgl i.e., the water logging situation has been less critical.

Fig. 16.1 Waterlogged area in the South of Village Ghagga

Fig. 16.2 Salt affected area adjacent to crop area

Fig. 16.3 Poor crop growth in salt-affected areas

Table 16.5 Change in groundwater level (m bgl) from 1980 to 2007

Village	1980	1985	1990	1995	2000	2007
Fakkarsar Theri	10.38	4.90	1.30	2.38	2.55	2.75(Jan 05)
Gidderbaha	NA	12.62	9.61	8.15	5.1	3.73
Malaut	11.7	9.85	5.2	0.23	0.64	2.11(Jan 05)

Source (Chaube 2008)

16.3.3 Causes of Water Logging in Sample Villages

1. The seepage from the damaged portion of the lining of Sirhind Feeder Canal and due to the construction of the canal in the filling.
2. Burrow pits were excavated during the construction of canal in filling. These burrow pits do not have adequate outlets.
3. Sirhind Feeder and Rajasthan Feeder Canals cut across the natural gradient, leading to the impoundment of monsoon runoff on the upstream side of structures. This aggravates the water logging problem.
4. Temporary flooding due to inadequate capacity and low maintenance of drains. The drains bring stormwater and seepage water from upstream beyond the project area.
5. Excessive application of irrigation water and poor on-farm water management practice. Farmers are adopting a fixed cropping pattern of rice wheat or cotton-wheat.
6. Non-extraction of groundwater for domestic and irrigation purposes due to poor water quality.

The following remedial measures were suggested to Irrigation Engineers:

i. The lining of damaged portions in the canal
ii. Filing of borrow pits dug up along the canal
iii. Tree plantation along canal banks, in borrow pits, in waterlogged areas
iv. Construction of more surface drains to avoid drainage congestion
v. Clearance of weeds and sediments in existing drains
vi. Linking of main drains with field drains

16.4 Components of Field Drainage

An agricultural drainage system consists of the following:

(a) A drainage outlet,
(b) Outfall drain,
(c) Some collector drains, and
(d) Field drains.

The outfall drain runs through the lowest parts of the agricultural area. It follows the main depression, and branch drains follow lesser depressions (Fig. 16.4). If, however, the land surface is essentially an inclined plane, the outfall drain is aligned to fall towards the lowest contour, as shown in Fig. 16.4.

In the case of the inclined plane, branch drains run close to the predominant slope, and collector drains run at a small angle to the contour. In these circumstances, a collector is adjacent to a watercourse on the upslope side (Fig. 16.5).

This has the advantage of intercepting seepage in that direction but has the disadvantage of promoting it by the steepness of the hydraulic gradient (Fig. 16.6). The system of minor channels shown in Fig. 16.5 can be rotated relative to the predominant slope to distribute the fall of ground surface most economically between field,

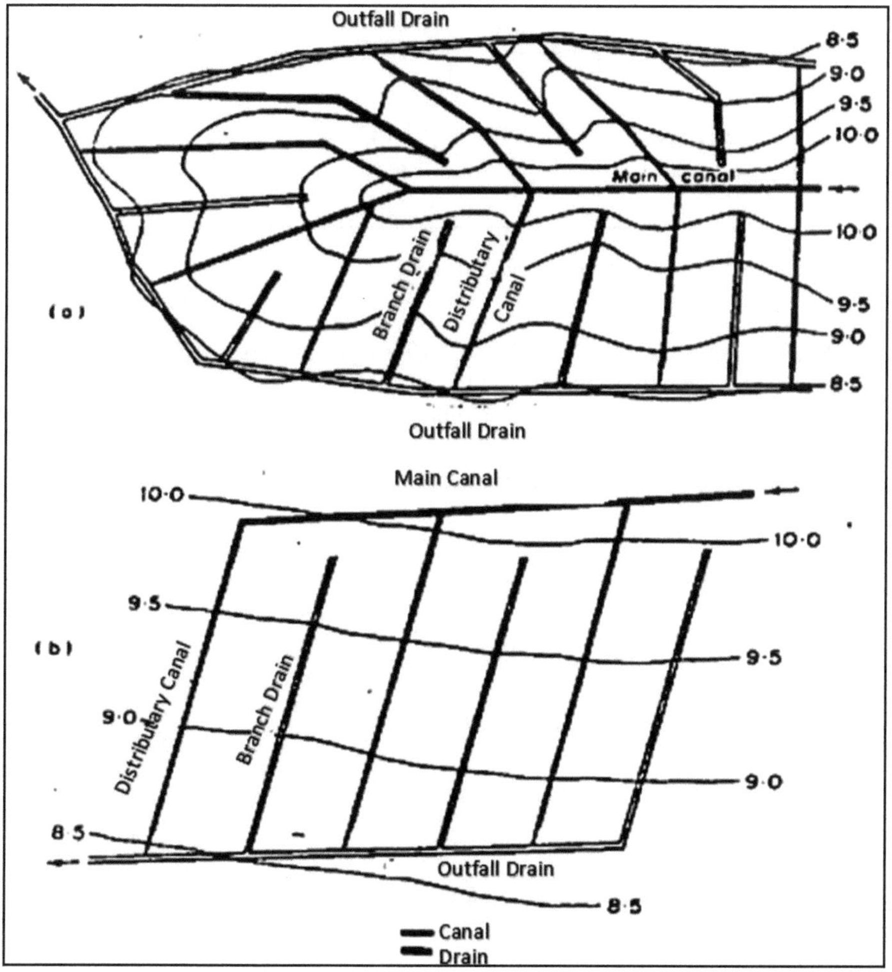

Fig. 16.4 Details of layout of minor canals and drains (WALMI 1987)

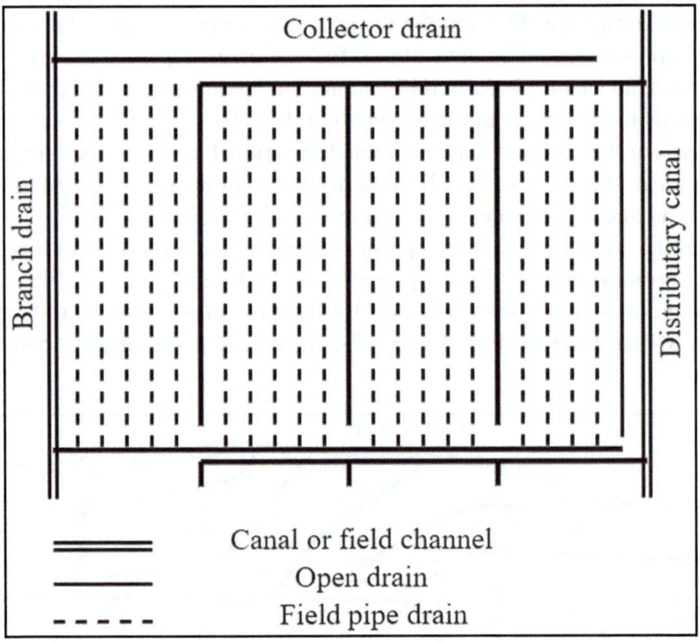

Fig. 16.5 Details of layout of minor canals and drains

collector, and secondary drains. Figure 16.7 shows the use of pipe and ditch in subsurface drainage of agricultural field.

The length and spacing of the field or lateral drains will be as uniform as applicable. Both collector and field drains can be open drains or pipe drains. They are determined by a wide variety of factors, such as topography, soil type, farm size, and the method of field drainage.

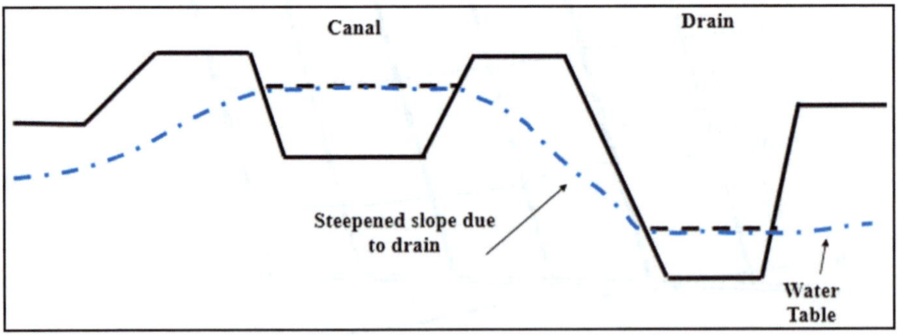

Fig. 16.6 Adjacent canal and drain

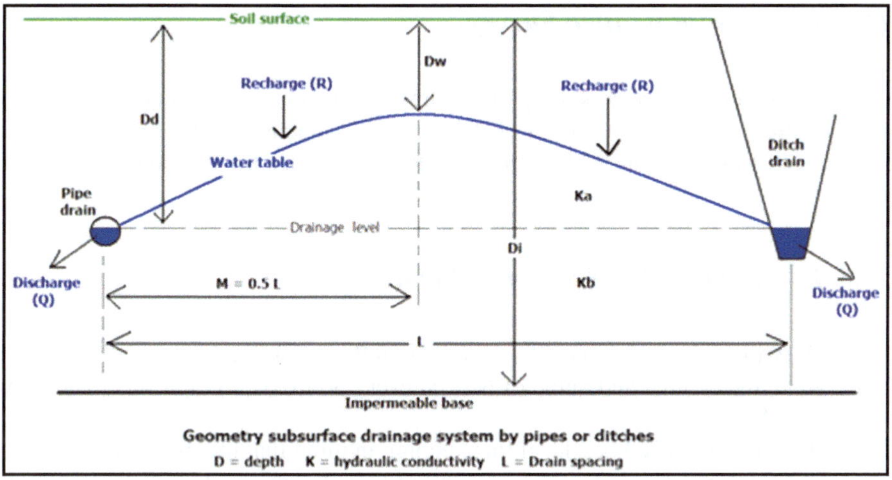

Fig. 16.7 Geometry of subsurface drainage by pipes and ditches

16.5 Implementation of Field Drainage

WALMI (1987) has brought out field guidelines for the implementation of field drainage.

16.5.1 Unit for Design

The smallest unit for design should be a minor or a distributary serving a group of outlets. This will ensure that field conditions of adjoining outlets will be reflected in the design. If the command area is very large, it should be divided into convenient subcommands, which will be taken as units for design.

16.5.2 Collection of Data

Data pertaining to the following needs to be collected.

(a) Distribution network plan and L section, cross-sections of distributary/minor, Shazra map, outlets location, sill level, and their CCA (Irrigation Department).
(b) The drainage plan, L section of seepage, carrier, and submain and main drains (Irrigation Department).
(c) Cultivator list, revenue map, and road map of the area (Revenue Department).
(d) Locations and values of permanent benchmarks.
(e) Cropping pattern with dates of sowing and harvesting.
(f) Meteorological data of the station nearest to the command.

16.5.3 Reconnaissance Survey

Carry out a reconnaissance survey of the area to find any variation between survey and planning, such as excavation of drains, canal capacity works, construction of structures, and power lines. Mark the variations, if any, on the grid map.

16.5.4 Prepare of L Section of Field Drains

i. Mark the points at every 30 m. interval on the drain line.
ii. Write the change and value of NSL at every 30 m. on a graph sheet and draw the NSL line.
iii. Draw the proposed bed level line keeping in mind that the difference between NSL and bed level should not be more than 1.5 m and less than 0.6 m. On average, the difference should be 1.1–1.2 m. so that a sufficient quantity of soil is available for the construction of service roads along both sides of the field drain.
iv. No fall structure should be proposed, as far as possible.
v. The bed line should be drawn in such a way that outfall is available at the outfall point. The maximum can be flushed with the bed of carrier drain at the cutfall point only if there is a main drain at the outfall.
vi. If outfall is not available, then a note of the same should be given for the deepening of the carrier/main/sub-main drain.
vii. The outfall level should be shown at the outfall point.
viii. The bed slope of the drain should normally be the same as the ground slope. However, it should not be flatter than 1:1500 and steeper than 1:500 as far as possible.
ix. Get the L Section approved by higher authorities.
x. If there is any change, then incorporate the same.
xi. Write the bed level values at every 30 m, calculate the cutting, and write the values at every 30 m.
xii. Write the discharge values, velocity, bed width, side slope, bed slope, etc. with the help of the Design Manual.
xiii. Draw a typical cross-section of the drain at a particular change.

16.5.5 Preparation of Layout Plan

i. Reduce the approved final planning at the scale 1:6000 by pentagraph.
ii. Prepare the layout plan showing contours at 30 cm. intervals, watercourses, field drains with roads, canals, main/submain/seepage/carrier drains, and all existing structures on canals and drains which have already been excavated.

iii. Complete the writing work, including the number of watercourses, its offtaking position, CCA, FSL designed and available in the canal, FSL required in the watercourse, number of drains, name of the canal, and RD values at all permanent structures.
iv. All structures, as per the mapping symbols approved, should be marked on the layout.
v. Show the north direction in the top right-hand corner.
vi. Show the dummy at the right corner.
vii. Prepare a chart of the drainage system showing different drains, their classification, length, outfall point, design discharge, catchment area, outfall level, and proposed culverts/inlet at the cultivator/Government account on the layout plan.
viii. Mark Class II and Class III land, if any, on the layout map.
ix. Calculate the area under class-II and Class-III land.

16.5.6 Calculation of the Right of Way

The area under watercourses, field drains, carrier drains, and approach roads is known as 'right of way,' and at the time of reallotment, it is deducted from each holding on a prorated basis. The right-of-way calculation is based on the strip width specified for various items of work, such as watercourses (3 m), field drains with both sides approach road (14 m), carrier drains with both sides approach road (14 m), field drains with one side approach road (9.8 m), field approach road along seepage drain (5.4 m), the distance between canal boundary and watercourses running along it (2 m), and village cart tracks (6 m).

Calculate the length and cost per ha for the watercourse, and field drain. Calculate the cost of structures. Calculate the cost of realignment and land leveling. Prepare cost abstract.

16.5.7 Execution of Field Drainage System

i. Arrange jungle clearance and Typha cutting, if needed, on the alignment of drains (to be done departmentally).
ii. Carry out dog-belling (to be done departmentally).
iii. Arrange/supervise the drains excavation to the given specifications, particularly for side bulging, berm clearance, and road formation.
iv. Check bed levels and ensure their conformation to the designed levels.
v. Construct drain crossings, if any.
vi. Ensure the outfalls to field drains and get the drain inlets constructed from Irrigation Wing on the carrier/sub-main/main drain.

vii. Arrange the fixing of field inlets as per specifications giving due care to the joining of pipes, grade, and proper earth cushion over these pipes.
viii. Arrange sodding of doob grass on 10 cm mesh as per specifications and periodical watering till the well establishment of doob grass is ensured.

16.6 Maintenance of Surface Drains

The maintenance of drains is as important as the maintenance of irrigation canals. Unfortunately, drains usually receive insufficient attention. Usually, there are no service roads along the drainage network, so 'out of sight, out of mind. Experience shows that drains receive the last priority for maintenance when funds for maintaining the irrigation system (unless heavy monsoon rains occur!) are not adequate, which usually is the case.

16.6.1 Monitoring of Groundwater Levels

Monitoring groundwater depth below the ground surface in any irrigation system is extremely important. If groundwater levels are deep, then perhaps only a single observation well in the lower reaches of the irrigation system is adequate. But the history of irrigation projects is that groundwater levels usually continue to rise with time, unless there is substantial groundwater pumping. Where groundwater levels are shallow, say less than 5 m; groundwater monitoring becomes extremely important for indicating the impending waterlogging and salinity problems and the effectiveness of the open drain network for subsurface drainage. Also, the fluctuations in the groundwater levels in response to cleaning of the drains can be observed, which is another indicator of the effectiveness of the network for subsurface drainage and the relative importance of properly maintaining open drains.

16.6.2 Erosion and Sedimentation

A major emphasis should be placed on minimizing the flow of sediments into the drainage network. This can begin with the practices used by farmers in releasing ponded water from their fields. Then, the flow of water in the tertiary systems should be advised, and farmers should be requested to take appropriate measures.

A 'walk-thru' along the drainage network will undoubtedly disclose many sites where water is allowed to run into the open drains in a manner that causes erosion. As drains are located in the lowlands, their gradient is usually very flat, so sediments are readily deposited in the bed of drains, reducing their draining capacity and eventually clogging them.

16.6.3 Aquatic and Vegetative Growth

The greatest difficulty in the maintenance of drains is the clearing of vegetative growth. Often, there is massive growth that becomes a tremendous job to clear. The problem is aggravated by stagnant ponds of water that support both aquatic and vegetative growth.

The best means for removing vegetative growth is uprooting the plants by hand labour. This can usually be done easily in the bed of a drain but will become difficult on the drain bank, where the vegetative growth will have to be cut by hand labour. The vegetative material should be stockpiled and removed or burned.

16.6.4 Hydraulic Performance

Rarely is the hydraulic performance of a drain network evaluated. Everyone assumes that drains are performing as designed. Actual field measurements will likely prove to be shocking. Almost assuredly, a field evaluation would result in a much greater emphasis on more frequent maintenance of drains.

Although many drainage networks have very few structures, invariably, there will usually be some culverts or bridge crossings. Consideration should be given to either calibrating these structures for discharge measurement, or constructing a few control structures that can be calibrated, or modifying existing structures to better serve as flow measurement structures.

Monitoring of discharge rates at just a few locations in the drainage network combined with monitoring groundwater levels at some locations and rainfall will provide valuable insights into the hydraulic performance of drains. This, in turn, will provide guidance regarding scheduling maintenance activities.

16.7 Lining of Drains to Maintain Hydraulic Efficiency

The main objective of the lining of field drains is to maintain hydraulic efficiency through the retaining section, the gradient of the drain, and control weed growth. The lining should particularly be done at locations such as (i) 5 m upstream and downstream of all structures, (ii) curves up to 5 m on either side in addition to the curve, (iii) banking more than 0.5 m, and (iv) other lengths where erosion is observed. Examples of different types of lining are given in Table 16.6. Requirements of good lining are impermeability, structural stability, durability, and economy (WALMI 1987).

Table 16.6 Example of different types of linings

Sl. No.	Type of lining and thickness	Work reported in Maharashtra		Seepage losses cumecs/ Mm^2	Structural strength	Coeff of rugosity for design (N)	Sid slope H:V	Remarks
		Project	Km					
1	Soil cement slabs 4 cm	Jayakwadi	3.6	2.4	Low	0.025	As for soil	Not recommended
2	PVC & LDPE (both of 200 gauge)	Bhima	2.5	1.2–3.1	–	As for soil (0.03)	As for soil	Not recommended for adoption without further experimentation
3	Murum/ CNS	–	–	–	Stabilises the slope	0.03	1:1	Recommended for black cotton soil
4	UCR Masonary	Mula	–	–	Medium	0.03	Vertical	Prohibitively costly
5	Shahabad stone	Bhima	1.8	1.2–3.5	Medium	0.025	1:1	Risk of theft
6	Brick	Bhima	2.9	1.8–2.8	Medium	0.030	1:1	Trapezoidal section susceptible to damage by cattle
7	Cast in situ concrete	Bhima	0.4	–	–	–	–	Not recommended
8	PCC slabs	Jayakwadi	0.9	1.9	Medium	0.025	1:1	Recommended
		Bhima	17.0	2.6	Medium	0.025	1:1	
9	Premoulded trapezoidal	Jayakwadi	6.0	2.3	High	0.025	¼:1	Recommended
		Bhima	37.0	2.4				
10	Premoulded half round	Bhima	25.0	0.5	High	0.022	–	Recommended

16.8 Example: Petlad Drainage Cooperative Society (Gujarat)

In the Mahi project command, 35,765 ha of land was affected by waterlogging and salinity. The Petlad drainage cooperative society covers 103 farmer members in a 50 ha service area of 4 outlets under Petlad branch canal of Mahi Right Bank Canal (Gujarat). The society is registered under the Gujarat Cooperative Societies Act. It is located in Dhabhou village of Kheda District, Gujarat. Three more similar drainage cooperatives were subsequently farmed. The main purpose of the society was to reclaim badly affected saline and waterlogged land by maintaining a system of underground pipes leading to a sump which is pumped out, and to make the land productive again. The society also managed irrigation with regard to the control of groundwater table.

The farmers also told that they were ready for their cooperative to take water on a volumetric basis and manage irrigation distribution, and to procure other agricultural inputs for the members.

For tube well water from private parties, farmers pay market rates, per hour, depending upon the discharge capacity of the pump (diesel or electric). These rates are the same irrespective of crops.

Maintenance: The main purpose of the drainage cooperative society was to maintain the sub-surface drainage system and monitor its functioning, which it took over in 1991. Under the cooperative system, landowners shared the costs of drainage, which worked out to be Rs. 300/ha/year at that time. The cost included diesel, the salary of the pump operator, and maintenance and repair of drainage pipes. Farmers pay this amount in addition to the regular charges for canal water. For canal water, the usual assessment is made by the irrigation department based on land area and crop.

When the cooperative completely takes over irrigation water distribution, it will have to determine ways to maintain field channels like any other water users' organization, and restrain waterlogging by over-irrigation.

Income to the Society: The fee of Rs. 300 per acre per year for drainage was collected by the cooperative to make the society economically viable in view of the restoration of the land to reasonable productivity. Farmers also paid a fee of Rs. 50 to join the cooperative. There were no other fees. Advantages: (i) Land has been made productive again. This cooperative has helped define practical measures to desalinate the salt-affected and/or water-logged heavy soils, and the water table has measurably declined.

(ii) Crop yields have increased considerably. There has been a 100% increase in the yield of kharif paddy and a 400% increase in the yield of Rabi wheat.

(iii) Through technical assistance, farmers have learned proper soil and water management practices.

16.9 Non-conventional Method of Drainage

Vertical drainage using shallow or deep wells or a shallow multiple well-point system, bio drainage, pump drainage, and mole drainage are some non-conventional drainage methods (Bhattacharya and Michael 2003). Construction of ring bunds to protect agricultural lands from inundation by flood waters (widely adopted in Kerala), washing out the dry season surface accumulated salts in the saline land by using the water from the initial monsoon rains, and reducing the depth of accumulated runoff by recharging it into groundwater are some of the other non-conventional methods (Bhattacharya and Michael 2003). The function of the non-conventional drainage methods is to achieve the same goals as those of the conventional drainage methods, i.e., control of excess water and excess salts in agricultural lands. However, their scope, working principle, and design methods are different.

In lowland regions, the natural drainage network is usually found non-functional due to either siltation or higher water level in the drains than in the field. This situation does not permit to use of gravity drainage through a conventional surface drainage network in the waterlogged fields. Renovation of the silted natural drains may be costly, and hence a feasible option in such situations is to construct a dyke and adopt pump drainage to pump out the accumulated runoff from the cultivated land upstream to the dyke and dispose of it through shallow carrier drains towards the natural drain or directly into the natural drain (Bhattacharya and Michael 2003). On the other hand, pumping groundwater in canal commands not only augments the surface water supply but also helps control the rise of the water table due to continuous irrigation.

In heavy soils (soils having very high clay content, which are also known as 'black cotton soils' in India), conventional subsurface drainage by pipe or tile drains is usually expensive. Hence, a mole drainage system alone or in combination with subsoiling proves to be effective and less expensive in heavy soils.

Different types of vertical drainage systems as well as one environment-friendly drainage technique known as 'bio drainage' are briefly discussed.

16.10 Vertical Drainage Using Tubewells

Tubewell drainage is a technique for controlling water table and salinity in agricultural areas. It consists of pumping an amount of groundwater equal to the drainable surplus using a series of wells (Figs. 16.8 and 16.9). Tubewell drainage is not new but has not been widely used.

The discharge capacity of a tubewell is governed by the aquifer properties, namely, transmissivity and storage coefficient or specific yield. The actual discharge determines the extent of the drawdown. The discharge and the aquifer properties together govern the shape and the spread of the cone of depression (i.e., the volume of soil being drained). The pumping schedule, non-pumping hours, and the density of the tubewell network determine the residual drawdown after selected intervals of time and its cumulative effect over the years in a certain region. It is theoretically possible to predict the combined effect of the above factors on the water-table scenario of a region. However, the best method is to regularly monitor the water table in a network of observation wells over a region and analyze the data to enable proper interpretation in terms of future possibilities of water-table rise or fall. In India, the Central Groundwater Board (CGWB) and the State agencies collect pre- and post-monsoon groundwater-level data in different basins or sub-basins. These data should be analyzed along with the data of past developments in the region regarding agriculture, urbanization, and industrialization to obtain a clear picture of the possible future scenarios of water table behavior. Also, suitable theoretical analyses are necessary to estimate a proper tubewell density and a desirable rate of groundwater pumping

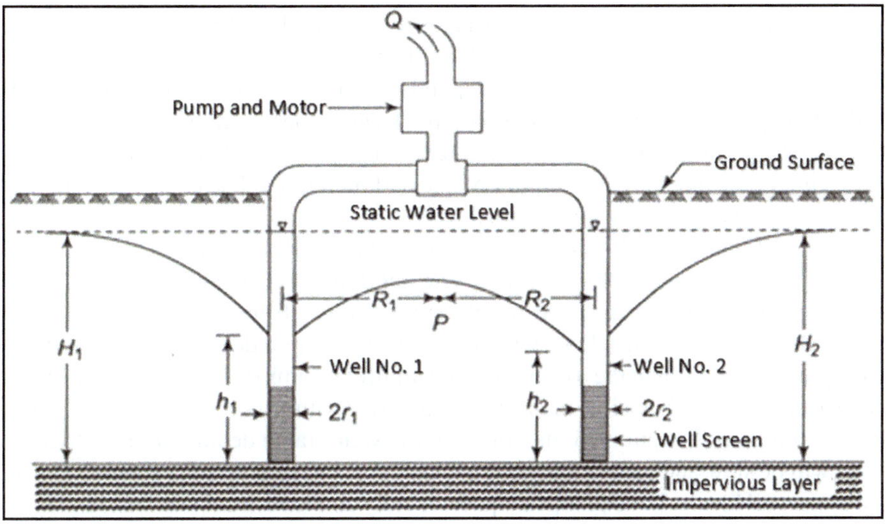

Fig. 16.8 Schematic of Tubewell drainage technique: two wells tapping an unconfined aquifer in series. (*Source* Michael et al. 2008)

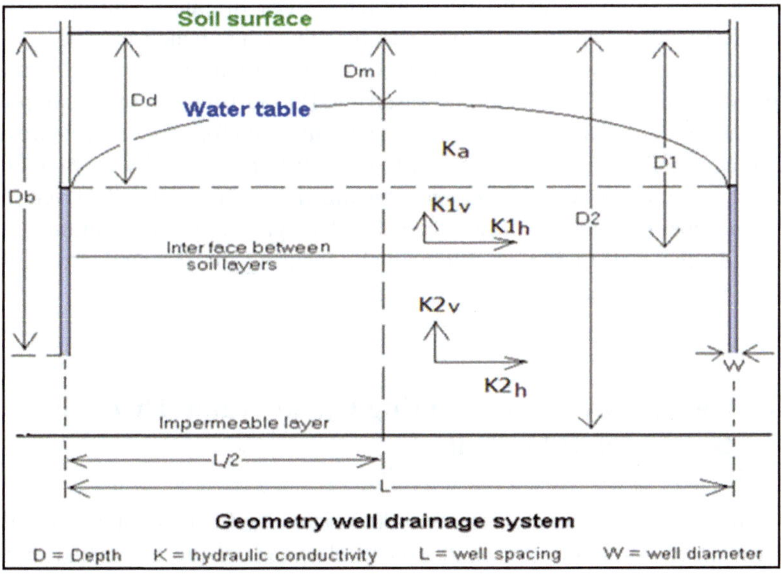

Fig. 16.9 Geometry of Well Drainage In Two Soil Layers

to maintain a favourable groundwater balance. Standard textbooks on groundwater hydrology may be referred to for detailed information on well hydraulics.

The success of tubewell drainage depends on many factors, including the area's hydrogeological conditions, the aquifer's physical properties to be pumped, and the physical properties of the overlying fine-textured layers. Another important factor is that skilled personnel are needed to operate and maintain tubewells, and to monitor water tables and the quality of pumped water.

Tubewells may be shallow or deep. A shallow tubewell is that which draws groundwater from the top unconfined aquifer. The depth of such an aquifer is highly variable from one region to another, but it usually ranges from a few metres to 30 or 60 m. Deep tubewells are those which draw water from a deep confined aquifer or multiple aquifers. When the top unconfined aquifer is pumped, its effect on the water table decline is clearly visible in the upper soil layer within a short period of pumping. If a deep tubewell is pumped, its effect on the water table decline in the upper soil layer will be slow and visible if the pumped aquifers have a hydraulic connection with the upper soil layers. If the pumped aquifer is fully confined, pumping from it may not affect the water table in the overlying unconfined soil system. It is worth mentioning that in most cases, the planned tubewell constructions have been for purposes other than drainage, i.e., to supply water for irrigation, domestic and industrial purposes.

Groundwater withdrawal in the command areas of purely surface irrigation projects (e.g., Kosi and Gandak projects in Bihar) has benefitted the command area by lowering the water table and thus saving the land from waterlogging and from possible chemical degradation (Prasad and Prasad 1996). In these two projects, groundwater withdrawal was not a planned activity, even for irrigation. However, the plentiful availability of groundwater at the low lift and generally inefficient functioning of irrigation projects led the enterprising and surface water-deprived farmers to sink and operate many shallow tubewells. This gave the desired drainage effect by lowering the water table, simultaneously ensuring the farmers' availability of irrigation water on demand.

16.11 Case Study of Augmenting Canal Water Through Battery of Tube Wells

A case study of planned groundwater withdrawal and its effect on water table behavior in the irrigation command (Hisar district in Haryana) of the Fatehabad branch of the Bhakra Canal System has been reported by Dua and Puri (1996). Along the branch canal, 50 strainer-type and gravel-packed tubewells of depth varying between 27 and 38 m were commissioned at a spacing of 120 m. The tubewell discharge varied between 10 and 15 L/s. The pumped water was fed into the branch canal to augment the water supply. Tubewell spacing, depth, and the pumping rate were fixed, keeping in mind the aquifer characteristics, presence of saline water at deeper depths, and for avoiding the upcoming of the saline water zone into the freshwater zone during pumping, respectively. Water table observations were taken at a number of piezometers installed perpendicular to the branch canal at several locations. The quality of pumped water was also monitored. These observations revealed a decline in the water table between 0.32 m and 3.09 m (average is 2.13 m, for 12 locations, covering a maximum distance of 600 m from the branch canal). The increase in the electrical conductivity (EC) of the pumped water after more than a year was slight (0.1 dS/m).

16.12 Use of Multiple Well-Point System

A multiple well-point systems consists of a network of closely spaced shallow tubewells to dewater a waterlogged area where the water table is close to groundwater or very close to the root zone, and pumping by a single or a few scattered deeper tubewells are not adequate to lower the water table. It is also suitable when the deep well pumping or pumping at a high rate from a single tubewell may be hazardous due to poor-quality water at deeper depths (Bhattacharya and Michael 2003). Thus, a multiple well-point systems of groundwater pumping is essentially a drainage method. In this system, shallow tubewells are closely spaced to produce an interference effect when they are simultaneously pumped. To cause an effective drawdown over a larger waterlogged area, should be several tubewells located within their radii of influence. It is important to note that well interference is not desirable for the tubewells constructed for water supply purposes.

The result of operating multiple well-point systems will be that all the individual cones of depression will superimpose on one another, giving a larger overall water table decline while pumping over a larger area. Pumping from such a network can be accomplished using a single pump installed on the main line after joining all the well points with air-tight pipes. Alternatively, individual rows of well points can be joined by a pipe that delivers water in a common sump well. By using bends, the delivery points of these pipes are kept sufficiently deep in the sump well. A single pump draws water from the sump well, causing a drawdown, adequately below the surrounding water table. This enables the pipelines connected to the well points to draw groundwater and deliver it into the sump well. A specific situation where such a system is useful is when the water table is shallow, and the soil is fine-textured and non-cohesive. Under these conditions, making stable open drains or trenches for the installation of a pipe subsurface drainage system become a difficult task due to the immediate collapse of drains or trenches. Further, to take full advantage of the system, the groundwater quality should be such that it can be used for irrigation or other purposes.

In **brackish (saline) groundwater regions**, groundwater extraction is low as the water cannot be used for irrigation. Use of only surface water for irrigation causes a gradual rise of poor-quality groundwater. **Conjunctive use of surface water and groundwater** can be adopted in such places for controlling water tables and increasing irrigation water availability. Gupta et al. (1987) studied this aspect in the

Faridkot region of south-west Punjab, where the EC of groundwater was about 8 dS/m, and the land had turned waterlogged and saline due to prolonged irrigation with surface water only. This study revealed that of the total water applied in irrigation, the groundwater and the surface water mixed in the proportion of 40:60 resulted in a maximum return from irrigation. Obviously, the water table would decline when 40% of the irrigation water is withdrawn from the aquifer rather than irrigated with only surface water.

16.13 Vertical Drainage Using Dug Well

Dug wells tap shallow unconfined aquifers. Therefore, pumping from dug wells immediately affects the shallow water table. However, since dug wells are shallow, have a small discharge capacity, and have a small zone of influence, the effect of water table decline is limited to a small area surrounding the wells. The constraint of low discharge can be partially avoided by withdrawing water at a high rate to cause adequate drawdown within a short period and stop pumping thereafter to allow recovery (Bhattacharya and Michael 2003). If the recovery rate is very slow, as in situations where the surrounding soil/porous medium has a low saturated hydraulic conductivity, the drawdown effect in the close vicinity of the well lasts longer. This period can be further lengthened by pumping a number of times in a day. The water so pumped can be used for irrigation in lieu of using the canal water if the area is canal irrigated.

Example: Pumping from Dug well
This example illustrates the effect of pumping from a dug well in a shallow water table farm holding in the Barna Command area of Madhya Pradesh. The crops were soybean in Kharif and wheat in rabi. It was found that the temporary water table declined due to pumping from dug wells and using the water for irrigation instead of the available canal water, which could give a 14.6% and 27.3% increase in yield, respectively, for the kharif soybean and rabi wheat as compared to the area where irrigation was only by canal water. In the rabi season, the unpumped dug wells in the canal-irrigated region recorded a rise in the water table by 2 cm, whereas the pumped-dug wells recorded a decline in the water table by 15 cm from the start to the closure of the canal during the rabi season. For the pumped-dug wells, the well water was used for irrigation, not the canal water. Water table observations made in a number of auger holes in the areas under canal irrigation and under well irrigation revealed a reduction in the SEW_{100} (number of days when the water table was shallower than 100 cm) values in the rabi season from 76 in the canal irrigated region to 24 in the dug well-irrigated region.

Based on the above information and discussion, it can be concluded that vertical drainage has many advantages. It is a means of controlling the water table in the unconfined aquifer and leaky aquifer regions as well as it is a vital component of conjunctive use of surface water and groundwater which is considered an appropriate mechanism of land and water management against degradation, especially in canal command areas (Bhattacharya and Michael 2003). A constraint, perhaps in several countries, in adopting vertical drainage at all places where it is needed is that it requires an assured supply of additional energy to operate a pumping system. In fact, the availability and cost of energy and the timely replacement of pumps and engines after their economic lifetime are the key determinants in selecting vertical drainage.

16.14 Biodrainage

Low-cost and environment-friendly drainage method like biodrainage provides several advantages over conventional drainage systems. According to FAO (2002): "Drainage engineers should no longer ignore the opportunities that biodrainage systems can offer. When planning for projects, the agricultural sector increasingly feels pressure from other users of the environment. For example, it is becoming increasingly unacceptable to set aside land exclusively for routinely designed irrigation and drainage projects. This illustrates the possible advantages of biodrainage systems".

16.14.1 Concept of Biodrainage

All living plants transpire water. The source of water is either irrigation water or groundwater. The transpiration capacity of a plant depends on its species' root depth and spread, canopy area, leaf area, and leaf structure, besides climatic conditions. When transpiration is met primarily by withdrawing groundwater, the process is known as biodrainage in drainage engineering. The total water transpired from the groundwater reservoir in a region, and hence its drainage effect (i.e., effect on water table decline) in a region is a function of plant density and other plant factors. Rice plants transpire quite heavily, but the process is not called biodrainage because the rice root systems are shallow (30–40 cm deep) (Bhattacharya and Michael 2003).

Medium to deep-rooted plants in a shallow water table region may act as small-capacity tubewells, constantly pumping groundwater to maintain transpiration rates (FAO 2002; Bhattacharya and Michael 2003). The difference between the two is that in the case of tubewell drainage, the area encompassed by the tubewell network is available for normal crop production, but in the case of biodrainage by a cluster of plants, the area within the cluster cannot be used for normal crop production.

16.14.2 Advantages and Disadvantages of Biodrainage

There are several advantages of biodrainage in comparison to the conventional drainage methods. The major advantages are that biodrainage is a low-cost measure, does not require gravity outlet and hence no physiographic constraint, operation, and maintenance are required only at the plant establishment stage, and there is no energy requirement for the operation and supply of fuel and fodder materials.

The disadvantages of biodrainage are that some area is required for growing plants, which cannot be made available for crop cultivation, and that good-quality water should be available for plant establishment. Therefore, despite the advantages outweighing the disadvantages, it is not yet a recognized drainage method, and it is necessary to conduct field studies on evapotranspiration from afforested areas, changes in chemical properties of the soil under the plants established for biodrainage, and the drainage effect within and outside the plantation area.

Questions

1. Explain the use of data on groundwater levels in deciding the maintenance needs of a drain network.
2. Explain the role of farmers in the maintenance of field drains.
3. Explain in brief the following terms: vertical drain, lateral, main drain, surface drainage, and subsurface drainage.
4. Explain various components of a subsurface drainage system and describe their maintenance needs.
5. Explain maintenance problems and methods of maintenance for surface drains.
6. What are the advantages of biodrainage and constraints in its implementation?

References

Bhattacharya AK, Michael AM (2003) Land drainage: principles, methods, and applications. Konark Publishers Pvt. Ltd., New Delhi

Chaube UC (2008) Area drainage study of proposed site for gidderbaha thermal power project. Unpublished Report of Sponsored consultancy project, IIT Roorkee, Roorkee

Dua SK, Puri TS (1996) Vertical drainage for removing drainage congestion in selected command area of Fatehabad branch: a case study. In: Technical papers, national workshop on reclamation of waterlogged, saline and alkaline lands and prevention thereof, Ministry of Water Resources, New Delhi, India

FAO (2002) Biodrainage: principles, experiences, and applications. IPTRID, Food, and Agriculture Organization (FAO), Knowledge Synthesis Report, No. 6, Rome, Italy

G.O.I. (2006) Assessment of water logging and salt and/or alkaline affected soils in the commands of all major and medium irrigation projects using satellite remote sensing. Technical report no

RCJ/TR/2006/5, report by regional Remote sensing Service Centre (ISRD) Jodhpur Govt of India
Gupta PK, Khepar SD, Kaushal MP (1987) Conjunctive use approach for management of irrigated agriculture. J Agricult Eng XXIV 3:307–316
ICID (1979) Amendments to the constitution, agenda of the international council meeting at rabat. International Commission on Irrigation and Drainage (ICID), Morocco, ICID, New Delhi, India, pp A-156–163
Michael AM et al (2008) Water well and pump engineering, 2nd edn. Tata McGraw Hill Education Pvt. Ltd, New Delhi
Prasad T, Prasad RS (1996) Conjunctive irrigation as a strategy for prevention/remedy of water-logging in humid alluvial plains for sustainably productive agriculture. In: Technical papers—supplementary, National Workshop on Reclamation of Waterlogged, Saline and Alkaline Lands and Prevention thereof, Ministry of Water Resources, New Delhi, India
WALMI (1987) On-farm development works(Second Edition), Publication number 12 of Water and Land Management Institute, Aurangabad, Maharashtra

Chapter 17
Diagnostic Analysis of Canal Irrigation System

Abstract Diagnosis of health (problems and their causes) of an irrigated agriculture system is the first step in seeking improvement in the operation and maintenance of the irrigation system. The objectives of this chapter are to increase our understanding of an irrigated agriculture system as it operates, recognizing its strengths (good features/values/benefits), as well as its constraints, and gain knowledge of field procedures for identifying the major constraints related to the main conveyance, on-farm, cropping, and socioeconomic subsystems that limit agricultural production. Reconnaissance survey is explained to understand the system as it presently operates. The reconnaissance help in designing detailed field surveys and experiments. Based on experience in supervising field-based diagnostic studies of minor canals of the Upper Ganga Canal system, the steps involved in planning and conducting a field study are discussed. A detailed list of surveys and experiments during the field study is given for each subsystem. A typical list of equipment required for the field study and some important Do's and Don'ts are indicated. Emphasis of a field study is not on the results of experiments or surveys but on generating meaningful information for analysis. Analysis should provide answers to checklist questions related to crop yield, utilization of irrigation potential, water availability at the head of the canal, water delivery above and below outlets, economic effectiveness, etc. A case study illustrates a diagnostic field study of the cropping subsystem and economic subsystem. Two minor (tank) irrigation projects (Khairana and Hinauta) in the Sagar district of Madhya Pradesh state in India are selected to illustrate the procedure for the diagnosis of these subsystems.

17.1 Concept and Knowledge Base of Diagnosis

17.1.1 Concept

Diagnosing the health (problems and their causes) of an irrigated system is the first step in seeking improvement in the operation and maintenance of the irrigation system. The diagnostic analysis provides a basis for developing and assessing

U. C. Chaube et al., *Canal Irrigation Systems in India*, Water Science and Technology Library 126, https://doi.org/10.1007/978-3-031-42812-8_17

solutions and then implementing these solutions to improve the operation and maintenance of an irrigation system. It requires understanding the difference between how a system should work and how it does work. Expectations regarding a system vary widely with the interests of stakeholders (farmers and institutions) (Lowdermilk et al. 1983).

A diagnosis identifies the system's performance and its problems and their causes according to the formally chosen reference: each reference will result in a different diagnosis, and the choice of reference may be subjective. Diagnosis of an irrigated system leads to a non-unique performance assessment. Because of the bias inherent in a diagnosis, it is important to be clear about the information needed to make an appropriate and accurate diagnosis.

17.1.2 Knowledge Base

All the technical fields concerned with the irrigated system come under the domain of Civil Engineering, Agricultural Engineering, Agronomy, Economics, and Social science. These fields do not constitute independent modules, but they are interlinked fields. The four technical domains are (Lowdermilk et al. 1983):

Civil Engineering: Meteorological data, hydrology, water resources, water storage, hydraulic and irrigation structures, aquifers, hydrological balance, losses, and discharge distribution in the channels (water storage and distribution).

Agricultuural Engineering: Weather data, field geometry, soil conditions, source of water, water quality, crops and evapotranspiration, consumptive use, conveyance channels, infiltration, irrigation methods, irrigation schedules, drainage network, and water recycling.

Agronomy: Cropping pattern, soil conditions, water scheduling, inputs, labour, agronomic considerations, household availabilities, needs and requirements, irrigation practices, and crop requirements.

Economics: Farm economics, market conditions, loans, and project economics.

Social science: Equity, local organization, political tensions, leadership, labor relations, tradition, kinship and group effects, laws and rules related to land and water, taxation conditions, and irrigation management.

Human factors account for much of the management of an irrigation system, and the collection of information regarding these technical domains is necessarily biased. The value of information given by an informant depends on his/her perception of the person asking the question. The answers are often qualitative and depend on the informant(s) and the person asking questions to interpret the situation correctly through these biased answers.

17.2 Objectives and Steps Involved in Diagnostic Analysis

Water and Land Management Institute, Aurangabad, Maharashtra, has brought out a publication (WALMI 1985) dealing with a procedure to conduct a diagnostic field study of minor irrigation schemes.

17.2.1 Objectives of Diagnostic Study

1. Understanding the irrigation system as it operates, recognizing its strengths and constraints. Instead of strengths, words such as values, useful features, and benefits are also used;
2. Identification of major physical, biological, economic, and organizational constraints to an irrigation system that limit agricultural production; and
3. Listing the above-mentioned constraints in order of priority based on stated criteria to assist the development and assessment of possible solutions.

The investigation team is responsible for establishing the criteria upon which priority will be assigned to constraints or the problem. The criteria must weigh the magnitude of the problem against the ability of any solution to affect the system immediately. Those problems that might take years before any solution would affect or remove them are not appropriate for a diagnostic analysis study.

17.2.2 Need for Identifying Right Parameters

The selected indicators must be relevant to the objectives of the diagnosis. Usually, the indicators have a focus on a particular aspect, such as:

1. The irrigation system (focused on hydraulic and hydrologic aspects);
2. The cropping system (crop yields expressed as a ratio per acre, per unit of water).
3. The agricultural economic system (economic returns, cost-benefit analysis);
4. The sociopolitical system (labour, equity, living standard, and quality); or
5. The ecological system (focused on long-term trends, irreversible changes, and degradations of climate, vegetal cover, water resources, and animal species balance).

Identifying the adequate indicators and right parameters that one should observe to explain poor (or good) performance is quite easy when there is a one-to-one correspondence between the performance measured and its causes. Unfortunately, this is seldom the case in irrigation systems; there are many loose linkages between the various factors, which make the backtracking of the root causes difficult. Very few disruptions in an irrigation system fully prevent any output from this system, and many different causes can be responsible for a similar performance. Therefore, quite often, the parameters identified as the causes of explaining a particular situation might also be mere consequences of this situation. This often leads experts into vicious circles during their reasoning process. They tend to cure symptoms but not causes due to their inability to find the right culprits.

17.2.3 Six Steps

The steps involved in diagnostic analysis are depicted in Fig. 17.1:

17.3 Reconnaissance Survey Procedure

Reconnaissance is the initial examination or survey of an irrigation system. The examination is not specific at this stage; rather, it is a sweeping overview of all dimensions of the irrigation system whether they are in the engineering, sociological, economic, and agronomic spheres. From the beginning, it is an interdisciplinary team effort.

Reconnaissance, when successfully executed, allows the team to identify some potential key constraints which will be more closely examined during the phase of detailed studies. Based on the reconnaissance, the team may refine the original

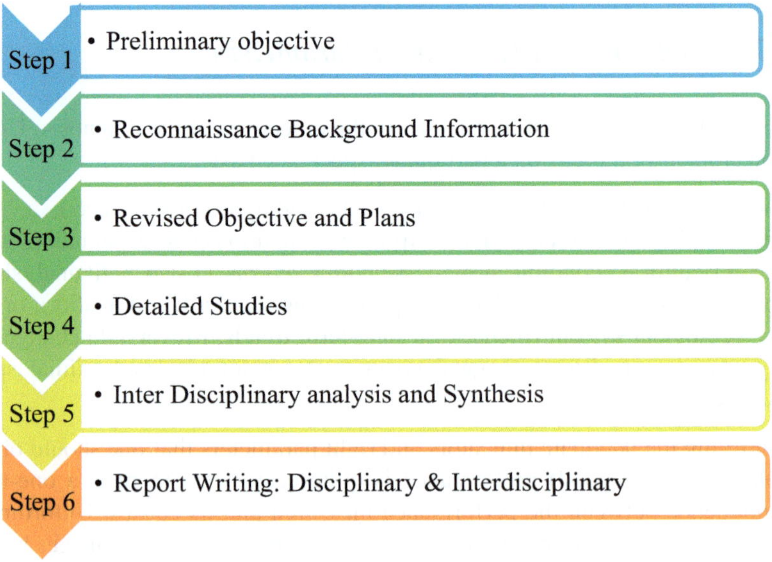

Fig. 17.1 Six steps of diagnostic analysis

objectives, questionnaire redrafted, and priorities recorded. Ultimately, reconnaissance permits the subsequent detailed studies phase to be planned according to the realities of the system rather than supposition or guess work-a common but ineffective means of investigation.

Note, however, that the aim of a reconnaissance is not the accumulation of unmanageable amount of data, but rather the systematic selection of data, however preliminary, that establishes perimeters, or limits, to the team task. In doing so, those constraints that most limit the viability and effectiveness of an irrigation system will change.

The actual field-level reconnaissance can be as short as one day or may last for a week. It is not the time frame but the careful planning which determines the effectiveness of reconnaissance. The flow chart in Fig. 17.2 illustrates seven steps for effective reconnaissance.

17.3.1 Preliminary Objectives of Reconnaissance

1. To understand the irrigation system as it presently operates.
2. To observe both its values and constraints
3. To order the observable constraints according to priorities based on predetermined criteria, and
4. To use the conclusion for planning detailed studies, including documenting the magnitude and causes of constraints.

These four objectives offer a fuller exploration of the purpose of reconnaissance.

17.3.2 Allocation of Responsibility

During the reconnaissance phase, however, responsibilities are shared informally among the members. The understood goal, at this point, is to go out into the system to whatever boundary level is appropriate in order to observe its operation and to interact with farmers and local officials.

The leader might ask the engineer to concentrate on the conveyance system or the sociologist to collect preliminary information on available institutional services, but both the engineer and the sociologist (and all the other members, for that matter) function collectively. This collaboration allows for an interdisciplinary approach from the very beginning of the investigation.

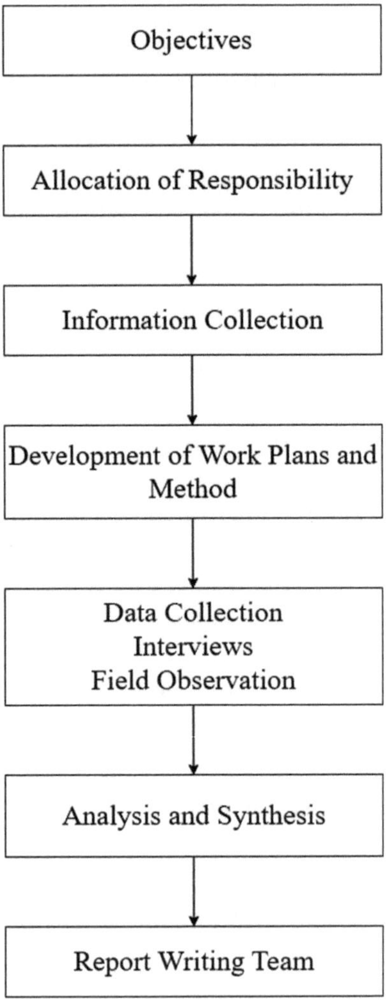

Fig. 17.2 Seven steps to an effective reconnaissance

17.3.3 *Information Collection*

The required information includes available monographs, research studies, project reports, and pertinent written materials about the area under investigation. This information should be condensed and combined, if possible, into a readily accessible form that is available to all team members. It is important to remember that while information may be collected along disciplinary lines, the entire thrust of the diagnostic analysis study is interdisciplinary. Mutual education, therefore, is a continual by-product of its procedures.

Visual Information: A photograph speaks much more than a written note on field observations. For example, during a reconnaissance survey, photographs are taken to depict the following:

Unauthorized use of pumps for water withdrawal from the canal, unauthorized outlets to take more water from the canal; and unauthorized use of canal water and the bank for laundry, placing obstruction to raise the canal water level for feeding the outlet. Similarly, the physical status of the canal (cracked lining of watercourse, damaged canal section, flood irrigation of field) is depicted in the photographs.

Photo: Unauthorized water withdrawal by pumping and by additional pipe outlets

Photo: Unauthorized outlet, laundry, obstruction in the canal to raise the water level

Photo: Crack in a lined water course

Photo: Damaged canal section and wrong method of disposing of the silt from the canal

Photo: Survey of the cross-section of the Lakhnauta minor

Photo: Field flooding is a common method of irrigation

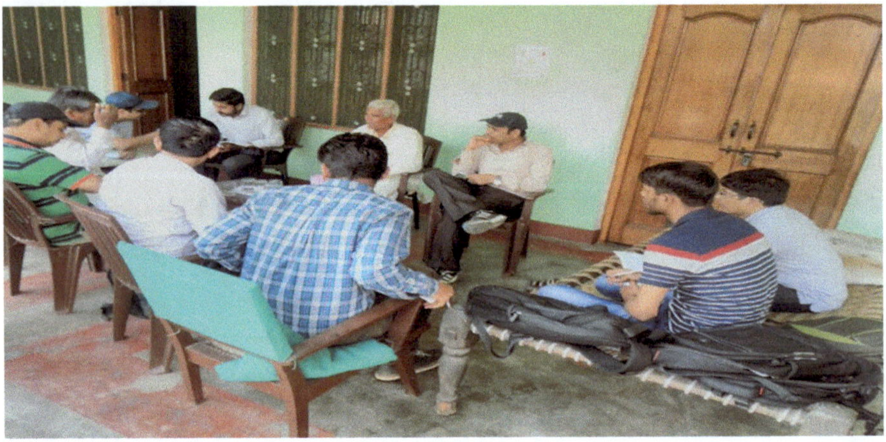

Photo: Evening/late evening is the appropriate time for interaction with farmers

17.3.4 Development of Work Plans and Methods

Flexibility is a primary characteristic of a functional work plan. Therefore, team leaders and members will review and change plans as needed. This revision is particularly appropriate with the completion of the information collection step.

It is also an appropriate time to review and/or finalize data collection methods. Simpler methods of diagnosis are needed, along with intelligent observation and team collaboration. A trained agronomist easily spots the visual signs of nitrogen deficiency; the engineer needs only to walk a distribution channel to know much about its conditions; and the sociologist or economist can determine general socio-economic conditions by simply viewing the size and quality of houses and roads. Collaboration among team members involves sharing data collection and observations, which allows the focus of reconnaissance to be revised to gain a better understanding of specific aspects of the system. Such collaboration is fundamental to any systematic, interdisciplinary undertaking

17.3.5 Data Collection

Data collection during reconnaissance is the simple gathering of certain facts based on informal conversations with project personnel and farmers and through visual

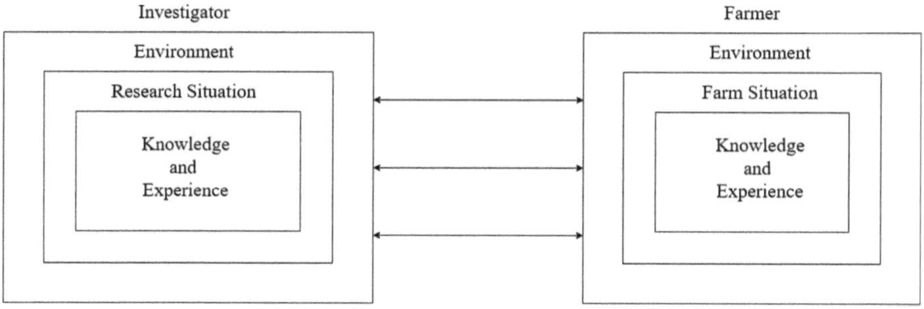

Fig. 17.3 Interdependence of investigator and farmer

observations. From these facts, tentative conclusions and perimeters will be drawn concerning the system. Informal, however, does not mean unsystematic. A sociologist can lean against the side of the community's well, chatting with villagers casually while learning much about the local water users group; or he can conduct a structured interview, based on the previously designed questionnaire and acquire the same information.

As mentioned above, reconnaissance uses a two-fold approach to data collection; field observation and interviewing.

Reconnaissance is field oriented. We cannot avoid going out of the system. At the end of reconnaissance, however, the team will have acquired, along with the soiled band and clothes, sufficient preliminary information upon which to develop a hypothesis, or tentative assumptions, about the primary values and constraints of the system. These values and constraints will be documented later in the detailed studies phase of the diagnostic analysis study.

This step of reconnaissance, therefore, is the initial link between the world of the investigator (or researcher) and that of the farmer. This link should stimulate mutual learning, communication, and undertaking, as seen in Fig. 17.3. If we had not done so already, we will recognize by the end of the reconnaissance how vital the farmer is to our understanding of the system.

17.3.6 Analysis and Synthesis of Reconnaissance

To analyze anything, we separate it into its parts or components to examine them critically. To synthesize means we assemble or bring separate parts into a whole. Synthesis indicates putting together the causes of different parts so that different relationships can be observed, causes and effects can be discovered, and new insights can be framed. Synthesis, by its very nature, crosses all disciplinary lines and results in an interdisciplinary understanding of the system and its possible improvement.

17.3.7 Report of Reconnaissance

The final step in the reconnaissance segment of the diagnostic analysis study demands that we put our thoughts and conclusions on paper. Having scanned the system, examined its components, and searched out new information, the reconnaissance is finished. The report will contain the results of the team's analysis and synthesis. Based on this, the objective and plans of diagnostic analysis will be revised.

17.4 Detailed Study Procedure

17.4.1 Five Steps

The detailed study phase has five activity steps, namely objectives, allocation of responsibility, formal field studies, data analysis, and disciplinary report writing (Fig. 17.4).

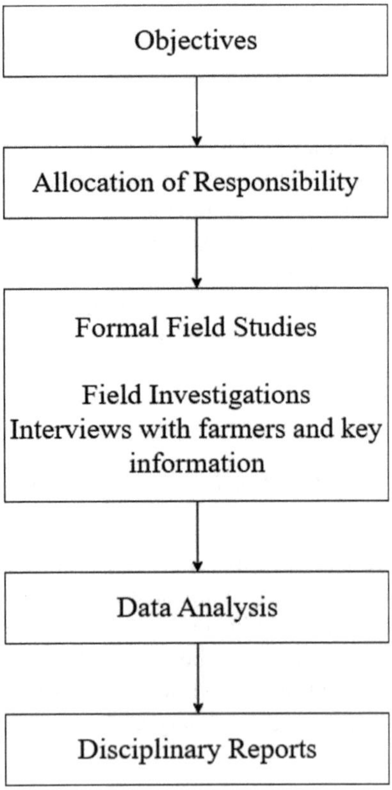

Fig. 17.4 Flow chart of detailed studies

The entire period of detailed study consists of field investigations and data analysis. All those activities that help with the evaluation of values and constraints of the irrigation command must be carried out in a scientific and integrated manner. Analysis and interpretation of results proceed simultaneously so that (if necessary) modification, repetition, multiplication of experiments, and field surveys could be done within the available time. It must be remembered that irrigation command problems are location-specific, and therefore specific fieldwork/experiments are required. Also, the temptation to conclude a diagnostic study regarding field surveys/experiment results is erroneous. The main objective of the fieldwork is to find the causes (in quantifiable terms) of the values and constraints of the system. Some aspects to be examined during the field study are indicated below.

17.4.2 Main Conveyance System Activities

(i) Collection of index map of the system.
(ii) Measurement of channel cross-section, long sections, bed material sizes, and silt concentration.
(iii) Estimation of discharge carrying capacity at designed F.S.L. of the existing cross-section, slope, and roughness coefficient.
(iv) Measurement of discharge at minor heads and outlet heads, along with operating water levels
(v) Measurement of seepage/conveyance losses in the channel and conveyance efficiency.
(vi) To check the S-D curve at the channel head
(vii) To determine the discharge capacity of the outlet at the operating head.
(viii) Evaluation of implementation and monitoring of main system regulation programme/roster.
(ix) Evaluation of the effect of entropic processes on the channel condition and the maintenance procedures.
(x) Evaluation of hydraulic behaviour of channel structures, such as adequacy of energy dissipation, presence of stream-lined flow without turbulence, and flow separation.
(xi) Evaluation of flow characteristics along the channel during the running days of the canal. Observe water level variations at specified points, and evaluate if flow is uniform/non-uniform, steady/unsteady.
(xii) Prepare canal water budget.

17.4.3 On Farm System

(i) Preparation of outlet command area map.
(ii) Watercourse carrying capacity measurement.

(iii) Conveyance loss measurement in watercourse/seepage losses, percolation, and spillage.
(iv) Surveying for land levelling of outlet command.
(v) Analysis of rotational water delivery schedule (designed, actual, time adjustment for filling, emptying, and seepage loss in canal).
(vi) Irrigation system performance evaluation (evaluation of field application efficiencies).
(vii) Soil and water quality analysis (surface and subsurface).
(viii) Evaluate the existing conjunctive use of surface and groundwater in outlet command and future potential.
(ix) Evaluation of on-farm structures.
(x) Pre- and post-monsoon water table and assessment of the waterlogged area.
(xi) Designed (planned) irrigation potential and actual utilization during the last ten years.

Data on the actual irrigated area (crop-wise) should be collected.

17.4.4 Cropping System

(i) Record acreage and yield of crops in outlet command.
(ii) Record the cultural practice of crops in the command.
(iii) Identification of cultivation problems in the command concerning insects and pests.
(iv) Agro-climatic production potential assessment of crops.
(v) Irritability and drain ability classification of land.
(vi) Taxonomic classification of land.
(vii) Crop production potential assessment of land.

17.4.5 Socio-economic Aspects

(i) To examine the social and organizational aspects of water availability, equity, and dependability.
(ii) To examine farmers' values and perceptions, level of knowledge, and decision-making processes related to irrigation, and the combination of enterprises' (crop-mix) use of technology.
(iii) To identify the level and use of institutional services and their significance to farmers in improving farm performance.
(iv) To study the costs and returns of important crops grown and resource use efficiency.
(v) To identify problem areas and potential decay constraints, including the viability of the system.

17.4.6 Interdisciplinary Analysis and Synthesis

At this point within the diagnostic analysis, time should be devoted to the interdisciplinary evaluation. Taking the whole apart, the team examines separate components and then synthesizes the results so that detailed studies can now substantiate the exact causes of constraints.

The primary constraints within the system would be identified for which future solutions could be developed. If there is more than one feasible solution to a problem, the recommendations should be placed in order of priority based on stated criteria. It is also important to recognize the values or benefits of the system as it functions. Such positive information can be just as valuable to system management.

17.4.7 Report Writing

The report should illustrate the systems perspective, the interdisciplinary approach, and the farmer's involvement. The report should not reflect simply parallel views of the system, for that is a multi-disciplinary approach. Rather, the report should present the team's findings in such a manner as to reflect the mutually interdependent components, the team's collective understanding of how it is functioning, and proposed recommendations.

17.5 Planning for Fieldwork

Preparation for fieldwork of diagnostic study covers field orientation activities, arrangement of equipment/facilities, laboratory experiments, collection of project data, etc. A bar chart showing the date, duration, and man-hour requirement for each activity should be prepared. Individual task assignment to each team member should be done.

17.5.1 Background Information

Background information to be collected is as under:

(i) The roster of the parent channel.
(ii) Maps and tables showing layout/details of channel, outlets, outlet command, and farmer fields.
(iii) History of irrigated cropping pattern.
(iv) History of actual roster followed in the previous five years.

(v) Design details, L-section, X-section, head inlet, S-D curve, Manning's n, culverts, falls, design-discharge and water supply level, outlets, discharge, outlet type, and position.
(vi) History of maintenance and repairs and inspection reports.
(vii) Budget allocation and actual expenditure on operation and maintenance activities.
(viii) History of the fragmentation of farm holdings in the command.
(ix) Rotational water delivery schedule.
(x) Water charges record for the past five years.
(xi) Existing reports on water theft and complaints of farmers.
(xii) Road map of the area.
(xiii) Water table fluctuations for the past five years.
(xiv) Medical reports on water-related diseases.
(xv) Marketing information.
(xvi) Community organization.
(xvii) Cultural pattern.
(xviii) Behavior pattern.
(xix) The welfare of farmers, and so on.

17.5.2 Equipment/Facilities

See Table 17.1.

The list of equipment required for laboratory tests is shown in Table 17.2.

17.5.3 Some Important Do's

(i) Prepare a flexible schedule to take care of contingency, for example, a canal may be suddenly closed, or a team member may fall sick.
(ii) Check all the instruments/equipment to be carried for the day's work.
(iii) Carry a medical first aid box, GPS equipment, and a road map of the area.
(iv) Review the progress of fieldwork every day.
(v) Carry index maps and other relevant information, such as water distribution schedule, and village population data.
(vi) Divide the day's work among team members to work on:

- Main system (from the head of the canal to outlet),
- On-farm system (below the outlets),
- Cropping system, and
- Socio-economic system.

(vii) Use polite language and local dialect for communication. Take the help of school teachers and students

Table 17.1 List of typical equipment for field work

S. No.	Equipment	Requirement	S. No.	Equipment	Requirement
1	Leveling instrument (with staff and equipment)	1 set per team	14	Wooden P	–
2	Current meter	1 per team	15	Copper ring	2 per team
3	Soil auger (screw type) and core auger	1 set/team	16	Staff gauge	2 per team
4	Double ring inflitrometer	1 set/team	17	Wooden plank (3 × 3 m)	2 per team
5	Flumes (can be Parshall, cutthroat or broad created weir)	2 per team	18	Piezometer	1 per team
6	V-notch weir	1 per team	19	Water bottle	5 per team
7	Tape measures	1 per team	20	Float (set)	1 per team
8	Hammers	1 per team	21	Calculators	2 per participant
9	Stopwatches	1 per team	22	Rulers (large metric)	2 per team
10	First aid kits	1 per team	23	Orifice	1 per team
11	Plastic buckets	3 per team	24	Camera photo	1 per team
12	Large poles	2 per team	25	Crawler	1 per team
13	Plastic weir	10 m per team	26	Spade	1 per team

Table 17.2 List of typical equipment for laboratory test

S. No.	Equipment	Requirement
1	Electric conductivity meter	1 per team
2	Measuring glasses	2 per team
3	Plastic mugs	1 per team
4	Pressure membrane apparatus	1 per team
5	Drying oven	1 per team
6	A core sampler	3 per team
7	Moisture cans	1 per team
8	Measuring cylinder	1 per team
9	Aluminum watertight moisture boxes	3 per team
10	Polythene sheets (large)	2 per team
11	Water bottle	5 per team
12	Plastic water can	1 per team
13	Calculators	1per member

(viii) Review day's progress at the end of every day of field visit.

17.5.4 Some Important Dont's

(i) Take discharge measurements when the water level fluctuates. Water level fluctuations occur during the filling and emptying of the canal. Fluctuations may also occur when an obstruction is placed in the canal or when outlets are opened or closed. Water level fluctuations may cause a hysteresis loop in the stage-discharge curve.
(ii) Conduct the ponding test for seepage analysis in the irregular section, which does not represent the entire canal system.
(iii) Use any current meter without a rated calibration chart. Each current meter has its own rating chart. The recent rating chart (within three months) should be used.
(iv) Take discharge measurements of the watercourse immediately below the outlet because the water level of a watercourse may rise, reducing the outlet's net operation head, which gives misleading discharge of the outlet.
(v) Conduct an infiltration test much before the advance and recession tests because the infiltration test has to represent the plot's infiltration characteristics at that particular time when the recession test is conducted.
(vi) Conduct any test on a field without the farmer's approval.
(vii) Take an interview when the farmer is preoccupied. A formal and technical approach should be avoided.

17.6 Analysis Using Checklist of Performance

The purpose of the diagnostic analysis is to evaluate the health (performance) of the project. Therefore, field experiments/surveys should be designed carefully to provide necessary data on the analysis (interpretation) and give information on the project's health. Emphasis is not on the results of experiments or surveys but on meaningful information. Data obtained from field experiments/surveys should provide answers for the performance checklist of questions indicated below (Podmore and Eynon 1983):

Check List on Crop Yield

(i) How do the existing crop yields (in the irrigated area) compare with crop yields before the project, crop yields in the adjoining unirrigated areas, and target crop yields?
(ii) What are the bottlenecks in achieving the desired yield?
(iii) What can be said for the future?

Check List on Utilization of Irrigation Potential

(i) How much of the planned irrigation potential has been utilized?
(ii) Compare crop areas irrigated with those proposed in the design cropping pattern.
(iii) Based on historical data, analyze the trend.
(iv) Is water shortage the only reason, or are there other reasons also?
(v) Has irrigation demand been met for each crop and each season?
(vi) Identify reasons for under utilizing irrigation potential by examining water utilization in each crop season.

Check List on Water Availability

(i) Assess 75% dependable yield using up-to-date hydrologic data.
(ii) Is it different from the design data?
(iii) What was the actual water supply in different years compared to the design yield?

Check List on Water Delivery

(i) Prepare water budget; compare water delivery with design targets for each season. Explain the difference.
(ii) Compare the water utilization factor with the potential values of the factor.
(iii) Evaluate the influence of actual conveyance efficiencies on water utilization factors.
(iv) Compare actual efficiencies with those adopted in the design of the project.
(v) Which reaches of the conveyance network show poor efficiencies?
(vi) Identify reasons for excessive losses.

Check List on Performance of Water Courses

(i) Is there excessive leakage?
(ii) What are the O&M problems?
(iii) Is it appropriately aligned?
(iv) Correlate seepage losses with undisturbed hydraulic conductivity, depth to confining layer, and estimated water table level.
(v) What are the problems being faced by farmers in the O&M of watercourses?
(vi) What is the level of farmers' awareness about the O&M of watercourses?

Check List on Outlet Command

(i) Determine the size of the outlet command.
(ii) What should be the appropriate size?
(iii) Fix the boundaries and check/propose alignment of field drains.

(iv) Verify and refix the outlet positions. Check the alignment of field channels.
(v) Check alignment of graded bunds etc.

Check List on Economic Effectiveness

(i) Prepare farm budget before the project, without project (in adjoining unirrigated area), and with project (in irrigated area) conditions.
(ii) Evaluate the composite net return (Rs. per ha) for crop patterns under three conditions.
(iii) Has the ultimate net return target been achieved?
(iv) Is the project still in the transition stage after construction to the full development stage?
(v) Have the irrigation area and agriculture technology transitions been achieved?
(vi) What is the impact of canal lining and conjunctive use on the water supply factor and the project's economic efficiency?

17.7 Case Study of Cropping Subsystem and Economic Subsystem

Two minor (tank) irrigation projects (Khairana and Hinauta) in the Sagar district of Madhya Pradesh state in India are selected to illustrate the procedure for the diagnosis of these subsystems. The study was carried out by IIT Roorkee as a field research project (Chaube 2006; Nihal 2007).

Cropping subsystem analysis consists of cropping pattern, cropping intensity, and irrigation intensity on seasonal and annual bases. Economic subsystem analysis consists of the cost of cultivation, crop return, gross farm income, net farm income, and incremental farm income related to the cropping subsystem.

Details of the diagnostic study of the cropping system and economic system are given in Annexure 17.A.

Appendix 17.A: A Case Study on Cropping and Economic Subsystems

General

A cropping system comprises unirrigated and irrigated crop areas in the kharif and rabi seasons. It is characterized by cropping pattern, cropping intensity, and irrigation intensity on seasonal and annual bases. Farm income covers the analysis of the cost of cultivation, gross farm income, net farm income, and incremental farm income with reference to the control situation, which is taken as prevailing before

Table 17.A.1 Salient features of the tank irrigation projects

	Unit	Khairana	Hinauta
Village		Khairana	Hinota
Block		Raheli	Banda
Distance from Sagar	km	56	47
CCA	ha	308	386
Kharif irrigation area	ha	20	63
Rabi irrigation area	ha	275	185
Total irrigation area	ha	295	248
Irr. intensity	%	95.78	617.25

the implementation of projects (balance works). Two minor (tank) irrigation projects (Khairana and Hinauta) in India's Sagar district of Madhya Pradesh state are selected for this case study. A detailed analysis of cropping system and farm income under various situations was carried out. Details of the study are given below (Chaube 2006 and Nihal 2007).

Tank Irrigation Schemes

The two minor (tank) irrigation projects are Khairana and Hinauta in India's Sagar district of Madhya Pradesh state. These tanks are designed for the storage of monsoon runoff of local streams. Irrigation is proposed in Kharif (wet) and Rabi (dry) seasons, emphasizing Rabi irrigation. The salient features are stated in Table 17.A.1.

Before Project, Designed, Existing, and Ultimate Situations

Before Project Situation: This corresponds to the situation prevailing at the time of NABARD sanction for implementing balance works.

Designed: This corresponds to the proposed cropping system and related parameters as given in the project proposals submitted by WRD for NABARD finance.

Existing Situation: This corresponds to the present situation in the command area of the project. The existing situation may differ over the projects due to the difference in the implementation status of projects. Part of the command area is under canal irrigation, in the case of Khairana and Hinauta Kharmau projects. Therefore, the existing situation in the command areas of these projects pertains to the transition stage to the ultimate situation.

Ultimate Situation: This refers to the expected situation when the projects will be in full operation stage. It is assumed that the existing rabi crops, which are irrigated from other sources as well as unirrigated rabi crops, will ultimately be irrigated by canal water. Further, the direct use of tank water for irrigation will not occur under the ultimate situation. However, it is debatable whether farmers already accustomed to the direct use of tank water will surrender such a benefit.

The cropping system and farm income are different under the above-mentioned four situations. Changes in the cropping system and farm income may occur over a period of time, even without a project due to improvements in agriculture technology (use of high-yielding variety seeds, farm machinery, pesticides, fertilizer, bank credit, and switch over to cash crops). Such impacts cannot be attributed to the irrigation project. For a realistic assessment of the impact attributable to the projects, it would be more logical to consider the change from "without a project in future" to "with a project in future" situations.

Exercise on the evaluation of project impacts gets further complicated due to the recognition of the fact that even where canals are not in operation or are partly in operation, the tank water is being utilized for irrigation (without using canals). Whether the cropping system based on such use of tank water should be considered as pertaining to "with project" or "without project" situation is debatable. In the present study, it has been assumed to be pertaining to "with project" situation.

In the following sections, cropping system and farm income are analyzed under different situations using sample data and other information.

Design Irrigation and Actual Canal Irrigation

The conventional procedure to determine the design irrigation area is based on 75% dependable water supply (with or without storage) and irrigation demand (usually monthly or fortnightly) within a normal year. Long-term simulation study incorporating variation in (i) cropping pattern, (ii) irrigation requirement, (iii) water availability, and (iv) irrigation efficiency over the years is desirable but usually not carried out. Table 17.A.2 shows the design irrigation area and canal irrigated area in different years as per the information supplied by WRD.

There is under-utilization of irrigation potential with significant fluctuations over the years in the Khairana and Hinauta projects. Annual irrigation shown in the table above is for canal irrigation only. It does not include direct irrigation from tank,

Table 17.A.2 Design and actual irrigation by canal

Name of the project	CCA (ha)	Designed irrigation (ha)			Actual irrigation (ha)		
		Kharif	Rabi	Annual	Year 1	Year 2	Year 3
Khairana	308	20	275	295	Nil	166 (54%)	60 (20%)
Hinauta	385	63	185	248	80 (32%)	Nil 15 (6%)	138 (56%)

streams, and wells. Tank bed cultivation is common in all projects. The cultivated tank beds and adjoining cultivated land on the periphery of the tanks are irrigated directly from the tank using pumps to lift water.

In addition, farmers are making use of water flowing in Nala in downstream, which is augmented due to seepage from dam, leakage from intake, and also unauthorized diversion of tank water into Nala by cutting the canal in head reach. In a way, this has resulted in the utilization of tank water, which otherwise was not possible due to incomplete/damaged canal network. In addition, tank storage and canal irrigation have increased the artificial recharge of ground water which is being utilized for irrigation.

Cropping Pattern and Cropping Intensity

A cropping pattern is expressed in terms of the percent of cultural command area (CCA) under each crop. Thus, there is a design cropping pattern (as reported in the project proposal), existing cropping pattern (as observed during a recent socio-economic survey), and a cropping pattern expected in the full operation stage (ultimate), which is based on a realistic assessment of crop and crop areas. Cropping intensity is defined as the total cropped area (irrigated and unirrigated) expressed as a percentage of CCA on an annual basis. Annual cropping intensity may not necessarily be equal to the sum total of seasonal cropping intensities if there are two seasonal or perennial crops. Seasonal irrigation intensity is the total of irrigated crop area expressed as a percentage of CCA in a season. Annual irrigation intensity is the sum of seasonal irrigation intensities. The concept of cropping intensity is appropriate to farm budgeting and assessing water charges etc. which are calculated on the basis of the annual cropped area. The concept of irrigation intensity is useful for assessing water requirements and how much land is to be irrigated.

Data have been collected from the farmers on the cultivated crops, crop areas (irrigated and unirrigated), and source of irrigation. The total land area and total crop areas owned by the farmers were worked out to assess the actual cropping pattern in the command areas, as shown in Table 17.A.3.

Soyabean is the most important Kharif crop. Pulse (Udad) is also an important Kharif crop in Maheri Project. Wheat, gram, and pulse are important crops (irrigated/unirrigated) in rabi season. Socio economic survey of the farmers has revealed that farmers have followed a similar cropping pattern for several years using improved agriculture technology and other sources of irrigation. Seasonal and annual cropping intensities are high in these projects. Irrigated crop areas and irrigation intensities are based on area irrigated from all sources of irrigation. Only rabi crops are generally irrigated. Soil type being black cotton, irrigation is generally not practiced in kharif season.

Table 17.A.3 Sample survey of crop areas (acres) and cropping pattern

Item	Khairana area (%)	Hinauta area (%)
Number of farmers surveyed	21	19
Area owned by the surveyed farmers(acre)	342	412
Item	Khairana area %	Hinauta area %
Number of farmers surveyed	21	19
Area owned by the surveyed farmers (acre)	342	412
CCA of projects (acre)	770	965
Crop areas (acres)	(308 ha)	(386 ha)
Kharif		
Soya	176 51.46	391 917.9
Paddy	1 0.29	–
Pulses	1 0.29	–
Vegetable	–	–
Udad	–	–
Rabi		
Wheat irrigated	92 26.9	127 30.83
Gram irrigated	144 42.1	148 35.92
Pulses irrigated	21 6.14	35 8.5
Vegetable irrigated	–	–
Wheat unirrigated	117.5 17.24	39 9.47
Gram unirrigated	25 7.31	51 12.38
Pulses unirrigated	38.5 11.27	–
Vegetable unirrigated	–	–
Irrigated Intensity		
Kharif	0	0
Rabi	75.14	75.25
Annual	75.14	75.25
Cropping Intensity		
Kharif	52.04	917.9
Rabi	97.96	97.1
Annual	150	192

Note—Irrigated crop areas include areas irrigated using various sources of irrigation (canal, tank, stream, and wells). Area owned by sample of farmers is distributed in canal command as well as in tank bed, tank periphery, and outside command area

Impact on Cropping System

Changes in Cropping Patten

Table 17.A.4 shows cropping patterns under different situations (designed and existing) for the projects. The existing cropping pattern is significantly different from the designed cropping patterns in all the projects. Unirrigated/irrigated paddy area which was existing before the project and which was proposed for canal irrigation (designed) are not exists now, as it is not a cash crop. Instead, soyabean has become an important Kharif crop in all the projects, irrespective of the availability of canal irrigation.

Wheat was the only crop proposed for rabi irrigation in all four projects. The existing pattern shows that gram has emerged as a more important rabi crop in all the projects. The irrigated wheat area is now significantly less than what was proposed in the Khairana project.

Ultimately, when the projects are in full operation stage, unirrigated wheat, and gram crops are expected to be converted into irrigated crops. The area under pulses and vegetables is also expected to increase.

Table 17.A.4 Cropping pattern under different situations (designed and existing)

Rop	Khairana		Hinauta	
	Designed	Existing	Designed	Existing
	(ha)	(ha)	(ha)	(ha)
Kharif				
Paddy (unirrigated)	200	1	144	–
Paddy (irrigated)	20	–	23	–
Soyabean	–	159	40	366
Vegetable/pluses	–	–	–	–
Udad	–	–	–	–
Rabi				
Gram (unirrigated)	–	23	–	48
Gram (irrigated)	–	130	–	139
Wheat (unirrigated)	–	13	–	37
Wheat (irrigated)	275	83	185	119
Pluses (unirrigated)	–	35	–	–
Pluses (irrigated)	–	19	–	33
Vegetable	–	–	–	–
Total	495	462	392	742
Cropping intensity %	160.7	150.0	101.6	192.2

Changes in Cropping Intensity

Table 17.A.5 shows the annual cropping intensity under different situations for the projects. Cropping intensity has increased significantly in all the projects. In the case of Khairana project, the cropping intensity was already high (137.7%) even before the project. It was designed to be 160.7%, but it is 150.3% at present, mainly due to the non-existence of paddy crop area. In the case of the Hinauta project, the change in cropping intensity is 192.2% compared to 817.7% before the project and 101.5% as designed (Fig. 17.1).

Table 17.A.5 Annual cropping intensity under different situations

Crop	Khairana (CCA-308ha)				Hinauta (CCA-386ha)			
	Before (%)	Designed (%)	Existing (%)	Ultimate (%)	Before (%)	Designed (%)	Existing (%)	Ultimate (%)
Kharif								
Paddy (Unirrigated)	71.43	617.94	0.325	0.00	47.41	37.31	0.00	0.00
Paddy (irrigated)	–	6.49	–	0.32	–	5.96	–	–
Soyabean	–	–	51.62	51.62	6.22	10.36	917.82	917.82
Vegetable/ pluses	–	–	–	–	–	–	–	–
Udad	–	–	–	–	–	–	–	–
Rabi								
Gram (unirrigated)	–	–	7.47	–	–	–	12.44	–
Gram (irrigated)	–	–	42.21	49.68	–	–	36.01	48.45
Wheat (Unirrigated)	52.60	–	17.22	–	31.09	–	9.59	–
Wheat (irrigated)	–	89.29	26.95	31.17	–	47.93	30.83	40.41
Pluses (unirrigated)	13.64	–	11.36	–	–	–	–	–
Pluses (irrigated)	–	–	6.17	17.53	–	–	8.55	8.55
Vegetable	–	–	–	–	–	–	–	0.00
Cropping intensity %	137.66	160.71	150.32	150.32	817.72	101.55	192.23	192.2

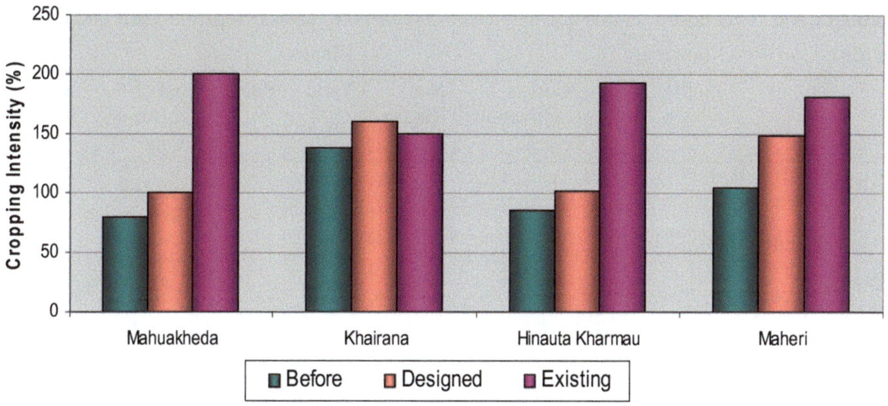

Fig. 17.A.1 Shows the comparison of cropping intensities in different situations

Unit Cost of Cultivation and Net Return for Crops

Cost of Cultivation and Net Crop Return under Existing Situation

Farmers have been interviewed to provide information on various farm expenditures and income items. These data were collected for unirrigated and irrigated crops during Kharif and Rabi seasons. For irrigated crops, the cost of water is different for canal water and water lifted from wells or streams using pumps.

There are wide variations in the reported cost of cultivation and farm income. Data provided by some of the farmers were either unrealistic or incomplete. The cost of self-employed family labor and the cost of owned machinery (tractor, thrasher, etc.) was not reported correctly by some farmers. Therefore, information has also been collected from other sources to have a realistic assessment of cost and benefits. Prices were ascertained by the survey of local markets.

Table 17.A.6 shows yield, prevailing market prices, unit cost of cultivation and net crop return for various crops during Kharif and Rabi seasons in the projects. As seen in the table, crop yields and cost of cultivation differ over the projects.

Comparison with Designed Data

Crop yield, rate, and cost of cultivation assumed during projects' appraisal are compared with data collected during the present study in Table 17.A.7. Crop yields for various irrigated/unirrigated crops are more or less the same except for wheat. The yield of irrigated wheat under the existing situation (25 quintals/ha) is only 71.4 % of the yield taken for project appraisal. Due to the overall increase in market prices,

Table 17.A.6 Unit cost of cultivation and net crop return for various crops in Kharif and Rabi

Crop	Khairana				Hinauta			
	Yield	Rate	Cost of cultivation per ha	Net Return	Yield	Rate	Cost of cultivation per ha	Net Return
Kharif								
Paddy (Unirrigated)*	–	–	–	–	–	–	–	–
Paddy (irrigated)*	–	–	–	–	–	–	–	–
Jowar	–	–	–	–	–	–	–	–
Soyabean	13	1200	6650	8950	11	1200	5300	7900
Rabi								
Wheat (unirrigated)	117.5	800	5250	6350	13	800	5200	5200
Wheat (irrigated)	25.5	800	5850	14550	25	800	5700	14300
Gram (unirrigated)	10.5	2000	6500	14500	7	2000	5300	8700
Gram (irrigated)	117.5	2000	6800	22200	13.5	2000	5550	21450
Pulses (unirrigated)	7.5	1500	5200	6050	–	–	–	–
Pulses (irrigated)	14	1500	5300	15700	15	900	5575	7925

the cost of cultivation has increased but at the same time the gross return has also increased.

Average parameters for the projects show that Soyabean provides highest returns (Rs. 9137/ha) in kharif season. The net returns with irrigation (gram, pulses and wheat) are significantly higher than those for without irrigation in rabi season. Irrigated gram crop provides the highest net return (Rs. 20,410/ha).

Incremental Farm Income

Farm income is assessed in terms of annual benefit per ha of land with a particular cropping pattern. It is equal to the sum total of net crop return multiplied by crop area per ha of CCA. Crop areas per ha of CCA are nothing but the cropping pattern expressed in fractional form. Cropping patterns are given in Tables 17.A.4 and 17.A.5. Net crop returns are given in Table 17.A.6. Farm incomes under different situations (before, designed, existing, ultimate) are shown in Table 17.A.8. It is to be noted that the net crop returns are for the existing situation. Thus, the incremental farm

Table 17.A.7 Comparison of appraisal data with recent data

Crop	Appraisal for NABARD assistance			Recent socio economic survey			
	Yield Qtls/ha	Rate Rs/ Qtls	Cost of cultivation Rs/ha	Yield Qtls/ha	Rate Rs/ Qtls	Cost of cultivation Rs/ha	Net crop Return Rs/ ha
Kharif							
Paddy (unirri)	20	500	3500	20	700	7400	6600
Paddy (irri)	25	500	4520	25	700	9900	7600
Jowar	18	450	1800	–	–	–	–
Soyabean	15	900	5200	13	1200	6463	9137
Rabi							
Wheat (unirri)	15	550	2700	14	800	5337	5863
Wheat (irri)	35	550	5500	25	800	7140	12,860
Gram (unirri)	–	–	–	9	2000	5800	12,200
Gram (irri)	–	–	–	14	2000	7590	20,410
Pulses (unirri)	6	1200	2730	7.5	1500	5200	6050
Pulses (irri)	–	–	–	14	1500	5300	15,700

income with reference to the before-project situation shows the impact of change in the cropping pattern on farm income.

Table 17.A.8 also shows the incremental farm incomes with reference to the control situation (before the project situation). The incremental farm income in the existing situation is significantly more than the envisaged (designed situation). It will further improve when projects are in the full operation stage (ultimate) and at present is high and will further improve in the ultimate situation.

Questions

1. Explain the following terms:

   (i) diagnosis, (ii) constraints, (iii) values, (iv) main conveyance system, (v) on-farm system, (vi) walk-through survey, (vii) minor canal, (viii) outlet, (ix) water course, and (x) field channel.

Table 17.A.8 Incremental farm income (Rs/ha)

Project	Annual farm income (Rs. per ha of CCA)	Incremental Farm income with reference to before project condition (Rs. per ha of CCA)
Khairana (308 ha)		
Before Project	8783	
Designed	17771	8988
Existing	20940	12157
Ultimate	22961	14178
Hinauta (386 ha)		
Before Project	5809	–
Designed	10588	4779
Existing	21882	16073
Ultimate	24340	18531

2. Draw similarities between the diagnosis of health of human body and diagnosis of health of an irrigated agriculture system.
3. Explain the field study procedure to evaluate the discharge capacity of a minor canal.
4. What precautions/care should be taken for conducting a socio economic survey of farmers.
5. What are the essential Do's and Don'ts before going for field work?
6. List various items related with agronomic survey.
7. List various items of engineering survey.
8. List various items related with economic survey.
9. Compare the cropping subsystem performance of the two tank irrigation projects (Khairana and Hinauta).
10. Compare the economic subsystem performance of the two tank irrigation projects (Khairana and Hinauta).

References

Chaube (2006) Report on evaluation of rural infrastructure (irrigation) projects in Sagar District of Madhya Pradesh. Sponsored Research by National Bank for Agriculture and Rural Development (NABARD), Mumbai, India

Government of India (1979) Manual on irrigation and water management. Ministry of Agriculture and Irrigation, New Delhi

Lowdermilk MK et al (1983) Diagnostic analysis of irrigation system, vol I. Concept and methodology. Colorado State University, U.S.A

Nihal KDS (2007) Financial and economic analysis of tank irrigation projects, report of M.Tech dissertation (unpublished) supervised by Prof U C Chaube, IIT Roorkee, Roorkee

Peri G, Skogerboe GV (1979) Evaluation and improvement of irrigation system. Colorado State University, Report 49A, U.S.A

Podmore CA, Eynon DG (1983) Diagnostic analysis of irrigation system, vol II. Evaluation technique. Colorado State University, U.S.A

WALMI (1985) Spl. course on diagnostic analysis of minor irri.schemes, Publication number 8 of Water and Land Management Institute, Aurangabad, Maharashtra

WRDM (2010–2019) Unpublished reports on diagnostic study of Lakhnauta Minor Canal and Gadarjudda Minor Canal,carried out by trainee engineers as part of their training at Department of WRDM, IIT Roorkee

Chapter 18
Soil and Water Quality Management

Abstract This chapter describes the study of soil and water quality management and some critical soil and water quality parameters concerning irrigated agriculture. The methods of determining and managing saline/sodic soils and water are presented. The rise in the water table within 1–2 m of the surface causes waterlogging and contributes significantly to evaporation from the soil surface, leaving the salts in the top layer of soil (see Chap. 16 for details). Management of soil water quality is the main concern of this chapter. The objectives of the chapter, therefore, are to: (a) understand various parameters of soil and water quality and procedures to carry out sampling and testing of soil and water; (b) understand hazards of salinity and sodicity; (c) appreciate the role of drainage, conjunctive use in the control of salinity; and d) irrigation method and its frequency in the control of salinity. Laboratory determinations that are needed to evaluate irrigation water salinity are given. Soil and water sampling procedures and the information sheets for each sample are explained. Hazards associated with salinity and sodicity are discussed. These hazards can be controlled through various measures, such as leaching, irrigation technology, conjunctive use, drainage, and chemical amendments. Leaching is an economical method for reclamation and maintaining salt balance during the irrigation season. The conjunctive use of fresh and saline waters is important for areas under saline groundwater or waterlogged saline lands. Numerical examples of (a) leaching requirement and (b) blending saline and fresh water for conjunctive use are given. Case studies conducted on soil and water quality testing are included. Physical and chemical parameters and heavy metals were examined and compared with acceptable and permissible values. The study area consisted of irrigated agriculture soils, surface water, and groundwater facilities near a village in District Bilaspur of Chhattisgarh state, India.

18.1 Introduction

Before irrigation, a balance exists amongst rainfall, runoff (surface and ground), evaporation, and transpiration. This balance is disturbed when large quantities of water are brought into an area and spread on the land. Groundwater level rises due to seepage from canals and field applications. The rise in the water table within 1–2 m of

U. C. Chaube et al., *Canal Irrigation Systems in India*, Water Science and Technology Library 126, https://doi.org/10.1007/978-3-031-42812-8_18

the surface contributes significantly to evaporation from the soil surface, leaving the salts in the top layer of soil. The salts depend upon the mineral content of irrigation water. The concentration and composition of dissolved constituents determine water quality for irrigation use. The soil and water quality is an important consideration in any appraisal of salinity or alkaline conditions in an irrigated area.

The accumulation of salts usually occurs in areas that receive salt from other locations, and water is the primary carrier. Historically, many civilizations that nearly always relied on irrigated agriculture had declined in the ancient past primarily due to the failure to cope with salinity, sodicity, and waterlogging problems.

Example: Gidderbaha tehsil in Muktasar district in the Punjab state of India has been facing an acute problem of water logging due to the rise in groundwater level on the west side of the Rajasthan and Sirhind feeder canals in Punjab. In low-lying areas, the land stands submerged and is totally unfit for cultivation (Chaube 2008).

This chapter discusses some critical soil and water quality parameters for irrigated agriculture. The methods of determining and managing saline/sodic soils and water are presented. Codes of practice published by the Bureau of Indian Standards describe methods of sampling, testing, and analysis of water and soil quality parameters. These have been referred to in the preparation of the chapter.

1. BIS (2012): Code IS 10500:2012 Drinking water specification, Bureau of Indian Standards, Manak Bhavan, 9 Bahadur Shah Zafar Marg, New Delhi.
2. BIS IS 1622:1981 Methods of sampling and microbiological examination of water (first revision), Bureau of Indian Standards, Manak Bhavan, 9 Bahadur Shah Zafar Marg, New Delhi.
3. BIS IS 3025(part 1): 1987 Methods of sampling for water and waste water, Bureau of Indian Standards, Manak Bhavan, 9 Bahadur Shah Zafar Marg, New Delhi.
4. BIS IS 3025(part 2): 2002 Determination of 33 elements by inductively coupled plasma atomic emission spectroscopy, Bureau of Indian Standards, Manak Bhavan, 9 Bahadur Shah Zafar Marg, New Delhi.
5. BIS IS 2720:Part 1to Part 26: Method of Test for soils, Bureau of Indian Standards, Manak Bhavan, 9 Bahadur Shah Zafar Marg, New Delhi.

18.2 Definition of Quality Parameters

Quality parameters are defined in BIS Codes and various textbooks such as in Biswas and Mukherjee (1987) and Michael (1990).

18.2.1 Electrical Conductivity

Electrical conductivity (EC) is expressed in micro-mhos per centimeter (mmho/cm) for high salinity and μmmho/cm for low salinity waters. The SI units are decisimens per meter (dS/m).

1 mmhos/cm at 25 °C = 1 dS/m, and.
1 μmmhos/cm = 1 μdS/cm.

The temperature corrections for determinations made at a temperature different from 25 °C are approximated by the following relationship:

$$EC_{25} = EC_t - 0.02(t - 25)EC_t$$

where EC_t is the electrical conductivity at temperature T^0 centigrade.

18.2.2 Sodium Adsorption Ratio (SAR)

SAR is defined as the ratio of concentration (in meq/1) of monovalent cations to the square root of the concentration of divalent cations in the soil water solution:

$$SAR = \frac{\mathrm{Na}^+}{\sqrt{(\mathrm{Ca}^+ + \mathrm{Mg}^+)/2}}$$

where Na and (Ca + Mg) are the values of soluble ions in meq/1. If the ion concentration is expressed in mmol/1, then

$$SAR = \frac{\mathrm{Na}}{\sqrt{(\mathrm{Ca} + \mathrm{Mg})}}$$

Exchangeable sodium ratio (ESR): It is defined as the concentration of exchangeable monovalent and divalent cations:

$$ESR = a + k(SAR)$$

where, a and k are the empirical constants for a given soil type. Also,

$$ESR = \frac{\mathrm{Na}}{\mathrm{Ca} + \mathrm{Mg}} = \frac{\mathrm{Na}}{CEC - \mathrm{Mg}}$$

where Na and Ca + Mg are the exchangeable ion values, and CEC is the exchangeable cation capacity. All the values are expressed in meq/100 gm.

18.2.3 Exchangeable Sodium Ratio or Percentage (ESR or ESP)

It is the degree of saturation of the soil exchange complex with sodium.

$$ESR = \frac{\text{Na}}{CEC} \times 100$$

The values of Na and CEC are in meq/100gm,

18.2.4 Adjusted SAR

The SAR of soil water, for steady-state conditions, is determined from irrigation water composition after it has been saturated with calcium carbonate ($CaCO_3$) precipitation or dissolution. Langelier proposed an index that he termed as saturation index (SI) to indicate the extent to which water flowing in a closed system (without any of C09) will precipitate or dissolve $CaCO_3$. The SI is defined as

$$SI = pH_a - pH_c$$

where pHa is the actual pH of water, and pHc is the theoretical pH that the water would have in equilibrium with $CaCO_3$:

$$Adj\ SAR = SAR(1.84 - pH_c)$$

18.2.5 Residual Sodium Carbonate (RSC)

The concept of residual sodium carbonate is used to evaluate the alkali hazards of water as

$$\text{RSC(meq/l)} = (\text{CO}_3 + \text{HCO}_3) - (\text{Ca} + \text{Mg})$$

The use of water containing high proportions of sodium (>75% of total cations) and the predominance of $CO2^-$ and HCO_3 ions relative to Ca^{2+} and Mg^{2+} lead to high ESP development and high pH in the soils. The following reaction takes place:

$$\text{Ca}^{++} + 2\text{HCO}_3^- = \text{CaCO}_3 + \text{H}_2\text{O} + \text{CO}_2$$

This reaction reduces the concentration of Ca and Mg, thereby increasing the proportion of Na. This increases ESP but reduces soil salinity. The residual bicarbonate reacts with sodium and forms $NaHCO_3$, which, upon evaporation, becomes Na_2CO_3.

$$Na^+ - HCO^3 = NaHCO_3$$

$$2NaHCO_3 = Na_2CO_3 + H_2O + CO_2$$

Na_2CO_2 reacts with normal soils and replaces Ca^{2+} with Na^+.

18.3 Measurement for Water Quality Evaluation

The measurement of salinity/alkalinity parameters in irrigation waters and soil is the key to developing management practices. Laboratory determinations that are needed to evaluate irrigation water quality are given in Table 18.1. Water quality represents the total sum of dissolved inorganic ions and molecules. The important components of salinity are cations of Calcium (Ca), Magnesium (Mg), and Sodium (N) and anions of Chlorides (CI), Sulphate (SO_4), and bicarbonate (HCO_3). The minor components may include Potassium (K) and Nitrate (NO_3). In quantitative terms, the salinity is measured by electrical conductivity (EC), which is closely related to the sum of major cations (or anions) as determined by chemical analysis. The EC does not depend upon the type of salt.

(1) mg/1 = miligram per litre = parts per million (ppm).

meq/1 = milliequivalent per litre (mg/1 – equivalent weight = meq/l); in SI units, 1 me/1 = 1 million/litre adjusted for electron charge.

(2) SAR is calculated from the Na, Ca and Mg reported in meq/l.

The EC is used to express the salinity of irrigation water (EC_W) and the salinity of soil water (EC_{sw}). Since the effects of applied water salinity are transmitted to the plants through the soil medium in which the concentration of salt (EC) changes with moisture, it may be useful to have salinity measurements at reference soil moisture content. USSL (1954) has developed a procedure to estimate soil salinity that uses soil moisture at saturation as a reference. The salinity so determined is known as salinity in saturation extract (EC_E). The EC_{SW} field capacity is about $2EC_e$. The relationship between EC and EC_{SW} may be useful in relating the salt tolerance of crops to EC_W.

The osmotic potential (OP) is related to EC by the following equation:

$$OP = -0.36\,EC$$

where OP is expressed in kilo Pascal (k Pa) and EC in dS/m at 25 °C.

Table 18.1 Laboratory determinations needed to evaluate common irrigation water quality problems

Water parameter	Symbol	Unit(1)	Usual range in irrigation water
SALINITY			
Salt Content			
Electrical conductivity (or)	EC_w	dS/m	0–3
Total Dissolved Solids	TDS	mg/1	0–2000
Cations and Anions			
Calcium	Ca^{++}	me/1	0–20
Magnesium	Mg^{++}	me/1	0–5
Sodium	Na^{+}	me/1	0–40
Carbonate	Co_3^{--}	me/1	0–1
Bicarbonate	$HC0_3^{-}$	me/1	0–10
Chloride	Cl^{-}	me/1	0–30
Sulphate	SO_4^{--}	me/1	0–20
SODIUM ADSORPTION RATIO (2)	SAR		0–15

The total soluble salts are related to EC by the following equations:

$$\text{Log TSS} = -0.9S0 + 1.055 \log \text{EC}$$

$$\text{TSS(mmolc/1)} = 10\,\text{EC(dS/m)}$$

where TSS is the concentration is mmolc/l.

If the water's salinity is 1–5 dS/m, it approximately equals 10 meq/1 total cation concentration (TCC) or 640 mg/1 of total dissolved solids (TDS). If the salinity increases beyond 5 dS/m, the corresponding TCC and TDS increase to 12 mg/1 and 950 mg/1, respectively. In the old literature, salinity was expressed in parts per million (ppm). This has now been changed to mg/1. The two units are identical for waters up to EC of 5 dS/m.

18.4 Soil and Water Sampling

Each sample must have a reference note as per the proforma given below.

Reference note on the soil sample

Name of the farmer:

Address:

Field/Keela No. sampled	Sample No.	Sampling Depth	Irrigated by tubewell No., if any
Crops grown	Yield obtained	Fertilizer used kg/ha	Irrigated non-irrigated

General field conditions

Permeability to water,Sub-soil hard layers , Leveling conditions :, Depth of water - table , Drainage conditions :

Crop symptoms and problems :

Any other relevant point one would like to mention:

Reference note on the water sample

Name of the farmer:
Address:

Source of water (check) canal drain well
If underground water sample, indicate - depth of well_____________
metres depth of filter opening _____________ metres

Use of water	Tubewell No.	Plot No. irrigated with tubewell Nos.

1. Acreage served
2. Crops grown
3. Change in crop or land symptoms observed, if any
4. Soil in which water is to be used — Texture: Permeability:
5. How long has the land been irrigated with this water?
6. What problems did you face with the use of this water
7. Give your opinion on water quality
8. Soil samples sent along with water samples yes/no

18.5 Salinity Hazards

The force with which the soil holds water is known as matric potential. If the soil is relatively dry, the metric potential is high, and the plant also has to exert more force. When salts are present in water, the osmotic potential is developed, and the plants have to exert more force against this additional potential force, commonly known as the osmotic effect. The potentials can be considered additive to compute the total potential (Michael 1990).

The waters having EC_W of less than 0.75 dS/m do not create a salinity problem and provide a full choice of crops. The build-up of salinity in the crop root zone depends upon EC_W, leaching fraction, and leaching efficiency (depending upon soil texture). The leaching fraction (LF) is defined as the fraction of water entering the soil that passes beyond the root zone. The average salinity of the crop root zone is closely related to EC and LF and the crops primarily respond to the average salinity of the root zone. The average salinity of the root zone can be estimated for known values of EC_W, LF, and assumed seasonal distribution of water uptake from the root zone. The most commonly used distribution of water removal by evapotranspiration is in the proportion of 40: 30: 20: 10 from the first, second, third, and fourth quarters of the root zone. Though soil moisture at field capacity (FC) does not change much, soil water concentration (CF = 1/LF) varies with LF. The degree of soil water concentration, relative to irrigation water, LF, and root zone depth, are given in Table 18.2. If it is desired to have CF at the soil moisture content equal to the saturation percentage, one has to simply divide the CF values given in Table 18.2 by two (because $2EC_e = EC_{SW}$). The CF values can be used to establish the relation between EC_e, and EC_W for a given LF.

Table 18.2 Relative solute concentration of soil water (field capacity) compared to that of the irrigation water

Root zone in quartiles[a]	Cu[b]	Concentration factor (CF = 1/LF) at LF values of					
		0.05	0.10	0.20	0.30	0.40	0.50
0	0	1.00	1.00	1.00	1.00	1.00	1.00
1	40	1.61	1.56	1.47	1.39	1.32	1.25
2	70	3.03	2.70	2.27	1.96	1.72	1.54
3	90	7.14	5.26	3.57	2.70	2.17	1.82
4	100	20.00	10.00	5.00	3.33	2.50	2.00
Average[c]	FC	5.58	3.76	2.58	2.06	1.74	1.54

[a]Assumes a water uptake of 40%, 30%, 20%, and 10% from the first, second, third, and fourth quarters, respectively
[b]Cumulative percentage of consumptive use above each indicated quartile depth
[c]The average for each quarter of the root zone is divided by four

18.6 Sodicity/Alkalinity Hazards

The sodicity/alkalinity problems are severe in irrigated soils. The long-term use of alkaline waters containing a high proportion of sodium (>75% of total cations) and predominance of CO_3 and HCO_3 ions relative to Ca^{2+} and Mg^{2+} lead to the development of high ESP and high pH in soil. The high sodicity and pH adversely affect the physical properties of soil. Infiltration in such soils reduces drastically, resulting in stagnation or high runoff following irrigation on rainfall.

18.7 Leaching for Salinity Control and Land Reclamation

Leaching means the removal of salts from the root zone by applying water (there could be natural leaching by rain) in excess of crop ET requirements. In leaching, salts get dissolved and removed beyond the root zone. Leaching is an economically effective practice for controlling salts brought in by irrigation water.

18.7.1 Water and Salt Balance in Soil Profile

The salt moves with water, and the major water inflows and outflows in the soil profile can be described by the water balance equation:

$$D_r + D_g + D_w = D_e + D_t + D_d + D_s$$

where D is the water depth, and subscripts r, g, w, e, t, d, and s stand for rainfall, upward groundwater movement, irrigation water, evaporation, transpiration, drainage, and change in storage, respectively.

The change in storage can be written as

$$D_s = D_r + D_g + D_w - D_e - D_t - D_d$$

The corresponding change in salt storage can be written as

$$S_s = D_r C_r + D_g C_g + D_w C_w + S_m + S_f - D_d C_d - S_p + S_c$$

where C with different subscripts stands for the concentration of salts in the rain (r), upward groundwater flow (g), irrigation water (w), and drainage water (d), respectively; S_m is the salt added due to mineral dissolution; S_f is the salt added by fertilizers and amendments; S_p is the salt precipitated; and S_c is the salt removed by the crops.

Generally, D_rC_r, S_m, S_f, S_p and S_c are relatively low compared to the other factors and hence ignored. The salt balance equation can, therefore, be written as

$$S_s = D_g C_g + D_w C_w - D_d C_d$$

If the water table is deep, the term $D_g C_g$ can also be neglected, and S_s can be written as

$$S_s = D_w C_w - D_d C_d$$

If the water inflow ($D_r + D_g + D_i$) is less than the outflow ($D_e + D_t$), the storage is reduced. The depletion of D_s dries up soil reducing D_e and $D_{t,}$ and this increases the concentration of soil water. If the concentration is beyond a certain level, the plants, depending on their sensitivity, will suffer growth reduction.

In the presence of a water table, the shortage of water supply in the soil profile may be met by D_g. In such a situation, no drainage takes place, and no salts are exported. Such a situation cannot continue indefinitely without severe effects on crop yield. Over a period of time, the salinity levels fluctuate in a range, and the purpose of leaching is to keep these fluctuations within limits.

18.7.2 Leaching Requirements

The leaching required to maintain an economic yield level depends upon the contents of soil, water, climatic conditions, and management practices. Keeping the net flow of water downward is the only way for sustainable cropping.

For no upward movement of groundwater and under steady-state conditions ($S_s = 0$),

$$D_g C_g + D_w C_w - D_d C_d = 0$$

If $S_s = D_r + C_r = D_a$ (applied water), then

$$\mathrm{D_a C_a = D_d C_d}$$

$$LF = \frac{D_d}{D_a} = \frac{C_d}{C_a} = \frac{EC_a}{EC_d}$$

where LR is the leaching fraction.

Determining C_a is very important to work out the minimum leaching requirement. The approximate method, which accounts for the salinity at the top (EC_t) and bottom (EC_b) of the soil profile to estimate the average salt concentration of the saturation extract, is in common use. According to this method,

$$EC_e = \frac{K_1(EC_t + EC_b)}{2}$$

The value of K_1 is taken as 0.8 for a low leaching fraction. The above equation is based on salinity at the approximate soil moisture content relationship at saturation and field capacity, which can be shown to be of the form:

$$EC_d = 5EC_e - EC_w$$

A more precise relationship to estimate the mean root zone salinity has been developed by van Genuchten and Hoffman (1983).

Now LR can be written as

$$LR = \frac{EC_w}{5EC_e - EC_w}$$

The total amount of water that needs to be applied is

$$D_a = \frac{ET}{(1 - LR)}$$

where LR is the leaching requirement as a fraction.

18.7.3 Example: Leaching Requirement Calculation

A crop of wheat sown in sandy loam soil is irrigated by border irrigation with an application efficiency of 70%. The salinity of tubewell water used for irrigation (EC_w) is 5dS/m. Seasonal evapotranspiration is 45 cm, total seasonal rainfall is 5 cm. The leaching requirement is computed to obtain a 100% yield.

Solution:

$$\begin{aligned} \text{Given } EC_w &= 5\text{dS/mw} \\ AE &= 0.70 \\ ET &= 45\,\text{cm} \end{aligned}$$

$$\begin{aligned} \text{Net water required} &= \text{ET} - \text{Rainfall} \\ &= 45 - 5 = 40\,\text{cm} \end{aligned}$$

Total water that must be applied to meet ET demands

$$= \frac{ET - Rainfall(effective)}{Application\ efficiency} = \frac{40}{0.70} = 55.55$$

Leaching requirement (LR)

$$LR = \frac{EC_W}{5EC_e - EC_w} = \frac{5}{5 \times 6 - 5} = 0.2$$

18.7.4 *Leaching for Reclamation*

When the salinity level greatly exceeds the tolerance of a crop, major leaching may be required before planting the crop. It may not be possible to accurately predict the water required to accomplish reclamation by leaching. It is reported in the literature that between 70 and 80% of the salts could be removed by passing a unit depth of water per unit depth of soil. The leaching requirements of heavy soils are more than light-textured soils. Only 0.40 units of water per unit of soil depth may be sufficient to leach 80% of the salts from the root zone. A basic understanding of the leaching concepts is required to design suitable leaching programmes.

18.8 Irrigation Timing, Frequency, and Method for Salinity Control

The salinity in the root zone fluctuates during the irrigation cycle. The lowest values prevail in the immediate post-application phase, whereas the salinity is at its peak just before irrigation. If the irrigation intervals are large, the variation in pre- and post-irrigation salinity values is also large. Since the plant water uptake is affected by the osmotic potential, it is reasonable to assume that the effect of high osmotic stress can be avoided by modifying the irrigation frequency (Michael 1990).

An increase in irrigation frequency without an unacceptable increase in water requirement is not always possible, particularly with surface methods. Of course, precision levelling and suitable design of surface irrigation layout may permit some control on the application depth. Sprinkler and drip systems are better suited to high-frequency irrigation.

Sprinkler irrigation, when used in moderate climates, has the advantage of maintaining better soil moisture and salinity regimes. Sprinkler irrigation can be beneficial in the pre-germination phase. However, when the high salinity waters are used during hot weather and evaporation rates are high, problems of leaf injury are encountered. Experimental results from Hisar show that yield reduction in cotton, which is grown during the hot season when irrigated with sprinkler using saline water, was more than wheat which is grown during the winter season. The problem can, however, be minimized by irrigating at night when temperatures are low or by running the sprinkler with fresh water toward the end of the irrigation application.

Drip or trickle irrigation is an ideal choice for using saline waters in row crops. This is because the system permits accurate delivery control. Irrigation frequency adjustments are also possible.

18.9 Conjunctive Use for Salinity Management

The conjunctive use of fresh and saline waters has special importance to areas underlain by saline groundwater or waterlogged saline lands.

18.9.1 Two Important Methods

The two important methods connected with the conjunctive use of saline and fresh waters are (a) the development and blending of saline water and (b) following in the rainy season (Gupta et al. 1987).

(a) **Blending of fresh water and saline waters**

The blending of water allows more area to be irrigated than would be possible with fresh water alone. The permissible proportions of fresh and saline waters can be evaluated by the following equation:

$$Max\ ECW_s = \frac{ECW_f VM_f + ECW_s VW_s}{VW_f + VW_s}$$

where Max ECW_s is the maximum permissible salinity for the acceptable yield level; ECW_f and ECW_s are the salinities in dS/m of fresh and saline waters, respectively; and VM_f and VW_s are the volumes (ha-cm) of fresh and saline waters, respectively.

(b) **Land fallowing in the rainy season**

In areas where there is no other source of water, there is no escape from the use of saline water alone. In such cases, the practice of fallowing the land in the rainy season to allow for the leaching of salts from the root zone is practiced. Sometimes, when the salinity levels are high, greater care has to be exercised in the choice of crops.

18.9.2 Example: Blending Saline and Fresh Waters for Irrigation of Wheat

Given that

Salinity of canal water $SCW_c = 0.5$ dS/m.

Salinity of tubewell water $(ECW_t) = 8$ dS/m.

Evapotranspiration of wheat $(ET) = 45$ cm.

Application efficiency $(AE) = 70\%$

Determine the proportion in which the two waters are to be blended for obtaining 90% of normal yield to wheat at Indore.

Solution

The maximum applied water salinity for 90% yield of wheat at Indore = 4.7 dS/m (from the literature).

$EC_e = 1.5EC_w = 1.5 \times 4.7 = 7.05\,dS/m.$

LR of canal water =

$$\text{LR of canal water} = \frac{ECW_c}{5EC_e + ECW_t} = \frac{0.5}{5 \times 7.05 - 0.7} = 0.014$$

LR of Tubewell Water =

$$\text{LR of Tubewell Water} = \frac{ECW_t}{5EC_e + ECW_t} = \frac{0.8}{5 \times 7.05 - 8.0} = 0.294$$

$$\text{Crop water requirement with existing irrigation practice} = \frac{\text{ET}}{1 - \text{LF}}$$

$$\text{WR} = \frac{45}{1 - 0.3} = \frac{45}{0.7} = 64.29\,\text{cm}$$

$$\text{Actual canal water requirement}(CWR) = \frac{\text{ET}}{1 - \text{LR}} = \frac{45}{1 - 0.014} = 45.64$$

$$\text{Actual tubewell water requiement}(TWR) = \frac{\text{ET}}{1 - \text{LR}} = \frac{45}{1 - 0.294} = 63.74$$

If canal water alone is used, the actual requirement is 45.64 cm. On the other hand, if tubewell water alone is used, the requirement is 58.44 cm. Since the application efficiency with which the water is being applied is 70%, the water applied is 64.29 cm. So, there is no need to use extra water for leaching.

The proportion of Blended Water:

If the proportion of canal water is a and that of tubewell water b,

$$a + b = 1 \text{ or } a = (1 - b)$$

Using Eq.

$$Max\ ECW_s = \frac{ECW_f \times VM_f + ECW_s \times VW_s}{VW_f + VW_s}$$

$$\text{ECW}_s = 0.5(1 - b) + 8b = 7.05$$

or b = 0.0867

Therefore,

Table 18.3 Soil sampling locations

Site No.	Name of site	Sample id	Photograph id
1	S-2, Near Ramp of Dike-1 Village-Hardadih	S-2	C2
2	S-5, Near Dike-1 Village Rank	S-5	C5

a = 1−0.867 = 0.133.
The proportion of canal water (a) = 0.133.
The proportion of tubewell water (b) = 0.867 Ans.

18.10 Case Study on Soil Quality Testing

This case study is based on IITR (2022). The following soil quality parameters were considered by Chaube (2022).

Soil Physical Texture, Soil type, pH, Conductivity, SAR, CEC, Na, P, K, Cation (Ca, Na, K), Anion (Cl, SO_4, CO_3), Heavy Metals (As, B, Cd, Co, Hg, Total Cr, Pb, Mo, Ni, As, Zn, Cu, etc.), Fluorides. (One season).

Analysis was carried out in view of the prevailing drinking water standards.

18.10.1 Soil Sampling Locations

The study area comprises irrigated agriculture soils near village in District Bilaspur of Chhattisgarh state, India. Soil samples were examined for physico-chemical parameters and heavy metals. There were several sampling sites. For illustration, only two sites are considered here. Soil sampling locations are described in Table 18.3.

18.10.2 Soil Sampling Method

Soil sampling and testing were carried out in October, 2021. The soil samples were drawn by soil sampler auger from different sites manually, as per the steps given below:

1. Remove the surface litter at the sampling spot.
2. Drive the auger to a plough depth of 15 cm and draw the soil sample.
3. Collect two more samples from 2 to 3 m depth, wherever possible considering the groundwater table, and place them in a tray.
4. Mix the samples thoroughly and remove the foreign materials like roots, stones, pebbles, and gravel.
5. Reduce the bulk to about half to one kilogram by quartering.

6. Divide the mixed sample thoroughly into four equal parts. Discard the two opposite quarters, remix the remaining two quarters, and repeat this process until the desired sample size is obtained.
7. Collect the sample in a zipper pouch and label it with site information.

18.10.3 Soil Test Method

The samples were analyzed as per the established scientific methods for physico-chemical parameters. The heavy metals were analyzed by using Inductive Coupled Plasma-MS and OES Analyzer. The soil quality classification is recommended in Handbook of Agriculture, Indian Council of Agricultural Research, New Delhi (ICAR 2021). Soil quality classification is indicated in Table 18.4.

18.10.4 Soil Quality Test Results

Soil quality test results are given in Table 18.5. The soil sample detection limit and reference document on the methodology adopted for each parameter are also mentioned in the table.

18.10.5 Impact of Trace Elements on Agro-Eco System

Some trace elements such as copper (Cu), zinc (Zn), manganese (Mn), iron (Fe), molybdenum (Mo), and boron (B) are essential for plant growth and are called micronutrients. Some trace elements, such as cobalt (Co) and selenium (Se) are not essential for plant growth but are required by animals and humans. Other trace elements, such as cadmium (Cd), lead (Pb), chromium (Cr), nickel (Ni), mercury (Hg), and arsenic (As), have toxic effects on living organisms and are often considered contaminants. Trace elements in an agroecosystem are either inherited from soil parent materials or inputs through human activities. Soil contamination with heavy metals and toxic elements due to parent materials or point sources often occur in a limited area and is, therefore, easy to identify.

Table 18.4 Soil quality classification (ref. handbook of agriculture, ICAR, New Delhi)

S. No.	Soil test	Classification
1	pH	< 4.5 Extremely acidic 4.51–500 Very strongly acidic 5.01–600 moderately acidic 6.01–6.50 slightly acidic 6.51–7.30 neutral 7.31–7.80 slightly alkaline 7.81–8.50 moderately alkaline 8.51–9.0 strongly alkaline > 9.01 very strongly alkaline
2	Salinity electrical conductivity (μS/cm) (1 ppm = 640 μS/cm)	Upto 1.00 average 1.01–2.00 harmful to germination 2.01–3.00 harmful to crops (sensitive to salts)
3	Organic carbon (%)	Upto 0.2 very less 0.21–0.4 less 0.41–0.5 medium 0.51–0.8 on an average sufficient 0.81–1.00 sufficient > 0.1 more than sufficient
4	Nitrogen (kg/ha)	Upto 50 very less 51–100 less 101–150 good 151–300 Better > 300 sufficient
5	Phosphorus (kg/ha)	Upto 15 very less 16–30 less 31–50 medium 51–65 sufficient > 80 more than sufficient
6	Potassium (kg/ha)	Upto 15 very less 16–30 less 31–50 medium 51–65 sufficient > 80 more than sufficient

Table 18.5 Soil quality test results

Sr. No.	Parameters	Method	Unit	Detection limit	Sample id	
					S-2	S-5
01	pH value (at 5% Slurry)	APHA $4500H^{+}B$	NA	NA	7.90	*7.23*
02	Conductivity (at 20% Slurry)	IS:3025 (P-14)	μS/cm	NA	205	*158*
03	Arsenic (as As)	APHA 3120B	mg/kg	0.05	6.61	7.34
04	Nickel (as Ni)	APHA 3120B	mg/kg	0.05	22.01	17.29
05	Lead (as Pb)	APHA 3120B	mg/kg	0.05	12.93	15.96
06	Cadmium (as Cd)	APHA 3120B	mg/kg	0.01	BDL	BDL
07	Total chromium (as Cr)	APHA 3120B	mg/kg	0.005	27.75	23.65
08	Copper (as Cu)	APHA 3120B	mg/kg	0.05	18.54	17.47
09	Zinc (as Zn)	APHA 3120B	mg/kg	0.02	19.13	**33.48**
10	Mercury (as Hg)	APHA 3120B	mg/kg	0.005	BDL	BDL
11	Phosphate (as P)	APHA 4500P-D	mg/kg	0.05	48.15	57.1
12	Sodium (as Na)	APHA 3120B	mg/kg	1.0	72.33	**105.54**
13	Potassium (as K)	APHA 3120B	mg/kg	1.0	748.66	438.87
14	Texture (Soil type)	NA	NA	NA	Loamy silt	Clay
15	Sodium absorption ratio	By Calculation	NA	NA	0.22	**0.33**
16	Boron (as B)	APHA 3120B	mg/kg	0.05	16.74	16.1
17	Cation exchange capacity	IS:2720 (P-24)	Meq/100gm	NA	*0.14*	**0.31**
18	Chloride (as Cl)	APHA 4500Cl-B	mg/kg	2.0	*393.5*	489.8
19	Sulphate (as SO_4)	APHA $4500SO_4$-E	mg/kg	1.0	200	200
20	Fluoride (as F)	APHA 4500F-D	mg/kg	0.05	2.61	2.29

(continued)

Table 18.5 (continued)

Sr. No.	Parameters	Method	Unit	Detection limit	Sample id	
					S-2	S-5
21	Carbonates (as CO_3)	APHA 2320B	mg/kg	1.0	Nil	Nil
22	Cobalt (as Co)	APHA 3120B	mg/kg	0.005	2.55	4.57
23	Molybdenum (as Mo)	APHA 3120B	mg/kg	0.05	BDL	BDL
24	Calcium (as Ca)	APHA 3120B	mg/kg	1.0	4433.6	4650.92

Conclusion
1: Bold terms shows the parameter's maximum concentration among all soil samples
2: Italic terms shows the parameter's minimum concentration among all soil samples
3.Both sites have low concentrations of trace elements

18.11 Case Study on Surface and Ground Water Quality Testing

This case study is based on IITR (2022). The following water quality parameters were considered by Chaube (2022).

Water Physical Parameters (pH, conductivity, TDS), chemical parameters (Alkalinity, hardness, Cl, SO_4, Na, K, Ca, Mg, Silica, Oil, and Grease), heavy metals (As, B, Hg, Cd, Co, Cr(6), Total Cr, Cu, Zn, Se, Fe, Al, Mn, Ni, Mo, Pb), Fluorides (One season).

The study area consists of water bodies and cultivated soils near village in District Bilaspur of Chhattisgarh state, India). Water samples were examined for physico-chemical parameters and heavy metals. The samples were analyzed as per the procedures specified in 'Standard Methods for the Examination of Water and Wastewater' published by the American Public Health Association (APHA) and Bureau of Indian Standard Code.

18.11.1 Methodology of Water Sampling and Testing

Grab sampling in a plastic bag was carried out. For each parameter, the required sample size and storage/preservation requirement are indicated in Table 18.6. The methodology for sample collection and preservation techniques were followed as per the Indian Standard IS:3025(P-01). Test methodology and detection limits for each parameter are as per Indian Standard IS:3025 (P-02, P-11, P-14, P-16, P-21, P-24, P-32, P-40, P-46 etc).

Table 18.6 Water sample size and preservation

S. No.	Parameters	Sample size	Storage/preservation
01	**Physical**		
I	pH	50 ml	At site analysis
II	Electric conductivity	50 ml	At site analysis
02	**Chemical**		
I	Total dissolved solids	100 ml	By refrigeration, can be stored for 7 Days
II	Total hardness	100 ml	Add HNO3 (pH < 2), Refrigeration, 6 month
III	Chloride,	50 ml	Not required, 1Month
IV	Sulphate	100 ml	Refrigeration, 1Month
V	Alkalinity	100 ml	Refrigeration, 1Month
VI	Oil and grease	500 ml	Acidification (pH < 2.0), keep in dark for 1 month
VII	Silica	100 ml	Not Required. Keep in dark
VIII	Hexavalent chromium	50 ml	Refrigeration for 24 h
IX	Fluoride	300 ml	Keep sample one month with neutral pH
3	**Metal**		
I	Heavy Metal (Fe, Cu, Mn, Cr, Zn, Cd, Pb, Hg, Na, Mo, Ni, Co, K, Ca etc.)	200 ml	Acidification (pH < 2.0) with HNO_3, keep in dark for 1 month

18.11.2 Water Sampling Locations

Water samples were collected in October 2021 from 6 locations, as shown in Table 18.7.

Table 18.7 Canal and ground water sampling locations

Site No	Sample id	Photograph id(Annexure 4.II)
1	SW-1	A1
2	SW-2	A2
3	SW-4	A4
4	GW-2	B2
5	GW-5	B5
6	GW-7	B7

18.11.3 Surface Water Quality Test Results

Test results are shown in Table 18.8. The Bureau of Indian Standard Code IS:10,500:2012 has specified **Acceptable** and **Permissible** limits for water sample quality as follows:

Acceptable Limit: It is recommended that the acceptable limit is to be implemented. The values in excess of those mentioned under 'acceptable' render the water unsuitable but may still be tolerated.

Permissible Limit: The permissible limits indicate that water can be used up to the permissible use in the absence of an alternate source, above which the sources will have to be rejected.

(a) The results indicate the pH values in the range of 7.49–7.93. The minimum value was observed at SW1 (Kurung Left Main Canal near village Gataura), and the maximum value at SW5, and both are well within the specified standard range of 6.5–8.5.
(b) All the above surface water samples at different sites have parameters well within the acceptable limit, except at SW-4 (Rank Village Pond). At the site, SW-4 (Rank Village Pond), Copper and Aluminum were within permissible limits.

18.11.4 Ground Water Quality Test Results

The analysis results indicate that

1. pH ranges from 7.24 to 8.00; the minimum pH of 7.24 was recorded at GW9, and maximum at GW5. All samples have pH values well within the standard of 6.5–8.5.
2. Total hardness was observed to be ranging from 204 to 660 mg/l. All samples had total hardness within the permissible limit.
3. Chlorides range from 25 mg/l to 177 mg/l.
4. Total Dissolved Solids (TDS) concentrations range from 304 to 1012 mg/l.
5. In groundwater, GW-1 site has TDS, Total Hardness, Calcium and Total Alkalinity is within the permissible limit.
6. In groundwater, sites GW-2 and GW-3 have TDS, Total Hardness, Calcium, Magnesium and Total Alkalinity in the Permissible limit.
7. In groundwater, sites GW-4 and GW-5 have Total Hardness within the permissible limit.
8. In groundwater, site GW-6 has Total Hardness is within the permissible limit.
9. In groundwater, site GW-7 has TDS, Total Hardness, Magnesium, and Total Alkalinity is the Permissible limit, but Manganese is out of the limit.
10. In groundwater, site GW-8) has TDS, Total Hardness, Total Alkalinity and Boron in the Permissible limit.

Table 18.8 Surface and ground water quality results

S. No.	Parameters	Unit	Detection limit	Acceptable Limit	Permissible limit in the absence of an alternate source	Results					
						SW-1	SW-2	SW-4	GW-2	GW-5	GW-7
01	pH Value	NA	NA	6.5–8.5	No Relaxation	7.49	7.67	7.66	7.37	8.00	7.70
02	Total Dissolved Solids	mg/L	5.0	500	2000	222	86	158	536	304	522
03	Total Hardness (as $CaCO_3$)	mg/L	5.0	200	600	136	43	89	466	209	325
04	Calcium (as Ca)	mg/L	1.0	75.0	200	33	12	23	120	53	74
05	Magnesium(as Mg)	mg/L	1.0	30.0	100	20	3	8	40	19	34
06	Chloride (as Cl)	mg/L	2.0	250	1000	30	12	16	69	30	39
07	Sulphate (as SO_4)	mg/L	1.0	200	400	4	BDL	5	20	8.0	12
08	Iron (as Fe)	mg/L	0.01	1.0	No Relaxation	0.014	BDL	0.21	0.02	0.03	0.12
09	Copper (as Cu)	mg/L	0.01	0.05	1.5	BDL	BDL	0.06	**BDL**	BDL	0.04
10	Manganese (as Mn)	mg/L	0.005	0.1	0.3	BDL	BDL	0.08	0.03	BDL	0.89
11	Fluoride (as F)	mg/L	0.05	1.0	1.5	0.26	0.20	0.22	0.38	0.36	0.44
12	Mercury (as Hg)	mg/L	0.0005	0.001	No Relaxation	BDL	BDL	BDL	BDL	BDL	BDL
13	Cadmium (as Cd)	mg/L	0.002	0.003	No Relaxation	BDL	BDL	BDL	BDL	BDL	BDL
14	Selenium (as Se)	mg/L	0.002	0.01	No Relaxation	BDL	BDL	BDL	BDL	BDL	BDL
15	Arsenic (as As)	mg/L	0.005	0.01	No Relaxation	BDL	BDL	BDL	BDL	BDL	BDL
16	Lead (as Pb)	mg/L	0.002	0.01	No Relaxation	BDL	BDL	BDL	BDL	BDL	BDL
17	Zinc (as Zn)	mg/L	0.1	5.0	15	BDL	BDL	BDL	BDL	BDL	0.21
18	Total Alkalinity (as $CaCO_3$)	mg/L	1.0	200	600	106	55	93	306	142	153
19	Aluminium (as Al)	mg/L	0.005	0.03	0.2	BDL	BDL	0.07	**BDL**	BDL	0.03

(continued)

Table 18.8 (continued)

S. No.	Parameters	Unit	Detection limit	Acceptable Limit	Permissible limit in the absence of an alternate source	Results					
						SW-1	SW-2	SW-4	GW-2	GW-5	GW-7
20	Boron (as B)	mg/L	0.1	0.5	2.4	0.13	BDL	BDL	BDL	0.14	0.79
21	Molybdenum (as Mo)	mg/L	0.01	0.07	No Relaxation	BDL	BDL	BDL	BDL	BDL	BDL
22	Nickel (as Ni)	mg/L	0.005	0.02	No Relaxation	BDL	BDL	BDL	BDL	BDL	BDL
23	Total Chromium (as Cr)	mg/L	0.005	0.05	No Relaxation	BDL	BDL	BDL	BDL	BDL	BDL
24	Cobalt (as Co)	mg/L	0.05	NA	NA	BDL	BDL	BDL	BDL	BDL	BDL
25	Oil and Grease	mg/L	3.0	NA	NA	< 3.0	< 3.0	< 3.0	< 3.0	< 3.0	< 3.0
26	Silica (as SiO_2)	mg/L	0.05	NA	NA	5.5	BDL	4.4	7.0	9.6	7.9
27	Conductivity	μS/cm	NA	NA	NA	361	140	255	856	490	844
28	Hexavalent Chromium (as Cr^{+6})	mg/L	0.05	NA	NA	BDL	BDL	BDL	BDL	BDL	BDL
29	Sodium (as Na)	mg/L	1.0	NA	NA	14.27	10.49	13.80	25.74	16.58	24.05
30	Potassium (as K)	mg/L	1.0	NA	NA	2.03	2.12	3.49	2.30	1.84	1.79

11. In groundwater, site GW-9 has TDS, Total Hardness, Calcium and Total Alkalinity within the Permissible limit.
12. In groundwater, site GW-10 has TDS, Total Hardness, Calcium, Magnesium Manganese and Total Alkalinity in the Permissible limit.

18.11.5 Conclusions

Water Quality Test Results are given in Table 18.8, along with the detection limit, acceptable limit, and acceptable limit. Summar and conclusions are given below.

Water quality has been interpreted considering acceptable and permissible limits defined below:

Acceptable Limit: It is recommended that the acceptable limit be implemented. The values in excess of those mentioned under 'acceptable' render the water not suitable but still may be tolerated.

Permissible Limit: The permissible limits indicate that water can be used up to the permissible use without an alternate source, above which the sources will have to be rejected.

Surface Water Quality Test Results

The pH values are within the standard range of 6.5–8.5.

All the surface water samples have parameters well within the acceptable limit, except at SW-4 (Rank Village Pond) where Copper and Aluminum are above the acceptable limit but still within the permissible limit.

Ground Water Quality Test Results

Most of the villages in the study area have hand pumps and dug wells, as most of these villagers use this water for drinking and other domestic purposes. The analysis results indicate that some water quality parameters exceed acceptable limits but are still within permissible limits, as follows:

1. All samples have pH values, total hardness, and chloride within the specified limits
2. Total Dissolved Solids (TDS) concentrations range from 304 to 1012 mg/l.
3. Site GW-2 (Handpump Water near Manja Gali Village- Hardadih) has TDS, Total Hardness, Calcium, Magnesium and Total Alkalinity above the acceptable limit but within the permissible limit.
4. Site GW-7 (Handpump Water near Dike-1 Village- Rank) has TDS, Total Hardness, Magnesium and Total Alkalinity in the Permissible limit, but Manganese is out of the limit.

Photograph B2 Photograph B5 Photograph B6

Questions

1. How does irrigation lead to soil salinity?
2. Define electrical conductivity, Sodium Adsorption Ratio, and Exchangeable Sodium Percentage.
3. Explain the relationship between

   a. Osmotic potential and electrical conductivity,
   b. Total soluble salts and electrical conductivity, and
   c. Exchangeable sodium ratio and sodium adsorption ratio.

4. Identify important cations and anions and indicate the usual range of these in irrigation water.
5. Discuss salinity hazard in terms of the electrical conductivity of irrigation water.
6. Irrigation application efficiency is 75%. The salinity of irrigation water is 5 ds/m. Seasonal evapotranspiration is 44 cm, and total seasonal rainfall is 6 cm. Compute the leaching requirement for obtaining 100% yield.
7. Explain the scope of conjunctive use of surface and groundwater in the management of salinity.
8. What is the meaning of acceptable and permissible limits on quality parameters?
9. Referring to the case study on soil quality testing in the a village area, comment on soil suitability for rice crop cultivation.

10. Referring to the case study on water quality testing in the a village area, comment on the suitability of surface water and groundwater for a).drinking and b) irrigation purposes.
11. Comment on the presence of heavy metals in agricultural soil, in surface water and in groundwater.

References

BIS IS 3025(part 1) : 1987 Methods of sampling for water and waste water, Bureau of Indian Standards, Manak Bhavan, 9 Bahadur Shah Zafar Marg, New Delhi.

BIS IS 3025(part 2) (2002) Determination of 33 elements by inductively coupled plasma atomic emission spectroscopy, Bureau of Indian Standards, Manak Bhavan, 9 Bahadur Shah Zafar Marg, New Delhi

BIS (2012) Code IS 10500: 2012 drinking water specification, Bureau of Indian Standards, Manak Bhavan, 9 Bahadur Shah Zafar Marg, New Delhi

BIS IS 1622:1981 Methods of sampling and microbiological examination of water (first revision), Bureau of Indian Standards, Manak Bhavan, 9 Bahadur Shah Zafar Marg, New Delhi

BIS IS 2720:Part 1 to Part 26: Method of Test for soils, Bureau of Indian Standards, Manak Bhavan, 9 Bahadur Shah Zafar Marg, New Delhi

Biswas TD, Mukherjee (1987) Textbook of soil science. Tata Mc Graw-Hill Publishing Company Ltd, New Delhi

Chaube UC (2008) Area drainage study of the proposed site for Gidderbaha Thermal power project. Unpublished Report of the Sponsored consultancy project, IIT Roorkee, Roorkee

Gupta PK, Khepar SD, Kaushal (1987) Conjunctive use approach for management of irrigated agriculture. J Agricult Eng XXIV(3):307–316

ICAR (2021) Hand book of agriculture published by Indian council of agricultural research, Krishi Anusandhan Bhavan, Pusa, New Delhi 110012

IITR (2022) Report of “consultancy work for drainage study (surface and sub-surface) around ash dykes of NTPC Sipat”. Indian Institute of Technology Roorkee, Roorkee

Michael AM (1990) Irrigation: theory and practice. Vikas Pubs. House Pvt. Ltd., New Delhi

Chapter 19
Soil Moisture and Its Measurement

Abstract Soil moisture measurements are needed to schedule irrigation (when, how, and how much), for experimental studies on crop water requirements, and to evaluate surface irrigation systems. The objectives of this chapter are to: (a) gain basic knowledge about soil moisture and its importance; (b) understand different methods and principles of measurement of soil moisture; and (c) exercise precaution in the use of different soil moisture instruments. This chapter compares various methods of measuring soil moisture based on the inputs required, cost, response time, and accuracy of results obtained. Different points must be considered while choosing the methods, such as application, resource availability, accuracy, calibration requirements, replicability, cost, and ease of using methods. Considering the pros and cons of individual methods, it can be inferred that the gravimetric method is the most accurate, simple, direct, and economical, but it is laborious, destructive, and does not allow repetitions at the same location. Excluding pressure plate apparatus, all the indirect methods provide immediate results.

19.1 Introduction

Soil moisture is defined as the moisture present in the topmost upper soil layer, i.e., water in the root zone depth. It is influenced by irrigation, crop type and stage of development, precipitation, temperature, groundwater contribution, and soil properties. Soil moisture Measurement is necessary for irrigation water management, evapotranspiration, soil erosion, hydrological modeling, floods, droughts, and hydrological investigations. Soil moisture measurements are needed to schedule irrigation (when, how, and how much), for experimental studies on crop water requirements, and to evaluate surface irrigation systems. Soil moisture measurements are essential in areas that receive less rainfall to assess, plan, and manage different moisture conservation practices. Soil moisture is mainly expressed by the amount of water present in a given amount or the stress or tension under which the soil holds water.

U. C. Chaube et al., *Canal Irrigation Systems in India*, Water Science and Technology Library 126, https://doi.org/10.1007/978-3-031-42812-8_19

19.2 Basic Concepts and Terminology Related to Soil Moisture

19.2.1 Soil Water

In general, pore spaces are partly filled with air and partly with soil water. Water is distributed around soil particles under cohesive and adhesive forces when rainfall or irrigation is applied to dry soil. Then, the air is displaced from the pore spaces filled with water.

The soil water is mainly classified under three classes, such as (a) hygroscopic water, (b) capillary water, and (c) gravitational water. Hygroscopic water is defined as water tightly held by adsorption forces to the surface of soil particles. The water present around soil particles and in the capillary spaces by the forces of surface tension and continuous films is called capillary water. Gravitational water is defined as water freely drained out from the soil under the influence of gravity.

19.2.2 Soil Water Content

The soil water content reflects the availability of water in the soil. It is mostly defined as a percentage of water by weight, a percentage of water by volume (mm of water depth present in the one-meter soil depth). The water content by weight is defined as the ratio of the weight of water in the soil upon the dry weight of the soil:

$$water\ content\ by\ weight = \frac{weight\ of\ water}{dry\ weight\ of\ soil}$$

The water content by volume is defined as the ratio of the volume of water in the soil upon the volume of the soil:

$$\begin{aligned}\text{water content by volume} &= \frac{\text{weight of water}}{\text{dry weight of soil}} * \text{bulk density}\\ &= \frac{\text{Volume of water}}{\text{Volume of soil}}\end{aligned}$$

19.2.3 Saturation Capacity

The soil pores are filled with water after rainfall has occurred or irrigation is applied. The soil is said to be at the saturation capacity or its maximum holding capacity when the entire pore space is filled with water. It means that there is no air left in

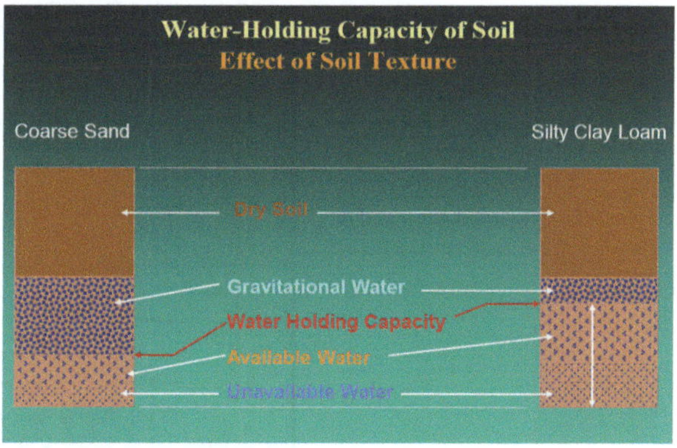

Fig. 19.1 Effect of soil texture on water holding capacity of soil

pore space. The plant needs both air and water for proper growth and development. Most crops, except rice, cannot stand saturated soil conditions beyond 2–5 days. The squeezing of saturated soil and the flow of muddy water is the most common practice to tell whether the soil is at saturation in the actual field condition.

19.2.4 Field Capacity (FC)

After the drainage of gravitational water, the amount of water that the soil against gravity can retain is known as field capacity. Soil is unsaturated, but still, it's quite wet. This situation usually exists 48–72 h after a significant irrigation or rainfall event. Field capacity is considered an ideal condition for crop growth. The macrospores are filled with air, and micropores are filled with air and water at field capacity. It is also known as the upper limit of available water and plant relations. The soil moisture tension at field capacity generally varies between 1/10 and 1/3 atm depending upon soil texture. It can depend upon several factors, such as soil layering, organic matter content, depth of wetting, and evapotranspiration. Also, it is considered the same throughout the growing season. The field capacity concept does not apply to soils that have shrinkage and swelling problems. The effect of Soil Texture on Water Holding Capacity of Soil is shown in Fig. 19.1.

19.2.5 Permanent Wilting Point (PWP)

It is defined as the moisture content at which a crop can no longer extract water to fulfill its crop water demand for growth and development and remain wilted unless water is added to the soil. At PWP, water available in the soil that plants cannot extract is known as hygroscopic water held by the adsorptive force. Most plants cannot recover from water stress, and finally, they die. It is also known as

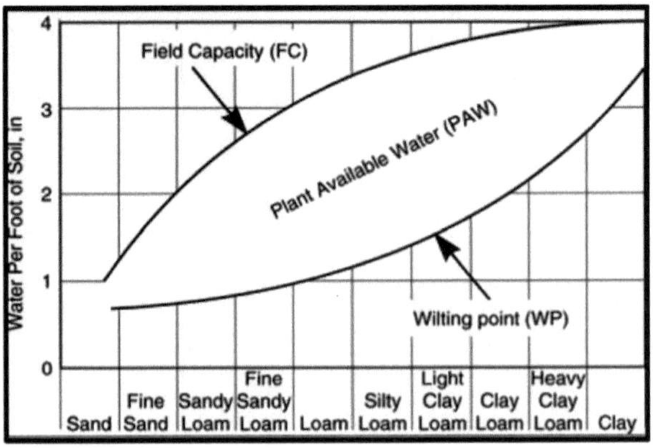

Fig. 19.2 Permanent wilting point of soil

a permanent wilting percentage and wilting coefficient. The soil moisture tension varies between 7 and 32 atm at the permanent wilting point. PWP is the plant's lower limit of available water (Fig. 19.2).

19.2.6 Available Water (AW)

The available water is expressed as the water retained by soil particles between field capacity and wilting point. It can also be defined as the water which is available for plants:

$$AW = (FC - PWP) \times D_{RZ}$$

where AW = the available water, FC = the field capacity in percent by volume, PWP = the permanent wilting point in percent by volume, and D_{rz} = the effective root zone depth.

19.2.7 Readily Available Water (RAW)

The readily available water is defined as the fraction of the available water which is readily available for plants to meet their need (Fig. 19.3). RAW is numerically considered as 50% of the available water.

$$RAW = 0.50 \times (FC - PWP) \times D_{RZ}$$

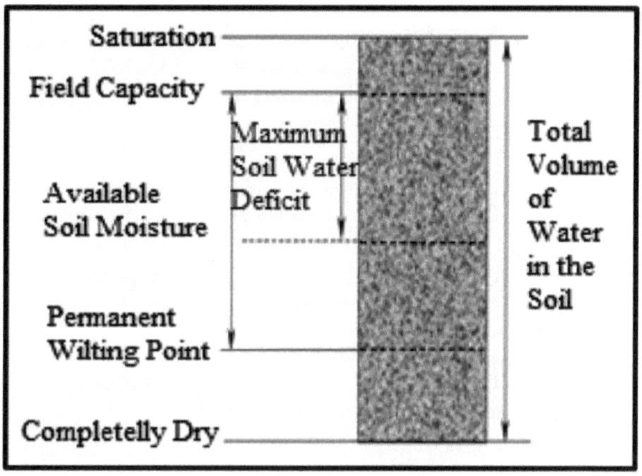

Fig. 19.3 Available water in the soil

19.2.8 Soil Water Potential

Soil water potential shows the energy status and describes the degree of firmness of attachment of water particles with soil. It is also known as soil tension.

19.2.9 Soil Porosity

It is defined as the ratio of the volume of pore spaces to the total volume:

$$\text{Porosity} = \frac{volume\, of\ pores\, space}{Total\ volume}$$

19.3 Soil Moisture Measurement Techniques

The soil moisture measurement techniques are classified under two categories: (i) direct methods and (ii) indirect methods. The soil moisture is calculated by indirect methods using the simple difference between soil sample weight before and after drying. In contrast, in indirect approaches, calculations are made by calibration against other measured factors (i.e., dielectric constant and electrical resistivity) that change with changing soil moisture content. Except for the gravimetric method, all other forms come under indirect methods. Several instruments are available commercially for measuring soil moisture. These instruments may not be accurate as the gravimetric method, but they are quick and less cumbersome. Each method has

Table 19.1 List of soil moisture measurement methods

Methods		Measured parameters	Criteria		
			Cost-effectiveness	Accuracy	Response time
Direct approach	Gravimetric method	Mass water content	Economical	High	24 h
Indirect approach	Tensiometers	Soil water potential	Economical	High	2–3 h
	Neutron probe	Volumetric soil moisture content	Expensive	High	1–2 min
	Time-domain reflectometer	Volumetric soil moisture content	Economical	High	Instantaneous (28 s)
	Capacitance and FDR	Volumetric soil moisture content	Expensive	High	Instantaneous
	Gamma-ray attenuation	Volumetric soil moisture content	Expensive	Low	Instantaneous (60 s)
	Gypsum block method	Soil moisture tension	Economical	Low	2–3 h
	Pressure plate method	Soil water potential	Expensive	Low	Soil dependent
	Feel and appearance	–	Economical	Low	1–2 min

its advantages and disadvantages. Some of the methods most commonly used are discussed below in Table 19.1.

19.3.1 Gravimetric/Oven Drying Method

The gravimetric method is the most accurate and standard. This method takes soil samples from desired depths at several locations in each soil type. Mainly soil samples are collected in airtight aluminum containers with a soil auger or soil sampler. The samples are weighed and dried at 105 °C for 24 h until all the moisture escapes. Then, the soil sample is removed, cooled at room temperature, and weighed again. The weight difference is the amount of water content in the soil, usually expressed as a dry weight percentage. More recently, infrared radiation is used to dry soil samples to reduce the oven drying time.

$$\text{Soil moisture}(\theta_\text{m}) = \text{Weight of water}(\text{Mw})/\text{Weight of dry soil}(\text{Ms})$$

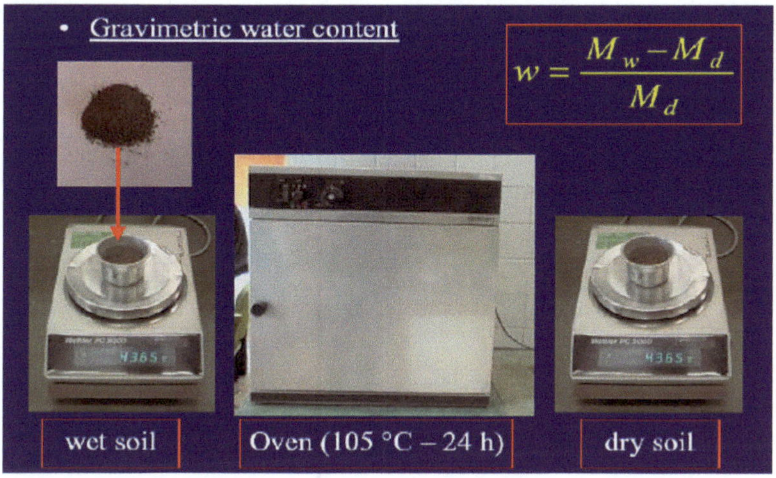

Fig. 19.4 Measurement of soil moisture by gravimetric method

This technique is quite simple, direct, cost-effective, accurate, and extensively used to measure moisture content, but it is time inefficient (at least 24 h), destructive, and difficult to use with rocky soils. It requires bulk density to convert moisture content from mass basis (θ_m) to volumetric basis (θ_v). Soils containing the maximum fraction of clay or organic matter can induce error (a considerable amount of adsorbed water in clay or at 105 °C oxidation of organic matter). The measurement of soil moisture by the gravimetric method is shown in Fig. 19.4.

Example

Weight of moist soil sample with container = 200 g.
Weight of dry soil sample with container = 165 g.
Container weight = 20 g.
Weight of moist soil (Mw) = 200 – 20 = 180 g.
Weight of dry soil (Ms) = 165 – 20 = 145 g.
Weight of water (Mw) = 180 – 145 = 35 g.
Therefore, moisture content (θ_m) = (35/145) × 100 = 24.14%

19.3.2 Tensiometers

A tensiometer directly measures the capillary or moisture potential with which soils hold water (Fig. 19.5). It can also be used for soil moisture estimation. The tensiometer consists of a porous cup containing water and a continuous water column to a vacuum measuring device, either a dial-type gauge or a mercury manometer. The cup is positioned in the soil at the desired depth. Because of the cup's porous nature, an equilibrium state is established between the water inside the cup and the water in

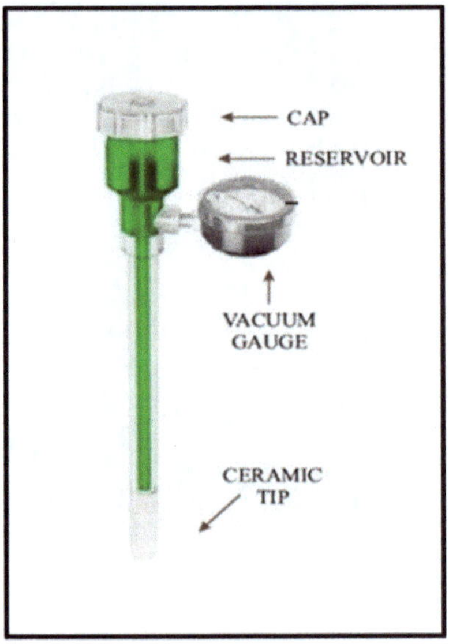

Fig. 19.5 Measurement of soil moisture tension by tensiometers

the soil outside. The water moves out or in the cup depending upon the soil water's tension increase or decrease. Fluctuations of the soil moisture tension are read above the ground on the vacuum indicator.

It is an indirect method as soil water is associated with soil water pressure potential (soil moisture characteristics curve). This technique is non-destructive and economical. It is capable of determining the moisture distribution for both saturated and unsaturated conditions. The equipment can be used for a long time if maintained properly. It provides continuous measurements of soil moisture without any disturbance to the soil. The usage of this instrument is obsolete now days due to high maintenance requirements. The tensiometers exhibit practical effects in the range of 0–0.85 bar of tension (a constraint on medium- and fine-textured soils), because the gauge will fail when air enters the ceramic tip, and the water in the tube separates.

19.3.3 Time Domain Reflectometry

Time-domain reflectometry (TDR) is a method that is launched along a waveguide formed by a pair of parallel rods embedded in the soil to determine the dielectric constant of the soil by monitoring the travel of an electromagnetic pulse.

The propagation velocity (v) = 2 l/t, where l is the length of the probe, and t is the travel time. The propagation velocity varies with moisture content due to the relatively large dielectric values of water. The accuracy of measurement is influenced by probe length.

Fig. 19.6 Measurement of soil moisture using TDR

The pulse is reflected at the end of the waveguide and its propagation velocity, which is inversely proportional to the square root of the dielectric constant. A lower propagation velocity shows wetter soil and vice-versa. Generally, the radius of influence is 30 cm. The most widely used relation between soil dielectrics and soil water content is given by Topp's equation as follows:

$$\theta_v = -0.053 + 0.029\varepsilon_b - 5.5 \times 10^{-4}\varepsilon_b^2 + 4.3 \times 10^{-6}\varepsilon_b^3$$

where

ε_b = the dielectric constant of the soil–water system, and
θ_v = the volumetric soil moisture content

TDR is portable, easy, and safe to operate (Fig. 19.6). It is a relatively less labor-intensive and non-destructive technique. It provides reliable volumetric moisture content within a short time and does not need any soil-specific calibration. TDR simultaneously measures several physical parameters, such as volumetric moisture content with ± 1% error, soil temperature, and soil electrical conductivity. The applicability of TDR in saline soil is limited.

19.3.4 Capacitance and Frequency Domain Reflectometry

Frequency Domain Reflectometry (FDR) is used to estimate the volumetric moisture content in the same way as TDR (Fig. 19.7). FDR works based on the variation of

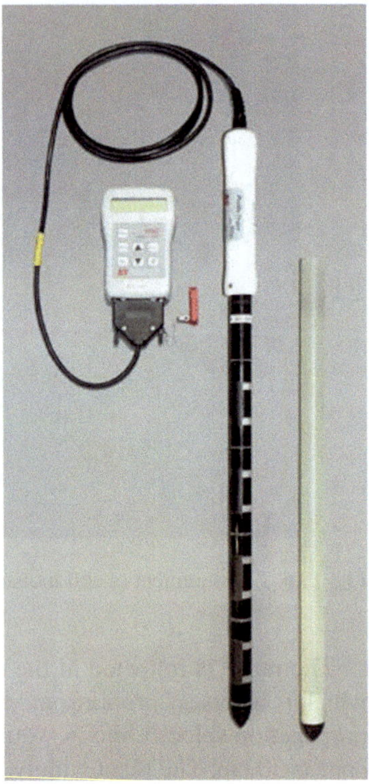

Fig. 19.7 Measurement of soil moisture using capacitance probe

frequency of a signal with dielectric properties of the soil. These techniques provide a reasonably accurate result with an accuracy of $\pm$ 0.01 ft^3/ft^3, but it needs site-specific calibrations. FDR is sensitive to air gaps, soil salinity, soil temperature, bulk density, and clay content and is restricted to the 1.6-inch radius of influence.

19.3.5 Gamma Ray Attenuation

This technique is used to determine the soil moisture content in the topmost soil layer (up to 1–2 cm). It is a radioactive method based on the assumption that diffusion and gamma rays absorption are correlated with the density of matter in its path. The gamma-ray attenuation method is unaffected by moisture in the soil and capable of estimating depth-wise moisture. It can be used to monitor the temporal change in soil moisture. It is a non-destructive technique but has a high installation cost and is difficult to use. It has been noticed that in highly stratified soils, these instruments show a large variation in bulk density and moisture content.

19.3.6 Gypsum Block Method

A gypsum block is an electrochemical cell with a saturated calcium sulfate solution that is placed at root zone depths and serves as an electrolyte. A small AC voltage is applied using a bridge circuit to determine electrode resistance. Gypsum blocks are buffers against changing soil salinity, as soil's electrical conductivity can significantly influence output. Electrical conductance is measured using the electrodes in the blocks, and the blocks absorb soil moisture. The technique requires minimal maintenance, is simple and economical, and uses a large volume (up to 8 cm radius), which is one of the most significant advantages in field studies. This technique's major limitation is the degradation and dissolution of the gypsum block and the need for recalibration with time. Salt content and soil temperature can affect the gypsum block, which shows an error of $\pm$ 0.01 ft^3/ft^3 in moisture estimation. This technique is not suitable where drainage is fast (sandy soils).

19.3.7 Pressure Plate Method

The pressure plate apparatus (Fig. 19.8) is mainly used to develop soil moisture characteristics curves (moisture content at different pressures) and to estimate soil's field capacity and wilting point. It is a laboratory method. The apparatus consists of two air-tight metallic chambers; first one is used for estimating field capacity, and the second one for wilting point with porous ceramic plates. To develop the soil moisture characteristic curve, first, saturate the ceramic plate and then put the soil sample on these plates. Both disturbed, and undisturbed soil samples can be used in the experiment. Then, soil samples are also saturated along with plates and are transferred to the metallic chamber. Special wrenches are used to tighten the nuts and bolts to seal and make it an airtight chamber. The required pressure is supplied and maintained through a compressor. The water starts to flow from the metallic chamber and continues to trickle till the equilibrium against applied pressure is achieved. After that, soil samples are taken out and oven-dried to determine the moisture content at 105 °C for 24 h. Similarly, the moisture content is determined for different pressure values (1/10, 1/3, 1, 3, 5, 7, 10 and 15), then a graph is plotted between moisture against pressure values.

19.3.8 Feel and Appearance

It is one of the oldest methods for determining soil moisture and has visual observation and feel of the soil, while the accuracy of judgment improves with experience. This method is only an estimate and lacks a scientific basis. Accurate measurement is not

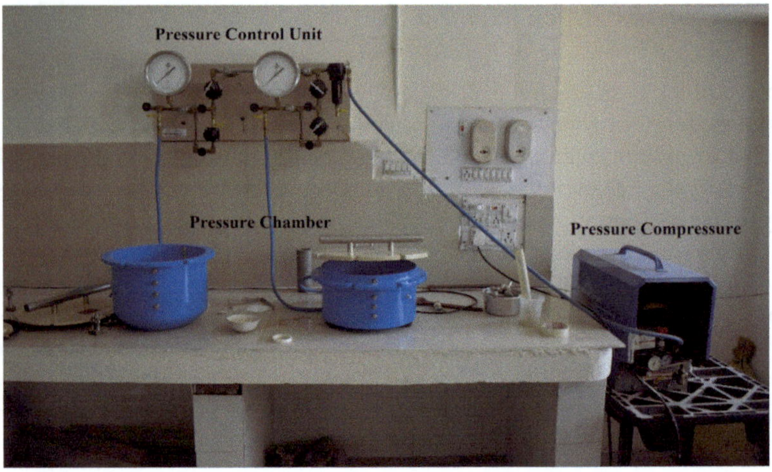

Fig. 19.8 Pressure plate apparatus

possible, but the method is an art developed over time with extensive use. Most of the farmers mainly adopt the feel and appearance method to irrigate their field through their experience.

In this method, first, soil samples are collected by the soil auger from different root zone depths. Then, with the help of a hand, formed soil into small balls. Based on the ball characteristics of squeezing, pressing, or appearance gives a rough idea about the soil moisture content in different layers of the soil. If, on squeezing, no free water appears on the soil, but the wet outline is left on the hand, soil moisture is considered field capacity.

This method is simple, indirect, and low-cost but requires huge experience and is not highly accurate.

Questions

1. Write down the factors which affect soil moisture under unsaturated conditions.
2. What are the principles and limitations of tensiometers?
3. What are the different applications of pressure plate apparatus?
4. Briefly explain the basic principles of neutron probes.
5. What are the different drawbacks of electrical resistance block?
6. What is the principal's method to express the soil moisture?
7. Explain the following terms:
8. Field capacity
9. Wilting point
10. Available water

11. Bulk density
12. Soil moisture characteristic curve
13. Particle density.

Chapter 20
Rehabilitation and Modernization

Abstract This chapter provides a brief study of the maintenance, rehabilitation, and modernization of the project. Maintenance is required to keep the project in working order and fulfilling its objectives, whereas rehabilitation is the process of renovating an existing project to meet its original objectives. Modernization is related to enhanced technical, social, or economic objectives. Possible deficiencies related to engineering and agronomic aspects in an existing project are stated. Rehabilitation is needed to remove these deficiencies. An existing project may undergo modernization by improvements in structural measures (canal lining, canal structures, extending irrigation service area, and drainage) and non-structural measures (conjunctive use management, improving organization structure, legal acts, participatory management, changing cropping patterns etc.). Due to a variety of constraints, it may not be possible to include all these measures in the rehabilitation and modernization (R&M) programme. The constraints could be related to time availability, finance availability, social feasibility, and site specific techno economic feasibility and so on. Based on experience, the relative importance of these measures for R&M is indicated. The Upper Ganga Canal, in its head reach, passes through difficult terrain. The UGC is being replaced by PUGC, and new canal structures are in this reach. This appendix explains various canal structures on old UGC and new PUGC. The need for modernization and replacement of old structures is explained. The discussion covers canal head works, silt ejector, cross regulator, super passage siphon, level crossing, and aqueduct. Cross-section, plan, layout, and schematic diagrams are provided as illustrations of the works.

20.1 Defining Maintenance, Rehabilitation, and Modernization

The definitions below indicate that rehabilitation and modernization include structural and non-structural measures (Chaube 2009).

Maintenance is the process of keeping irrigation, drainage, and any other infrastructural facilities in good repair and working order, fulfilling the objectives for which

U. C. Chaube et al., *Canal Irrigation Systems in India*, Water Science and Technology Library 126, https://doi.org/10.1007/978-3-031-42812-8_20

they were originally designed. It would also involve improvements of a relatively minor nature, which can be performed during the normal maintenance process.

Rehabilitation is the process of renovating an existing project whose performance fails to meet its original or enhanced technical, social, or economic objectives. It embraces improvements of the physical infrastructure, on-farm and production systems, operation, management, and institutional aspects, including policy measures influencing the overall project, which are designed and implemented to improve the economic and social benefits of the project.

Modernization is the process of updating and improving an existing project which otherwise is meeting its original objectives in order to meet enhanced technical, social, or economic objectives. It embraces changes to the physical infrastructure, on-farm and production systems, and all operation, management, and institutional aspects, including policy measures influencing the overall project, which are designed and implemented to enhance the economic and social benefits of the project.

20.2 Need for Rehabilitation/Modernization

During the past three decades or so, although performance evaluations of a number of irrigation systems in India have been carried out, these evaluations have not been comprehensive due to several reasons, including limited financial support. However, literature shows that the main deficiencies in irrigation projects are related to the following aspects (Chaube 2013).

20.2.1 Engineering Deficiencies

1. Reappraisal of the available surface and groundwater and return flow from irrigation.
2. Silting of reservoirs; and conducting sedimentation surveys and measures to prevent silting.
3. The insufficient number of hydrological and meteorological stations in catchment and command areas.
4. Excessive seepage and need for determining it. The lining of the canal is required.

The commonly accepted figures for transit losses in the alluvial plains of North India are 17% for main canals and branches, 8% for distributaries, and 20% for water courses which give a total loss of 45% of the water entering the canal head. Then, there are further losses in the field itself, and these have been estimated at 30% of the supply reaching the field, or 17% of the head discharge.

Thus, the total loss is 62% of water at the headgate, for a total project efficiency of 38%. With channel lining and good management, an overall project efficiency of

60% is economically attainable in India, or about 50% more land can be irrigated from the same initial water supply.

1. The tail reaches do not get enough irrigation water. There is over-irrigation in the head reaches of the distribution network.
2. Absence of conjunctive use of surface water and groundwater.
3. Salinity in soils and groundwater.
4. Inadequate drainage system
5. Bad maintenance of canal system. Untrained staff, limited maintenance grants.
6. Improper operation of reservoir and canal system. Decisions are ad-hoc and are not based on the systems approach.
7. The canal cannot carry design discharge.
8. Insufficient canal structures and improper maintenance.
9. Lack of communication facilities in the command.
10. Lack of field channels and proper maintenance.
11. Improper water management.
12. Waterlogging in irrigated areas.

20.2.2 Agronomy Related Deficiencies

1. Improper cropping calendar and cropping patterns.
2. Lack of research to determine the water requirement of crops.
3. Lack of research to determine suitable types of crops to suit the soils in command.
4. Poor extension services, lack of pilot projects, demonstration farms, etc.
5. Lack of detailed soil surveys of the command area.
6. Excessive application of irrigation water to crops.

An irrigation scheme can only function properly if design, construction, operation, and maintenance are adequate. A default in any of these phases may require rehabilitation work. Inadequate operation and maintenance is perhaps the most frequent cause of the need for rehabilitation, especially in developing countries. In particular, insufficient maintenance may render an irrigation scheme completely obsolete and unable to meet the project objectives.

Moreover, an irrigation scheme is not just a technical facility but a complex socio-technical mechanism compatible with environmental aspects.

A top-down approach has the indisputable advantage of speeding up the decision-making process as well as design and construction. However, the top-down approach emphasizes the technological component and disregards the human factor. Moreover, large schemes fit better the investment criteria of funding agencies.

A bottom-up approach can take into account the historical and human factors, such as the average cultural standards of the rural population, its readiness to learn new techniques, the collectivism existing in the rural community, and especially the endurance of sometimes very old traditions. However, a pure bottom-up approach can afford satisfactory results only in projects composed of many small-scale systems.

It cannot easily face the problems of large irrigation schemes, including design, construction, supervision, and operation and maintenance aspects of major structures and canal systems.

20.3 Components Requiring Improvements

The objective of the modernization of canal irrigation is to bring a significant increase in crop production per unit of available water and land at an economic cost. This requires improvements of the following significant components. The proposed improvements must be preceded by a diagnostic field study to find the values and constraints of the existing project.

20.3.1 Canal Lining

High-yielding varieties of crops need an assured water supply. India has limited water resources, so conservation of the resource by lining the conveyance system is necessary. Therefore, it is necessary not to confine lining only up to the canal and distributary system, but to extend it in the canal commands in blocks of up to 8 hectares so as to contain seepage losses on average of 35% in the fields itself beyond the outlets.

20.3.2 Conjunctive Use

The availability of groundwater in the command areas of existing surface irrigation projects has increased over the years. The use of groundwater, canal irrigation, and field drainage should be planned properly. Exploitation of groundwater resources requires careful planning so that the groundwater table is not lowered beyond the economic pumpage limit. Power is usually a big constraint, and its reliability needs to be improved while planning for the use of groundwater. The means of making combined or conjunctive use of surface water and groundwater are as follows:

(1) Irrigation of pockets exclusively with groundwater in a canal command, especially where the terrain is uneven.
(2) Augmentation of canal water by putting tube wells along the canal.
(3) Conjunctive use of groundwater during the period of low canal supply or canal closures.

20.3.3 Modernization of Structures

Modernization of canal structures, viz. head regulators, cross regulator, cross drainage, works, falls, bridges, escapes etc., will lead to the following benefits of the canal system:

1. Capability to pass designed discharge at every point in the canal system.
2. Replacement of old obsolete plank-controlled regulators by steel gated regulators will ensure better operational efficiency and saving of water.
3. Development of the capability to remove sediment in the canal system so as to maintain the uniform capacity of big as well as small channels in the monsoon season.
4. The introduction of water measuring devices will ensure the timely and equitable distribution of supplies and will be useful for the evaluation of distribution efficiencies.
5. It would be possible to adopt a user-oriented canal operation policy.

20.3.4 Remodeling and Construction of Additional Escapes

It is necessary to provide adequate escapes in the major canal in suitable locations to enable canals to run with full discharge even during the monsoon season without any danger of a breach in the case of suddenly no demand due to rainfall. The escaping capacity on a major canal system has got to be at least 75% of the head discharge of a canal. The provision of new escapes for remodelling of the existing inadequate escapes has to be done on the above principle.

20.3.5 Improvement of Drainage in the Command

The drainage system in the canal command has got to be reshaped drastically, with the introduction of a large number of artificial drains and the construction of link drains to improve the capacity of natural drainage. Constructing carriers and linking drains up to natural drainage will also be necessary to control canal seepage effectively.

20.3.6 Improvement of Tele-Communication on Canal Systems

Communication is an important factor in the efficient running of the canal system. The communication system needs modernization. A dependable communication system is a prerequisite to improve the canal system operation and management efficiencies.

20.3.7 Canal Service Roads

The canal service roads are generally earthen (kaccha) and are unserviceable, particularly in the monsoon season. At present, motorized transport needs the metalling of canal service roads to render all-weather communication, leading to efficient inspection and better operation and maintenance of works.

20.3.8 Engineering Infrastructure

The existing status of engineering infrastructure, such as communication systems, inspection houses, residential and non-residential buildings, vehicle facilities, etc., is inadequate to meet the present-day requirement of effective management and operational efficiency. These have to be adequate.

20.3.9 On-Farm Development Works

To ensure equitable and reliable water supply distribution, some of the essential features of the modernization of On-Farm Development (OFD) works are as below:

(i) **The Lining of Water Courses**:

Lining up to 8-hectare blocks in the command will save up to 35% of water in the field, thereby ensuring the uniform availability of water to the farmers. The additional cost of lining water courses will be balanced by the value of water saved and the number of other indirect benefits accruing to the farmers.

(ii) **Land Leveling:**

This will make possible efficient and effective utilisation of irrigation in the command. A detailed survey of the areas, preparation of contour maps, and other relative activities should be undertaken simultaneously with implementing an Irrigation Project.

(iii) **Farm Drainage:**

The programme will entail the construction of adequate artificial drains linked to natural drainage in the canal command. The farm drainage will not allow surface water to accumulate in fields to retard crop growth.

(iv) **Field Application Methods:**

Introduction of modern field's application methods like Sprinkler Irrigation should be examined as per countryside conditions, type of crop, and soil to reduce water wastage and bring more land under irrigation within the limited available supplies.

(v) **Outlets:**

The outlets have to be so designed that these cannot be easily tampered with or manipulated by the cultivators to ensure the equitable distribution of water to the farmers. The Adjustable Proportionate Module (A.P.M.) has been used in Haryana and Punjab and is functioning satisfactorily.

(vi) **Credit and Marketing Facilities:**

Intensive farming places a great demand on the resources of farmers. Appropriate Institutional Finance Framework in the cooperative sector has to be put up to meet the enhanced requirement of irrigated farming. Similarly, relevant agencies should be set up in cooperative sectors to handle the increased productivity of farmers and to ensure remunerative price to them.

20.3.10 Training

Modern irrigation has developed into a complex multi-disciplinary technology, involving engineering hydrology, agronomy, agricultural, economic, and socio-economic and management sciences. At present, irrigation engineers lack experience working with an interdisciplinary approach to identify problems in the ongoing systems and develop appropriate remedies. The project manager should have adequate knowledge of improved varieties of seeds, balance, and adequate use of fertilizers, use of pesticides and micro-nutrients, improved water application methods and transfer of technology from laboratory to field to ensure maximum agricultural produce and boost agricultural economy. Therefore, Irrigation Engineers must be given multi-disciplinary training in Modern Irrigated Agriculture Management before being inducted into regular service.

20.3.11 Culturable Command Area

Generally, it is found that the actual CCA below the outlets is lower than what was envisaged in the original project. Depending upon the topography, a detailed contour map of the project command with contour intervals of 0.25 m, should be prepared and the project command reassessed.

20.3.12 Crop Planning

Trends in cropping patterns indicate that farmers switch to cash crops under irrigated agriculture. Crop planning must be realistic, reflecting the aspirations of farmers as well as a strategy to bring the required changes in the cropping pattern. Accurate knowledge of crop behavior and crop water requirements is necessary.

20.3.13 Economic Viability

Modernization projects must envisage achieving yields 2 to 3 times the yield under average irrigated conditions. The cost per unit of additional water saved or provided must be less than the cost per unit of irrigation water for a new project.

20.3.14 Staff

The staff of the irrigation department should be capable of monitoring the construction programme, particularly for OFD works. Much of the success of a modernization project depends upon management efficiency during the operation stage. The jurisdiction of staff should be as given below:

Section officer	2000–3000 ha
Sub-divisional officer	8000–12,000 ha
Executive engineer	40,000–60,000 ha
Superintending Engineer	200,000 ha

20.4 Relative Importance of Measures During Rehabilitation and Modernization

There are a large number of structural and non-structural measures that could possibly be considered for the rehabilitation and modernization (R&M) of the following components of an existing irrigation project (Varshney 1992):

(i) Water conveyance and distribution network
(ii) On-farm irrigation
(iii) Drainage
(iv) Operation and management
(v) Agricultural aspects.

Due to various constraints, it may not be possible to include all these measures in the rehabilitation and modernization (R&M) programme. The constraints can be related to time availability, finance availability, social feasibility, site-specific techno-economic feasibility, and so on.

Table 20.1 Conveyance and distribution network—relative importance between rehabilitation and modernization

Items	Rehabilitation	Modernization
	Relative importance	Relative importance
Canal losses	4	3
Canal lining	4	3
Canal remodeling	3	2
Control/regulation structures	4	4
Water measurement	4	5
Remodeling the distribution system	3	4
Delivery scheduling	2	4
Automation	1	2
Pumping stations	2	4
Conjunction use of ground and surface water	1	2
Operation and maintenance	4	4

Scale: 1—low importance to 5—high importance

20.4.1 Conveyance and Distribution Network

The rehabilitation and modernization of conveyance and distribution systems always have structural components. In many cases, work is only directed toward the rehabilitation of channels. Other matters of great interest or concern to management are:

- The adoption of new delivery scheduling methods;
- The automation of conveyance and distribution networks, including the automation and remote control of diversion, regulation, and delivery structures and the use of computers in the operation of systems; and
- The automation of the pumping station.

The relative importance of different items is given in Table 20.1, which shows that there are other priorities of important questions between rehabilitation and modernization.

20.4.2 On-Farm Irrigation

In the past, importance has been placed on the R&M of the conveyance and distribution systems with too little emphasis on the R&M of the on-farm irrigation systems. The information given in Table 20.2 shows that there is now an awareness of the very high importance which should be given to on-farm irrigation in the R&M process.

Table 20.2 On-farm irrigation-relative importance between rehabilitation and modernization

Items	Rehabilitation	Modernization
	Relative importance	Relative importance
Proper irrigation techniques	4	5
Changing the irrigation methods	3	3
Coupling irrigation techniques with agricultural practices	4	4
Land reclamation	3	4
Improvement of soil use	4	5
Soil management	4	5
Land levelling	3	3
Irrigation scheduling	4	4
Improvement of irrigation efficiencies	5	5
Automation	1	2
Farm management	3	3

Scale: 1—low importance to 5—high importance

20.4.3 Drainage

The need to rehabilitate or modernize irrigation projects often arises from drainage and salinity problems. Concepts and policies and their relative importance are given in Table 20.3.

Table 20.3 Drainage-relative importance between rehabilitation and modernization

Items	Rehabilitation	Modernization
	Relative importance	Relative importance
Water table/groundwater control	3	2
Waterlogging/surface water	4	3
Disposal drainage systems	5	3
On farm drainage	4	3
Completion of surface water drainage systems	2	3
Operation and maintenance	5	4
Salinity related to the use saline waters	2	1
Salinity arising from excess irrigation losses at farm level	2	2
Salinity arising from water losses in the irrigation distribution system	2	2
Management of saline and alkali soils	3	2
Coupling irrigation and drainage	4	3

Scale: 1—low importance to 5—high importance

Table 20.4 Operation and management-relative importance between rehabilitation and modernization

Items	Rehabilitation	Modernization
	Relative importance	Relative importance
Operation and management systems	5	3
Rural infrastructure (roads, energy, workshops, storage, facilities, etc.)	4	3
Monitoring and evaluation	4	3
Training facilities	3	3
Farmer participation	5	3
Farmer education	4	3
Coupling water management services with extension services	4	4
Social administration and communication	4	4
Policies on water/irrigation management	4	4
Policies on water prices	4	5
Available technologies	3	3
Research on irrigation/water management	4	5

Scale: 1—low importance to 5—high importance

20.4.4 Operation and Management

The need to consider non-structural measures in rehabilitation and modernization is shown in Table 20.4.

20.4.5 Agricultural Aspects

The rehabilitation and modernization of irrigation and drainage are essentially improvements in the infrastructure for the benefit of agricultural production. The relative importance of agricultural aspects is shown in Table 20.5.

20.5 Upper Ganga Canal Modernization Project

The major cross-drainage works constructed with thumb rule designs had outlived their lives, which have been replaced without closing the irrigation water supply. Risk aversion of old canal system is one of the major components of the modernization project. Risk aversion works comprise replacing four major cross drainage works

Table 20.5 Agricultural aspects-relative importance between rehabilitation and modernization

Items	Rehabilitation	Modernization
	Relative importance	Relative importance
Appropriate technologies	5	4
Extension	5	4
Support by research and experimentation	5	5
Land tenure/land consolidation	3	3
Improvement of farm structure	5	4
Agricultural policies	5	5
Credit policies	4	4
Social and cultural aspects (family, women, etc.)	5	3

Scale: 1—low importance to 5—high importance

on the main canal and constructing a lined parallel canal (PUGC) connecting these structures from 6 to 36 km.

In the modernization project, Ranipur super passage has been replaced by a siphon, Pathri super passage has been replaced by a new super passage, Dhanauri Level Crossing has been replaced by the Ratmau aqueduct, and the Solani aqueduct has been replaced by a new aqueduct.

Appendix provides details of canal works in the head reach of UGC and the Parallel UGC (PUGC) as part of the modernization project (Govt. of UP 1991).

Appendix: Modernization of Upper Ganga Canal Structures

Introduction

This Appendix briefly describes the canal structures in the head reach of the Upper Ganga Canal (UGC) and the Parallel Upper Ganga Canal (PUGC) modernization project.

Headworks of UGC

The UGC is fed through a link channel with a capacity of 14,500 cusec. The old supply channel can carry about 40,000 cusec during floods. Bagh–Rao drainage adds about 15,000 cusec to the old supply channel. The excess water and the bulk of the silt load coming into it is made to escape Hardwar dam opposite Har-ki-Pauri. The reach between Har-ki-Pauri and Mayapur, canal head regulator, has a width of about 400 m and length is about 2 km. In this reach, Lalta Road drainage has an estimated flood of 12,000 cusec also meets. This reach serves as a sediment trap.

Mayapur regulator is the head regulator of UGC. This regulator is used to escape excess water through the Mayapur escape and regulate supply into UGC. Therefore, PUGC offtake (under the modernization project) was possible only from UGC some distance downstream of Mayapur. UGC passes through urban built-up area from Mayapur to Kankhal (about 3 miles). A silt ejector is located in this reach, with its escape channel having an outfall in the Ganga. PUGC off-take with a cross regulator has been fixed at miles 3–4 of UGC, considering various problems relating to land acquisition, construction cost, and administration problems.

Old Upper Ganga Canal Structures

Canal

Initially, the Ganga Canal was designed to carry 6750 cusecs (191 cumecs). Its bed width was kept at 140 ft (42.70 m) and its water depth was 10 ft. (3.05 m). A bed slope of 1.5 ft (0.29 m) per mile was adopted. The area commanded by it was 14.75 lac acres (5.97 lac ha). The irrigation supply was restricted from 30 to 45% of the commanded area. Subsequently, the canal was remodeled, and it is now capable of passing 10,500 cusecs (297.20 cumecs). The system comprises 570 miles (912 km) of the main canal and branches and 3560 miles (5696 km) of distributaries and minors. The area statistics are as below:

1. Gross commanded area: 50.60 lacs acres (20.48 lac ha.)
2. Culturable commanded area: 39.34 lacs acres (15.92 lac ha.)
3. Area irrigable at present: 19.12 lacs acres (7.74 lac ha.)
4. Area irrigated in Kharif: 8.38 lacs acres (3.39 lac ha.)
5. Area irrigated in Rabi: 7.70 lacs acres (3.12 lac ha.)

Silt Ejector at 2.2 km

Ganga Canal got silted up in 1970 due to heavy silt load. Therefore, a decision was taken to construct a silt ejector at 2.2 km downstream of the Mayapur head regulator to eliminate excessive silt entering the canal. Regular silt observation studies are being carried out during the monsoon period to check the excessive silt entry into the main canal to avoid the recurrence of the Alakhananda tragedy 1970.

Fig. 20.1 Permanent head works of UGC

Inlets

The Ganga canal, from its head to mile 19 (30.4 km) passes through a difficult terrain. In the head reach of 3 miles (4.8 km), the canal was aligned to pass through a very deep cutting and three inlets were provided on the right bank of the canal at (i) Lunda Leniwala, (ii) Kankhal, and (iii) Jawalapur to pass the flow of local drainage into the canal during monsoon season (Fig. 20.1).

Ranipur Super Passage

At its 5th mile (8th km), a hill torrent Ranipur Rao crosses the canal over the super-passage having a waterway of 196 ft (59.74 m). It carries a maximum flood flow of 28,000 cusecs (793 cumecs). Its bed consists of coarse sand of 0.45 mm diameter mixed with shingle. Its initial bed slope was 15.5 ft (2.95 m/km) per mile above and below the super-passage. Consequent to the operation of this super-passage since 1845, the bed slope of the torrent has flattened to 14 ft (2.67 m/km) per mile above the super-passage and steepened to 18 ft. per mile (3.43 m/km) below it, but it has not disturbed the regime of the torrent to induce any excessive scouring near the structure. With the utilization of the 8' (2.44 m) fall at Ranipur, in the canal, in Pathri Power House, the super-passage is now functioning as a canal siphon. The structure is functioning satisfactorily.

Pathri Super Passage

At its 9th mile (14th km) the canal meets another hill stream viz. Pathri torrent. As in the previous case, the torrent is taken over the canal through a super-passage, 296 ft (90.22 m) wide. Its maximum flood discharge has been estimated to be 36,000 cusecs (1020 cumecs). Its bed consists of fine sand of 0.25 mm diameter. It had an initial bed slope of 23.6 ft. per mile (4.50 m/km) above the super passage and 31 ft. per mile (5.90 m/km) below it. The accretion is taking place in its bed with the result that the slope is now 16 ft. per mile (3.05 m/km) upstream of the super passage and 12 ft per mile (2.29 m/km) downstream. The accretion on Pathri super passage started during 1868 and after 1880, considerable discharge of Pathri torrent spilled over into Ganga Canal during high floods. This accretion is due to the torrent fanning out in the swampy land after crossing the canal. This area has since risen and has become cultivable. There is no defined channel and water still flows over the area in a thin sheet leaving its bed load. Channelization of Pathri torrent downstream of Pathri super passage was attempted during 1885 to control the accretion problem. During the floods of 1894, the entire silt load was washed away. However, after 1938, the accretion had started again and at present there is 2 m silt load over the super passage. Various solutions including the diversion of torrent to Ratmau Rao, which has a defined bed, have been thought over.

Danauri Level Crossing

At the 13th mile (21st km) near the village Dhaunauri, Ratmau torrent crosses the canal at the same level and a level crossing has, therefore, been provided to pass its maximum flood flow of 80,000 cusecs (2265 cumecs). The torrent bed, which consists of sand of an average diameter of 0.35 mm had an initial bed slope of 10.3 feet per mile (1.41 m/km) downstream. In 1924 floods, the downstream apron (of boulder masonry laid on a network of wooden piles and boulder-filled crates) below the level crossing was damaged in a length of 300 feet (91.50 m) as a result of continued retrogression of the torrent bed by 15.0 feet (4.57 m). The damaged apron was, therefore, replaced by three vertical falls with impervious floors. In addition, five end spans on either side of the level crossing were closed. The retrogression, however, continued further. In the year 1948–1949, an arrangement of dissipaters in the form of baffle walls and staggered blocks evolved as a result of model studies was provided on the downstream floor of the last two falls. During July 1966, a flood of 80,000 cusecs (2265 cumecs) passed through the level crossing. Due to this flood talus downstream of impervious floor was washed away in 200 feet (61 m) length on the right side and impervious floor on compartment A by No. 2 and No. 3 was undermined and settled in 100 feet (30.5 m) length in bay No. 3 and 15 m length in bay No. 2. Due to the regular retrogression downstream of escape dam, the river bed in the reach below the dam has gone down by 25 feet (7.15 m). As a result of such a high drop in the bed, the water levels below the dam were substantially lowered with the result that the energy dissipaters suggested in 1948–1949 also did not function efficiently and the shooting flow caused damage of a large magnitude in the floods of 1966, when a flood of high intensity was received. The problem could only be solved by the provision of fall below the third floor.

Solani Aqueduct

From mile 17–7 to 18–6 (28.6 to 30 km), the canal passes through low-level areas. As its eighteenth mile (twenty ninth km) the most important and last cross drainage work is constructed over Solani River in the form of a 980 feet (298.70 m) long aqueduct which consists of fifteen spans of 50 feet (15.25 m) each separated by 10 feet (3.05 m) wide piers, the trough 1 being 175 feet (53.34 m) wide. The Solani River bed consists of sand of an average diameter of 0.25 mm and carries a maximum flood flow of 80,000 cusecs (22,265 cumecs). The channel in this reach runs in filling. The embankments are 2 miles 7 furlongs 507 feet (4 kms 781.4 m) long, and are held on one side by masonry walls, 55 feet (16.76 m) high, and supported on walls. These are provided on the water face with a flight of steps up to 3 feet (0.92 m) above the canal bed level. During the remodeling of the canal in 1956, the bed was lowered by 6 inches (0.15 m) by replacing the old clay bed overlaid with brick on edge by 3 inches (0.75 mm) cement concrete 1:2:4 which enabled the passing of an increased discharge of 10,500 cusecs (297.20 cumecs).

Modernization Project

The UGC was basically designed to provide protective Rabi irrigation to as large an area as possible. The original design discharge was 189 cumec (6750 cusec) but this has been gradually increased to 297 cumec (10,500 cusecs) to meet ever increasing demand for intensive irrigation throughout the year. Diversion of flood waters into UGC resulted in continuously injecting a high dose of sediment load. It has adversely affected the health of the system.

The major cross drainage works constructed with thumb rule designs had outlived their lives, and these have been replaced without closing the irrigation water supply. Risk aversion of the old canal system is one of the major components of modernization project. Risk aversion works comprised of the replacement of four major cross drainage work on main canal together with the construction of a lined parallel canal (PUGC) connecting these structures from 6 to 36 km.

In the modernization project, Ranipur super passage has been replaced by a siphon, Pathri super passage has been replaced by a new super passage, Dhanauri Level Crossing has been replaced by Ratmau aqueduct, and Solani aqueduct has been replaced by a new aqueduct.

Modern Structures on PUGC

Ranipur Syphon

At the time of construction of UGC, there was a 9 feet fall in UGC at this site, and a super passage was constructed. This fall was later shifted to the Pathri Power House site for power generation. Hence, the flow under the arches was not free, and the structure started acting partially as a siphon putting uplift pressure on arches. Syphon has now been proposed in PUGC at this crossing. The salient features of the siphon are as under:

(i) Type of structure: RCC structure having 6 rectangular Barrels and 2 triangular end barrels.
(ii) Length of barrels: 60 m.
(iii) Carrying capacity of canal: 310 cumec, with additional provision of 45 cumec for the operation of silt ejector.
(iv) Maximum flood discharge: 40,000 cusec (1133 cumec) in the torrent.

Ratmau Aqueduct at Dhanauri

At Dhanauri, the bed levels of drainages and also the UGC are practically the same, and it is for that reason that a level crossing was provided at this point. Sir Cautlay estimated the discharge of Ratmau River as 35,000 cusec (about 1000 cumec) and in most years it brought in flows of that order only. In 1947 it brought in a flood of about 80,000 cusec. This caused extensive damage, particularly in the escape structure. There were some damage to the escape structure in 1924 too. Again in 1966, the structure suffered damages. The position was that a good part of the sediment load was carried into the canal, and comparatively clean water escaped the drainage. Mainly for this reason, the drainage downstream of the escape continued to retrograde over the years.

Construction of a series of masonry weirs, each having a drop of 4–5 feet and interval of 200 feet, and later even construction of falls proved ineffective, and retrogression and damages have since been occurring. Experience of difficulties in level crossing called for some trouble-free design for this crossing.

Since Ratmau torrents bring flashy floods with heavy silt load and an appreciable quantity of this load passes down into the main canal through the cross regulator. A dam is provided on the drainage with 25 lift gates and 22 hinged-type falling gates, all manually operated. The present contemplated arrangement is that during flood season, beldars posted about one mile upstream of Ratmau torrents are supposed to inform the staff of level crossing about the impending flood with the help of gunshots. Who, in turn, is supposed to start operating the escape gates? The number of shots gives an idea of the quantum of the incoming flood. Even escaping through manually operated gates was quite efficient but restoration of these gates after every flood takes a long time. Assuming that the staff acts very promptly and operates the gates as required, again erecting the gates and bringing them in position every time a flood comes are quite a task, and all these years, there have been operation difficulties.

Keeping all the aspects and the retrograded bed of the torrent in view, it has been found that about 600 m downstream of the present crossing of an aqueduct is possible.

Hence, the alignment of PUGC in this reach from 19.50 to 21.50 km has been detoured to facilitate the construction of an aqueduct in place of the level crossing. The construction work was still going on in 1999.

The salient features of the aqueduct are as follows:

(i) Type of structure: Prestressed concrete through on RCC piers and wells.
(ii) Spans: 5 spans of 36.6 m each.
iii) Width of the trough (inside): 29.4 m.
(iv) Carrying capacity of canal: 310 cumecs.
(v) Maximum flood discharge: 1,50,000 cusec (4250 cumecs) in river.

Questions

1. Explain the difference between maintenance need, rehabilitation need, and modernization need.
2. For irrigation projects in your region, give examples of rehabilitation needs and modernization needs.
3. Give examples of structural and non-structural measures for the rehabilitation of an irrigation project.
4. Give examples of structural and non-structural measures for the modernization of an irrigation project.
5. Why the head regulator of UGC is not located adjacent to the barrage. Explain its location at Mayapur.
6. Explain the reasons for locating off-take of PUGC at km. 6.04 on the UGC.
7. Why level crossing in Ratmau River is not possible for the PUGC to cross the Ratmau.
8. Explain the reasons for aligning PUGC on the left side of UGC.
9. Explain the reasons for replacing UGC and its canal works in the head reach by PUGC and new works.
10. Write a brief note on cross drainage works on UGC.
11. Compare the cross drainage works on Parrallel Upper Ganga Canal (PUGC) with the cross drainage works on UGC.

References

Chaube UC (2009) Training course on rehabilitation and management of irrigation schemes, conducted by Prof U C Chaube at Kabul (Afghanistan)sponsored by Ministry of Energy and Water Resources, Govt of Afghanistan, Kabul

Chaube UC (2013) Lecture note on rehabilitation and modernization: need and scope, short term training course on modern irrigation practices. Centre for Continuing Education, IIT Roorkee, March 4

Govt. of UP (1991) Upper Ganga modernization project. I&P (World Bank) Govt. of Uttar Pradesh, Dept. of Irrigation Roorkee

IWRS: Modernization of Canal Irrigation. In: Proceedings volume. Indian Water Resources Society, WRDTC, University of Roorkee, Roorkee 247667 (U.P)

Varshney RS(1992) Rehabilitation and modernization of irrigation projects—concepts and policies. Write-up in Spl. Course 661 on Modernization of Canal Irrigation, Continuing Education Department, University of Roorkee, Roorkee, (UP) March/April

Chapter 21
Rehabilitation: A Case Study

Abstract Four minor (tank) irrigation projects in the Sagar district of Madhya Pradesh state in India are selected to diagnose deficiencies and identify the irrigation system's rehabilitation needs. The evaluation study of these tank irrigation projects was carried out as a field research project sponsored by the National Bank for Agriculture and Rural Development (NABARD) of India (Chaube, Report on evaluation of rural infrastructure (Irrigation) projects in Sagar District of Madhya Pradesh; sponsored research by National Bank for Agriculture and Rural Development (NABARD) 2006). These tanks are designed to store monsoon runoff of local streams used for irrigation in the Kharif (wet) and Rabi (dry) seasons, emphasizing Rabi irrigation. The objectives of this chapter are to: depict deficiencies in structural components of tank irrigation schemes, understand various physical and social phenomena causing maintenance problems, and recommend possible improvements to rehabilitate the projects. Various physical and social phenomena causing maintenance problems are depicted using field observations. For rehabilitation, these projects have been financed by a national bank. Financing of cost and expenditure for the four projects are compared. The cost overrun for each of the four projects is analyzed. Time and cost overruns are abnormally high in these four tank irrigation projects. The success/risk factors and learning points in project management are highlighted from the study of various aspects.

21.1 Salient Features of Tank Irrigation Projects

Four minor (tank) irrigation projects in India's Sagar district of Madhya Pradesh state are selected as case studies. The evaluation study of these tank irrigation projects was carried out by Chaube (2006). Further studies of these tank irrigation projects have been carried out by Nihal (2007) on financial and economic evaluation and by Nissanka (2007) on operation and maintenance aspects.

These tanks are designed for the storage of monsoon runoff of local streams. Irrigation is proposed in Kharif (wet) and Rabi (dry) seasons, emphasizing Rabi irrigation. The salient features are stated in Table 21.1. The headworks consist of earthen embankments with canal intakes and waste weir. A waste weir canal is

U. C. Chaube et al., *Canal Irrigation Systems in India*, Water Science and Technology Library 126, https://doi.org/10.1007/978-3-031-42812-8_21

Table 21.1 Salient features of tank irrigation projects

Item	Unit	Mahua Kheda	Khairana	Maheri	Hinauta Kharmau
Village		Mahua Kheda	Khairana	Basari	Hinota
Block		Jai Singh Nagar	Raheli	Khurai	Banda
Distance from Sagar	km	22	56	30.5	47
Stream		Local	Local	Local	Local
Catchment area	km^2	3.15	4.142	2.032	21.3
Annual rainfall	mm	1369	1177	1186	1163
Live storage	MCM	1.2	1.466	0.948	1.35
Silt reserve	MCM	0.08	0.096	0.07	0.13
Submergence at FRL	ha	–	50.70	52	36.8
Length of dam	m	660	1025	1320	670
Max. height	m	9.94	9.87	9.15	13
Top width	m	3	3.00	4.5	3
Length of waste weir	m	69.19	84.0	80.22	47.25
Canal head discharge	m^3/s	0.12 LBC+ 0.12 RBC–	0.1415	0.113	0.184
Length of canal	km	LBC 2.25 (CCA 126 ha)	3.51	3.21	1.8
Length of distr. and minor	km	RBC 1.32 (CCA 74 ha)	1.65	1.29	1.45
Irrigation area					
CCA	ha	180	308	160	386
Soil type		Clay loam	Clay loam Sandy loam	Clay	Clay
Kharif	ha	20	20	40	63
Rabi	ha	180	275	91	185
Annual irrigation	ha	200	295	131	248
Irr. intensity	%	111	921.78	81.8	64.25

excavated to carry surplus water from the tank to the stream downstream. Schematic diagrams of the project layout are shown in Figs. 21.1, 21.2, 21.3, and 21.4.

21.2 Field Observations

Physical verification and visual documentation of the completed works are carried out, and an Overview of the Implementation of Works is provided in Table 21.2.

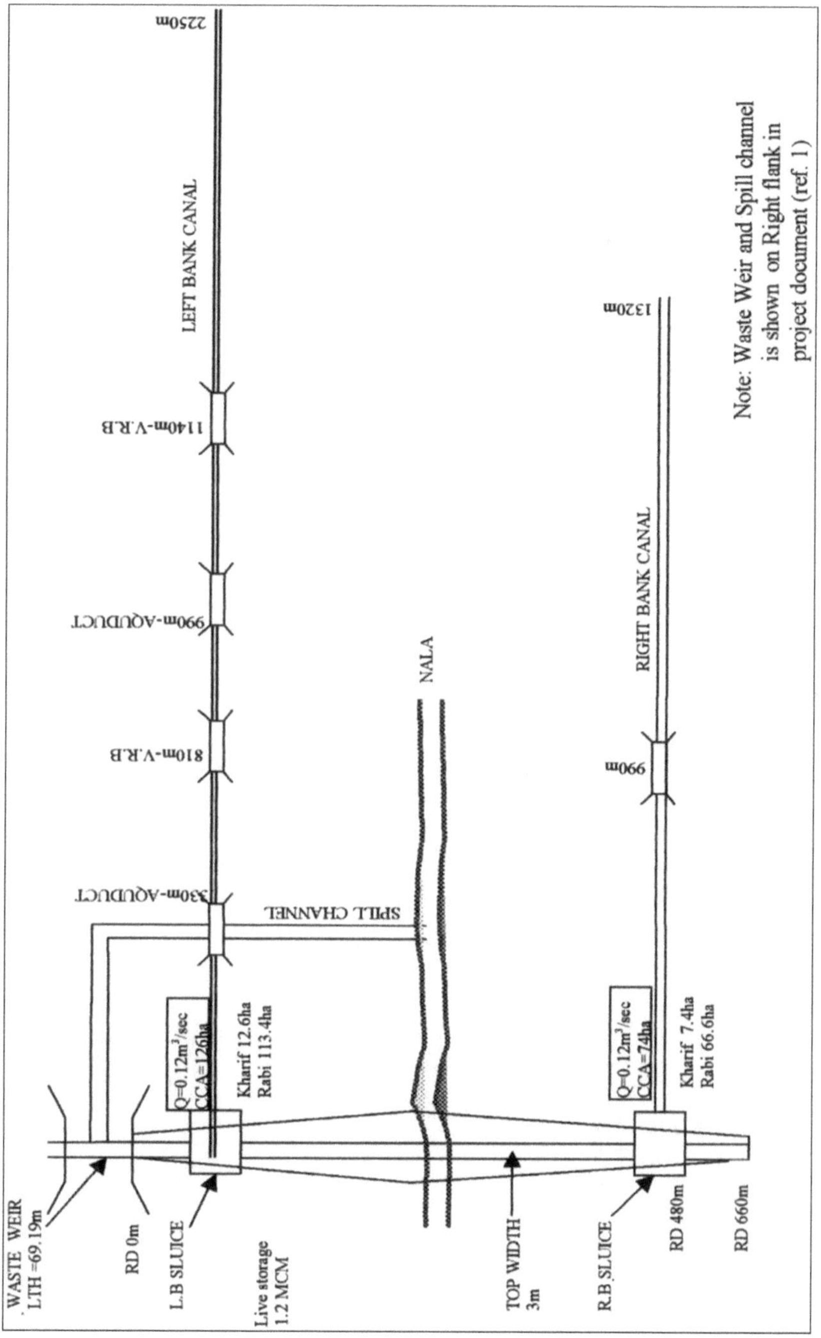

Fig. 21.1 Mahuakheda tank-line diagram

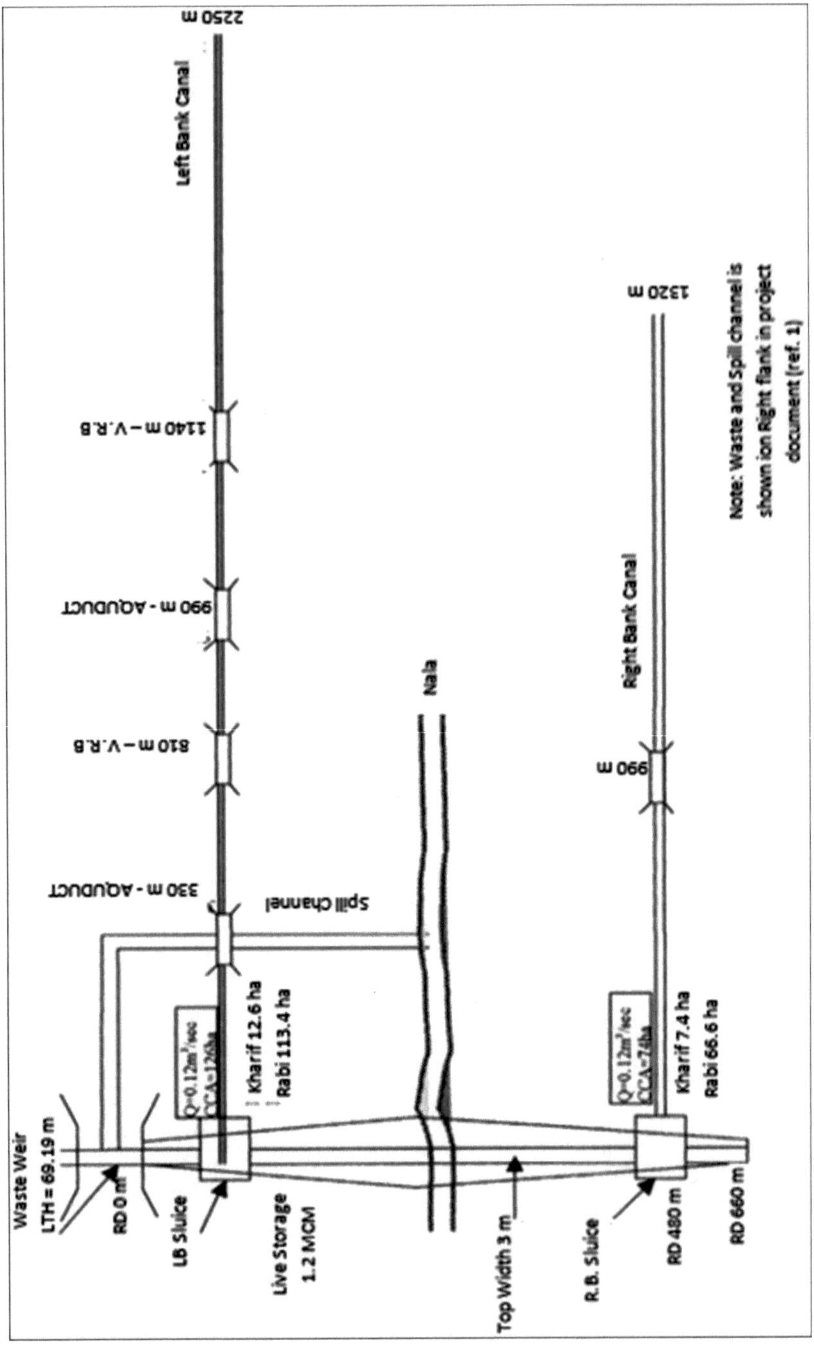

Fig. 21.2 Khairana tank-line diagram

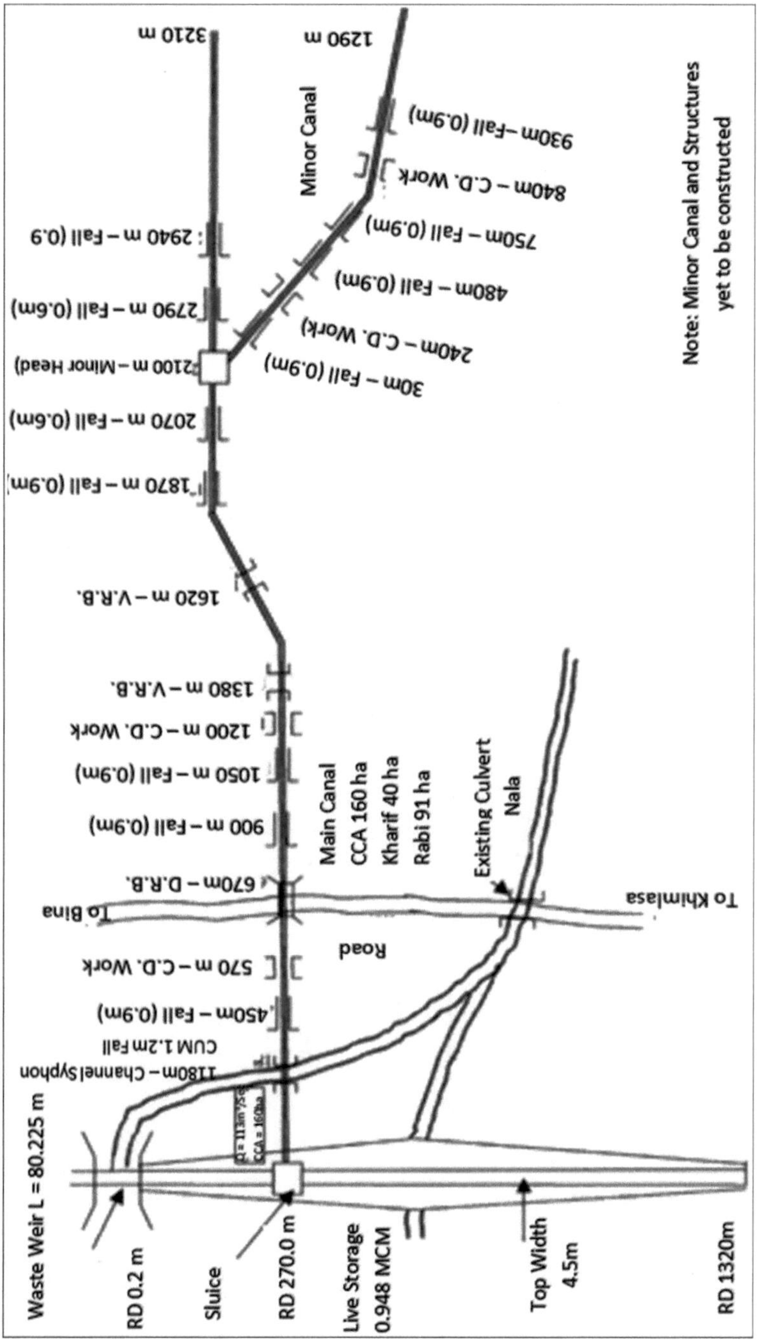

Fig. 21.3 Maheri tank-line diagram

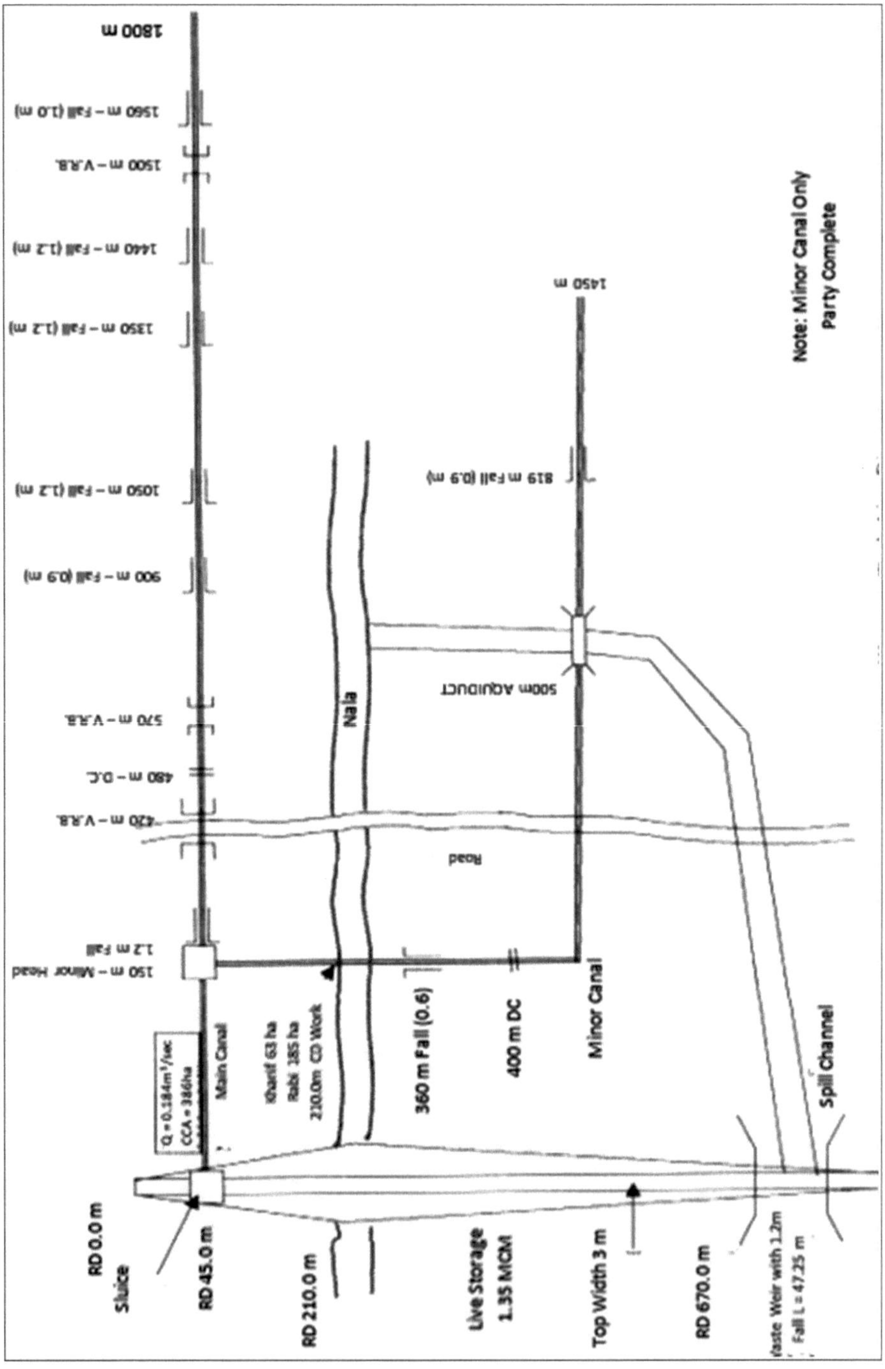

Fig. 21.4 Hinauta Kharmau tank-line diagram

Table 21.2 An overview of the implementation of works

	Mahuakheda	Khairana	Maheri	Hinauta Kharmau
Implementation of headworks				
Earthen embankment	Yes	Yes	Yes	Yes
Intake structure	Yes 2 (RBC + LBC)	Yes 1 (LBC)	Yes 1 (LBC)	Yes 1 (LBC)
Sluice gate	Yes	Yes	Yes	Yes
Waste weir	Flush bar	Flush bar	Flush bar	Fall structure
Spill channel	No	Yes	No	Yes
Seepage collection drains	Yes	Yes	Yes	Yes
Implementation of canal works				
Land acquisition	Yes	Yes	Yes	No
Main canal	Yes	Yes	Yes	Yes
Minor canal	–	–	No	Partly incomplete
Structures	No	Yes	No	Yes
Tail escape channel	Not found	Not found	Not found	Not found
On-farm dev. works				
Outlets	No	Yes (partly)	No	Only in lined portion
Watercourse	No	No	No	No
Field channels	No	No	No	No

21.2.1 Common Observations

1. The canal cross-section is non-uniform and irregular. The service road along the main canal is either non-existent or poorly maintained. The boundary and chainage stones are not provided.
2. Outlets are not existing, and on-farm development works have not been carried out. Intake sluice gates are poorly maintained. The water level scale either does not exist or is not readable.
3. The reservoir bed is used for the cultivation of rabi and summer crops.
4. There is no arrangement to measure the canal water level and discharge in the canal.

21.2.2 Observations on Mahuakheda Project

1. **Headworks**

    (i) The agricultural land immediately downstream of the dam is waterlogged due to heavy seepage from the dam body.

Fig. 21.5 Tank bed cultivation and embankment material taken from Mahuakheda reservoir bed

(ii) The soil has been extensively excavated from the reservoir bed adjacent to the dam and has been used as an embankment material (Fig. 21.5). According to the norms/rules of WRD; no material can be excavated from the bed up to 10H distance from the embankment.

(iii) The leakage is observed near the intake on the left bank, even though the gate is closed.

(iv) The scales for recording tank water levels are not marked on intake structures.

(v) The waste weir is located on the left bank. It consists of stone pitching with a flush bar.

(vi) The reservoir bed and land on the periphery of the reservoir are extensively cultivated.

2. **Canal system**

(i) The right bank canal had a deep pool of water downstream of intake with a heavy growth of weeds in the canal bed all along (Fig. 21.6). The wheat cultivation was observed in the canal bed during the second visit.

(ii) The left bank canal looked to be in an abandoned condition with a heavy growth of weeds and plants. The canal has since been cleaned but only in the head reach.

Fig. 21.6 Heavy weed growth in right bank canal of Mahua Kheda project

21.2.3 Observations on Khairana Project

1. **Headworks**

Heavy seepage occurs from the dam foundation, particularly in the Nala closure portion. Fields immediately downstream of the dam have become waterlogged (Fig. 21.7). Some fields have been rendered unsuitable for cultivation, and in some of the fields, rabi cultivation has been delayed abnormally.

The intake sluice gate has not been functioning properly since October 2002. The scale for the measurement of tank water level does not exist.

Fig. 21.7 Waterlogged fields downstream of Khairana dam

Fig. 21.8 Khairana tank: Canal is cut, and water diverted to Nala

2. **Canal system**
   (i) A cut on the canal's right bank at its head has been made to divert uncontrolled release from intake (due to its malfunction) to the Nala (Fig. 21.8).
   (ii) Too much weed growth in the canal bed and on banks near road crossing to Khairana village.
   (iii) The main canal, from 0 to 2.61 km in length (on contour) was damaged at many places due to heavy rains in the monsoon season 2021. An adequate drainage system for storm runoff from upstream of the contour canal does not exist.
   (iv) The bank height of the canal is inadequate near Khairana village. The canal banks and crop fields are damaged due to water overtopping the banks.

21.2.4 Observations on Maheri Project

1. **Headworks**
   (i) The intake gate is not functioning. The tank water level scale is not visible.
   (ii) Extensive rabi cultivation on the tank bed and its periphery. Electric wires were observed lying on the embankment and bamboo poles for unauthorized energization of pumps.
2. **Canal system**
   (i) A minor canal of 1.29 km in length is still not constructed.

(ii) The design of the canal syphon at the waste weir drainage crossing is poor. Transitions have not been provided. The canal at this location has been cut to divert canal flows into spill channel for use in the downstream Nala.

(iii) The canal section is almost non-existent at 420 m chainage. Further downstream, there is 1 m fall in the bed level, but no fall structure has been provided.

(iv) One cross drainage work and two fall structures are still not constructed. The canal has never been operated.

(v) The canal length between waste weir drain crossing and Bina road crossing is almost completely damaged. Therefore, during a rabi season, the Maheri tank water was released into Parasari Nala for lift irrigation by the farmers.

21.2.5 Observations on Hinauta Kharmau Project

Due to seepage from the dam foundation to the adjacent area, the downstream of the dam is waterlogged. The road crossing the drain from the waste weir is damaged.

Canal system

(i) Part of the minor canal in its head reach is lined and has outlets (Fig. 21.9).

The farmers of Hinauta village are benefiting, but Kharmau village farmers are deprived of irrigation benefits due to incomplete minor canals.

(ii) The left bank of the minor lined canal is heavily damaged at the Nala crossing. The unlined portion of the minor canal is poorly maintained and damaged.

Fig. 21.9 Hinauta Kharmau tank project: lined minor canal with silt and damage

21.3 Operation and Maintenance Status

21.3.1 Implementation Status

At present, the operation and maintenance of the projects are being managed by the government. The process is on to transfer the operation and maintenance of canal networks to Water Users Associations for which elections have already been held.

Water is yet to be released in the canals offtaking from Maheri and Mahuakheda tanks. Khairana and Hinauta Kharmau tank canals have been operated for the past 2–3 years, but the water supply has been inadequate to achieve the irrigation potential. The canal water supply was made for the first time in 2004–2005 from Khairana tank and in 2003–2004 from the Hinauta Kharmau tank. The Operation and Maintenance Status of Canals is provided in Table 21.3.

Even though gated intake structures are provided to regulate water supply into canals, the gates could not be operated to regulate supply into canals according to

Table 21.3 Operation and maintenance status of canals

Item	Mahuakheda	Khairana	Maheri	Hinauta Kharmau
The amount required for O & M	Rs. 0.1 lacs	Rs. 0.15 lacs	As per Deptt. rules	Rs. 0.28 lacs
The amount provided by Government for O & M	Nil	Nil	Nil	Nil
Canal water supply	Nil	Partly	Nil	Partly
Arbitrary use of tank water	Yes (tank bed)	No	Yes (tank bed)	No
Shortage at tail end	NA	Yes	NA	Yes
Water rate being charged	NA	Nil	Penalty charges for tank bed cultivation	Kharif Rs.215/ha Rabi (wheat) Rs. 235/ha
Adequacy of water rates	NA	NO	NA	NO
O & M by WUA	Not yet	Not yet	Not yet	Not yet
Training of farmers	No	No	No	No
Training of staff	Partly	Partly	Partly	Partly
Field drainage	Nil	Nil	Nil	Nil
Outlets	Nil	Yes	Nil	Partly
Special repair need	Urgent	-	Urgent	Urgent
Funds for special repair	Nil	Nil	Nil	Nil

water demand due to a combination of various factors, such as (i) improper maintenance, (ii) damage by miscreants, (iii) inadequate storage, and (iv) incomplete/damaged canal network.

The maintenance needs of an irrigation system arise because of physical and social phenomena. The bank erosion and cuts due to high-intensity rains, burrowing animals, vegetation growth, and seepage losses are some of the observed physical phenomena, while deliberate cutting of banks, putting boulders underneath intake gates so that water may continue to flow, putting obstructions in the canal bed to raise the water level, unauthorized/oversized outlets, bank encroachment, bathing, and washing unauthorized passage across canals are the observed social phenomena. It is expected that problems relating to social phenomena should significantly reduce after transferring O & M to WUAs.

There were heavy rains during July 2005, due to which canals suffered heavy damage requiring special repairs. These damages have not yet been repaired.

Canal damages have occurred due to a variety of reasons, such as (i) unauthorized conveyance of tank water into Nala at the head of the canal, (ii) taking canal water into fields in the absence of outlets, (iii) damage by storm runoff in the initial reach where the canal is aligned on contour, and (iv) creating the passage across the canal.

21.3.2 Analysis of Time Overrun

There is a significant time gap between the date of financial sanction by the funding agency and the date of administrative approval by the Government. The implementation period for balance head works (mainly nala closure) ranges from 43 months (Maheri) to 67 months (Mahuakheda). The implementation period for balance canal works ranges from 17 months (Khairana) to 59 months (Mahuakheda). Figures 21.10 and 21.11 show the time overrun in the construction of head works and canals and the overall time overrun, respectively.

Common Reasons for Time Overrun: Delay in the administrative approval and tendering process after receiving financial sanction and thin spreading of available funds over the years are common reasons for time overruns in all four projects.

21.3.3 Main Reasons for Time Overrun

A study of project documents shows that the following are the main reasons for time overrun.

Mahuakheda Project: Contractors who submitted the tender documents did not meet the sponsoring agency requirements necessitating the repetition of the tender

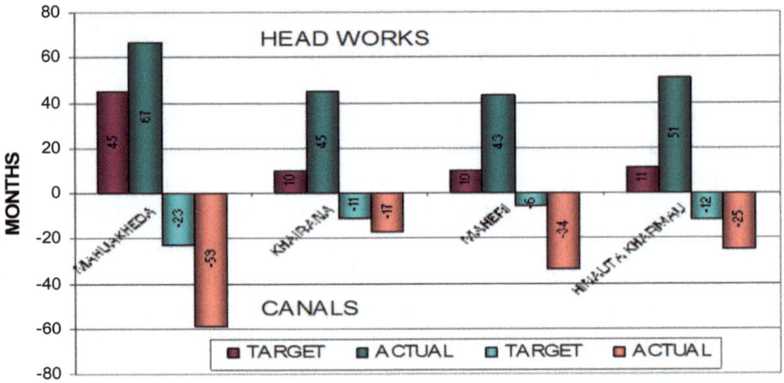

Fig. 21.10 Time overrun in the construction of head works and canals

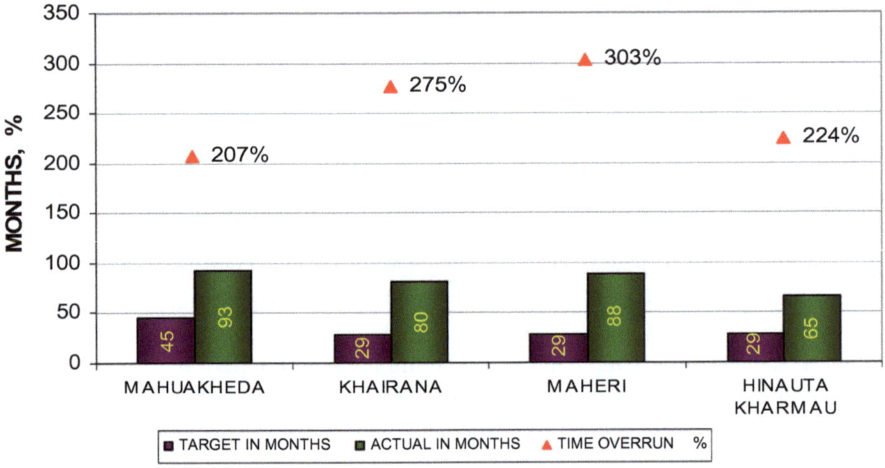

Fig. 21.11 Overall time overrun

calling process five times. Canal work has been delayed mainly due to incomplete land acquisition. Day-to-day hindrances were also caused by landowners in the construction work.

Khairana Project: The main reason for delay is the excessive time taken in according to technical sanctions and the necessity to frame a revised cost estimate. Further delay occurred due to the time taken to obtain the administrative approval for the revised estimate. Construction work progressed slowly due to the thin spreading of funds.

Hinauta Kharmau: The delay occurred in the tendering process, and work could start two years after the financial sanction. Work was again stopped for some time as per the instructions of the Chief Engineer as per policy matter of M.P Govt.

Maheri Project: The main reasons are delay in technical sanctions and higher authorities' administrative approval. The tendering process took sufficient time.

21.4 Finance and Expenditure on Rehabilitation

The amounts of cost finance and expenditures for the four projects are compared for (i) original cost and expenditure before finance sanction by the bank (NABARD), (ii) State Government share in the balance of cost and up-to-date expenditure, and (iii) bank share in the balance of cost and up to date expenditure (Fig. 21.12). The cash flow of bank loans and expenditures are graphically depicted.

Details on annual budget provisions are not available in the case of Maheri and Hinauta Kharmau TIPs. The annual project budget and disbursement figs do not match. There has been a thin spreading of the available funds over the years resulting in time and cost overruns. Time taken to complete the remaining works ranges from five years to seven years against the envisaged period of two to three years. Still, works remain incomplete in the case of the Maheri and Hinauta Kharmau projects (Ref. completion report). Cost history and cost reasons overrun for each of the projects are analyzed in the following paragraphs.

21.4.1 Analysis of Cost Overrun

Cost Overrun of Mahuakheda Project: Expenditures incurred to complete the remaining works are 300% of the estimated cost of balance works. The government share in this expenditure is 896% of the stipulated share of Rs. 21.05 lacs. The estimated cost of the balance of head works was Rs. 19.68 lacs, whereas expenditure

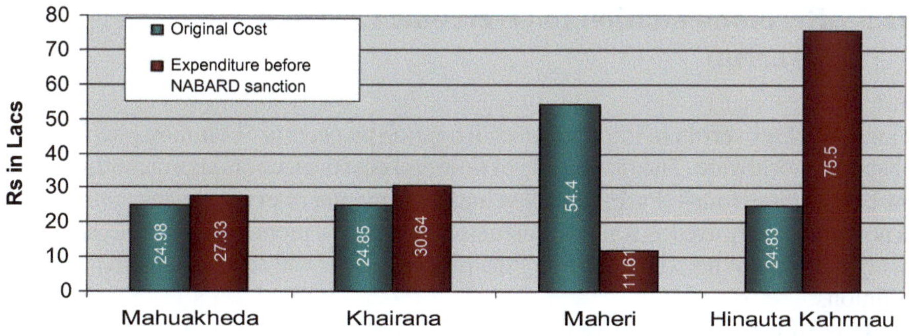

Fig. 21.12 Cost finance and expenditure comparison improving operation and maintenance budget

on this item was Rs. 39.58 lacs. The estimated cost of the balance of canal works was Rs. 2.8 lacs, whereas expenditure on this item is Rs. 24.39 lacs.

Cost Overrun of Khairana Project: The balance of Rs. 140.91 lacs was proposed to be financed with Rs. 127.98 lacs as the NABARD share and Rs. 12.93 lacs as the Government contribution. As per the completion report, the estimated cost of the balance of headwork was Rs. 122.19 lacs, whereas the actual expenditure on this item was Rs. 97.83 lacs. On the other hand, the estimated cost of the balance of canal work was Rs. 18.72 lacs, whereas the actual expenditure on this item was Rs. 38.5 lacs; the total expenditure being Rs. 138.61 lacks (including Rs. 2.276 lacs as fixed charges).

Cost Overrun of Maheri Project: Its original cost in the year 1991 was Rs. 54.4 lacs. It was revised to Rs. 98.19 lacs in the year 1997. From 1991 to 1997, only Rs. 11.61 lacs was spent on this project, unlike other projects where the amount spent before the NABARD sanction had been more than the original cost. The balance cost of Rs. 86.58 lacs was proposed to be financed with Rs. 76.58 lacs as the NABARD share and Rs. 10.0 lacs as the Government contribution.

The expenditure incurred on the remaining works is Rs. 123.18 lacs. Minor canals and related structures are yet to be constructed. The actual GOMP contribution in up-to-date expenditure is 466% of the stipulated expenditure of Rs. 10 lacs.

Cost Overrun of Hinauta Kharmau Project: A balance of the cost of Rs. 104.2 lacs was proposed to be financed with Rs. 86.39 lacs as the NABARD share and Rs. 17.81 lacs as the Government contribution. As per the information provided, the expenditure is Rs. 1021.3 lacs, out of which Rs. 86.39 lacs is the NABARD share and Rs. 18.94 lacs is the Government share. The estimated cost of the balance head work was Rs. 79.85 lacs, whereas the actual expenditure was Rs. 72.33 lacs. The estimated cost of the balance of, canal works was Rs. 24.35 lacs, whereas the actual expenditure is Rs. 33.00 lacs.

21.5 Recommendation to Overcome Cost and Time Overrun

Time and cost overrun in implementing irrigation projects is a common phenomenon observed worldwide. The magnitude of time and cost overruns is significantly large in the case of these four TIPs. Financial valuation at regular intervals during construction is now commonly used as a management tool. Cash flow monitoring and impact on the benefit–cost ratio help in identifying the problem areas and appropriate management solutions.

Before funding by NABARD, the projects had already been in the construction stage for a long period as indicated by the year of the original project start. It is seen that annual budget allocations for project construction have been inadequate and inconsistent with annual disbursements of loans by NABARD. Funds must be transferred to the project executive level swiftly for timely implementation.

Administrative approvals for the revision of project costs should be provided well in time. The NABARD criteria for accordance of administrative approval before the release of assistance should be strictly enforced.

The tender percentage being a function of various complex factors, NABARD did not adopt any hard and fast limit for cost escalation, and the percentage of higher tenders provided in the projects had been accepted as reasonable. Yet, the reasons for an unusually large increase in actual cost and time overrun are high percentage tender, which means that costs were either not assessed realistically or there were some other unexplained reasons beyond the scope of this chapter.

The department carries out an annual inspection and normal repair works as per the availability of funds. The budget requirement for yearly maintenance varies from Rs. 10,000 to Rs. 28,000 for these projects.

Budgeting for O & M is a mandatory requirement as per the Budget Manual in each state. Budget allocation signifies the approval of funds, while budget allotment is the formal authorization for incurring the expenditure.

The procedure for the receipt of fund allotment at the beginning of the year (say by April end) needs to be enforced so that the normal maintenance is carried out before the start of monsoon.

The establishment charges (part of O & M) are relatively high. Staff strength, particularly that of the work charged category, can be reduced, as WUA will look after O & M after transferring the canal system.

The budget proposal should include details of irrigation revenue for the Kharif and Rabi seasons (actual collection in the previous financial year with shortfalls, if any) and target revenue collection. Additional revenue should be correlated with irrigation supply.

Flood damages, which occurred in the monsoon season 2005, are yet to be repaired. There should be sufficient reserve (margin money) for urgent repairs in the case of flood damage.

The implementing agency's budgetary system emphasizes the financial aspect (salaries, cost of material, etc.) with a strong bias towards units of administration and objects of expenditure. It is not helpful to judge the progress towards the attainment of the objectives of the project. It does not serve as an adequate basis for informed decision-making.

It is recommended that the performance approach to budgeting be followed with emphasis on (i) accomplishments rather than on the means of accomplishment and (ii) precise definition of works to be done or service to be rendered. It will help in (i) measuring progress towards objectives, (ii) facilitating better appreciation and review of the management of the project, and (iii) make possible more effective performance audits.

21.6 Improving Monitoring and Evaluation

Monitoring is a process of data collection and compilation to review performance status at regular intervals and compare it with the targets set. Evaluation, i.e., concurrent review analysis of the collected data under monitoring, is carried out to know how the targets could be achieved. Data are the observed attributes and information resulting from data processing in a meaningful form.

The Water Resources Department is not doing regular monitoring and evaluation (in each crop season). An independent agency (outside the department) must also monitor and evaluate the performance, say once in three to four years. The monitoring data should be collected in prescribed formats to facilitate their computerized processing for evaluation.

(a) **Monitoring and quality control during construction**

Compaction tests were performed to check the degree of compaction in earth dams. However, waterlogging downstream of the dam body clearly indicates the low quality of seepage control measures adopted during construction. The quality control of canal construction is poor, as is evident from visual documentation.

(b) **Monitoring and evaluation during the O & M stage**

Reservoir water levels, canal operation schedule (target and actual), head and tail canal water levels, the progress of irrigation (adaptation), and regular maintenance activities should have been monitored regularly for meaningful evaluation exercise. No internal evaluation of the performance has been carried out so far.

(c) **Inspection visits**

Farmers' interviews revealed that irrigation officers have seldom visited the headworks and canal works. Further, farmers rarely had the opportunity to interact with irrigation officers and inform them about their problems (complaints). Inspection notes during site visits can provide valuable information on the projects. In the olden days, irrigation engineers were required to be on field visits for at least 20 days a month, and inspection registers containing technical notes on each visit were maintained at sites. Unfortunately, irrigation officers are not making regular and frequent site visits now. Many of the O & M problems can be solved effectively, economically, and promptly if the executive engineers regularly visit and interact with field-level staff and farmers to understand interface problems.

Officers can issue corrective instructions (in the inspection register) to the field staff and keep them alert in discharging their duties.

21.7 Recommendations for Improved Maintenance

(a) **Improvement in management**

1. Adequate provision and timely release of the amount (latest by April) for normal maintenance of headworks and canal before the onset of the monsoon season.
2. Immediate release of adequate funds for the special repair of canals damaged in the monsoon season of 1992.
3. Transfer of O & M of canals to Water User Associations only after necessary repairs and training of farmers in technical and financial management
4. Employment of one Junior Engineer by each WUA for O & M.
5. Construction of masonry structures and the remaining part of the canal at the earliest.
6. Splitting of O & M grants into (a) grant for operation, (b) grant for maintenance, and (c) grant for establishment. Increase the O & M grant.
7. Involving NGOs in the evaluation and monitoring units.
8. Segregating O & M units and establishment units in the Water Resources Department.

(b) **Technical improvement**

Cultivation on the periphery and in tank beds should be permitted only if these lands can be protected from soil erosion. Lifting arrangements for intake gates should be oiled, greased, and tested. Pulling vegetative weeds from canal beds and banks by wading laborers is more effective than cutting by hand blade as the regrowth rate can be reduced. The vegetative growth in lined channels should be eliminated by removing sediment from the channel bed. Joints in the lined portion should be sealed with asphalt.

Sediment removal from the canal should begin at the lower end of a channel and proceed upstream; if begun upstream and not completed before canal operation, the excavated material serves as a sediment trap. For animal crossings, a brick ramp on a slope of 3:1 should be provided on the canal at suitable locations (particularly where the canal passes through the village). Bathing steps should be made of bricks. Without these bathing steps, erosion will continue to occur in people's pathways, as observed in the canal of the Hinota tank irrigation project.

(c) **Farmers' perception, problems, and use of improved technology**

Techno socio–economic survey of a large number of farmers in command areas of the four projects was carried out with the assistance of the National Institute of Hydrology, Sagar. A number of farmer interviews had to be rejected/or repeated to obtain realistic and reliable information. In addition to the impact on the cropping system. The other findings of the socio-economic survey are given below:

1. Farmers are making better use of farming technology (farm machinery, fertilizer, pump sets, etc.) even without irrigation water.

2. Farmers have been practicing irrigated agriculture with groundwater and stream flow as a source of water. They have a fair knowledge and perception of field irrigated agriculture technology.
3. Farmers need training to improve field application efficiency of water, but on-farm development works are also necessary simultaneously.
4. Farmers' attitude towards the project implementation and irrigation service is not encouraging. Their problems relate to poor maintenance of canals, water logging, inadequacy and inequity in canal water supply, and damage to fields.

(d) **Issues for policy intervention**

Some important issues for policy intervention are:

(i) Block (clustered irrigation fields) irrigation systems can reduce the capital cost of the canal network, canal operation losses, and evapotranspiration demand of crops. This policy may conflict with the existing irrigation policy, which favors extensive irrigation.
(ii) Close monitoring of each project (the term 'minor' is misleading) is necessary so that public investments are used economically and efficiently. Simple and standard yardstick, procedures need to be evolved for rapid evaluation exercises (computerized) by independent agencies on behalf of NABARD.
(iii) In a tradition-bound society in rural India, religious beliefs/values can be used effectively for water conservation and management. Religion is an integral part of social life for the people in villages. In Bali island (Indonesia), irrigation water committee duties are considered religious duties. The challenge lies in promoting such religious beliefs/traditions.
(iv) Irrigation service area may include cultivated land on the tank bed and periphery which are already being irrigated from tanks. Erosion and reservoir sedimentation will need to be controlled
(v) The upper limit of acceptable cost of balance of works is based on Rs.90,000/ha of annual irrigation or service area, whichever (area) is more. A distinction is possible between the acceptable cost for headworks and canal works. Further, the realization of irrigation benefits (by the farmers) is possible only if on-farm development works are also financed and implemented by the government or by WUAs.

21.8 Success/Risk Factors and Learning Points

Four minor (tank) irrigation projects in the Sagar district of Madhya Pradesh state in India are selected for the diagnosis of maintenance problems study. These tanks have been designed to store monsoon runoff of local streams and provide irrigation water in Kharif (wet) and Rabi (dry) seasons, emphasizing Rabi irrigation. The structural components of tank irrigation schemes that are commonly implemented considering tropical monsoon hydrology have been illustrated through schematic diagrams of the four tank irrigation projects. Field observations have been described and depicted through photographs to understand various physical and social phenomena causing maintenance problems.

Bank erosion and cuts due to high-intensity rains, burrowing animals, vegetation growth, and seepage losses are some of the observed physical phenomena, while deliberate cutting of banks, putting boulders underneath intake gates so that water may continue to flow, putting obstruction in the canal bed to raise the water level, unauthorized/oversized outlets, bank encroachment, bathing, washing unauthorized passage across canals, etc. are the observed social phenomena. Problems relating to social phenomena are expected to significantly reduce after transferring O & M to WUAs.

Recommendations have been made on possible improvements in the O & M budget, monitoring of the status of O & M, and maintenance technology. Emanating from the study of various aspects, the following are the success/risk factors and learning points in project management.

Success Factors

1. There is no shortage of staff in the implementing agency.
2. Farmers are keen to receive canal water. It is cheaper than lift irrigation being practiced by farmers.
3. Farmers have a fair knowledge of field irrigation technology.
4. The use of farming technology (HYV seeds, fertilizer, farm machinery) has improved.
5. Stream flow is augmented due to seepage from the dam, leakage from intake, and diversion of tank water into nala. This has resulted in the utilization of tank water, which is otherwise impossible due to an uncompleted/damaged canal.
6. Tank storage has increased groundwater recharge. Farmers are using increased recharge for well irrigation and drinking water supply.
7. Due to the change in cropping pattern (shift to cash crops) and increase in cropping intensity, direct and indirect benefits have significantly improved.

Risk Factors and Learning Points

1. Delay in implementation of remaining incomplete works will not only cause an increase in cost but also a reduction in the present worth of benefit and thus will adversely affect the financial performance of projects.
2. Centralized decision-making and inadequate delegation of authority to lower levels are partly responsible for the delay in projects' implementation. Redundant hierarchical levels should be excluded from the decision-making process.
3. Thin spreading of funds over the years results in time and cost overrun.
4. On-farm development works (outlets, watercourses, field channels, field drainage, tail-end escape channel) are necessary (i) to achieve equity in water distribution, (ii) to improve project efficiency, (iii) to increase project output, and (iv) to minimize the risk of waterlogging.
5. Improvement in the capabilities of junior and middle-level staff is necessary as staff performance directly impacts project output.

6. The infrequently supervised field staff may develop a relationship with farmers and act at least to a degree as farmers' agents. An executive engineer should frequently make field inspections. The inspection register should be maintained at the site.
7. Conflicts may arise due to different perceptions and different approaches to solving farmers' problems, mainly relating to compensation for land acquisition.
8. High establishment charges reduce the amount available for actual maintenance. The strength of junior engineers and work-charged staff should be reduced.
9. Maintenance problems aggravate if not solved in time. A contour canal gets damaged due to inadequate drainage of runoff from the upper elevation.
10. Realistic cropping patterns and cropping intensity need to be adopted in project design. These have a major impact on direct benefits and other financial parameters.
11. Farmers' perception of irrigation projects and irrigation service is not encouraging. They need to be involved in implementing and maintaining works (particularly the water distribution network).
12. A key to the success of WUAs lies in ensuring their financial viability. Further, farmers will have to show a stronger affinity to WUAs as a social group than to other social groups to which they belong. WUAs should have representation from the weaker sections of society.
13. Strict enforcement of accountability for time and cost overrun is necessary. Responsible media, NGOs, and social workers should be encouraged to highlight values and constraints.

Questions

1. Explain the following terms: (i) tropical monsoon hydrology, (ii) tank irrigation, (iii) Rabi irrigation, (iv) Kharif irrigation, (v) irrigation intensity, (vi) CCA, and (vii) silt reserve.
2. Study the schematic diagrams depicting the Mahuakheda and Hinota tank irrigation project layout. Make a list of structural components in these projects and discuss the purpose of these components.
3. Write a note on the status of the implementation of the Maheri and Khairana tank irrigation projects.
4. What lessons can be learned from the observed deficiencies in the case of the Maheri tank irrigation project?
5. What are the deficiencies in the existing O & M budget provisions?
6. Compare time over run in the four projects.
7. Compare cost over run in the four projects.
8. What are the measures to reduce time over run?
9. What are the measures to reduce cost overrun?
10. Is it practically possible to have zero time over run and zero cost overrun?

References

Chaube UC (2006) Report on evaluation of rural infrastructure (irrigation) projects in Sagar district of Madhya Pradesh; sponsored research by National Bank for Agriculture and Rural Development(NABARD). Mumbai, India

Following documents on website www.nabard.org of National Bank for Agriculture and Rural Development(NABARD),Mumbai, India (a) NABARD: Financial Sanction Orders containing Schedule I, Schedule II and Schedule III on terms and conditions for implementation of the projects. (b) NABARD: RIDF-I for Irrigation Projects in Madhya Pradesh- Reports of the Appraisal Mission for Madhya Pradesh, (c) NABARD: RIDF-III: Financing of Irrigation Projects in Madhya Pradesh Phase V

Nihal KDS (2007) Financial and economic evaluation of tank irrigation projects, M.Tech Dissertation, supervised by Prof. Chaube UC, IIT Roorkee, Roorkee

Nissanka N (2007) O & M aspects of small tank irrigation projects—some case studies, M.Tech Dissertation, supervised by Prof. Chaube UC, IIT Roorkee, Roorkee

Chapter 22
Conjunctive Use Management

Abstract The implementation of a conjunctive use management plan is a process influenced by several factors, technical, legal, socio-economic, and organizational. The objectives of this chapter are to: understand the socio-legal issues in the implementation of conjunctive use plan; understand problems relating to irrigation water rates for surface water and groundwater, and illustrate the issues/problems and possible solutions through a case study. The implementation of a conjunctive use management plan involves finding a solution to several problems/conflicting issues. Issues/problems relating to irrigation water rates, interface problems between the various organizations and individual farmers and legal issues have been explained with an illustrative example of the Lakhawati branch command area of Madhya Ganga Canal in U.P. Due to large disparity in water rates for surface and groundwater, farmers prefer to make maximum use of surface water. Irrigation water charges are compared and rationalized. Several agencies are involved in the implementation of conjunctive use management. These are Irrigation Department, Agriculture Department, Minor Irrigation Dept., Tube Well Corporation, and Command Area Development Authority. Better coordination is necessary at field level among the government agencies on one hand and between farmers and Govt. agency on the other hand. Water user Associations have a very important role in conjunctive use as private use of groundwater by individual farmers needs to be controlled as a common property resource. Lacunae in surface and groundwater legal acts are briefly explained. There could be several conflict interfaces among farmers on one hand and implementation agencies on other hand. Acts are prepared to resolve various conflicts which arise at several interfaces. It is necessary to recognize these interfaces and understand conflicts. Possible measures for conflict resolution are discussed.

22.1 Issues in the Implementation of Conjunctive Use Management

Strategies for the use of groundwater in canal command area have been discussed in an earlier Chap. 12. A case study on groundwater recharge and design of wells and pumping facility for making use of groundwater has also been presented in Chap. 12.

U. C. Chaube et al., *Canal Irrigation Systems in India*, Water Science and Technology Library 126, https://doi.org/10.1007/978-3-031-42812-8_22

The implementation of conjunctive use management plan implies unified control, regulation, and efficient use of surface water and groundwater for achieving specific objectives. Implementation is a process influenced by several factors, technical, legal, socio-economic, and organizational. It can become a complex and long drawn-out process of adjustment, if the issues involved are not properly dealt with. It is necessary to emphasize that the implementation of a conjunctive use management plan requires an interdisciplinary approach (CWC 1995).

Some of the issues in the implementation of the conjunctive use management plan are (Khan 1993):

(i) A large disparity in water prices from canal and tubewell sources would discourage using more expensive groundwater. The disparity is source wise, crop-wise, and season-wise, which discourages the implementation of design cropping pattern and delivery of irrigation water to respective crops.
(ii) For proper implementation of conjunctive use plan, two tiers of management organization viz. government organizational structure and farmers organization need to be fully functional. The main functions of a government organization are to provide an efficient and equitable water distribution service up to the distributary level, while farmers organizations are responsible for the equitable distribution of water among farmers and its efficient application in the field.
(iii) The lack of agreement on respective roles and the resulting inadequate coordination and cooperation among government agencies may seriously hamper the implementation of conjunctive use plan. In most existing cases, funding, design and construction, and operation responsibility are assigned to different agencies.
(iv) The landholdings and fragmentations, and the choice of crops and agricultural practices have a firm hold on the farmers' minds. Traditions dictate them to adopt certain attitudes, which are difficult to change. This breads certain prejudice against any change that is sought to be brought about. Under the influence of these prejudices, farmers may reluctantly follow these only so long as there is a compulsion.
(v) In the absence of groundwater acts, the government has no control over the development and use of groundwater through private tubewells. Similarly, in the absence of sufficient legal powers, irrigation managers are handicapped to properly exercise their duties in the settlement of disputes.
(vi) In some of the states, as many as four Government Departments may be involved to manage irrigation schemes and agriculture production (Irrigation Department, Ground Water Organization, Agriculture Department, State Electricity Board, and Command Area Development Authority (CADA)). Sometimes, there is a lack of coordination among the activities of these government departments.
(vii) A farmer belongs to several social groups. There can be many conflict interfaces, such as between WUA-Individual Farmer; WUA-Project Authority; WUA—Religions/Cast Group; WUA—Electricity Department; and Project

Table 22.1 Recommendation of irrigation commission on water charges

Form of conjunctive use	Water charges
Pumped water from tubewells sunk alongside of a canal for augmenting canal supplies	Normal irrigation rate for the areas irrigated by channels as the two waters cannot be separated and quality of service is same
Water from shallow tube wells sunk as an anti-waterlogging measure, put into irrigation channels	Normal irrigation rate for the areas irrigated by channels as the two waters cannot be separated and quality of service is same
Private tubewells or filter points sunk in canal commands for irrigating crops when canal water is not available or is available inadequately	Normal canal water rates for use of canal water. But no charge where frmer uses water only from his own sources
State tubewells sunk in a canal command to irrigate pockets which cannot be served with canal water	Normal tubewell rates
Tubewell water for a second crop and canal water for the first crop	Canal water rate for the first crop and tube well rate for second crop
Tubewell and canal water for irrigating the same area in a crop season	Both canal and tubewell charges should be levied

Authority-private tube well owners. Legal Acts are enacted to resolve various conflicts that arise at several interfaces. Therefore, in framing legal Acts, it is first necessary to recognize multiple interfaces and the type of conflicts.

22.2 Irrigation Water Charges

22.2.1 Surface Water and Ground Water Charges for Crops

Water charges for the conjunctive use of surface water and ground water depends upon the form the conjunctive use may take. Irrigation Commission (GOI 1972) has provided the following guideline (Table 22.1) for fixing water charges for various forms of conjunctive use.

22.2.2 A Case Study on Water Rates

A study of the Lakhawati Branch Canal (Upper Ganga Canal System) in the western part of Uttar Pradesh was carried out by IIT Roorkee (Goel 2003). The study area receives canal water during the monsoon period. The project envisages irrigation of 24.5% paddy during the kharif season. However, calculation of water requirement of the paddy crop and availability of surface water indicates that 19.5% of maize area can be brought under paddy. During this period, surface water is insufficient to meet the water requirements of crops; therefore, groundwater is pumped to supplement available supplies.

In order to minimize the cost of supplying both surface water and groundwater and to control water table position, more surface water is allocated to areas where the water table position is deeper and more groundwater is pumped in the areas where the water table position is shallower. The command area was divided into ten zones to allocate source and groundwater. Most of these zones are commands of a distributary or a group of distributaries of minors.

Table 22.2 compares surface water and groundwater charges. For canal water supply, farmers were charged at nominal rates on the basis of the area irrigated under different crops. For the groundwater supplied from state tube wells, the farmers were charged on a volumetric basis, which worked out to be higher than the surface water irrigation charges. Moreover, the charges for groundwater from state tubewells were double for Rabi crops as compared to those for Kharif crops.

Such differentials in irrigation charges present a hurdle to implementing the optimum conjunctive use plan of surface water and groundwater. Therefore, it is desirable to rationalize irrigation water charges for the various sources of water (Chaube 2000).

There are different types of disparities in water rates. These are source wise, season-wise, crop-wise, and unit wise. These are discussed as follows:

Crop wise disparities

Irrigation charges for paddy and wheat crops are the same, i.e., Rs. 143.26/ha, but the depth of water required by paddy is just double that needed for wheat (Table 22.2). Similarly, the depth of water required by sugarcane and paddy is more or less the same, but irrigation charges for sugarcane are about twice that for paddy.

Table 22.2 Comparison of surface water and ground water charges

S. no.	Crop	Delta m	Vol. of water in ha-m	Surface water rate		Groundwater rate		Difference in rate	
				Rs/ha	Rs/ ha-m	Rs/ha	Rs/ ha-m	Rs/ha	Rs/ha-m
1	Rice	0.8611	0.8611	143.26	166.36	227.33	264	084.0	097.64
2	Wheat	0.4146	0.4146	143.26	345.50	218.9	528	075.0	182.5
3	Arhar	0.2017	0.2017	106.21	527.00	053.25	264	−52.75	−263.0
4	Mustard	0.2445	0.3445	106.21	434.4	129.10	528	022.89	093.6
5	Berseem	0.2910	0.2910	049.42	169.8	153.65	528	104.23	358.2
6	Guar	0.1048	0.1048	086.45	824.9	027.67	264	−58.78	−560.9
7	Maize	0.1947	0.1947	086.45	444.0	041.4	264	−45.0	−180.0
8	Sugarcane	0.9610	0.9610	237.12	246.47	504.4	528	267.28	281.30

Source wise disparities

Surface water charges are very much less than groundwater charges. For example, for paddy crop surface water irrigation charges are Rs.143.00/ha and for groundwater it works out to Rs. 227 per hectare for an irrigation depth of 0.8611 m. The irrigation charges for different crops, for surface water and groundwater with present rates, have been calculated and are presented in Table 22.2.

Season wise disparity

Season-wise disparity is also there in the present irrigation system where charges for groundwater in Rabi season are just double of those for Kharif i.e. for Rabi it is Rs. 528/ha-m and for Kharif it is Rs. 264/ha-m.

Region-wise disparity

Rates of irrigation are different for different regions i.e., in the Sarda system, the irrigation charges are different from those in the Ganga Canal System.

22.3 Rationalization of Water Charges

22.3.1 Principles to Be Followed

The following principles should be kept in view while fixing the rates for irrigation water.

1. Water charges should be adequate to cover the annual maintenance and operation charges and part of fixed costs. Efforts should be made to reach this idea over a period.
2. Water charges should be generally levied on a crop basis, giving due consideration to crop water requirements except in irrigation from tubewells and to water user associations.
3. The crop water charge should be related to the increased income from crops. It should range between 5 and 12% of the gross income, the lower limit being applicable to food and fodder crops, and the upper limit to cash crops.
4. The rates should be uniform for the same class of supply over a region and the disparity in charges for the same class of supply between regions should be minimum.
5. Where lift irrigation is done at the farmer's cost, water is generally used economically, and wastage is reduced. The resulting saving in water justifies lower water rates.
6. Water rates should be revised after every five years.
7. While promotional water charges may be necessary for projects, where cultivators are not familiar with irrigated agriculture and the water demand is not keen, concessions on a long term basis are undesirable.

Table 22.3 Example on rationalization of water charges for various crops

Crop	Water charge Rs/ha	Water charge as percent of gross receipt	Water charge as percent of net income
Rice	533.0	4.00	7.45
Wheat	655.0	4.35	8.43
Arhar	250.0	2.89	6.25
Mustard	386.0	4.39	8.97
Guar	130.0	2.29	4.88
Berseem	460.0	5.21	8.89
Sugarcane	1424.0	4.43	7.80

22.3.2 Example: Rationalization of Water Charges in Lakhauti Branch Command

As part of a sponsored research scheme, the University of Roorkee (now IIT Roorkee) carried out an exercise on uniform irrigation charges for surface and groundwater so as to recover the total O & M cost of surface water and groundwater (Chawla 1991). The average O & M cost of surface water and groundwater for Kharif crops worked out to be Rs. 1238.0/ha-m. In Rabi, only groundwater is to be used in the Lakhawati command area. The O & M cost of groundwater for Rabi crops worked out to be Rs. 1580.0/ha-m. Irrigation charges for various crops were estimated, taking into account the average O & M cost of irrigation water and crop water requirement.

In the case of paddy crop, about 57% of the applied water percolates and recharges the groundwater reservoir; therefore 50% of the water cost is proposed to be recovered from the beneficiaries as irrigation charges. Based on these considerations, the irrigation charges for various crops were worked out as given in Table 22.3.

The proposed water charges vary from 2.29 to 5.21% of gross income, which are well within the range of 5% to 12% of gross income, recommended by the Irrigation Commission of the Government of India (GOI 1972).

22.4 Improvements in Organization Structure

22.4.1 Deficiencies in Existing Organization

At present five Government Departments (Irrigation Department, Ground Water Organization, Minor Irrigation Department, and Agriculture Department, and Command Area Development Authority (CADA)) work for management of irrigation water and agriculture production. In the present organization structure, there are so many deficiencies as discussed below. But the main deficiency is the lack of proper coordination among various agencies at the field level. It is necessary to remove the following existing deficiencies for the successful implementation of conjunctive use plan:

(a) Field officers and staff are not reinforced with sufficient powers; therefore, they are not able to make decisions at the spot, which causes delay in maintenance works.
(b) Irrigation department's responsibility is only up to the outlet below which it is totally left to the farmers for the management and distribution of irrigation water (groundwater and canal water) among themselves.
(c) Supply of canal water is not reliable so people have a tendency to over flood their fields resulting in low field application efficiency, besides water logging and salinity problems. Also, in the absence of flow measuring devices correct flow is not maintained.
(d) There is no Ground Water Regulation Act, so people use groundwater indiscriminately, resulting in groundwater depletion and land subsidence.
(e) As far as state tubewells are concerned, electric supply is very irregular and fluctuation of voltages is very common, so cases of motor burning and other mechanical defects are being reported frequently.
(f) Staff and officers working in the command area development authority (CADA) are on deputation from other departments.

22.4.2 Example: Model Organization Structure for Conjunctive Use Management

The following organizational structure has been proposed for Lakhawati branch command project (Khan 1993). For correspondence and accounts there will be one section of accounts and correspondence.

The organization structure has been proposed (Table 22.4) on the following assumptions:

(i) All the state tubewells in the command area will be handed over to farmer's associations and management of canal irrigation system at tertiary level will also be given to the farmer's associations.
(ii) Dedicated electricity supply lines will be provided to the tube wells, to avoid interruption in power supply.
(iii) Ground water development by individual farmers will be controlled by the project manager.
(iv) Legal Acts will be enacted to control private development of ground water.

Table 22.4 Model organization structure for conjunctive use management

Designation	Number	Jurisdiction	Controlling officer
Project manager (executive engineer)	1	Lakhawati branch command project 49,500 ha	Superintending engineer of Madhya Ganga Canal Circle
Liaison officer	1	49,500 ha	Project manager
Hydrologist	1	49,500 ha	Project manager
Hydrology surveyors, mechanical supervisors, data analyst	2	24,750 ha each	Hydrologist
Area manager (assistant engineer)	3	Approx. 16,500 ha of CCA each	Project manager
Area supervisor (junior engineer)	5	Approx 3300 ha of CCA each	Area manager

22.4.3 Water User's Association (WUA) for Conjunctive Use Management

Farmers association or water user's association (WUA) will be formed on distributory level/minor level. All the presidents, outlet committees and Tubewell committees will be its members. These members will elect the following office bearers themselves.

1. President
2. Secretary
3. Treasurer

These association will be registered under irrigation department or under "Societies Registration Act". The duties and responsibilities of WUA will be as follows:

(i) Monitoring the flow in distributories.
(ii) Monitoring power supply to the tube wells.
(iii) Resolving conflicts that may arise among the outlet committees/tubewell committees.
(iv) Preparation of contingent plans for cropping pattern during scarcity conditions in consultation with the Agriculture Department officials.
(v) Having Liaison with Liaison officer regarding different issues.

Farmer's Federation at Lakhawati Project Level

For whole lakhawati command area there will be a single federation of farmers. The members of the federation will be presidents and secretaries of all 16 farmers association. These members will elect the following office bearers:

1. President
2. Vice President
3. General Secretary

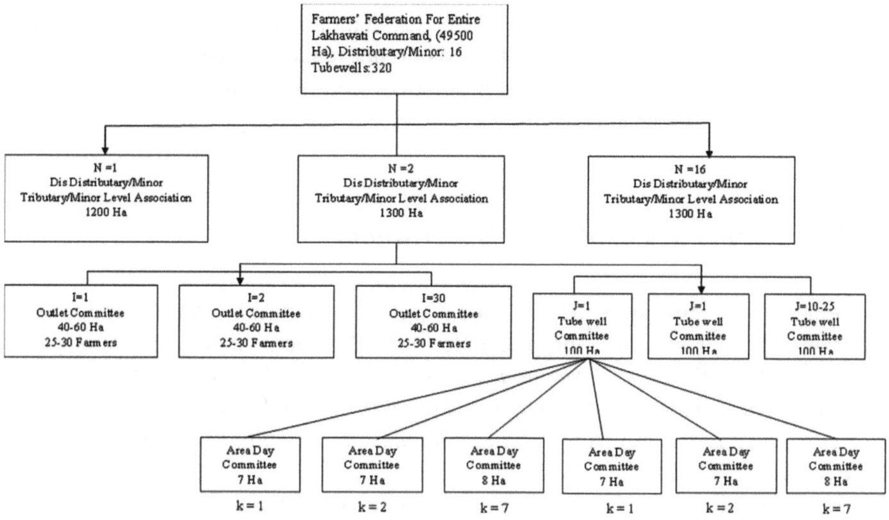

Fig. 22.1 Organization structure of farmers associations

4. Treasurer
5. Joint Secretary

Due representation will be given to different reaches and various strata of the society.

Duties and responsibilities of farmers' federation will be as follows:

(i) Discussing Irrigation policy matters with the project manager and other officers of the Lakhawati command project and taking decisions.
(ii) Formulation of cropping patterns in consultation with Agriculture Department officers.
(iii) Making suggestions for the release and stoppage of irrigation water/electric supply in the main system and main feeders, respectively.
(iv) Maintaining a cordial relationship among the farmers associations.

The physical structure of the farmers organization of Lakhawat command project is given in Fig. 22.1.

22.5 Surface Water Rights and Legal Issues

In India during the pre-independence period, several irrigation laws were enacted proclaiming the government's power over the use of water. The government has carried on with these irrigation laws even after independence without considering the water rights of people. The existing water rights of people and powers of the rights of the states are sometimes in conflict. These laws are very old and have many lacunae, that is why government agencies are handicapped to deal with the legal problems of irrigation water. Since there is no act regarding regulation and control on groundwater development, so the problems faced in groundwater regulation are very common and of serious nature.

22.5.1 Rights of People and Government

The concept of the nature of the state's rights in natural water is embodied in different acts. For instance, the preamble to the Northern India and Drainage Act 1873 provides: *Whereas, throughout the territories to which this Act extends, the provincial government is entitled to use and control for public purposes the water of all rivers and streams flowing in natural channels, and of all lakes and other natural collections of still water ...*

Most of the irrigation Acts empower the state government to issue a notification when water of rivers, streams etc. is to be applied for public purpose. For such use or allocation of water, the canal officers are empowered to enter into any land, remove any obstruction, close any channels, and exercise necessary powers. Similar provision to this effect exists in the Bombay Irrigation Act, the Rajasthan Irrigation Act and Drainage Act, J & K Canal and Drainage Act, the Bihar Lift Irrigation Act, and Mysore Irrigation Act. The Mysore Irrigation Act specifically provides that private irrigation works are to be constructed only with the sanction of the state government and subject to conditions which it may impose. The right of government to control the supply and distribution of irrigation water is not merely a proprietary right but is a sovereign right, as was expounded by the Madras High Court in Secretary of State V. Nageswara Iyer where it was observed.

A right by prescription can be acquired as against the proprietary right of another but not as against the sovereign right which under the Indian law the state possesses to regulate the supply of water in public streams so as to utilize it to the best advantage.

A comparative study of rights of people and government regarding surface water has been made in Table 22.5. For private wells and private tube wells people have unlimited right to draw water and state has no right to regulate except in Gujarat. Physical laws and legal Acts are in conflict with respect to the occurrence and utilization.

22.5.2 Lacunae in Northern India Canal and Drainage Act

Under this Act irrigation rights are transferable with the permission of government. But confusion arises where a person, owning land and entitled to get supply for a specific duration, sells some land but contract to allow more time for getting water than is proportionate to the land sold.

Table 22.5 Surface water rights of people and state

S. no.	Source	Rights of people	Rights of the state
1	Rivers and streams	Customary, reparian and other rights recognized by the courts and under the easement act	Absolute rights under irrigation and other laws
2	Canals	No rights, permission to use on payment of fees under warabandi, osrawandi and other schemes under irrigation laws	Absolute rights of ownership and for sale on fees
3	Tanks, lakes artificial	Individual rights of land owners customary usufruct rights of people	No rights if tank on private land power of the govt. to regulate use of private tanks in some states. Rights vested in the panchayats or municipality if tank on public land
4	Tanks, lakes natural	Customary rights of the people recognized by the courts, and under easement act	Absolute rights of ownership and use

Example: Court Case of Smt. Ishar Kaur Vs Harnam Singh. There the respondent sold eight bighas and eighteen biswas of his land to appellant and allowed her to take water supply for three hours which was more than ordinary allocation to the area that was transferred to her. But respondent later objected and filed a suit for injunction. It was held that there was a valid contract between the parties.

Under the Act the D.C.O. has not got sufficient powers to decide disputes at various levels. The dispute settlement machinery is not confined to the irrigation department. In the settlement of disputes the involvement of collector and civil court makes it very a lengthy and complicated process, thus disputes are not resolved timely.

For the construction and maintenance of w/c, Gano Sabha is responsible under the Act. D.C.O. is not empowered to acquire land for the construction of watercourses. In practice it is found that the requisitioning authority i.e. Tehsildar did not hand over the land for a long period further under section 50-c. It is the responsibility of Gano Sabha to get the land acquired or requisitioned but they have practically no means to prepare the land requisition or acquisition papers.

In the Act confusion arises regarding the conviction of offences because of the duplication of penal provision of irrigation statute on one hand and Indian penal code on the other hand. The case on this point is Mewa Ram vs. Emperor. In this case the accused persons were convicted under section 430 of the IPC for committing the offence of mischief. What was provided in this case was that accused persons forcibly opened the canal distributary and diverted the flow of water. But there was nothing to show that they permanently diminished the supply of water. In these circumstances the question was under which provision of law accused persons should be penalized, under section 430 of IPC or under section 70 of the Northern India Canal and Drainage

Act. The question was material in view of the fact that if conviction were to be under the former, the punishment up to five years could be given where as if conviction were to be under the later the maximum punishment could be only rupees fifty or one month punishment in default.

The Gano Sabha or irrigation panchayats are not having sufficient powers regarding collection of irrigation charges from individuals and also getting supply from the irrigation system. They are not reinforced with sufficient powers to acquire land for the construction of w/c and to resolve disputes among cultivators.

22.6 Ground Water Rights

22.6.1 Existing G/W Rights in Different States

Rights for groundwater belong to land owners, since they form part of the dominant heritage and the state's tenancy laws govern land ownership. There is no limitation on how much ground water a particular landowner may draw. Table 22.6 shows ground water rights of people and the state.

The consequence of such legal framework is that only the landowner can own groundwater in India. It leaves out all the landless and tribals who may have (community) rights over land but not private ownership. It also implies that the rich landlord can be water lords and indulge in openly selling as much water as they wish.

To ensure proper and equitable distribution of ground water it is necessary to separate water rights from land rights. The only state to have a groundwater law is Gujrat (Applicable in only one district). There is no separate law. Sections have been added to Bombay Irrigation Act 1976 (79). It merely tries to regulate water harvesting and marketing by restricting the depth of tubewells and introduces licensing procedures. Section 94 prohibits the construction of tubewells beyond 45 m depth without permission and section 99 of the Act regulates the wastage of groundwater. Evidently

Table 22.6 Ground water rights of people and state

S. no.	Source	Rights of people	Rights of the state
1	Wells private	Absolute rights of the land owners	No rights
2	Wells public	Customary rights of groups, castes or communities, but rights for all under the constitution and the civil liberties act	Rights to regulate
3	Tube wells private	Unlimited right to draw water from tube wells on private land	No rights to own or regulate except in one state Gujarat so far
4	Tubewells public	Usufruct. rights granted by the state	Rights to regulate

even this introductory regulation is welcome. However, a great deal of thinking and research need to be done to come up with appropriate groundwater rules.

In this context, the recent Kerala high court decision in Attakoya Thangal vs. Union of India, becomes relevant. In this case the residents claim that the excessive pumping of groundwater by the rich farmers was threatening the very availability of groundwater for all. Under Article 21 of the constitution, they claimed that their life opportunities were being threatened since the depleting groundwater resource was likely to become saline. The court upheld their claim. Such a decision once again makes the right to water natural or fundamental rights under Article 21 rights to life.

The Mysore Irrigation Act contains provisions for control by the state over the construction of wells where public irrigation works are constructed or proposed to be constructed. In such notified areas, no person can construct any well without prior permission of the state government.

Some states have enacted legislation regulating irrigation by tubewells. For instance, the Uttar Pradesh legislature has enacted the State Tube Well Act 1936 and the Punjab legislature has enacted the Punjab State Tube Well Act 1954.

22.6.2 *Lacunae in Ground Water Act*

All the state tube well Acts applicable in different states provide the construction and repair of state tube wells and supply of water from them to private land owners for irrigation purposes. There is no provision in any statute to regulate the digging, spacing, and depth of wells by individuals. We still lack any comprehensive legislative measure to regulate and control groundwater development. The model irrigation bill also has a major defect that it provides for separate management of surface water and groundwater, while conjunctive use management plan requires an integrated approach to the problem.

22.7 Conflict Interfaces

Acts are prepared to resolve various conflicts which arise at several interfaces. Therefore, in the framing of legal Acts, it is first necessary to recognise various interfaces and the type of conflicts.

In the management of irrigation system at micro level, Water User's Association (WUA) has been considered as a central body. It will act as representative body of farmer's interests. Interface size between individual farmer and other agencies can be significantly reduced. Conjunctive use project, various agencies, and groups are involved so conflicts are natural. There can be so many conflict interfaces, such as group, religious group, and agriculture department. Also project authority may have conflicts with individual farmers and electricity department. These conflict interfaces have been described in brief with possible remedial measures.

WUA—Individual Farmer

Water User's Association will be the first formal committee in the farmers organization. It will be involved in the management of tubewells and tertiary system and distribution of water at micro level. WUA will have conflicts with the individual farmers in the matter of supply of surface and ground water, O & M of watercourse and field channels, collection of irrigation charges, protection of irrigation works and illegal irrigation, and cropping pattern.

(i) **Supply of water**: Water User's Association will supply water to farmers on a rotation system under the written contract between WUA and farmers. The irrigation rights will be transferable with the transfer of land by the permission of project authority. The duration of supply for transferred land should not be more than allocated hours in proportion of the transferred land. Conflicts will develop in the case of interruption or failure of supply. WUA will own state tube wells besides canal water supply. The advantage of dual source will be that if association fails to supply water from one source the other source can be utilized to maintain the supply. In the contract, provision should be made for compensation in the case of failure of supply due to causes which are not mentioned in the contract. Similarly, if any farmer fails to receive water on his turn, his turn will automatically will be over.

(ii) **Maintenance of water courses and field channels**: Conflicts will arise when WUA will not get labour and material timely and sufficiently. If any farmer fails to provide his share, WUAs will provide his share for the time being, cost of labour and material provided by WUA will be recovered from the defaultor with sufficient penalty so that repetition of default could be checked. WUA should be empowered to acquire land, material and labour for emergency repair of water courses and field channels but it should not be required to do normal maintenance work.

(iii) **Collection of irrigation charges**: Conflicts will arise if farmers do not pay irrigation charges timely or if there is any mismanagement in areas irrigated. As far as dues from farmers are concerned, time of payment should be specified in the contract; preferably it should be just after the harvesting of each crop. If a farmer fails to pay his dues within specified time, irrigation charges should be recovered with late fees which may be 20–25% of normal charge. If a farmer fails to pay irrigation charges of two consecutive seasons's crops legal action should be taken against him besides the stoppage of supply. To avoid conflicts, measurement of area irrigated should be done in the presence of concerned farmer.

(iv) **Protection of irrigation works**: Conflicts will arise when individual farmers will not care for the irrigation work. So some responsibility should also lie with individual farmers. WUAs should be empowered to impose fine up to Rs. 1000/- for irrigation offenses, damaging irrigation works. There should not be duplication of conviction of offences. Offences should be convicted under the amended Northern Indian Canal and Drainage Act.

(v) **Illegal irrigation**: Illegal irrigation through cuts or bunds is one of the main sources of low irrigation efficiency. Therefore, to check illegal irrigation WUA should be empowered to recover penal rates, five times more than normal irrigation charges, for illegal and unauthorised use of irrigation water. In case if it is not possible to identify the actual offender penal charges will be recovered from the owner of the land irrigated.

(vi) **Cropping pattern**: If farmers choose crops other than designated others, ultimately conflicts will arise. So to avoid conflicts the name of crops to be irrigated should be specified in the contract. WUA should give water supply to only those crops which are specified in the contract.

WUA—Project Authority

Project authority will be responsible for the operation and maintenance of irrigation system and delivery of water supply only up to distributory and minor level, while O & M of tertiary system and state tubewells and distribution and management of irrigation water below distributory/minor level will be the responsibility of water user associations. WUAs may have conflicts with the project authority in the matter of water supply, O & M of watercourse, field channels and statetube wells, collection of irrigation charges, measurement of flow etc.

(i) **Supply of irrigation water**: Quantity, schedule, and duration of water supply should be decided in the beginning of each crop season. If the project authority fails to supply irrigation water due to a reason which is not mentioned in the contract WUA can claim compensation according to provisions made in the contract; on the other hand, if WUA fails to receive water on its turn, the turn will automatically be over and project authority can claim compensation for any loss due to spillage.

(ii) **O & M water course and field channels**: Though O & M of water course and field channels will be the responsibility of WUA, in the case of poor maintenance of water course project officers can give warning to WUA and in the case of total failure project authority will repair the water courses and field channels at its own cost. The cost will be recovered from WUA with a penalty ranging Rs. 1000 to Rs. 5000/-.

(iii) **O & M of state tube wells**: All state tube wells and their goals will be owned and maintained by the WUAs with the help and guidance of project officers and staff. In the case of poor maintenance, project authority will have a right to intervene and repair them at the cost of repair to be recovered from WUA with penalty. Project officer and staff can suggest and help farmers in regard to repair and maintenance of state tubewells. If WUA fails to repair, project authority will interfere and will repair at its own cost which will then be recovered from WUA with penalty according to the provision made in contract.

(iv) **Collection of water charges**: Project Authority will sell water to WUA on a volumetric basis. Conflicts may arise in the case of failure of payment of water charges by WUA. Therefore, the time of payment should be mentioned in the contract and there should be provision for late fees if WUA fails to pay water charges in time.

(v) **Measurement of flow**: This is also a point where conflict may arise between project authority and WUA. So far avoiding conflicts, metering devices should work correctly and efficiently. Project staff should take measurement of flow in the presence of representative of WUA.

WUA—Electricity Department

As far as electricity department is concerned, it will only supply electricity to the tubewells. Conflict may arise between WUA and electricity department in the matter of supply and theft of power, and electricity charges.

(i) **Supply of power**: Electricity department will supply power to the project (tubewells) under the written contract between electricity department and WUA with liaison officer as a mediator. In the contract, provision should be made for the flexibility of demand. Conflicts will arise in the case of interruption in supply. To avoid conflicts, the extent, schedule, and duration will be decided in the beginning of each crop season. In the case of failure of supply, WUA can claim for compensation. Since dedicated lines are proposed for tube wells, chances of interruption are less.

(ii) **Theft of power**: The theft of power is common which results in heavy revenue loss to the electricity department. Conflicts may arise on this matter. Checking of theft of power will be the responsibility of WUA. Under the contract, the electricity department will have a right to claim for any loss due to the theft of power.

Project Authority—Individual Farmer

Conflicts between individual farmer and project authority will arise mainly in the matter of indiscriminate use of groundwater, spacing, and depth of private tube wells which will result in ground water depletion besides the inequitable distribution of groundwater. To avoid this the project authority should declare the whole project command as notified area. In such a case, any person desiring to sink a well in the area will have to apply to the project authority for grant of a permit for the purpose. As far as existing users in the area are concerned they will have to apply to the authority for grant of a certificate. The holder of agricultural land where a well is situated will not use groundwater for any purpose other than agriculture or drinking and will not waste water for any reason. WUA will help project authority to regulate the pumping or use of both of ground water especially through private tubewells.

WUA—Agriculture Department

Conflict may arise regarding the time of supply or quality of agriculture input. Therefore, supply of seeds, pesticides, and fertilizer should be done under the written contract with WUA. In the case of untimely supply and unsatisfactory output or adverse effect of agriculture input on crops, WUA can claim for compensation.

Similarly, in the supply of machinery on a hire basis if department fails to provide machinery in time or provides it in bad order WUA can claim for compensation for any loss due to delay or disorder of machinery. On the other hand, if there is any loss or damage to machinery in the custody of WUA, WUA will have to pay for cost of repair/cost of the machinery.

WUA—Religions Group

Like in other parts of India, cultivators of Lakhawati command also belong to two main religions i.e. Hinduism and Islam. Conflict may arise between WUA and religious groups in the matter of schedule of water supply and poultry and pig farming.

(i) **Schedule of water supply**: Schedule of supply is the main point where conflict with religious groups may arise. For example, if a Hindu farmer is asked to receive water on Holi, Deepawali and Ramnavmi he will object to it, similarly a Muslim farmer will object to take water on Eid and, from 12 noon to 2.0 PM on Fridays. Therefore, to avoid conflicts, WUA should fix turns of individual farmers by keeping in view the religion of farmers.

(ii) **Poultry and pig farming**: Poultry farming and pig farming will be highly objectionable in the localities of Hindus and Muslims, respectively. Therefore, to avoid conflicts pig farming and poultry farming should be done in only those localities where this farming is not objectionable.

WUA—Cast Group

Cultivators within their religious groups belong to different cast sections. WUA may have conflicts with cast groups mainly in the matter of cropping pattern, dairy farming, and water supply. For example, oil men cast will be interested in oil seed production, among Muslims Ansaris will be oriented towards cotton production, Khachhis will be more interested in vegetable farming. Also, Yadavas among Hindus will be more interested in dairy farming, while other casts may have objection due to probable harm to their crops by cows and buffaloes, similarly Balmiki community will be more interested in pig farming but it may be objectionable to other casts.

Conclusions on Conflict Interfaces

It is seen that the present system of irrigation management is not efficient. Generally farmers are unaware of Irrigation Acts, therefore, provision of legal Act is meaningless for them. Penalties and fines for irrigation offences, such as damage to irrigation works and unauthorized irrigation, are nominal in comparison to benefits of additional water getting through illegal acts. A fine of Rs. 50/- can easily be paid by farmers if they get more water. Moreover, there is no provision for social sanctions against offenders. Also, irrigation officers are not reinforced with sufficient powers to exercise their duties, especially in the matter of settlement of disputes and requisition of land for the construction of water courses. There is no ground water Act to control and regulate its development especially through private tube wells.

Therefore, for efficient management of irrigation system it is proposed to assign more powers to irrigation officials especially in the matter of land requisition. Fine and penalties for unauthorized irrigation have been revised in order to make them really penal. To control and regulate the development of ground water through private tube wells, the whole area is required to be notified. In such a case every user of the area will have in get a certificate from project authority for using ground water. Various conflict interfaces and causes of conflicts have been identified and possible remedial measures have been suggested.

Questions

1. Explain why the charges for supply of canal water and ground water for irrigation are different?
2. Why the charges are different for different crops?
3. What is the basis for the rationalization of irrigation charges?
4. Discuss interface problems between
   (i) farmers and electricity departments, and
   (ii) farmers and canal irrigation department.
5. Identify various government agencies involved in the conjunctive use of surface water and ground water.
6. Explain needs for formations of water use associations (outlet committee, tubewells committee).
7. Explain role of water users association in the conjunctive use management.
8. Explain the need for a comprehensive Ground water Act.
9. Discuss the existing lacunae in Northern India Canal Act.
10. Discuss the existing ground water rights.
11. Identify the conflict interfaces relating to the use of groundwater.
12. Discuss the conflicts arising due to religion and cast feelings.
13. Discuss the usefulness of Water Users Association in managing various types of conflicts.
14. Explain improvements in organization structure for conjunctive use management in Lakhauti canal Command of Madhya Ganga Canal system.
15. Explain improvement in organization structure of WUA for conjuctive use management.

References

Chaube UC (2000) Conjunctive use management of surface water and ground water Course package (unpublished). Centre for Continuing Education IIT Roorkee

Chawla AS (1991) Evaluation and improvement of irrigation technology in tube well commands: a report. University of Roorkee, Roorkee, WRDTC

CWC (1995) Guidelines for planning conjunctive use of surface and ground water in irrigation project. Central Water commission, New Delhi

Goel MMK (2003) Spatially distributed simulation of an irrigation system, PhD research thesis under the supervision of Prof. U. C. Chaube, Indian Institute of Technology, Roorkee

GOI (1972) Report of the irrigation commission. Ministry of Irrigation & Power, Government of India, New Delhi

Khan S (1993) Implementation of conjunctive use management plan—a case study. M.E. dissertation report, supervised by Pro U C Chaube,WRDTC, University of Roorkee, Roorkee, December

Chapter 23
Economics of Irrigation and Flood Control

Abstract This chapter provides a detailed description of irrigation and flood control economics. During the implementation of an irrigation project, economic analysis may be employed as an effective management tool. Economic analysis is also required to evaluate economic performance during the O & M stage of an existing project and to justify rehabilitation and modernization need. Concept of 'before project', 'with project in future' and 'without project in future' in economic analysis are explained. The existing procedure for the estimation of the benefit of irrigation projects is based on the 'before project' and 'with the project in future' concepts. The annual cost is made up of interest on capital costs, depreciation, and operation, maintenance, and repair (OMR) costs. The interest is worked out on the total investment, not on decreasing investment due to depreciation. It is more appropriate to estimate benefits on a 'with project' and 'without project' basis. Data should be collected on cropping patterns, irrigated and un-irrigated areas under the present condition and 'without project' (Wo/P) and 'with project' (W/P) future conditions for each season. Farm budgets should show the net economic returns per hectare for each crop under the present, without the project, and with project conditions. The transition periods from the present to projected areas and cropping patterns to a projected level of agricultural technology should be realistically assessed. The objectives of this chapter are to learn the current method of economic evaluation and understand the information required for the improved method of economic evaluation. Procedures for economic analysis of (i) canal irrigation; (ii) groundwater development; (iii) economics of water losses, groundwater, and lining; and (iv) economics of sprinkler irrigation are discussed. Numerical examples are given to illustrate the procedures. In deltaic regions, drainage and flood control are also part of the irrigation project. Therefore, methods for the economic evaluation of flood control projects with possible improvements are explained and illustrated with examples. The existing methodology for the economic appraisal of flood control projects in India is explained. Improvements in the estimation of benefit and cost and in the comparison of benefits with costs have been suggested by Rashtriya Barh Ayog (1980). These have been briefly explained with illustrative examples. A case study of the Mhaisal Lift Irrigation Scheme (Mhaisal LIS) is presented to illustrate the use of CWC guidelines (CWC, Central water commission (CWC) Government of India, Ministry of Water

U. C. Chaube et al., *Canal Irrigation Systems in India*, Water Science and Technology Library 126, https://doi.org/10.1007/978-3-031-42812-8_23

Resources, 2010) for the computation of cost, benefit, and benefit–cost ratio of canal irrigation scheme.

23.1 Economic Evaluation Criteria: Irrigation Water Charges

23.1.1 *International Bank Criteria*

It has been a standard practice for International Banks to use the Economic Internal Rate of Return (EIRR) criterion. The project is considered economically viable if its EIRR exceeds the economic opportunity cost of capital in the country concerned. Because it is difficult, in practice, to estimate precisely what this value should be for each country, 10–12% is used for all countries as the minimum rate of return for projects for which an EIRR can be calculated, and the rate at which to choose the least-cost options (EDRC 1997).

The projects may have some benefits or costs that cannot be quantified or valued. The minimum rate of return within the range of 10–12% could be interpreted to take account of these factors. The banks follow the guideline given below:

- accept all independent projects and subprojects with an EIRR of at least 12%;
- accept independent projects and subprojects with an EIRR between 10 and 12% for which additional unvalued benefits can be demonstrated and where they are expected to exceed unvalued costs;
- reject independent projects and subprojects with an EIRR between 10 and 12% for which no additional unvalued benefits can be demonstrated or where unvalued costs are expected to be significant; and
- reject independent projects and subprojects with an EIRR below 10%.

The Economic Internal Rate of Return (EIRR) or Economic Net Present Value (ENPV) is calculated using the most likely values of the variables incorporated in the cost and benefit streams. Future values are difficult to predict, and the project results will always be uncertain. The effects of different values should be investigated.

Sensitivity analysis is a simple technique to assess the effects of adverse changes on a project. It involves changing the value of one or more selected variables and calculating the resulting change in the NPV or IRR. Changes in variables can be assessed one at a time to identify the key variables. Possible combinations can also be assessed. Sensitivity analysis should be applied to project items that are numerically large or for considerable uncertainty.

23.1.2 Historical Changes in Evaluation Criteria in India

The criteria followed in India for considering whether a project is economically or financially viable have changed with time and location (GOI 1972).

Period/ year	Criteria
Before 1947	Under the British rule in India, the criteria initially followed was to classify a project as 'productive' or 'unproductive'. The productivity was defined in terms of the government receipts from the sale of water in excess of expenditure. However, this criterion was found to be too harsh, as the irrigation water rates could not be fixed high to recover the cost of project due to social constraints. Then, the rate of return was followed as the criterion which continued to be followed even after independence
Year 1949	After independence, the provision of irrigation was deemed to be a responsibility of the government for social welfare. The minimum acceptable rate of return on capital investment was lowered from the pre-independence level of 6–3.75%
Year 1960	The minimum acceptable rate of return on capital investment was again raised to 4.5% in 1954 and further to 5% in year 1960
Year 1964	Gadgil committee asserted that the minimum acceptable rate of return on capital criterion for sanctioning irrigation projects was highly inappropriate from a social point of view. It recommended a BC ratio of 1:5 for considering the project viable
Year 1972	The Irrigation Commission decided that the B/C ratio of 1 should be considered as acceptable for drought prone areas and B/C ratio of 1.5 or more for other areas
Year 1983	The B/C ratio criterion was replaced by the internal rate of Return (IRR) criterion. To qualify, projects were required to yield a minimum IRR of 9%. For the irrigation project located in drought prone areas, hilly track and in areas where 75% of the dependable flow of basin had already been trapped, a lower IRR of 7% was allowed. However, the B/C ratio criterion recommended by Irrigation Commission continues to be followed in most of the states even now

23.1.3 Current Method of Economic Evaluation in India

The criterion initially followed for selecting an irrigation project was to classify a project as 'productive' or 'unproductive'. The Irrigation Commission appointed by the Government of India also examined the aspect of benefit–cost analysis and recommended the following procedure for calculating the benefit–cost ratio (B/C) (GOI 1972) is provided in Table 23.1.

The above method is currently being followed for appraising irrigation projects by the Planning Commission/Central Water Commission (CWC 2010). Only such irrigation projects that have a B–C ratio of 1.5 or more, are approved.

The Irrigation Commission also recommended the lower limit of B/C ratio as 1 to cover projects intended to provide protective irrigation in drought-prone or backward/ tribal areas and schemes for the modernization of existing works to improve their

Table 23.1 Procedure for calculation of benefit cost ratio

A	Before introduction of irrigation	B	After irrigation
(a) Gross receipts			
1	Gross value of farm produce	1	Gross value of farm produce
2	Dung receipts at 30 of fodder expenditure	2	Dung receipts at 30% of fodder expenditure
Total gross receipts			
(b) Expenses			
1	Expenditure on seeds	1	Expenditure on seeds
2	Expenditure on manure and fertilizers	2	Expenditure on manure and fertilizers
3	Expenditure on hired labour (human and bullock)	3	Expenditure on hired labour (human and bullock)
4	Fodder expenses[a]	4	Fodder expenses[a]
5	Depreciation of implements[a]	5	Depreciation of implements[a]
6	Share and cash rent[a]	6	Share and cash rent[a]
7	Land revenue	7	Land revenue
		8	Interest on land leveling cost
(c) Total expenses			
	Net direct annual benefit (Total gross value minus total expenses) before introduction of irrigation		Net direct annual benefit (Total gross value minus total expenses) after introduction of irrigation and full development
	Net annual benefit attributed to irrigation		= Difference in the net (B–A)

Annual costs:

1. Interest on capital at 10% (annual interest rate)
2. Depreciation (straight-line depreciation) of works over a project lifetime 75 or 100 yrs
3. Administrative expenses (operation and maintenance cost, Rs.25/ha of CCA)

Benefit cost ratio $= \frac{\text{Net annual benefit}}{\text{Annual cost}}$

[a] As a percentage of the gross value of produce

efficiencies or provide carry-over storages over the years to improve the dependability of water.

The Government of India occasionally revises the land development cost per hectare (part of the capital cost) and O & M cost per hectare (recurring annual cost). Therefore, recent government notifications need to be referred to.

23.2 Limitations of the Current Method

Concept of Before Project, Without Project, and With Project: To identify project costs and benefits, the situation "without the project" should be compared with the situation "with the project". The "without-project" situation is not the same as the "before-project" situation. The "without-project" situation can sometimes be represented by the present levels of productivity of the relevant resources. However, the present levels of productivity would frequently change without the project, and this should be taken into account in defining the "without-project" situation (USAID 1982).

As is evident from the foregoing, the current method is based on the 'before and after' concept. Even without the implementation of irrigation projects, some changes in agricultural production are likely to occur with the passage of time. By thinking in terms of 'before and after', even such changes wrongly get attributed to the irrigation project, although they are not actually caused by it. The proper way would be to analyze the future situation on a 'with and without project' basis.

The annual cost, which occurs in the denominator of the benefit–cost ratio, comprises interest on capital costs, depreciation, and operation, maintenance, and repair (OMR) costs.

Interest Rate: The interest rate is worked out on the total investment, not on decreasing investment due to depreciation. In other words, it amounts to the recovery of cost more than investment, which does not appear to be justified. On the other hand, if interest is calculated only on the balance cost at any point in time, then the present value of all recovered costs plus interest is found to be equal to the initial investment, no matter what the pattern of investment recovery costs over the life of the project is assumed.

There has been a tendency in the past to underestimate costs and overestimate benefits so that the B.C. ratio becomes favourable for getting clearance for the project. It is necessary to provide safeguards to ensure the reliability of estimates. Therefore, estimates of benefits and costs for the project under consideration should be checked at least on a sample basis by an outside independent agency. Also, actual benefits and costs for other projects in the O & M stage can provide useful information for a realistic assessment of benefits and cost.

23.3 Information Required for Economic Analysis

The following information is needed for the economic analysis of an irrigation project:

(i) **Cropping Patterns**: Under the present condition and in the future 'without project' (Wo/P) and 'with project' (W/P) conditions for each season and both for irrigated and rainfed conditions.

(ii) **Irrigated and rainfed areas are** present without project in the future and with project in the future and for each season. Areas to be irrigated should be based on average water supply rather than design supply.
(iii) Farm budgets show net economic returns per hectare for each crop under the present without project and with project conditions.
(iv) Annual operation and maintenance costs for irrigation works, land development, pumping, and any other costs for water use management.
(v) Capital costs of major works, land development, and any other required investments necessary to achieve the objectives of the project.
(vi) Transitions from present to projected areas and cropping patterns to a projected level of agricultural technology.

23.4 Net Value of Crops

The ultimate net returns occur only after full development and adaptation of project. These are assumed to occur at a uniform annual rate and continue up to the end of project life. The net value of crops (Rs. per ha. of crop area) and composite net value (Rs. per ha of command area) are to be estimated first to work out the ultimate net return.

Economic crop budgets for principal crops grown in the region are prepared based on projected economic prices. The crop budget is used to evaluate each crop's net value under three conditions, namely (a) before project, (b) without project, and (c) with project. An example of finding the net value of the crop is given in Table 23.2. The data used is only to explain the procedure and is not representative of the current market condition.

23.5 Composite and Ultimate Net Return

The net value of crops based on projected economic prices for the crops in the region and cropping patterns in the project area under three conditions, namely (a) before project, (b) without project and (c) with project, are used to find the composite net return per hectare of the cropped area for a particular cropping pattern. The ultimate net return for each condition is found by multiplying the composite net return per ha. with the cropped area. For project condition, the average water supply factor should also be taken into consideration. The method is illustrated with the following example of the Karwappa Nalla Irrigation Project in Maharashtra, for which the irrigation command area is 3890 ha (Table 23.3) (Chaube 1997). The monetary values in the example are not representative of the present-day market conditions. They are used here only to illustrate the procedure for the computation of composite and ultimate net return. The ultimate net return is evaluated for before project, with project and without project in future condition.

Table 23.2 Economic crop budget and net values of crops: Karwappa Nalla (Maharashtra)

Crops	Yield Qtls/ha	Price Rs./Qty	By product Rs./ha	Inputs		Net value Rs./ha
				Labour Rs./ha	Other Rs./ha	
1. Before project						
Kharif rainfed						
Paddy	15.0	137	160	270	775	1170 (150 × 137 + 160–270-775)
Rabi rainfed						
Wheat	5.0	127	64	165	455	79
2. Without project						
Kharif rainfed						
Paddy	17.0	161	176	281	978	1654
Rabi rainfed						
Wheat	5.8	153	77	180	560	224
3. With project						
Kharif irrigated						
Paddy	40	144	400	420	2101	3638
With project						
Rabi irrigated						
Wheat	25	153	256	330	1558	2193

23.6 Annual Net Returns over Different Years

The ultimate net returns occur only after full development and are assumed to continue up to the end of project life. Time is required to place the additional area under irrigation and implement new irrigated agriculture technology. Thus, the annual net returns during the transition and development period will be lower than the ultimate net returns. The transition periods will differ for 'with project' (W/P) and 'without project' (WO/P) conditions. Under the W/P conditions, the transition period would be required for improved agricultural technology and area transition from rainfed to irrigated conditions, whereas under the Wo/P conditions, the transition period is required only for adopting improved agricultural technology.

A review of the development history of irrigation projects indicates that, in general, there has been a shortfall in achieving the design areas for irrigation i.e., the ultimate irrigated area is observed to be lower than the designed irrigation area. However, the area ultimately irrigated tends to be reached in three or four years. In some projects, the seasonal irrigation water demands have shifted heavily toward higher-water-using Rabi over Kharif compared to design projections. Normally, the following transitions could be adopted for different conditions:

Table 23.3 Cropping pattern and net economic returns: Karwappa Nalla (Maharashtra)

S. no.	Crop	Net value of crop Rs./ha	Cropped area as % of command area	Composite net return (Rs./ha)	Ultimate net return
1. Before project-rainfed					
i	Paddy	1170	65	760	Ultimate net return is equal to the composite net return x cropped area. For example, if the irrigation command area is3890 ha, then
ii	Sorghum	419	6	25	
iii	Pulses, other	327	14	46	
iv	Fallow	0	15	0	
	Total			831	
2	Without project rainfed			1172	2.:Without project 3890 × 1172 i.e. Rs.4,559,080
3	With project-irrigated				3. With project: (a) Irrigated: 3890 × 5163 × 0.97 i.e. Rs.19,482,000 (0.97 is irrigation water supply factor)
i	Perennials			–	
ii	Two seasonal			–	
iii	Kharif			–	
iv	Rabi			–	
	Total			5163	
v	Unirrigated			334	(b) Unirrigated: 3890 × 334 i.e. Rs. 129,900 So total for with project condition is Rs. 20,781,000

(i) The transition from 'Before Project' to Wo/P condition: 8 years (technology transition).
(ii) The transition from 'Before project' to W/P condition: 8 years.

   (a) Irrigation area transition: 5 years (i.e., 20%, 40%, 60%, 80%, and 100% of ultimate area are brought under irrigation in each successive year).
   (b) Agricultural Technology Transition: 8 years (on the incremental area brought under irrigation).

A study of the productivity of present irrigation compared to projected productivity as measured by the effect on the net W/P returns suggests an agricultural technology transition beginning at 0.5 level i.e.; present productivity is only one-half of full development productivity. Thus, after the construction period, the transition begins at 0.5 level and in each successive year incremented by 0.0625, i.e. [(1–0.5)/8] up to eight years to reach the level of 1.0.

During the transition period, the net returns grow according to the proportion of area brought under irrigation and according to the proportion of area brought under improved agricultural technology. A composite transition factor indicating the impact of these two transitions can be worked out for each year of the transition period.

The annual net returns are then calculated for W/P and Wo/P conditions, and net returns attributable to the project are determined. The procedure is illustrated with the following example (Table 23.4). The ultimate net return with project is 20,781,000 Rs. and the ultimate net return without the project is Rs.4,559,000.

Composite transition factor for a particular year = Sum of the product of area transition factor and appropriate Agricultural Technology transition factor for each year after construction.

Table 23.4 Annual crop net economic returns: Karwappa Nalla (Maharashtra) (irrigation command area 3890 ha)

Transition			Net returns Rs. 000				
Year	Composite area	Old area	W/P			WO/P	Net total (6–7)
			New irrigation	Present transition	Total		
1	2	3	4	5	6	7	8
4	0.0	1.0	0.0	3233	3233	3233	0
5	0.11	0.8	2286	2719	5005	3399	1606
6	0.24	0.6	4987	2139	7125	3565	3560
7	0.38	0.4	7897	1500	9397	3730	5667
8	0.52	0.2	10,806	774	11,580	3869	7711
9	0.69	0	14,340	0	14,340	4052	10,288
10	0.75	–	15,586	–	15,586	4228	11,358
11	0.81	–	16,833	–	16,833	4393	12,440
12	0.87	–	18,079	–	18,079	4559	13,520
13	0.93	–	19,326	–	19,326	4559	14,767
14	0.96	–	19,950	–	19,950	4559	15,391
15	0.99	–	20,573	–	20,573	4559	16,014
16	1.00	–	20,781	–	20,781	4559	16,222
24			20,781		20,781	4559	16,222
25			20,781		20,781	4559	16,222

- Col (3) = indicates the areas not yet brought under irrigation
- Col (4) = ultimate net return x composite transition factor
- Col (5) = Col. (3) × ultimate net return Wo/P condition
- Col (6) = Col (4) + Col (5)
- Col (7) = Agricultural Technology transition factor x ultimate net return for without project (Wo/P) condition
- Col (8) = Col (6)−Col (7). This column gives the net return in different years (transition period and post-transition period up to project life)

23.7 Estimation of Cost and Benefit/Cost Ratio

23.7.1 Initial Cost Estimate

The initial cost includes construction cost, engineering and administration cost, right-of-way cost, and other minor costs. The construction cost is the amount spent on completing works outlined in the plans and specifications. Engineering and administrative costs are the expenditure on preparing the necessary plans and specifications and supervising the construction work. The right-of-way cost is the opportunity cost of using the land required for project installation and maintenance.

23.7.2 Land Development Cost

The government has recently proposed the land development cost @ Rs.1000–3000 per ha. The land development capital costs include the development of communal field channels, distributaries, and drainage below 40 hectares (Part-I) and land leveling and drainage on individual farms (Part-II). It is assumed that lands under 0.6% and over 3% slopes would not be graded. Part I costs (field channels and drains below the outlet) are applied to 100% of CCA, and Part II costs (land development on individual farms) to 50% of CCA.

23.7.3 Annual Operation and Maintenance Costs

These are estimated at 50 Rs./ha plus one percent of headworks cost. Operation and maintenance costs for land development are estimated at 5% of capital costs. All capital costs and prices should correspond to the same base year. Where sunk costs or earlier or later estimates are involved, these should be corrected using all India wholesale price index (WPI) and labour cost indices (LCI) or an appropriate combination of these indices (weighted construction index).

$$\text{Weighted Construction Index(WCI)} = \text{WPI} \times (\text{Material fraction}) + \text{LCI} \times (\text{Labour fraction of cost})$$

For example,

$\text{WCI} = 1.384 \times 0.6 + 1.188 \times 0.4 = 1.3$

and base year cost = (Construction Cost Estimate)/WCI.

Further construction cost factor (CCF) is to be applied to convert financial cost into economic cost, i.e.

$$\text{Economic Capital Cost} = (\text{Base Year Cost}) \times (\text{CCF})$$

Similarly, land development and O & M's financial costs should be converted to economic costs by applying the construction cost factor. The salvage value of the project components and land development should be taken into account. The salvage value of land development may be assumed as 100%.

23.7.4 Calculation of Benefit/Cost Ratio

The benefit and cost streams are converted into the present worth of benefit and cost by applying a present worth factor. An appropriate discount rate has to be assumed for this purpose.

23.8 Economic Analysis of Groundwater Development

23.8.1 Economic Cost of Ground Water Development

The cost estimation of groundwater development is not as difficult as the cost estimation of surface water development. Structural components of a tubewell or dug well are few, and these are explicitly defined. The main problem in groundwater development is to design a proper capacity of the well from cost considerations. The cost function of a tubewell or dug well should provide a solution to find the optimal capacity which caters to the groundwater need at the lowest cost.

The procedure to evaluate the economic cost of groundwater development is explained below. The cost structure of a tubewell is depicted in the chart given below (Fig. 23.1).

Financial investments on various items are first worked out based on market prices for materials and equipment and the technical data. The financial investments are converted to economic costs by adjusting economic prices. Works and equipment have a different life. The capital costs of works and equipment are annualized, considering the life span and the discount rate. The total annual cost is the sum of annualized capital cost and the O & M cost.

The cost per unit of pumped water delivered is obtained by dividing the annualized cost of tubewell by the annual effective water output. The procedure of annual cost estimation is illustrated with the following example (Table 23.5):

1. Power requirement 0.067 KWh/m^3 @ Rs.0.3/kWh
2. Pumping 1760 hr @ 2.01/hr @ Rs.0.90
3. 1% per annum
4. 6% per annum
5. 8% per annum.

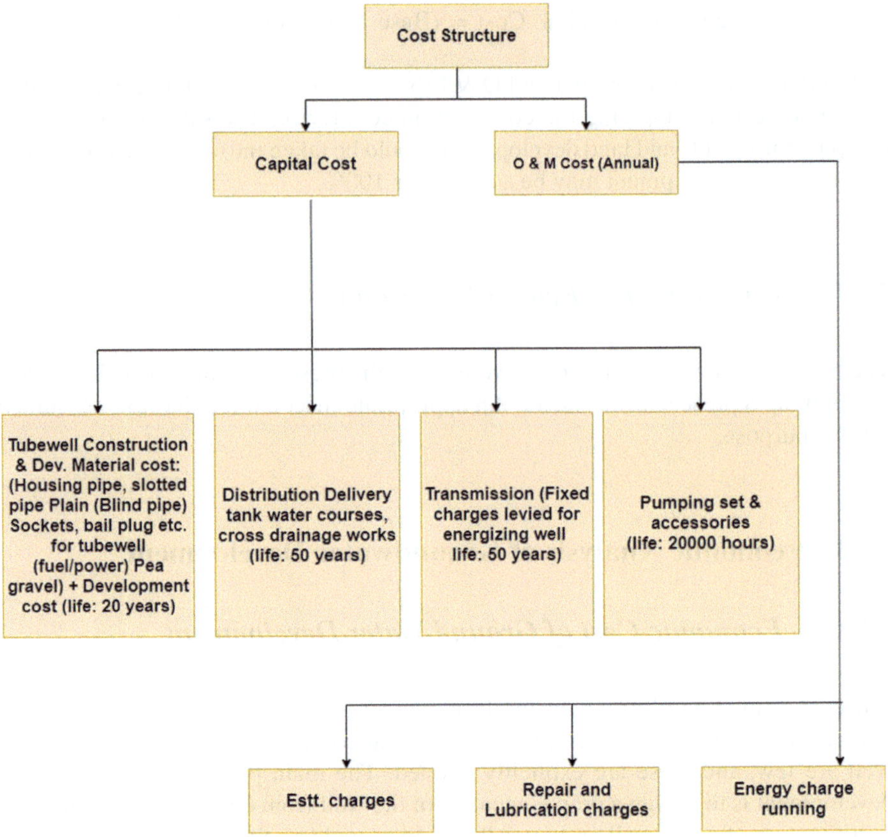

Fig. 23.1 Flowchart to develop economic cost of groundwater development

23.8.2 Benefits of Ground-Water Development

Benefits are calculated per 1000 m^3 using the information on cropping patterns and estimates of irrigation requirements at the pump outlet. The ultimate crop return is calculated based on crop prices and crop patterns for 'with project' conditions. The procedure is the same as described earlier for the surface irrigation project. 10-year irrigation area transition and 5-year agricultural technology transitions are assumed to work out composite transition factors and the economic crop returns during the transition period. The economic crop returns (during the transition and ultimate) are worked out for 1 ha size.

The pump outlet's annual water requirements (m^3) are worked out for the cropping pattern under W/P conditions. Thus, the benefit stream for the W/P condition is found. The difference of annual economic crop returns (W/P condition-Wo/P condition) for 1000 m^3 per year groundwater supply is then evaluated and converted to the present

Table 23.5 Economic cost of groundwater development

	Tubwell in fractured rock			Dug-cum-bore well in deep alluvium		
	Works	Equipment	Total	Works	Equipment	Total
Well yield per year (m^3)	–	–	20,000	–	–	99,000
Life (years)	30	5	20,000	25	5	
Financial investment (Rs.)	17,800	3800	21,600	14,100	11,900	26,000
Conversion factor	0.75	1.00		0.75	1.00	
Economic rate (Rs.)	13,350	3800	17,150	10,580	11,900	22,480
Annual capital cost at 12% (Rs.)	1660	1050	2710	1310	3300	4610
O & M cost						
Fuel/power (Rs.)		400	400		3170	3170
Lubrication and repairs (Rs.)	130	230	360	110	950	1060
Sub-total	130	630	760	110	4120	4230
Total annual cost		3470				8840
Cost per 1000 m^3 at well (Rs.)		179				89
Field irrigation efficiency		0.77				0.73
Cost per 1000 m^3 at field (Rs.)		232				122

worth at a proper discount rate. The equivalent uniform annual benefit can then be found.

23.9 Economics of Water Losses, Groundwater, and Lining

The purpose of an economic study of water losses is to assess the economic effects of seepage losses on potential irrigated areas of lining and potential for groundwater development and illustrate an analytical methodology for such studies.

The estimate of the irrigation potential of a project is based on the design water supply and assumed efficiencies for the main canal, distribution canal, and field channels. The actual available crop water supply will depend upon the actual operational, conveyance, and application losses. Even with moderate seepage rates, losses from long unlined canals can significantly reduce areas that can be irrigated with available water supplies. Thus, the actual economic crop returns could significantly differ from those anticipated for the irrigation potential created or planned.

The recovery of water losses by groundwater development or improving the crop water supply by lining canals can restore the benefits which otherwise are lost. However, groundwater development or lining alternative involves additional cost.

Table 23.6 Canal water budget

	Item	Unlined canal	Lined canal
1	Available to the canal (mm^3)	26.711	26.711
2	Operational loss	−5.34	4.00
3	Seepage loss	−9.23	2.31
4	Sub-total	12.14	20.40
5	Distribution and field channel loss	−2.43	−4.08
6	Sub-total	9.71	16.32
7	Field application losses	1.94	3.26
8	Available for crop	7.77	13.06
9	Overall efficiency	29.1	48.9
10	Total losses	18.94	13.65
11	Available for pumping	9.47	6.82

Note
Line 1: Water available in the reservoir for the canal = gross storage−dead storage−evaporation loss−allocation for other use
Line 2: 20% of line 1 for unlined canals and 15% for lined canals
Line 3: Seepage loss (mm^3) = (Sectional wetted perimeter, m) x (length, m) x (loss rate cumec per million meter square) × days of operation) × 24 × 3600 × 10–6
Line 4: Line (1)−Line (2)−Line (3)
Line 5: 20% of line 4
Line 6: Line 4−Line 5
Line 7: 20% of line 6
Line 8: Line 6−Line 7
Line 9: Line 8/Line 1
Line 10: Line 2 + line 3 + Line 5 + Line 7
Line 11: 50% of line 10. It is assumed that 50% of losses could be recovered through groundwater development

23.9.1 Canal Water Budget

The first step in the economic study of water losses, groundwater development, and the lining is to prepare a water budget for the irrigation canal under unlined and lined conditions to find the actual water availability for crops and water available for pumping. The following example (Table 23.6) illustrates the procedure for the canal water budget.

23.9.2 Irrigated Area Correction

The area that can actually be irrigated will be according to the actual water available, not according to the design water supply. The irrigated area correction under unlined and lined conditions can be estimated by finding the water supply factor as explained below:

$$\text{Water supply factor} = \frac{\text{Available water for crops after all actual losses}}{\text{Design water supply for crops}}$$

Available water: Water supply factor = design water supply.

Available water: The water available for crops after deducting all actual losses.

Design water supply: Designed seasonal allocation to the canal × delivery efficiency.

e.g. if allocation for Rabi is 31.47 mm^3 and delivery efficiency is 0.48, then the water available for the crop (as per designed supply) is $31.47 \times 0.48 = 15.1$ mm^3.

Water supply factor (unlined) 7.77/15.1 = 0.51.

Water supply factor (lined) 13.06/15.1 = 0.86.

Water supply factor (pumping of seepage water) 17.24/15.1 = 1.14.

These water supply factors are used for finding actual irrigated areas and the corresponding crop returns under various conditions. The procedure for irrigated area correction for crop returns has been explained earlier.

23.9.3 Benefit–Cost Evaluation

23.9.4 Cost of Recovering Seepage Water

The procedure for the estimation of groundwater development cost has been explained earlier. The cost is evaluated per 1000 m^3 annual groundwater supply. Thus, the cost of pumping the seepage canal water (50% of the losses are normally assumed as recoverable through pumping) can be estimated.

23.9.5 Benefits from Recovery of Seepage Water

The net economic returns of the conjunctive surface and seepage canal water use depend upon the improved water supply factor due to the recovery of seepage water.

23.9.6 Cost of Canal Lining

The lining of the canal increases canal cost but reduces seepage losses. The procedure to evaluate the project's cost with lined canals is the same, except that the additional lining cost is added to the capital cost.

The economic study of the Shivna Irrigation Project (Maharashtra) has shown that lining becomes economical if seepage losses exceed about 6–8 CFS/Msf for lining costing Rs.30 to Rs.40/m^2 regardless of whether or not losses are recovered by groundwater development.

23.9.7 Benefits of Canal Lining

The net economic returns will depend upon the improved water supply factor due to reduced seepage losses.

23.10 Economics of Sprinkler Irrigation

Economic analysis of sprinkler irrigation based on the current method (as recommended by Irrigation Commission) is discussed through an example of the Khera Kheri distributaries command area in Haryana. Table 23.7 shows benefit–cost analysis, and Table 23.8 shows calculation details to find the net value of produce. These tables are self-explanatory.

23.11 Existing Benefit Cost Analysis of Flood Control Projects

Benefit–cost analysis (B-C analysis) of flood control projects helps select the optimum level of adjustment to floods and decide the optimum combination of measures for the purpose. The erstwhile Ministry of Irrigation and Power, in their communication of 12th October 1955 to State Governments, prescribed a detailed format showing the manner in which benefits and costs should be analysed and evaluated for flood control projects before submission to Central Government for clearance. The rates prescribed regarding interest on capital and maintenance have been revised from time to time.

Costs

The capital cost of a project is compiled by adding costs of various items, such as investigation and planning, land, building, works, tools, and plants, work charged

Table 23.7 Benefit–cost analysis of sprinkler irrigation in Chaks of Khera Kheri distributary command area in Haryana

Description		Chak RD. 16,145/R 101 acres		Chak RD.18,000/R101 acres	
		Rabi	Kharif	Rabi	Kharif
		1978–1979	1979	1978–1979	1979
Cost of sprinkler set (Rs.)	Operation and maintenance	77,137	77,137	81,768	81,768
Electric power charges (Rs.)		1200	1200	960	960
Repairs (Rs.)		500	500	500	500
Labour charges for maintenance (Rs.)		3850	3850	3850	3850
Depreciation @ 10% (Rs.)		3857	3857	4088	4088
Interest @ 10% (Rs.)		3857	3857	4088	4088
Total expenditure (sum of line 2–6) (Rs.)		13,264	13,264	13,486	13,486
No. of working hours		787	860	591	940
Rate per hour (Rs.)		16.80	15.40	22.80	14.35
Net value of produce (Rs.) (Procedure is explained in Table 17.7)		73,477	28,849	71,455	38,862
Benefit–cost ratio Col.10–Col.7		5.54:1	2.18:1	5.30:1	2.88:1
Total area irrigated (ha)		41.4 + 12.3[a] = 53.7	24.6	41.4 + 7.8[a] = 49.2	29.9
Rate of sprinkler irrigation per ha Rs. Col.7–Col.12		247.0	539.2	274.1	451.0
Average cost for Rabi and Kharif		393.1		362.5	

[a] Additional area irrigated adjoining to the main chak

(a) Since the cost per ha irrigated by sprinkler is much higher than ordinary gravity flow irrigation, sprinklers' use is recommended only for those portions of a chak which are higher than the gravity flow command and are at present dependent only on rainfall.

(b) The net value of production shows an 8 to ninefold increase if sprinkler irrigation is introduced in previously rain-dependent areas.

staff, etc., as per prevailing standards in the Irrigation and Flood Control Department of State Governments. From this, an estimate of the annual average cost is obtained by adding annual interest, depreciation, and maintenance costs, each calculated as some prescribed percentage of total capital cost. These rates have changed from time to time. In 1972, the annual cost was recommended to be calculated at 16% of the capital cost in the case of embankment and drainage schemes and 17% in the case of anti-erosion and anti-seacoast-erosion schemes.

Benefits

Annual benefits of flood control works are estimated by finding out the average monetary value of annual flood damages based on at least 10 years of data before the

Table 23.8 Estimated net value of produce (CHAK RD. 16,145-R—Kharif 1979)

Crop	Area irrigated (q/ha) (q)	Produce per ha in qtls	Total produce in qtls	Rate per QTL (Rs./q)	Value of total produce (Rs.)	Inputs				Hired Rate (Rs./ha)	Labour total (Rs.)
						seed		Measures			
						rate (Rs/.ha)	Total (Rs.)	rate (Rs./ha)	Total (Rs.)		
Bajra	14.4	14.58	210	115	24,150	48.6	700	48.6	700	437.5	6300
Jawar	5.7	7.37	42	115	4830	49.1	280	–	–	550.79	3139
Cotton	4.5	14.67	66	350	23,100	36.7	165	464.4	2090	488.9	2200
Total					52,080		1145		2790		11,639
Say											11,640

Gross receipt		Expenses		Net value of produce	
Gross value of produce	52,080	Expenditure on seeds	1145	Total gross receipt	53,642
Dung receipts @ 30% of Fodder expenditure	1562	Expenditure on manure	2790	Minus total expenditure	24,793
Total	53,642	Expenditure on hired labour	11,640	Net value of produce	28,849
		Fodder expenditure on 10% of the gross value of produce	5208		
		Depreciation on implements @ 2.7% of the gross value of produce	1406		
		Share and cost rent @ 3% of the total gross value of produce	1562		
		Land Revenue 2% of the total gross value of the produce	1042		
		Total	24,793		

project's construction. From this, an estimate of the average annual damage after the construction of the project is deducted. There is a provision for the adjustment for the beneficiary value of silt deposition, if any. The benefit takes into account expenditure on relief and rehabilitation, revenue remission, agricultural loans, etc.

For flood embankments (most common method of flood control), the following procedure has been conventionally followed:

The annual cost of flood control component:

(i) 12% of allocated cost of dam (10% interest + 1% depreciation (100 years life) + 1% maintenance)
(ii) 16% of allocated cost of embankment (10% interest + 2% depreciation (50 years life) + 4% maintenance)
(iii) Total annual cost (i + ii)

(iv) Average annual damage computed on the basis of at least last 10 years of data
(v) Average annual damage anticipated after the execution of project
(vi) Savings in annual damage (item 5 - item 6)
(vii) B.C. ratio = item 7/item 4 (iii).

While costs are estimated at current prices, benefits, being regarded as the average of flood damage, are calculated at the respective current prices of past several years.

23.12 Improvement in Cost Estimation of Flood Control

Quantum of each work item, including labour, should be estimated in a realistic manner. National level guidelines issued by Central Water Commission and Neeti Ayog should be followed for the preparation of cost estimates. Table 23.9 shows CWC guidelines for estimating certain costs.

Additional capital works, such as anti-erosion measures (spurs, revetments), are often undertaken for stabilising the benefits of embankments. The cost of these additional capital works could easily amount to a significant proportion of the original capital cost (example Puthimari embankment in Assam, Kosi River embankment in

Table 23.9 Norms for estimating certain cost items

Items	Norms
1. Preliminary expenses	1% or more of cost of I-works. In the case of big projects costing more than Rs. 30 crores it could be up to 5% (1–2% for a diversion scheme and 2–4% for a storage scheme)
2. Cost of buildings	3–5% of I-works. 15% of cost of temporary and semi-permanent buildings shall be taken under V-receipts and recoveries
3. Miscellaneous (electrification, water supply security etc	4% of the cost of I-works. Resale value to be taken under receipts and recoveries
4. Maintenance during construction	1% of the cost of I-works less A - Preliminary, B-Land and Q-special T and P
5. Losses on stock	0.25% of the cost of I-works less A -preliminary, P-land and Q-special T&P
6. Establishment (for works let out on contract)	8–10% for concentrated works and 10–12% for scattered works (say canals)
7. Establishment (for works done departmentally)	15%
8. T and P	1% of the cost of I-works
9. Audit and account charges	1% of the cost of I-works
10. Abatement of land revenue	either at 5% of land cost or 20 times of annual revenue lost

Source GOI (1976, 1980)

Bihar). Provision for such works (if necessary due to the meandering nature of the river) should be made in original cost estimates.

Information on the construction schedule and time phasing of estimates should be provided so that proper time value of money is taken into consideration. The interest rate during construction should be considered, as it affects the cost if the project is delayed.

The annuity method (simple interest at 10% and straight-line depreciation at 2%) results in a higher than economically justified figure. The compound interest rate is more appropriate than simple interest. The sinking fund method of depreciation is better than straight-line depreciation. Instead of working out annual cost and annual benefit, the process of determining the present worth of cost and benefit through discounting would take care of annual interest and depreciation.

Maintenance cost should be computed at a certain percentage of the cost of works but not of the entire capital cost.

Cost allocation of multipurpose projects should be done, following the method of separable cost remaining benefit as recommended by Central Water Commission.

23.13 Improvement in Benefit Estimation of Flood Control

Flood control benefits are complex in nature. A better system of reporting and evaluation can be useful in removing part of the difficulty in the quantification of benefits. In brief, improvements are required in the following (NFC 1980; Chaube 1997):

- Assessment of the area to be benefitted with the help of contour maps should be made with respect to the design flood. Low-lying areas which are always submerged should be excluded.
- Longer the period of past annual damage data, the more reliable is the estimate of average annual damages. However, changes in prices, landuse, cropping patterns, development activities occur over a long period. Floods of similar type of magnitude can produce a different order of damage today. Damages should be evaluated in terms of the current year's prices. Damage data for 15–20 years should be used for deriving the average annual damage.
- Transfer payments (relief, rehabilitation, loans, and remission of land revenues) should not be considered in benefits.
- Additional areas made available by the project should be included in benefits by considering their productivity and other attributes.
- Benefits from the protection of land should be reckoned either in terms of an increase in income which is measured by damage prevented, or in terms of rises in the value of land but not both
- Effect of the fertilising value of silt brought by flood may be determined by comparing data on the yield of a representative sample of flood-affected farms with similar farms in nearby flood-free areas.

- Post-project damages continue to take place. Sometimes the damage may be produced by both flood and drainage congestion. The problem of drainage congestion may remain even after protection is provided against flood.

23.14 Improvement in B.C. Analysis of Flood Control

Benefits and costs need to be expressed in comparable terms. Therefore, benefits and costs should be expressed in terms of the same year's prices, and the time value of money should be considered.

While costs are estimated at current prices, benefits being regarded as the average of flood damages, are calculated at the respective current prices of the past several years. This procedure is defective. Flood damage data from different years should be evaluated in terms of prices of the base year, which are used for cost estimation.

Cost and benefits data of different years, as usually given in an unprocessed form, are not comparable. These should be properly discounted to represent the time value of money with respect to a particular year and then the B.C. ratio should be computed. Discounting should be done at the prescribed interest rate (or social discount rate).

The following formulae may be used for calculating discounting factors for any interest rate.

(i) Single payment present worth factor $= \frac{1}{(1+i)^n}$
(ii) Uniform payment series present worth factor $= \frac{(1+i)^n - 1}{i(1+i)^n}$

where i is the rate of interest, and.
n is the number of year.

Tables 23.10 and 23.11 illustrate the methodology with examples. It will be noticed that, according to the present methodology, the B.C. ratio remains constant irrespective of the time schedule of construction. When calculated by the improved method, the ratio goes down progressively as the period of completion is increased. The examples also show that the B.C. ratio could be higher or lower, depending upon the period of construction.

23.15 A Case Study of Mhaisal Lift Irrigation Scheme

A study carried out by Purandare and Bajaj (2017) has been reviewed to illustrate the existing procedure for the economic study of an irrigation project in India. The study is relevant as it is based on recent data.

Table 23.10 Example 2-B/C ratio as per prevailing and proposed methodology

Assumptions
Life of project—55 years
Rate of interest—10%
Years of completion—5 years

Year	Cost (Rs.)	Benefit (Rs.)	Present worth factor	Costs (Rs.) (2 × 4)	Benefits (Rs.) (3 × 4)
1	2	3	4	5	6
1	20,000		0.9091	18,182	
2	20,000		0.8265	16,530	
3	20,000		0.7513	15,026	
4	20,000		0.6830	13,660	
5	20,000	20,000 (for uniform series from 6 to 55th year)			
6	4000	20,000			
7	4000	20,000			
8	4000	20,000	6.1563	24,625	123,126
9	4000	20,000			
10	4000	20,000			
55	4000	20,000			
Total	100,441	123,126			

Notes B/C ratio as per prevailing methodology = 20,000/16,000 = 1.25
B/C ratio as per proposed methodology = 123,126/100,441 = 1.23

23.15.1 The Mhaisal Lift Irrigation Scheme (Mhaisal LIS)

The Mhaisal Lift Irrigation Scheme (Mhaisal LIS) is part of the Krishna Koyna Lift Irrigation Scheme (KKLIS) in Maharashtra state of India. Water is lifted from River Krishna in the rainy season and from Koyna reservoir viz. Shivajisagar by releasing water in River Krishna, as and when required during the other two seasons. The whole area falls under the Mhaisal LIS, falls under drought area. The total Irrigable command area of Mhaisal is 81697 ha. While cropped area expected is 82922 ha. By the end of 2014, 25,061.00 ha potential has been created. The work of Mhaisal was started in the year 1986.

23.15.2 Assumption for Calculation of Benefit Cost Ratio

"Benefit–Cost Analysis" is based on the CWC guidelines (CWC 2010).

Table 23.11 Example 4-B/C ratio as per prevailing and proposed methodology

Assumptions
Life of project—65 years
Rate of interest—10%
Years of completion—15 years

Year	Cost (Rs.)	Benefit (Rs.)	Present worth factor	Costs (Rs.) (2 × 4)	Benefits (Rs.) (3 × 4)
1	2	3	4	5	6
1	7000		0.9091		
2	7000		0.8265		
3	7000		0.7513	5259	
4	7000		0.6830		
5	7000		0.6209		
6	7000		0.5645		
7	7000		0.5132		
8	7000		0.4665		
9	7000		0.4241		
10	7000		0.3856		
11	6000		0.3505		
12	6000		0.3186		
13	6000		0.2897		
14	6000		0.2633		
15	6000		0.2394		
16	4000	20,000 (for uniform series from 16 to 65th year)			
…	…	…	…	…	…
65	4000	20,000	2.3735	9494	47,470
Total	61,277	47,470			

Notes B/C ratio as per prevailing methodology = 20,000/16,000 = 1.25
B/C ratio as per proposed methodology = 47,470/61,277 = 0.77

(a) The base year referred to is 2013–2014. The life of Mhaisal LIS project is 100 years.
(b) The data regarding the cost of the pumping system and raising main of Mhaisal LIS is 75% of that of KKLIS.
(c) The agriculture prices of year 2013–2014 are referred to. The output price values for without and with project situations for all agriculture crops are considered as given by the office of KKLIC. In addition to the direct agricultural income, the income through the by-product is also taken into consideration. The total agricultural income is estimated in the proportion 9:1 of direct agriculture benefits with market prices and with minimum support prices, respectively.

(d) For benefits through drinking water supply from Mhaisal to villages during scarcity time.

23.16 All Costs are Converted to Year 2013–2014 by Compounding the Past Values and Discounting the Future Values

(e) For the estimation of the gross value of farm produce, the table is referred to as mentionedin the guidelines of CWC.
(f) For Benefit–Cost Ratio estimation, the format is referred to as mentioned in the guidelines of CWC.

23.16.1 Input–Output Values

The input–output values without project (Table 23.12) and the potential created so far, with the project (Table 23.13) are estimated using the circular of the Joint Director Agriculture of the region and data of M.P. Krishi Vidyapeeth Rahuri.

23.16.2 Gross Values of Farm Produce

The gross values of farm produce without project (Table 23.14) and with project (Table 23.15) are estimated.

23.16.3 Annual Cost Calculations

The annual cost calculations are provided in Table 23.16

23.16.4 Estimation of Benefit Cost Ratio

The Estimation of Benefit Cost Ratio as per the Present Procedure (as Prescribed by CWC) is given in Table 23.17.

The Benefit Cost Ratio estimated is only 0.288 for the year 2013–2014 with around 30% potential creation, so with 100% potential creation this can be estimated as 0.864 which shows economic infeasibility of the project.

Table 23.12 The input–output values for the potential created without project

Seasons	S. no	Crops	The area under the crop		Output prices of produce			Input costs							
					Yield (Q/Ha)	Rate 2013–2014	Total income	Cost of seeds (2013–2014)	Total seed cost	Manure and fer. (2013–2014)	Total Mannure and fer. cost	Pesticides (2013–2014)	Total pesticide cost	Irrigation	Labour (2013–2014)
						Rs./Q	Rs. in lakhs	Rs./Ha	(Rs. Lakh)	Rs./Ha	(Rs. Lakh)	Rs./Ha	(Rs. Lakh)	Rs./Ha	Rs./Ha
	1	2	3	4	5	6	7	8	9	10	11	12	13	14	16
			%	Ha			= 6 × 4 × 5		= 8 × 4		= 10 × 4		= 12 × 4		= 7 × 0.2
Kharif	1	Sugarcane	4	3267.9	500.0	233.7	3818.5	15,216.0	566.9	13,159.5	430.0	1124.2	36.7	0.0	763.7
	2	Kh. Pulses	16	13,071.5	11.5	3164.0	4756.2	2400.0	354.5	295.6	38.6	0.0	0.0	0.0	951.2
	3	Groundnut	16	13,071.5	4.2	3520.0	1932.5	7033.0	1176.7	4237.8	553.9	0.0	0.0	0.0	386.5
Rabi	1	Rabi jawar	44	35,946.7	5.1	2022.4	3707.6	412.0	189.6	648.6	233.2	0.0	0.0	0.0	741.5
	2	Bajara	16	13,071.5	3.5	1408.0	644.2	72.0	12.0	1888.3	246.8	0.0	0.0	0.0	128.8
	3	Wheat	4	3267.9	9.5	2240.0	695.4	3272.0	136.9	3597.2	117.6	0.0	0.0	0.0	139.1
		Grand Total	100	81,697.0			15,554.4		2436.6		1620.1		36.7	0.0	3110.9
						Weightage assigned			**0.8**		**1.0**		**1.0**		**0.8**
						Revised input values are			**1949.3**		**1296.1**		**29.4**		**2488.7**

[a] Only pesticides and fertilizer (fer.) are tradable, so weights assigned are one. And the weightage given to all non tradable commodities is 0.8

Table 23.13 The input–output values for the potential created with project

Output prices of produce							
Seasons	S. no	Crops	Area		Yield (Q/Ha)	Rate 2013–2014 Rs./Q	Total income Rs. in lac
	1	2	3	4	5	6	7
			%	Ha			$= 6 \times 4 \times 5$
Perenial	1	Sugar cane	9.9	1638.2	1500.0	233.7	5742.57
	2	Pamogranate	6.8	1127.1	86.2	8131.0	7902.18
	3	Grapes	32.1	5317.4	140.0	3325.0	24,752.44
Kharif	1	Hy. Jowar	3.0	498.0	7.4	1792.0	65.59
	2	Maize	8.0	1317.3	18.0	1472.0	349.24
	3	Soyabean	3.8	629.1	17.0	2700.0	288.74
	4	Udid	5.4	891.2	6.6	3951.6	233.12
	5	Gr. Nut	0.6	91.1	4.2	3520.0	13.47
		Bajra	6.6	1087.7	24.0	1488.0	388.45

(continued)

Table 23.13 (continued)

Output prices of produce							
Seasons	S. no	Crops	Area		Yield (Q/Ha)	Rate 2013–2014 Rs./Q	Total income Rs. in lac
Rabbi seasons	2	Shalu Jawar	19.4	3208.2	5.1	2022.4	330.90
	1	Wheat	2.5	419.4	13.2	2240.0	123.81
	2	Jawar	1.3	209.7	8.4	1848.5	32.44
	3	Gram/Tur	0.6	104.8	11.5	3164.0	38.15
	2	Chili (Polyhouse)	0.2	26.2	25.0	9450.0	61.92
				16565.3			40323.0

(continued)

Table 23.13 (continued)

Input costs								
Cost of seeds (2013–2014)	Total seed cost	Manure and fer. (2013–2014)	Total Mannure and fer. cost	Pesticides (2013–2014)	Total pesticide cost	Irrigation	Total irrigation charges	Labour (2013–2014) (20% * col 90
Rs./Ha	(Rs. lakhs)	Rs./Ha	(Rs. lakhs)	Rs./Ha	(Rs. lakhs)	Rs./Ha	Rs. In lakhs	Rs./Ha
8	9	10	11	12	13	14	15	16
	$= 8 \times 4$		= 10 * 4		$= 12 \times 4$		$= 14 \times 4$	$= 7 \times 0.2$

(continued)

Table 23.13 (continued)

Input costs								
17,346.2	284.2	13,159.5	215.6	1124.2	18.4	6290.0	103.0	1148.5
20,404.0	230.0	24,847.6	280.0	1274.0	14.4	4720.0	53.2	1580.4
57,190.0	3041.0	63,160.3	3358.5	38,112.8	2026.6	4720.0	251.0	4950.5
1139.2	5.7	1959.6	9.8	0.0	0.0	240.0	1.2	13.1
4390.4	57.8	2902.6	38.2	0.0	0.0	240.0	3.2	69.8

(continued)

Table 23.13 (continued)

Input costs								
4202.0	26.4	2376.2	14.9	425.0	2.7	240.0	1.5	57.7
1704.2	15.2	1646.7	14.7	245.8	2.2	240.0	2.1	46.6
9002.2	8.2	4237.8	3.9	0.0	0.0	240.0	0.2	2.7
933.0	10.1	1612.0	17.5		0.0	240.0	2.6	77.7
527.4	16.9	648.6	20.8	0.0	0.0	0.0	0.0	66.2
4188.2	17.6	3597.2	15.1	0.0	0.0	470.0	2.0	24.8
586.3	1.2	1179.8	2.5		0.0	397.0	0.8	6.5
2712.0	2.8	295.6	0.3	0.0	0.0	350.0	0.4	7.6
6750.0	1.8	3420.0	0.9	2530.0	0.7	1042.0	0.3	12.4
	3718.9		3992.7		2064.9		421.5	8064.6
For Social Benefit calculations only: Weightage assigned	0.8		1		1		0.8	0.8
Revised input values are	2975.2		3992.7		2064.9		337.2	6451.7

Table 23.14 Gross value of the farm produce (without project)

		Crop			Gross value of farm produce				
Season	S. No	Particulars	Area of Mhaisal (in Ha) 81,697		Yield in (Q/ Ha)	Area (' 000 ha.)	Yield in (T/ ha)	Production (Th.Tonnes)	By Production (Th.Tonn
	1	2	3	4	5	6	7	8	9
			Crop %	area in Ha		= 4/ 1000	= 5 × 0.1	= 6 × 7	
Kharif	1	Sugarcane	4	3267.9	500	3.27	50	163.39	62.09
	2	Kh. Pulses	16	13,071.5	11.5	13.07	1.15	15.03	19.54
	3	Groundnut	16	13,071.5	4.2	13.07	0.42	5.49	12.6
Rabi	1	Rabi jawar	44	35,946.7	5.10	35.95	0.51	18.33	25.7
	2	Bajara	16	13,071.5	3.5	13.07	0.35	4.58	6.41
	3	Wheat	4	3267.9	9.5	3.27	0.95	3.10	5.12
		Grand Total	100	81,697.0					

Gross value of farm produce							Converting prices for SBCR	
Market price (WOP)	Minimum support	By product price	Market value	Minimum support value	By product value	Gross value of farm produce	Coversion factor for traded and non traded	Converted gross value of FP
In Rs./Q			(Rs. in lakhs)			(Rs. in lakhs)		(Rs. in lakhs)
10	11	12	13	14	15	16	17	18
			$=\frac{(8\times10\times0.9)}{10}$	$=\frac{(8\times11\times0.1)}{10}$	$=\frac{(12\times9)}{10}$	=13+14+15		=16×17
234	210	125	3436.7	343.13	74.5	3854.30	1	3854.30
3164	4300	55	4280.6	646.39	0.0	4296.97	0.8	3941.58
3520	4000	277	1739.2	219.60	157.8	2116.68	1	2116.68
2022	1520	361	3336.9	278.66	141.2	3756.99	0.8	3005.35
1408	1250	188	579.7	57.19	177.4	814.36	0.8	651.48
2240	1400	123	625.9	43.46	184.9	854.25	1	854.25
						16323.24		**14423.64**

Table 23.15 Gross value of the farm produce (with project)

		Crop			Gross value of farm produce				
Seasons	Sr. No	Crops	Area of Mhaisal (in Ha)		Yield (Q/Ha)	Area ('000 ha.)	Yield in (T/ha)	Production (Th.Tonnes)	By Production (Th.Tonn
			25,061						
	1	2	3	4	5	6	7	8	9
			%	Ha					
Perenial	1	Sugar cane	9.9	1638.2	1500.0	1.638	150.0	245.724	93.3751
	2	Pamogranate	6.8	1127.1	86.2	1.127	8.6	9.71859	0
	3	Grapes	32.1	5317.4	140.0	5.317	14.0	74.4434	0
Kharif	1	Hy. Jowar	3.0	498.0	7.4	0.498	0.7	0.36603	0.51244
	2	Maize	8.0	1317.3	18.0	1.317	1.8	2.37253	3.81978
	3	Soyabean	3.8	629.1	17.0	0.629	1.7	1.06939	7.96
	4	Udid	5.4	891.2	6.6	0.891	0.7	0.58995	0.22635
	5	Gr. Nut	0.6	91.1	4.2	0.091	0.4	0.03827	0.08801
	6	Bajra	6.6	1087.7	24.0	1.088	2.4	2.61057	5.54164
Rabbi seasons	1	Shalu Jawar	19.4	3208.2	8.4	3.208	0.8	2.68524	7.89852
	2	Wheat	2.5	419.4	24.3	0.419	2.4	1.01865	0.09771
	3	Jawar	1.3	209.7	8.4	0.210	0.8	0.17551	0.51624
	4	Pulses (Gram)	0.6	104.8	11.5	0.105	1.2	0.12057	0.00898
T.S	1	Chili (Polyhouse)	0.2	26.2	25.0	0.026	2.5	0.06553	0
				25,061.3					

Gross value of farm produce							Converting prices for SBCR	
Market price	Minimum support	By product price	Market value	Minimum support value	By product value	Gross value of farm produce	Coversion factor for traded and non traded	Converted gross value of FP
In Rs./Q			(Rs. In lakhs)			(Rs. In lakhs)		(Rs. In lakhs)
10	11	12	13	14	15	16	17	18
			$= \frac{(8\times10\times0.9)}{10}$	$= \frac{(8\times11\times0.1)}{10}$	$= \frac{(12\times9)}{10}$	= 13 + 14 + 15		= 16 × 17
233.7	210	126	5168.3	516.0	1176.53	6860.86	1	6860.86
8131.0	0	0	7902.2	0.0	0.00	7902.18	1	7902.18
3325.00	0	0	24,752.4	0.0	0.00	24,752.44	1	24,752.44
1792.0	1500	269	59.0	5.5	13.78	78.31	0.8	62.65
1472.00	1310	236	314.3	31.1	90.15	435.54	0.8	348.43

(continued)

Table 23.15 (continued)

Gross value of farm produce							Converting prices for SBCR	
Market price	Minimum support	By product price	Market value	Minimum support value	By product value	Gross value of farm produce	Coversion factor for traded and non traded	Converted gross value of FP
In Rs./Q			(Rs. In lakhs)			(Rs. In lakhs)		(Rs. In lakhs)
2700.0	2560	130	259.9	27.4	103.48	390.72	1	390.72
3951.6	4300	180	209.8	25.4	4.07	239.25	0.8	191.40
3520.0	0	145	13.5	0.0	1.28	14.75	1	14.75
1488.0	1250	166	349.6	32.6	91.99	474.23	0.8	379.38
2022.4	1520	269	488.8	40.8	212.47	742.04	0.8	593.63
2240.0	1400	123	205.4	14.3	1.20	220.82	1	220.82
2022	1500	360	31.9	2.6	18.58	53.16	0.8	42.53
3164	3100	148	34.3	3.7	0.13	38.20	0.8	30.56
9450.0	0	0	61.9	0.0	0.00	61.92	1	61.92
						42,264.43		**41,852.28**

Table 23.16 Annual cost calculations

(a)	Interest on the total cost of the project @ 10% on 378,328	37,833
(b)	Depreciation @ 1% of the cost of the project excluding ETP & c & d is 343088	3431
(c)	Depreciation of pumping system (assuming life of machinery 12 years) at 8.33% on Rs. 1209	1007.7
(d)	Depreciation of rising main (assuming life of rising main 30 Years) @ 3.33% on Rs. 20,169	503.7
(e)	Power charges for lift irrigation (year 2013–2014)	1873.57
(f)	O & M charges Rs. 455.24 per Ha. For croped area = 81,697 Ha	371.9
(g)	Maintenance cost @1% of cost of head works Rs. 228,140	2281

Table 23.17 Estimation of benefit cost ratio as per present procedure (as Prescribed by CWC)

	Calculation of benefit cost ratio				
					Rs. in lakhs
	Particulars			Without Project	With Project
A	**Gross receipts**			A.1	A.1
1	Gross value of farm produce			14,423.64	41,852.28
2	Dung receipts @ 30% of the fodder expenditure			519.25	1004.45
3	Total (A): Gross receipts (1 + 2)			14,942.89	42,856.73
B	**Expenditure**				
1	Expenditure on seeds			1949.25	2975.15
2	Expenditure on manure, Fertiliser, pesticides etc.			1325.50	6057.59
3	Expenditure on hired labour			2488.71	6451.68
		Without project	With project		
4	Fodder expenses	15% of Item A.1	10% of item A.1	1730.84	3348.18
5	Depreciation on implements	2.7% of item A.1	2.7% of item A.1	389.44	1130.01
6	Shared and cash rent	5% of Item A.1	3% of item A.1	721.18	2092.61
7	Land revenue	2% of Item A.1	2% of item A.1	288.47	837.05
8	Irrigation water Charges			0.00	337.20
9	Total (B): expenses (1–8)			**8893.39**	**23,229.48**
C	**Net value produce**				
1	Total Gross receipts (Total A.3)			14,942.89	42,856.73
2	Minus total expenses (Total B.9)			8893.39	23,229.48
3	Net Benefit of produce: (1–2)			6049.50	19,627.26
D	**Annual agricultural benefits**				
1	Net benefit with project (C.3)			19,627.26	
2	Net benefit without project (C.3)			6049.50	
3	Loss of agriculture benefits due to sumergence			0.00	
4	Net annual benefits (Net incremental benefits) (D): (1–2)			13,577.75	
	Total agriculture benefit			13,577.75	
		Revenue through water charges		46.25	
		Grand total of benefits		**13,624.00**	
		Total annual costs		**47,301.87**	
		Benefit cost ratio		**Net annual benefits** **Net annual cost**	
		Benefit cost ratio		**0.288**	

Questions

Questions on irrigation economics

1. Explain the current method of economic evaluation of government sponsored irrigation projects.
2. What are the limitations of the existing procedure of economic analysis of irrigation projects?
3. What do you understand by the three conditions of analysis: (i) before irrigation project, (ii) with irrigation project, and (iii) without irrigation project in future?
4. What information is required for an improved method of economic evaluation?
5. Explain the procedure to be followed for the estimation of annual net return over the years under three conditions i.e., i) before project, ii) without project in future, and iii) with project in future.
6. Collect the data for an existing canal irrigation project and illustrate the procedure for economic analysis of the project.
7. Collect the data for a state tube well/or a cluster of state tube wells and work out the annual equivalent cost of groundwater development.
8. Explain the cost and benefit of (i) water loss recovery for irrigation, (ii) canal lining, and (iii) sprinkler irrigation.

Questions on economics of flood control

Q.1 Explain in brief the following terms (i) Capital cost, (ii) depreciation, (iii) annual cost, (iv) annual benefit, (v) discounting, (vi) cost allocation in a multipurpose project, (vii) transfer payment, and (viii) present worth.

Q.2 Identify percentage rates used in computation of annual interest, depreciation, and maintenance cost for a flood embankment project.

Q.3 Explain deficiencies and suggest improvements in the estimation of annual cost.

Q.4 Explain the current procedure for the estimation of annual benefit of a flood control benefit. Identify deficiencies and suggest improvements.

Q.5 Give examples explaining how period of construction affects the B.C. ratio.

Q.6 Explain how damage-frequency analysis is useful in economic appraisal of a flood control project.

Q.7 Explain the need to employ an independent agency in the evaluation of a flood control project.

Project work on Mhaisan lift irrigation scheme

i. Explain each column and row of the tables.

ii. Redo the entire benefit–cost analysis for ultimate stage (design irr area and design cropping pattern in place of data pertaining to year 2013).

iii. Search the literature (or visit the Mhaisan lift irrigation scheme) to obtain data on recent status (cropping pattern, area equipped with irrigation facility, crop yields, unit costs, prices of input and output) and work out benefit cost ration.

References

Chaube UC (1997) Economic analysis, budget planning and project management. Course package (Unpublished). Centre for Continuing Education, IIT Roorkee, Roorkee

CWC (2010) Central water commission (CWC) Government of India, Ministry of Water Resources. Guidelines for preparation of detailed project reports (DPR) of Irrigation & Multipurpose Projects

EDRC (1997) Guidelines for the economic analysis of projects. Economic and Development Resource Center (EDRC), Website: www.vita.virginia.gov

GOI (1972) Report of the irrigation commission 1972, Vol I, Ministry of Irrigation, and Power, Government of India, New Delhi

GOI (1976) Broad guidelines for preparation of project estimates for major irrigation and multipurpose projects. Government of India: Central Water Commission, July

GOI (1980) Guidelines for preparation of detailed project reports of irrigation and multipurpose projects. Govt. of India, Min. of Irrigation

NFC (1980) Report of the National Flood Commission (Rashtriya Barh Ayog), Vol I, Chapter XII, Published by Govt. of India, March

USAID (1982) Maharashtra irrigation technology and management project paper volume I and II. U.S. Agency for International Development, New Delhi, April

Chapter 24
Operation and Maintenance Budgeting and Financing

Abstract Budgeting refers to the estimation of irrigation revenue and expenditure on O & M in advance of the ensuing year. The objectives of this chapter are to discuss the procedure for preparing budget proposals and possible improvements. Shortcomings in the conventional budget are discussed. The importance of performance-based budgeting is highlighted. While allocating the budget, the emphasis should be on the accomplishments rather than on the means of accomplishments. Not only the financial aspects but physical accomplishment and performance also need to be monitored and considered in evaluating the financial performance. Charges for the water supply generally take the form of a water rate, which varies with the crop grown and is levied on the basis of the area irrigated and matured. The water rate has no relation to the cost of supplying water but is fixed by each State Government from time to time, taking into consideration a number of factors. There is considerable variation in the methods followed in different states for assessing water charges. Existing water rates (i) in respect of flow vis-a-vis lift irrigation and the dates when the rates were last revised; (ii) crop-specific water rates in respect of flow Irrigation; and (iii) crop-specific water rates in respect of lift Irrigation are discussed, and state level water rates are presented. Norms prescribed for the O & M budget are given, followed by examples of O & M budget for two irrigation divisions of the Upper Ganga Canal. Data regarding budget and expenditure for O & M are analyzed. Norms for the allotment of funding for O & M are not generally followed. This has affected the overall performance of the canal system. It is observed that there is disproportionately more expenditure on operation compared to maintenance. A major portion of the O & M budget is spent on establishment charges (mainly salary component). Guidelines are given for working out the annual O & M budget.

24.1 General Aspects of Budgeting and Financing

Budgeting refers to estimating irrigation revenue and expenditure on operation and maintenance activities before the ensuing financial year. Financial year in south Asian countries is conventionally from 1st April of the current calendar year to 31st March

U. C. Chaube et al., *Canal Irrigation Systems in India*, Water Science and Technology Library 126, https://doi.org/10.1007/978-3-031-42812-8_24

of the next calendar year. The following are the general aspects of budgeting and financing (Chaube 1997):

(i) Estimation of irrigation revenue and expenditure on programmers and works and activities of O & M in advance for the ensuing (budget) year (April–March) is referred to as the budgeting year. Such estimation for the future year(s) beyond the budget year is termed as 'forecast' budget.

(ii) Budgeting for O & M is a mandatory requirement as per the Budget Manual in each state to ensure timely receipt of formal authorization/allotment of funds at the beginning of the year under consideration. The allocation should be distinguished from allowance; the latter is the formal authorization for incurring expenditure, while the allocation signifies the approval of funds, not the authorization to spend.

(iii) A procedure for receiving firm allotment at the beginning of the year, say by the end of April (fiscal year), must evolve. So far as the O & M activity is concerned, if the system of LOC (Letter of Credit) is followed, the credit should hold good for at least six months.

(iv) The budget allocation should be periodically reviewed in tune with the escalation in labour and material prices and submitted to the appropriate authority, intimating the revised need for budget allocation and fund allotment accordingly for the proper maintenance of the I&D System.

(v) Any work that is carried out to create an asset is a 'plan work'. In the O & M programs, rehabilitation and modernization works are considered as "Plan Works" and are accordingly classified under 'Capital Budget' or 'Plan Expenditure.'

(vi) The capital cost of a major/medium project includes the construction cost of project components and facilities and the construction of water courses up to 5–8 ha blocks.

(vii) O & M costs during the construction period are generally met from the capital budget for the completed portion of work, yielding partial benefits.

(viii) The O & M funds include the establishment (staff) charges and are generally allotted on an adhoc basis based on the potential created/C.C.A. per ha. Alternatively, the per km basis of channel length (length standardized for channel size) could also be considered for the allotment of funds. The physical condition of work also needs consideration.

(ix) Due to the uncertainty of allotment from the revenue (or non-plan) budget for O & M, the overall resources available for capital and revenue expenditure of the department/organization may be considered together for O & M so that a view may be taken about the availability of resources for O & M.

(x) An adequate allotment needs to be made for the proper upkeep of the existing systems rather than creating new irrigation potential to get better returns from the investment already made. This calls for a balanced view of the deployment of available funds for planned projects and for meeting O & M costs of existing works.

(xi) The establishment charges (part of O & M charges) have been mounting, particularly to continue the work-charged staff even after the completion of a project, according to the Supreme Court (SC) directive. Staff costs for 1991–1992 in J&K for medium and major projects was 76.67%. Suitable measures need to be evolved to limit staff costs, e.g., transferring surplus staff to new investigation works or construction sites.

(xii) Cost Allocation in Multi-Purpose Projects: Different methods have been applied to allocate costs and revenues. Cost allocation is required when more than one purpose is being served by the project. For example, a project may be generating hydropower, providing flood control, and providing irrigation. Even in the case of single-purpose project, the question of cost allocation in a project serving only one state is not relevant as the expenditure is met from the state revenues. However, in cases where the project serves more than one state and the Central Government contributes to one of the project purposes, cost allocation becomes necessary.

24.2 Guidelines for Preparation of Budget Proposal

Guidelines for budget preparation are generally available in the Budget Manual of the state irrigation departments. Rules and procedures are also contained in irrigation manuals/financial handbooks.

(i) The budget cycle extends over formulation and approval processes, implementation of the approved budget during the financial year, and final review and evaluation of budget performance at the end of the year of implementation. The formulation and approval period is in accordance with fixed deadlines as given in the Budget Manual. Accordingly, the organization units will need to have a disciplined approach to this activity.

(ii) Irrigation Revenue: Most of the receipts come from the irrigation charges which are estimated separately for Rabi and Kharif seasons. A small portion of revenue comes from the sale of water to cities and towns for domestic and industrial purposes and receipts for the cultivation of canal lands, water mills, and plantations.

Dates for the submission of budget estimates by Executive Engineer (EE) to Superintending Engineer are prescribed. For example, in one of the northern states, the dates are:

- Estimate for the ensuring year: 1 October
- Revised estimate for current year: 15 October
- Final revised estimate: 15 December.

The revised estimate would include actual receipts for Rabi and a forecast for Kharif. The final revised estimate would consist of actual receipts for both crop seasons—Rabi and Kharif. The revenue receipts are collected from the Office of the Board of Revenue in each state. All estimates are consolidated and submitted to the government through the Chief Engineer/Engineer-in-Chief.

The revenue assessment should be simplified by evolving a flat rate system for Kharif and Rabi seasons and collection at the outlet head. It would spare revenue staff for watching and supervising the system, thereby reducing the O & M expenditure.

(iii) Expenditure: Expenditure is worked out for the works for which authorization has been issued.

Dates for submission of estimates for expenditure by EE to SE similarly are:

- Expenditure for the ensuing year: 15 September
- Preliminary forecasting of expenditure for the current year: 15 October
- Final forecast of expenditure: 15 December
- Supplementary forecast (demand): 1 February.

The expenditure will include ordinary repairs, special repairs (including reserves for SE and CE) for emergency repairs and improvements, depreciation reserve for distress conditions (small percentage, say 1% of capital cost to O & M); tools and plants; provision for stores; establishment charges including salaries, allowances, contingency; and collection charges paid to the Revenue Department. After consolidation, the estimates are submitted by the Chief Engineer/Engineer-in-Chief to the government.

(iv) Presentation of Budget Proposals: The budget proposals submitted to the government by the Chief Engineer/Engineer-in-Chief are scrutinized and finalized by the State Finance Department for presentation to the Legislature, which usually passes:

  - Supplementary budget for the current year in February
  - Budget for the ensuing year (1st April) in March.

Allocations under different heads are communicated to the Head of the Department (Chief Engineer/Engineer-in-Chief) and are followed by formal allotment/authorization of funds. The Head of the Department makes allocations soon thereafter to the various Irrigation Divisions.

24.3 Financing of Operation and Maintenance Works

Almost all irrigation and drainage projects in India are government sponsored and owned by the government. The operation and maintenance charges of irrigation and drainage projects are met out of the general revenue of the state, which includes irrigation revenues realized from irrigation projects. The annual expenditure incurred in O & M of major and medium projects in nearly all states has been very low. The

O & M grants need to be enhanced in tune with the need of proper upkeep, based on per ha rate approved by the government from time to time.

Example: Norms for O & M grants are different in different states and change with time over the years due to inflation. Recommendations given by Ninth Finance Commission are given below as an illustration. This grant is exclusive of interest charges otherwise payable by the State Government on loans advanced by the Central Government.

24.3.1 Major and Medium Surface Irrigation Projects

(i) Rs. 180 per ha per annum of gross irrigated area for O & M grant, taking the base year as 1988.
(ii) Out of this, the allocation for headworks should be to the extent of Rs. 30 to 40 per ha, depending on the type of the headworks.
(iii) While working out the gross irrigated area of any project, the two seasonal crops and perennials are counted only once along with Kharif and Rabi crops.
(iv) Rs. 65/- per ha to Rs. 90/- per ha per annum of C.C.A. for the component of the regular establishment.
(v) An amount of at least Rs. 25/- per ha of the protected area is to be provided for maintaining the drainage system in the command area.
(vi) 1/3rd of the norms in para 1.1 above should be provided for the unutilized potential.
(vii) 20% of para 1.1 above should be provided for special repairs over and above the normal maintenance grants as and when required.

24.3.2 Minor Surface Irrigation Schemes

(i) For minor surface irrigation schemes in hilly areas of the Himalayan region, the O & M grant should be at least Rs. 900 per ha of gross irrigated area, including the cost of regular establishment.
(ii) In addition, 20% of the above O & M grant should be provided for special repairs as and when required.
(iii) For hilly regions of other states, the grants may be increased by 30% of para 1.1 above for the extra maintenance requirements in such systems.

24.3.3 Lift Irrigation Schemes

(Inclusive of Electricity Charges and Establishment).

(i) Lift Irrigation schemes by pumping, the rate in Rs. per ha of actual irrigation from river and storage.
  Group A up to 0.15 cumec: 770.00.
  Group B above 0.15 to 0.75 cumec: 620.00.
  Group C above 0.75 to 3.00 cumec: 500.00.
  Group D above 3.00 cumec: 475.00.
(ii) Lift irrigation from canals.
  Group A up to 3.00 cumec: 550.00.
  Group B above 3.00 to 15.00 cumec: 520.00.
  Group C above 15.00 cumec: 500.00.
(iii) Irrigation from augmentation tubewells: 735.00.
(iv) Irrigation from direct state tubewells: 665.00.
(v) The O & M grant should be updated annually for the escalation in the costs of labour, material, and equipment based on the overall increase in the All India Consumer Price Index.

24.3.4 Variation in Cost and Revenue

The costs associated with an irrigation system's operation and maintenance have fixed cost elements (salaries, etc.) and variable cost elements. Frequently, costs are incurred on droughts and floods over which management has no control. The construction type, texture, structure of soils, and meteorological factors affect the maintenance effort.

On the other hand, the anticipated revenue may vary from the "normal". The revenue is based on the supply of water. The occurrence of a drought reduces supply, higher than normal rainfall reduces demand, and both these situations will result in lower than normal revenue, even though actual operating costs may be higher because of these abnormal circumstances. Any drastic cut in O & M costs would adversely affect the system performance, thereby reducing revenue/benefit in the future.

24.4 State-Wise Water Charges (Rates)

Charges for the water supply generally take the form of a water rate, which varies with the crop grown and is levied on the basis of the area irrigated and matured. The water rate has no relation to the cost of supplying water but is fixed by each State Government from time to time, considering several factors.

There is considerable variation in the methods followed in different states for assessing water charges. The presently existing system of water charges in the States/UTs has been detailed in the report of Central Water Commission (CWC 2017): *Pricing of Water in Public Systems in India.*

The CWC report provides state-wise crop-specific water rates for some important crops (viz. paddy, wheat, sugarcane, cotton, oilseeds and pulses) in respect of flow and lift irrigation. This apart, an analysis of capital expenditure and working expenditure along with gross receipts over 2000–2001 to 2013–2014 for major & medium projects has been provided. Finally, it details the revenue gap assessed and realized for the same period for all States and UTs. Details of water rates followed by state governments in India are given in tabular form as below CWC (2017).

Table 24.1 presents state-wise the overall range of water rates in respect of flow vis-a-vis lift irrigation and the dates when the rates were last revised. Table 24.2 presents state-wise crop-specific water rates in respect of flow irrigation.

Table 24.3 presents state wise crop-specific water rates in respect of lift irrigation.Table 24.2 presents state-wise crop-specific water rates in respect of flow irrigation.

24.5 Conventional Versus Performance Budget

Chaube (1997) has discussed the shortcomings of conventional budgets and highlighted the importance of adopting performance-based budgeting for O & M activities. The present 'conventional budget' (also referred to as 'administrative budget') has a strong bias towards organizations and objects of expenditure. The budget shows item-wise details of provisions towards salaries and cost of materials under each administrative unit. In this form, it has a number of shortcomings, such as:

(i) Difficult to analyze the impact of government transactions on the total economy;
(ii) Difficult to relate the purposes and objectives for which resources are being allocated. The functional classification should be used to remedy this
(iii) Difficult to judge the progress towards the attainment of long- and short-term objectives and societal goals; and
(iv) Does not provide an adequate basis for informed decision-making.

The emphasis in performance budgeting is on the accomplishments rather than on the means of accomplishments. The performance approach to budgeting is based principally on the use in the budget management of three inter-related considerations. First, a meaningful classification structure in terms of programmes and activities is established under each of the functions entrusted to an organization to show precisely the objectives of various agencies, the work done by them, and the organizational responsibility. Second, the system of accounts and financial management is brought in line with this classification. Third, under each programme and activity, action is taken to establish work unit norms, standards, and other performance indicators for the appraisal and evaluation of performance. The above constitutes the three basic steps in the introduction of performance budgeting, each serving the management needs in part and together forming an important tool for review and analysis.

Thus, the main purpose of performance budgeting is to:

(i) provide meaningful classification;
(ii) correlate the physical and financial aspects of programmes and activities;

Table 24.1 Water rates in respect of flow vis-a-vis lift irrigation and the dates since applicable

State / UT	Flow Irrigation Range		Lift Irrigation Range		Date since applicable
	Max	Min	Max	Min	
1	2	3	4	5	6
Andhra Pradesh	864.50	148.20	NA		01-07-1996
Arunachal Pradesh	No water rates				29-12-2008
Assam	751.00	150.00	751.00	150.00	30-03-2000
Bihar	370.50	74.10	NA		Nov-2011
Chhattisgarh	741.00	123.50	741.00	123.50	15-06-1999
Delhi	148.20	34.03	148.20	33.35	2009
Goa	360.00	72.00	720.00	144.00	01-04-2013
Gujarat	300.00	160.00	100.00	53.33	01-01-2007
Haryana	197.60	24.70	98.80	12.35	27-07-2000
Himachal Pradesh	49.92	49.92	99.81	99.81	01-04-2015
	298.87	121.03	2998.58	298.87	01-04-2015
Jammu & Kashmir	370.50	74.10	370.50	74.10	26-11-2001
Jharkhand	988.40	37.00	1976.80	74.00	13-07-2000
Karnataka	99.00	37.00	148.50	93.00	18-09-1974
Kerala	960.00	50.00	960.00	50.00	31-12-2005
Madhya Pradesh	6297.00	119.00	5405.00	20.00	01-07-2003
	602.00	184.00	602.00	184.00	24-08-2013
Maharashtra	No water rates				
Manipur	No water rates				
Meghalaya	No water rates				
Mizoram	930.00	60.00	NA		05-04-2002
Nagaland	123.50	123.50	123.50	123.50	12-11-2014
Orissa	286.52	29.64	573.04	14.82	24-05-1999
Punjab	250.00	10.00	NA		2002
Rajasthan	61.78	2.77	NA		06-11-1987
Sikkim	312.50	312.50	312.50	312.50	01-10-2003
Tamil Nadu	474.00	30.00	237.00	15.00	18-09-1995
Tripura	474.00	30.00	237.00	15.00	18-09-1995
Uttarakhand	123.50	37.06	2015.52	251.94	01-07-2003
Uttar Pradesh	No water rates				
West Bengal	NA				
A & N Islands					
Chandigarh[a]	830.00	110.00	275.00	75.00	29-01-1996
Dadra & Nagar Haveli	286.00	286.00	286.00	286.00	2007
Daman & Diu	No water rates				
Lakshadweep	NA				
Puducherry					

[a] In rural areas of Chandigarh, the water rate is Rs. 23/- per hour with effect from 01.01.2010. NA: Not Available

Table 24.2 Water rates for crops utilizing flow irrigation (Unit Rs./hectare)

State/UT	Paddy		Wheat		Sugarcane		Cotton		Oilseeds		Pulses	
	Max	Min	Max	Min	Max	Min	Max	Min	Max	Min	Max	Min
1	2	3	4	5	6	7	8	9	10	11	12	13
Andhra Pradesh	494.00	370.50	NA	NA	864.50	864.50	864.50	864.50	247.00	148.20	NA	NA
Arunachal Pradesh	No water rates											
Assam	751.00	281.24	562.50	562.50	222.00	222.00	NA	NA	562.50	562.50	562.50	562.50
Bihar	247.00	108.40	185.25	138.32	370.50	370.50	NA	NA	98.80	74.10	98.80	74.10
Chhattisgarh	494.00	200.07	NA	NA	741.00	741.00	NA	NA	247.00	123.50	247.00	123.50
Delhi	148.20	148.20	66.63	66.63	NA	NA	NA	NA	44.46	44.46	NA	NA
Goa	180.00	180.00	NA	NA	360.00	360.00	NA	NA	120.00	120.00	NA	NA
Gujarat	160.00	160.00	160.00	160.00	300.00	300.00	160.00	160.00	160.00	160.00	160.00	160.00
Haryana	148.20	123.50	123.50	111.15	197.60	172.90	123.50	111.15	123.50	111.15	98.80	86.45
Himachal Pradesh	49.92	49.92	49.92	49.92	49.92	49.92	49.92	49.92	49.92	49.92	49.92	49.92
Jammu & Kashmir	298.87	298.87	150.67	150.67	298.87	298.87	NA	NA	150.67	150.67	121.03	121.03
Jharkhand	217.36	108.68	185.25	138.32	370.50	370.50	NA	NA	98.80	74.10	98.80	74.10
Karnataka	247.10	247.10	148.25	148.25	988.40	988.40	148.25	148.25	148.25	148.25	86.50	86.50
Kerala	99.00	37.00	NA	NA	NA	NA	NA	NA	NA	NA	NA	NA
Madhya Pradesh	155.00	85.00	125.00	75.00	960.00	960.00	70.00	70.00	75.00	50.00	75.00	50.00
Maharashtra	476.00	119.00	476.00	476.00	6297.00	6297.00	1924.00	724.00	1438.00	476.00	357.00	357.00
Manipur	602.00	305.00	305.00	305.00	NA	NA	NA	NA	184.00	184.00	184.00	184.00
Meghalaya	No water rates											
Mizoram	No water rates											
Nagaland	No water rates											
Orissa	NA	NA	170.00	170.00	500.00	500.00	280.00	280.00	170.00	60.00	170.00	60.00
Punjab	123.50	123.50	123.50	123.50	123.50	123.50	123.50	123.50	123.50	123.50	123.50	123.50
Rajasthan	197.60	49.40	148.20	64.22	286.52	103.74	177.84	88.92	113.62	64.22	79.04	49.40
Sikkim	100.00	60.00	NA	NA	NA	NA	NA	NA	NA	NA	NA	NA
Tamil Nadu	49.42	5.56	NA	NA	55.60	5.56	61.78	16.86	8.35	2.77	8.35	2.77
Tripura	312.50	312.50	312.50	312.50	NA	NA	312.50	312.50	312.50	312.50	312.50	312.50
Uttarakhand	287.00	40.00	287.00	128.00	474.00	99.00	114.00	35.00	NA	NA	NA	NA
Uttar Pradesh	287.00	40.00	287.00	40.00	474.00	99.00	114.00	35.00	NA	NA	NA	NA
West Bengal	123.50	37.06	49.40	49.40	NA	NA	NA	NA	NA	NA	NA	NA
A & N Islands	No water rates											
Chandigarh[a]	NA											
Dadra & Nagar	140.00	140.00	NA	NA	830.00	830.00	NA	NA	NA	NA	NA	NA
Daman & Diu	286.00	286.00	NA	NA	NA	NA	NA	NA	NA	NA	NA	NA
Lakshadweep	No water rates											
Puducherry	NA											

[a] In rural areas of Chandigarh, the water rate is Rs. 23/- per hour with effect from 01.01.2010

(iii) improve budget formulation and review decision-making at all levels of management;
(iv) make possible more effective performance audits;
(v) measure progress toward objectives as envisaged; and
(vi) bring annual budget and development plans closely together through a common language.

Table 24.3 Water rates for crops utilizing lift irrigation (Unit Rs./hectare)

State/UT	Paddy		Wheat		Sugarcane		Cotton		Oilseeds		Pulses	
	Max	Min	Max	Min	Max	Min	Max	Min	Max	Min	Max	Min
1	2	3	4	5	6	7	8	9	10	11	12	13
Andhra Pradesh	NA											
Arunachal Pradesh	No water rates											
Assam	751.00	281.24	562.50	562.50	222.00	222.00	NA	NA	562.50	562.50	562.50	562.50
Bihar	NA	NA	NA	NA	NA	NA	NA	NA	NA	NA	NA	NA
Chhattisgarh	494.00	200.07	NA	NA	741.00	741.00	NA	NA	247.00	123.50	247.00	123.50
Delhi	148.20	148.20	66.69	66.69	NA	NA	NA	NA	44.46	44.46	NA	NA
Goa	360.00	360.00	NA	NA	720.00	720.00	NA	NA	240.00	240.00	NA	NA
Gujarat	53.33	53.33	53.33	53.33	100.00	100.00	53.33	53.33	53.33	53.33	53.33	53.33
Haryana	74.10	74.10	61.75	55.58	98.80	86.45	61.75	55.58	61.75	43.23	49.40	43.23
Himachal Pradesh	99.81	99.81	99.81	99.81	99.81	99.81	99.81	99.81	99.81	99.81	99.81	99.81
Jammu & Kashmir	1499.96	1499.96	748.41	748.41	2998.58	2998.58	NA	NA	449.54	449.54	298.87	298.87
Jharkhand	217.36	108.68	185.75	138.32	370.50	370.50	NA	NA	98.80	74.10	98.80	74.10
Karnataka	494.20	494.20	296.50	296.50	1976.80	1976.80	296.50	296.50	296.50	296.50	173.00	173.00
Kerala	148.50	93.00	NA	NA	NA	NA	NA	NA	NA	NA	NA	NA
Madhya Pradesh	155.00	85.00	125.00	75.00	960.00	960.00	70.00	70.00	75.00	50.00	75.00	50.00
Maharashtra	357.00	357.00	535.00	535.00	5405.00	3600.00	843.00	20.00	1200.00	20.00	476.00	476.00
Manipur	602.00	305.00	305.00	305.00	NA	NA	NA	NA	184.00	184.00	184.00	184.00
Meghalaya	No water rates											
Mizoram	No water rates											
Nagaland	No water rates											
Orissa	NA											
Punjab	123.50	123.50	123.50	123.50	123.50	123.50	123.50	123.50	123.50	123.50	123.50	123.50
Rajasthan	395.20	24.70	296.40	32.11	573.04	51.87	355.68	44.46	177.84	32.11	158.08	24.70
Sikkim	NA											
Tamil Nadu	NA											
Tripura	312.50	312.50	312.50	312.50	NA	NA	312.50	312.50	312.50	312.50	312.50	312.50
Uttarakhand	143.00	20.00	143.00	20.00	237.00	49.00	57.00	18.00	NA	NA	NA	NA
Uttar Pradesh	143.00	20.00	143.00	20.00	237.00	49.00	57.00	18.00	NA	NA	NA	NA
West Bengal	2015.52	503.88	503.88	503.88	1259.70	1259.70	NA	NA	503.88	251.94	NA	NA
A & N Islands	No water rates											
Chandigarh[a]	NA											
Dadra & Nagar Haveli	275.00	75.00	275.00	275.00	100.00	75.00	NA	NA	NA	NA	275.00	275.00
Daman & Diu	286.00	286.00	NA	NA	NA	NA	NA	NA	NA	NA	NA	NA
Lakshadweep	No water rates											
Puducherry	NA											

[a] In rural areas of Chandigarh, the water rate is Rs. 23/- per hour with effect from 01.01.2010

24.6 Prescribed Norms for Maintenance Grant

There are prescribed norms for maintenance grants on the canal system in India, which vary from State to State. A case study on the analysis of annual maintenance grants for a canal division in the Upper Ganga Canal (UGC) has been carried out (Singh 1998; Chaube 1997).

At present, maintenance grants on canal systems are typically worked out on the following basis.

(a) Rs. 100/- per ha (approximately) on actual irrigated areas;
(b) Rs. 45/- per ha (nearly) on un-irrigated area in canal command; and

(c) On the basis of the normal sanctioned estimate of divisions/systems.

The overall maintenance grant was up to Rs. 133/- (nearly) per ha for the Rabi and Kharif area during 1996–1997. The normal estimates of different divisions are sanctioned based on norms issued vide E-in-C Office Memo. 6738-IB/83B-100-A/grant dated 30.11.1959, revised from time to time. Estimates are revised and sanctioned by Chief Engineers with administrative control.

Chief Engineer allots funds for annual maintenance, including special repairs based on sanctioned normal estimates of a particular division. In practice, the total funds available to the divisions are generally limited to 70–80% of the sanctioned normal estimate on average.

An example of sanctioned Normal Estimate of a Division controlling a particular branch system of Ganga Canal system is given below (Table 24.4).

In addition, Engineer-in-Chief, U.P. Irrigation Department, has issued specific instructions for the utilization of maintenance funds on different parts and elements of a canal system (Ref. Office Memo. No.2884/18/108B/AA/March-77/Section Revenue/87–88 dated 30.5.87). The prescribed percentages are as below:

(i) Silt clearance on channel 25%
(ii) Repairs to Masonry Works 35%
(iii) Strengthening banks 10%
(iv) Escapes and connected drains 10%
(v) Headworks 10%
(vi) Miscellaneous and Vehicle maintenance 5%
(vii) Inspection Houses and other buildings 5%
Total: 100%

These days, in view of the continuous rising pressure in feeding the maximum number of tails of channels, a high percentage of budgets are utilized on silt clearance and strengthening of channels with nominal expenditure on masonry work.

Besides the above, the Chief Engineer controlling the administration of different canal systems also exercises necessary controls from time to time for better and more useful utilization of available grants on various items by issuing specific orders for his systems with detailed instructions to S.E. and E.E.s. The Superintending Engineer in U.P. Irrigation Department is primarily the competent officer to exercise and maintain proper financial controls over the expenditures of irrigation divisions. They carry out frequent inspections in the field in every crop season. The Executive Engineer sends comments in a prescribed detailed format to the Chief Engineer. In addition, monthly monitoring is also carried out on the progress, achievements, and expenditures in all the divisions at the Monitoring Cell established in the office of the Engineer-in-Chief, and headed by a Chief Engineer.

Also, to keep the system in satisfactory condition, the Chief Engineers keep a certain reserve (nearly 10% of the maintenance budget) with them, which is normally utilised for special and urgent repairs of essential works and reaches of canals after careful examination and approval of Superintending Engineer and Chief Engineer. This is to keep the system functional even with inadequate budgets by taking timely

Table 24.4 Example on normal estimate of a canal division (year 1995–1996)

Sl. no	Name of item	Quantity	Rate	Amount (Rs.)
A. II.M.C. & branches				
1	B-land	1 no. (each divn.)	100.00	100.00
2	D-regulator 500 to 200 cusecs	5 nos	3815.00	19,075.00
3	E-falls 500 to 2000 cusecs	5 nos	3179.00	28,611.00
4	X-drainage works 500 to 2000 cusecs	24 nos	3139.00	75,336.00
5	G-bridges 500 to 2000 cusecs	42 nos	4504.00	189,168.00
6	H-escapes 500 to 2000 cusecs	6-0-0 mile	7935.00	47,610.00
7	J-mills	2 nos	6030.00	12,060.00
8	K-buildings (i) Constructed before 1950 (ii) Constructed after 1950	As per statement		109,500.00
9	L-earth work Total length of A.B.	82-0-0	12,233.00	1,003,106.00
	a. Mileage bases	–	–	–
	b. Acre bases	-16.1 acre	-540.00	-8694.00
10	O-miscellaneous millage bases	82-0-mile	735.00	60,270.00
11	R-communication mile 82-0-0	121.00 kms	13,500.00	1,768,500.00
			Total	33,22,030.00
	Special repair 25% on Rs. (33,22,030–1,09,500)			803,133.00
			Total	41,25,163.00
B. III. dys & minors				
1	K-buildings (i) Buildings constructed before 1950 (ii) Buildings constructed after 1950	As per statement		15,586.00
2	L-earthwork below below 20 cusecs	147.1 -556	6668.00	981,734.00
	20 to 200 cusecs	227-0-0116	9139.00	2,074,581.00
	200 to 500 cusecs	62-4-312	12,347.00	772,417.00
3	H-escape	4-3-446	5023.00	22,400.00
		Total III Dys. & Minors		3,866,718.00
	Special repairs 25% on (3,866,718–15,586)			962,783.00
			Total	48,29,501.00
	Total of II.M.C. & branches + III dys + Mrs (41, 25, 163 + 48, 29, 501)			89,54,664.00
	Losses on stock 0.1%			8955.00
	Compensation 0.1%			8555.00
	S.E's. reserve 10%			895,466.00
	CE's reserve 5%			447,733.00
	Grand Total			1,03,15,773.00
	Say Rs.			103.16 lacs

action to carry out special repairs on vulnerable reaches to avoid heavy damages and breakdowns in the system. Special repairs to the main canal are normally attended to during well-planned annual short closure periods of nearly 2 to 3 weeks on alternate years or after a few years gap.

24.7 Examples

24.7.1 Allotment, Expenditure, Revenue Over Ten Years Muzaffarnagar Division

Actual data on annual O & M allotment (establishment), actual expenditure, irrigated areas in Kharif and Rabi seasons, and revenue collection during Kharif and Rabi seasons were collected from the Ganga Canal Division, Muzaffarnagar. The data were collected for the ten financial years 1987–1988 to 1996–1997. The CCA in the jurisdiction of the above Division is 1,17,388 ha, while the proposed irrigation intensity for a year is 107%. Analysis of the data revealed the following:

(i) **Maintenance allotment and expenditure**: The maximum maintenance expenditure per unit of the irrigated area was Rs. 80.73/ha from 1990 to 1991. It was lesser in the remaining nine years. The norms for maintenance grant on the canal system in U.P. is approximately Rs. 100.00/per ha (approximately) of actual irrigated area, while the actual maintenance grant in the said Division was always less than Rs. 80.73 per ha during the last 10 years. This reveals that insufficient maintenance grant has been a constraint causing a gradual but definite and substantial deficiency and ultimately deteriorating the health of canal systems.

(ii) **Operational expenditure**: The annual operational (establishment) expenditure has been found to increase gradually. The operational expenditure during 1987–1988 was Rs. 76.25 lacs, while it was Rs. 280.62 lacs during 1996–1997. The expenditure on operation has increased four times, while maintenance grant has increased less than two times (nearly) over the past 10 years. The annual operation expenditure per unit irrigated area was maximum (Rs. 202.57/ha), during 1996–1997, which was 3.1 times (nearly) more than maintenance expenditure per unit irrigated area during 1996–1997.

(iii) **O & M expenditure**: Overall annual O & M expenditure gradually increased during the last 10 years. Though the O & M expenditure had increased 3.0 times (nearly) over the last 10 years, the total irrigated area during Kharif and Rabi had remained the same.

(iv) **Revenue**: Revenue collected during Kharif was generally 3 times (nearly) of the Rabi revenue. The overall revenue collected during 1996–1997 was maximum (Rs. 413.413 lacs).

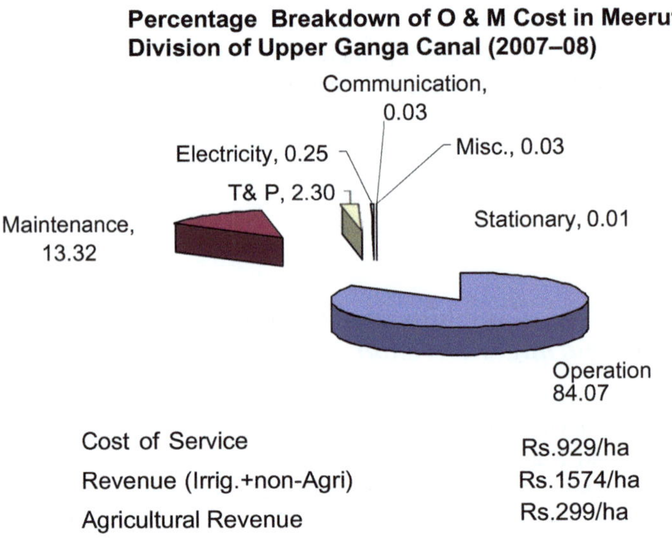

Fig. 24.1 Breakdown of the annual O & M budget in Meerut division (Upper Ganga Canal)

24.7.2 Percentage Breakdown of Annual Budget in Meerut Division

Break down of annual O & M budget (1st April, 2007 to 31st March, 2008) in the Meerut division of the Upper Ganga Canal project in Uttar Pradesh is depicted in Fig. 24.1. There is significantly higher provision for operation (84.07%) compared to that for maintenance (13.32%).

The agriculture revenue is only Rs. 300/ha. In contrast, the total revenue reaches Rs. 1574/ha, meaning that agriculture does not even contribute to 20% of the revenue (the bulk of the revenue is provided by domestic water sold to cities).

The cost of the current operation and maintenance organisation has been estimated to be at Rs. 929/ha, significantly below the revenue, which means that there is room to improve management, operation, and maintenance within the revenue limits.

Appendix: Questions

1. Explain in brief the following terms:

   (a) Plan work, establishment charges, allocation, allotment, special funds, irrigation revenue, capital cost, and supplementary budget.

2. Compare the norms recommended by the Ninth Finance Commission with those followed in Utter Pradesh.

3. Compare the norms for O & M grant prescribed by Irrigation Department of Uttar Pradesh with the allocation, allotment and actual expenditure per hectare of irrigated area in (a) Meerut Division of UGC and (b) Muzaffarnagar Division of UGC.
4. Explain the procedure for the preparation of O & M annual budget.
5. Identify the dates for the preparation/submission of various statements of revenue and expenditure.
6. Explain controls exercised by the Chief Engineer for the proper utilization of budget allotments.
7. Discuss the role of Superintending Engineer and Chief Engineer in the preparation, allotment, and utilization of budget grants.
8. Compare the water rates in Odisha and Andhra Pradesh for rice (paddy) crop.
9. Compare the water rates for lift irrigation in Tamil Nadu and Punjab.
10. Compare the water rates for wheat crop in Punjab and Uttar Pradesh.
11. Critically review the dates on which the water rates were last revised in various states.

References

Chaube UC (1997) Economic analysis, budget planning and project management. Course Package (Unpublished). Centre for Continuing Education, IIT Roorkee, Roorkee

CWC (2017) Report on pricing of water in public systems in India. Central Water Commission, Government of India, New Delhi

Guide for preparation of plans of operation and maintenance of irrigation systems in India. Indian National Committee on Irrigation and Drainage, New Delhi, March 1994

Internet Sources; Websites of (i) Ministry of Water Resources (mowr), Government of India (Jal shakti-dowr.gov.in; www.mowr.gov.in), (ii) Central Water Commission (www.cwc.gov.in), (iii) Irrigation Department of Government of Uttar Pradesh (https://idup.gov.in), (iv) Maharashtra Water Resources Regulating Authority (https://mwrra.maharashtra.gov.in)

Singh JP (1998) Operational practices in Upper Ganga Canal system—constraints, problems. M.E. dissertation supervised by Prof UC Chaube, unpublished report, WRDTC, University of Roorkee, Roorkee, India, January

MIX
Papier aus verantwortungsvollen Quellen
Paper from responsible sources
FSC® C105338

If you have any concerns about our products,
you can contact us on
ProductSafety@springernature.com

In case Publisher is established outside the EU,
the EU authorized representative is:
Springer Nature Customer Service Center GmbH
Europaplatz 3, 69115 Heidelberg, Germany

Printed by Libri Plureos GmbH
in Hamburg, Germany